Earth System Modeling, Data Assimilation and Predictability

Second Edition

Since the publication of the first edition of this highly regarded textbook, the value of data assimilation has become widely recognized across the Earth sciences and beyond. Data assimilation methods are now being applied to many areas of prediction and forecasting, including extreme weather events, wildfires, infectious disease epidemics, and economic modeling. This second edition provides a broad introduction to applications across the Earth systems and coupled Earth–human systems, with an expanded range of topics covering the latest developments of variational, ensemble, and hybrid data assimilation methods. New toy models and intermediate-complexity atmospheric general circulation models provide hands-on engagement with key concepts in numerical weather prediction, data assimilation, and predictability. The inclusion of computational projects, exercises, lecture notes, teaching slides, and sample exams makes this textbook an indispensable and practical resource for advanced undergraduate and graduate students, researchers, and practitioners who work in weather forecasting and climate prediction.

Eugenia Kalnay completed her PhD at the Massachusetts Institute of Technology (MIT) under Jule Charney and became the first woman on the faculty in the Department of Meteorology. In 1979, she moved to NASA's Goddard Space Flight Center, where she developed the fourth-order global numerical model and led experiments in the new science called "data assimilation." In 1984, she became Head of NASA's Global Modeling and Simulation Branch. In 1987, she became Director of the National Oceanic and Atmospheric Administration's Environmental Modeling Center, where many improvements of models and data assimilation were developed for the National Weather Service forecasts. Her paper "The NCEP/NCAR 40-year reanalysis project" (Kalnay et al., 1996) is the most cited paper in geosciences. In 1997, Kalnay became Lowry Chair at the University of Oklahoma and in 1999 became Atmospheric and Ocean Sciences Department Chair and professor at the University of Maryland, where she was later elected a Distinguished University Professor.

Safa Mote is Assistant Professor of Computational and Applied Mathematics at Portland State University and Visiting Assistant Professor of Atmospheric and Oceanic Sciences at the University of Maryland who has worked on a wide range of challenging interdisciplinary problems. He has two PhD degrees in Physics and in Applied Mathematics and Statistics, and Scientific Computing (AMSC) from the University of Maryland. He designs mathematical models to propose and assess holistic policies that lead to sustainability in interconnected environmental, economic, climate, and health systems. He develops computational methods based on Dynamical Systems, Machine Learning, and Data Assimilation to forecast extreme weather and climate events, improve subseasonal to seasonal predictions, and create projections for the coupled energy–water–food nexus.

Cheng Da works on Coupled Data Assimilation as a postdoctoral research associate at the University of Maryland and the Global Modeling and Assimilation Office at NASA's Goddard Space Flight Center. Supported by the NASA Earth and Space Science Fellowship, he earned his PhD degree under the supervision of Professor Kalnay at the University of Maryland, focusing on the assimilation of precipitation and nonlocal observations in the ensemble data assimilation system and coupled data assimilation. Before this, he earned his bachelor's and Master's degrees in Meteorology at Florida State University, working on radiance assimilation from spaceborne sensors.

Earth System Modeling, Data Assimilation and Predictability

Atmosphere, Oceans, Land and Human Systems

Second Edition

EUGENIA KALNAY
University of Maryland, College Park, MD, US

SAFA MOTE
Portland State University, Portland, OR, US
University of Maryland, College Park, MD, US

CHENG DA
University of Maryland, College Park, MD, US
NASA Goddard Space Flight Center, Greenbelt, MD, US

Shaftesbury Road, Cambridge CB2 8EA, United Kingdom

One Liberty Plaza, 20th Floor, New York, NY 10006, USA

477 Williamstown Road, Port Melbourne, VIC 3207, Australia

314–321, 3rd Floor, Plot 3, Splendor Forum, Jasola District Centre, New Delhi – 110025, India

103 Penang Road, #05–06/07, Visioncrest Commercial, Singapore 238467

Cambridge University Press is part of Cambridge University Press & Assessment, a department of the University of Cambridge.

We share the University's mission to contribute to society through the pursuit of education, learning and research at the highest international levels of excellence.

www.cambridge.org
Information on this title: www.cambridge.org/9781107009004

DOI: 10.1017/9780511920608

First edition © Eugenia Kalnay 2003
Second edition © Eugenia Kalnay, Safa Mote and Cheng Da 2024

This publication is in copyright. Subject to statutory exception and to the provisions of relevant collective licensing agreements, no reproduction of any part may take place without the written permission of Cambridge University Press & Assessment.

When citing this work, please include a reference to the DOI 10.1017/9780511920608

First published 2003, reprinted with corrections 2004
Second edition 2024

A catalogue record for this publication is available from the British Library

Library of Congress Cataloging-in-Publication Data
Names: Kalnay, Eugenia, 1942– author. | Mote, Safa, author. | Da, Cheng (Meteorologist), author.
Title: Earth system modeling, data assimilation and predictability : atmosphere, oceans, land and human systems / Eugenia Kalnay, University of Maryland, College Park, Safa Mote, University of Maryland, College Park, Cheng Da, University of Maryland, College Park.
Other titles: Atmospheric modeling, data assimilation, and predictability.
Description: Second edition. | Cambridge, UK ; New York, NY : Cambridge University Press, 2024. | Earlier edition published in 2003 as: Atmospheric modeling, data assimilation, and predictability. | Includes bibliographical references and index.
Identifiers: LCCN 2023057631 (print) | LCCN 2023057632 (ebook) | ISBN 9781107009004 (hardback) | ISBN 9780511920608 (ebook)
Subjects: LCSH: Numerical weather forecasting.
Classification: LCC QC996 .K35 2024 (print) | LCC QC996 (ebook) | DDC 551.63/4–dc23/eng/20240324
LC record available at https://lccn.loc.gov/2023057631
LC ebook record available at https://lccn.loc.gov/2023057632

ISBN 978-1-107-00900-4 Hardback
ISBN 978-1-107-40146-4 Paperback

Additional resources for this publication at www.cambridge.org/kalnay2e

Cambridge University Press & Assessment has no responsibility for the persistence or accuracy of URLs for external or third-party internet websites referred to in this publication and does not guarantee that any content on such websites is, or will remain, accurate or appropriate.

This book is dedicated to
the Grandmothers of Plaza de Mayo
for their tremendous courage and leadership in defense of human rights
and democracy.

Front Cover Legend: Importance of ensemble forecasting and data assimilation for the forecast of the Hurricane Sandy landing in the New York metro area in 2012 (The actual trajectory of Sandy is shown with a thick black line with a black circle every 6 hours indicating when a new "assimilation of the observation of Sandy's location" took place). This figure, created by Clark Evans, Professor of Atmospheric Sciences at the University of Wisconsin–Milwaukee, (evans36@uwm.edu), shows the National Weather Service ensemble prediction of the devastating Sandy Hurricane landing.

Note that the ensemble trajectories of the NCEP 10-day ensemble forecasts of the Sandy hurricane trajectory were started on October 23, 12UTC (shown with blue lines at the first identification of Sandy), and 6 and 12 hours later (shown with green and red lines respectively). The majority of the earliest (in blue) ensemble forecasts miss the hurricane being "captured" by a strong atmospheric trough (see the inside cover of the book) and continue moving eastwards, driven by the Atlantic westerly winds, as the majority of the Atlantic hurricanes normally do. The green trajectories, that after 6 hours underwent one additional data assimilation, clearly turned west, indicating the influence of the trough that captured the hurricane. The red trajectories, started October 24 00UTC, from the next data assimilation just 12 hrs after the blue trajectories, have a majority that correctly turns west for the landfall.

Why did Sandy turn west around 00Z/29/10/2012?
It was captured by a deep trough!

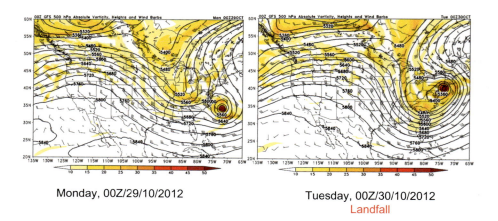

Monday, 00Z/29/10/2012   Tuesday, 00Z/30/10/2012
                                                             Landfall

500 hPa NCEP analysis of absolute vorticity, winds, and heights

Contents

Foreword to the First Edition	*page* xiii
Preface to the Second Edition	xv
Reviews and Comments on the First Edition	xviii
Acknowledgments from the Second Edition	xxiv
Acknowledgments from the First Edition	xxv
List of Variables	xxvii
List of Abbreviations	xxix

1	**An Overview of Numerical Weather Prediction**	1
1.1	Introduction	1
1.2	Early Developments	7
1.3	Primitive Equations. Global and Regional Models, and Nonhydrostatic Models	13
1.4	Data Assimilation: Determination of the Initial Conditions for the Computer Forecasts	15
1.5	Operational NWP and the Evolution of Forecast Skill	23
1.6	Weather Predictability, Ensemble Forecasting, and Seasonal to Interannual Prediction	27
1.7	The Future	33
2	**The Continuous Equations**	35
2.1	Governing Equations	35
2.2	Atmospheric Equations of Motion on Spherical Coordinates	39
2.3	Basic Wave Oscillations in the Atmosphere	40
	2.3.1 Pure Types of Plane Wave Solutions	42
	2.3.1.1 Pure Sound Waves	42
	2.3.1.2 Lamb Waves (Horizontally Propagating Sound Waves)	43
	2.3.1.3 Vertical Gravitational Oscillations	43
	2.3.1.4 Inertia Oscillations	44
	2.3.1.5 Lamb Waves in the Presence of Rotation and Geostrophic Modes	44
	2.3.2 General Wave Solution of the Perturbation Equations in a Resting, Isothermal Atmosphere	45
	2.3.2.1 External Waves	48
	2.3.2.2 Internal Waves	48

viii **Contents**

	2.3.3	Analysis of the FDR of Wave Solutions in a Resting, Isothermal Atmosphere	49
2.4		Filtering Approximations	50
	2.4.1	Quasi-geostrophic Approximation	50
	2.4.2	Quasi-Boussinesq or Anelastic Approximation	51
	2.4.3	Hydrostatic Approximation	52
2.5		Shallow Water Equations, Quasi-geostrophic Filtering, and Filtering of Inertia-Gravity Waves	55
	2.5.1	Quasi-geostrophic Scaling for the SWE	57
	2.5.2	Inertia-Gravity Waves in the Presence of a Basic Flow	60
2.6		Primitive Equations and Vertical Coordinates	61
	2.6.1	General Vertical Coordinates	61
	2.6.2	Pressure Coordinates	64
	2.6.3	Sigma, Eta, and Hybrid Coordinates	65
	2.6.4	Isentropic Coordinates	67
2.7		Introduction to the Equations for Ocean Models	68
	2.7.1	Primitive Equations for the Oceans	68
	2.7.2	Ocean Boundary Conditions and Coupled Atmosphere–Ocean Models	70
2.8		Kelvin Waves and Equatorially Trapped Waves	71
	2.8.1	Kelvin Waves	71
	2.8.2	Equatorially Trapped Waves	72
3		**Numerical Discretization of the Equations of Motion**	76
3.1		Classification of Partial Differential Equations	76
	3.1.1	Reminder about PDEs	76
	3.1.2	Well-posedness, Initial and Boundary Conditions	77
3.2		Initial Value Problems: Numerical Solution	80
	3.2.1	Finite Differences Method	80
	3.2.2	Truncation Errors and Consistency	82
	3.2.3	Convergence and Criteria for Computational Stability	83
		3.2.3.1 Criterion of the Maximum	84
		3.2.3.2 Von Neumann Stability Criterion	85
		3.2.3.3 Leapfrog Scheme Initialization	90
		3.2.3.4 Robert–Asselin and Williams Time Filters for Leapfrog	90
	3.2.4	Implicit Time Schemes	94
	3.2.5	Semi-implicit Schemes	97
3.3		Space Discretization Methods	100
	3.3.1	Space Truncation Errors, Computational Phase Speed, Second- and Fourth-Order Schemes	100
	3.3.2	Galerkin and Spectral Space Representation	103
	3.3.3	Semi-Lagrangian Schemes	106
	3.3.4	Nonlinear Computational Instability, Quadratically Conservative Schemes, and the Arakawa Jacobian	108
	3.3.5	Staggered Grids	115
	3.3.6	Finite Volume Methods	118
3.4		Boundary Value Problems	120

	3.4.1	Introduction	120
	3.4.2	Direct Methods for Linear Systems	122
	3.4.3	Iterative Methods for Solving Elliptic Equations	123
	3.4.4	Other Iterative Methods	124
3.5		Lateral Boundary Conditions for Regional Models	126
	3.5.1	Introduction	126
	3.5.2	Lateral Boundary Conditions for One-Way Nested Models	127
	3.5.3	Other Examples of Lateral Boundary Conditions	130
	3.5.4	Two-Way Interactive Boundary Conditions	131
3.6		Nonhydrostatic Models	132
3.7		Need to Replace Spectral Models: Experiments to Choose the Next Generation Global Model at NCEP	135
3.8		How to Validate NWP Models That Are Based on Machine Learning and Artificial Intelligence?	137

4 Introduction to the Parameterization of Subgrid-Scale Physical Processes 139

4.1 Introduction	139
4.2 Subgrid-Scale Processes and Reynolds Averaging	141
4.3 Overview of Model Parameterizations	144
4.4 The SPEEDY Model and Documentation	147
4.5 Cumulus Parameterizations and "Superparameterization"	147

5 Data Assimilation 150

5.1		Introduction	150
5.2		Empirical Analysis Schemes	151
	5.2.1	Early Approaches to Objective Analysis	151
	5.2.2	Successive Correction Method	153
	5.2.3	Nudging	155
5.3		Introduction to Statistical Estimation Methods through the Use of Toy Models	156
	5.3.1	Sequential (or Least Squares) Method	156
	5.3.2	Variational (Maximum Likelihood) Approach	160
		5.3.2.1 Bayes Theorem Applied to Data Assimilation	161
	5.3.3	Analysis Cycle Equations for the "Stone in Space" Toy Model	162
5.4		Multivariate Statistical Data Assimilation Methods	163
	5.4.1	Multivariate Analysis Cycle: Equations and Their Interpretation	164
	5.4.2	Derivation of OI and 3D-Var Analysis Equations	166
		5.4.2.1 Some Mathematical Remarks	167
		5.4.2.2 Statistical Assumptions and Derivation of OI and 3D-Var Formulas	168
	5.4.3	Numerical Solutions of OI and 3D-Var	170
		5.4.3.1 Remarks: Errors of Representativeness, Error Correlations, and Super Observations	171
		5.4.3.2 Optimal Interpolation	171
		5.4.3.3 3D-Var	173
		5.4.3.4 Computation of $\mathbf{A}$, $\mathbf{C}$, and $\mathbf{V}$ from the "NMC Method" Background Error Covariance $\mathbf{B}$	176

	5.4.4	Estimation of the Background Error Covariance **B**	178
		5.4.4.1 Introduction	178
		5.4.4.2 Estimations of **B** Used in OI before the "NMC Method"	179
		5.4.4.3 The "NMC Method"	182
	5.4.5	Physical-Space Statistical Analysis Scheme, and Its Relationship to 3D-Var and OI	184
5.5		Advanced Data Assimilation Methods with Evolving Covariance: 4D-Var	187
	5.5.1	Introduction: "Errors of the Day"	187
	5.5.2	4D-Var Extension of 3D-Var and Its Relationship to Kalman Filter	189
	5.5.3	Numerical Solution of 4D-Var: Inner and Outer Loops	192
	5.5.4	Further Remarks on 4D-Var	194
	5.5.5	Introduction to the Construction of the Tangent Linear and Adjoint Models	196
5.6		Advanced Data Assimilation Methods with Evolving Covariance: Ensemble Kalman Filter	199
	5.6.1	Introduction	199
	5.6.2	Introduction to the Kalman Filter and Extended Kalman Filter Equations	200
	5.6.3	Introduction to Ensemble Kalman Filtering Methods: Stochastic and Square Root Filters	202
	5.6.4	Example of a Square-Root EnKF: Local Ensemble Transform Kalman Filter	204
		5.6.4.1 Analysis Weights Interpolation	207
	5.6.5	Hybrids of Ensemble and Variational Data Assimilation	208
		5.6.5.1 Covariance Hybrid	208
		5.6.5.2 Gain Hybrid	209
		5.6.5.3 4D-Var and 4DEnVar	211
	5.6.6	Running in Place: A No-Cost Smoother	213
		5.6.6.1 4D-LETKF and No-Cost Smoother	214
		5.6.6.2 Use of the No-Cost Smoother to Accelerate the Spin-Up (Running in Place and Quasi Outer Loop)	215
	5.6.7	Ensemble Forecast Sensitivity to Observations and Proactive Quality Control	216
		5.6.7.1 FSO, EFSO, and HFSO	216
		5.6.7.2 Brief Derivation of EFSO, and a "Bridging" Example with a Low-Resolution GFS Model and PrepBUFR Observations	217
		5.6.7.3 Results of EFSO/PQC with a Low-Resolution GFS Model and PrepBUFR Observations	218
	5.6.8	Particle Filter	219

6		**Atmospheric Predictability and Ensemble Forecasting**	224
6.1		Introduction to Atmospheric Predictability	224
6.2		Brief Review of Fundamental Concepts about Chaotic Systems	226
6.3		Tangent Linear Model, Adjoint Model, Singular Vectors, and Lyapunov Vectors	229
	6.3.1	Tangent Linear Model and Adjoint Model	230
	6.3.2	Singular Vectors	232
	6.3.3	Lyapunov Vectors	237
	6.3.4	Simple Examples of Singular Vectors and Eigenvectors	240
6.4		Ensemble Forecasting: Early Studies	243

	6.4.1	Stochastic-Dynamic Forecasting	244
	6.4.2	Monte Carlo Forecasting	245
	6.4.3	Lagged Average Forecasting	247
6.5	Operational Ensemble Forecasting Methods		250
	6.5.1	Breeding	253
	6.5.2	Singular Vectors	259
	6.5.3	Ensembles Based on Multiple Data Assimilation	262
	6.5.4	Multisystem Ensemble Approach	263
6.6	Growth Rate of Errors and the Limit of Predictability in Mid-latitudes and in the Tropics		263
6.7	The Role of the Oceans and Land in Monthly, Seasonal, and Interannual Predictability		268
6.8	Decadal Variability and Climate Change		272
6.9	Historical Development of Earth System Models: Progressive Coupling of New Components		274
6.10	Domination of the Climate System by the Human System		275
6.11	Developing Data Assimilation Methods for Improving Human System Modeling		278
6.12	Controlling Chaos in Control Simulation Experiments		280

Appendix A Coding and Checking the Tangent Linear and the Adjoint Models   283

A.1	Verification	286
A.2	Example of FORTRAN Code	287

Appendix B Postprocessing of Numerical Model Output to Obtain Station Weather Forecasts   292

B.1	Model Output Statistics	292
B.2	Perfect Prog	295
B.3	Adaptive Regression Based on a Simple Kalman Filter Approach	296

Bibliography	299
Index	337

Foreword to the First Edition

During the 50 years of numerical weather prediction, the number of textbooks dealing with the subject has been very small, the latest being the 1980 book by Haltiner and Williams. As you will soon realize, the intervening years have seen impressive developments and success. Eugenia Kalnay has contributed significantly to this expansion, and the meteorological community is fortunate that she has applied her knowledge and insight to writing this book.

Eugenia was born in Argentina, where she had exceptionally good teachers. She had planned to study physics but was introduced to meteorology by a stroke of fate; her mother simply entered her in a competition for a scholarship from the Argentine National Weather Service! But a military coup took place in Argentina in 1966 when Eugenia was a student, and the College of Sciences was invaded by military forces. Rolando Garcia, then Dean of the College of Sciences, was able to obtain for her an assistantship with Jule Charney at the Massachusetts Institute of Technology. She was the first female doctoral candidate in the department and an outstanding student. In 1971, under Charney's supervision, she finished an excellent thesis on the circulation of Venus. She recalls that an important lesson she learned from Charney at that time was that if her numerical results did not agree with accepted theory, it might be because the theory was wrong.

What has she written in this book? She covers many aspects of numerical weather prediction and related areas in considerable detail, on which her own experience enables her to write with relish and authority. The first chapter is an overview that introduces all the major concepts discussed later in the book. Chapter 2 is a presentation of the standard equations used in atmospheric modeling, with a concise but complete discussion of filtering approximations. Chapter 3 is a roadmap to numerical methods providing students without a background in the subject with all the tools needed to develop a new model. Chapter 4 is an introduction to the parameterization of subgrid-scale physical processes, with references to specialized textbooks and papers. I found her explanations in Chapter 5 of data assimilation methods and in Chapter 6 on predictability and ensemble forecasting to be not only inclusive but thorough and well presented, with good attention to historical developments. These chapters, however, contain many definitions and equations. (I take this wealth as a healthy sign of the technical maturity of the subject.) This complexity may be daunting for many readers, but this has obviously been recognized by Eugenia. In response, she has devised many simple graphical sketches that illustrate the important relations and definitions.

An added bonus is the description in an appendix of the use of *Model Output Statistics* by the National Weather Service, its successes, and the rigid constraints that it imposes on the forecast model. She also includes in the appendices a simple adaptive regression scheme based on Kalman filtering and an introduction to the generation of linear tangent and adjoint model codes.

Before leaving the National Centers for Environmental Prediction (NCEP) in 1998 as Director of the Environmental Modeling Center, Eugenia directed the *Reanalysis Project*, with Robert Kistler as Technical Manager. This work used a 1995 state-of-the-art analysis and forecast system to reanalyze and reforecast meteorological events from past years. The results for November 1950 were astonishing. On November 24 of that year, an intense snowstorm developed over the Appalachians that had not been operationally predicted even 24 hours in advance. This striking event formed a test situation for the emerging art of numerical weather prediction in the years immediately following the first computations in 1950 on the ENIAC computer discussed in Chapter 1. In 1953, employing his baroclinic model, and with considerable "tuning," Jule Charney finally succeeded in making a 24-hour forecast starting on November 23, 1950, of a cyclonic development, which, however, was still located some 400 kilometers northeast of the actual location of the storm. This "prediction" played a major role in justifying the creation of the Joint Numerical Weather Prediction Unit in 1955 (Chapter 1). By contrast, in the Reanalysis Project, this event was forecast extremely well, in both intensity and location – as much as three days in advance. (Earlier than this the associated vorticity center at 500 mb had been located over the Pacific Ocean, even though at that time there was no satellite data!) This is a remarkable demonstration of the achievements of the numerical weather prediction community in the past decades, achievements that include many by our author.

After leaving NCEP in 1998, Eugenia was appointed Lowry Chair in the School of Meteorology at the University of Oklahoma, where she started writing her book. She returned to Maryland in 1999 to chair the Department of Meteorology, where she continues to do research on a range of topics, including applications of chaos to ensemble forecasting and data assimilation. We look forward to future contributions by Professor Kalnay.

Norman Phillips

Preface to the Second Edition

It has been two decades since Cambridge University Press published the first edition of my book on data assimilation in 2003. Its title was *Atmospheric Modeling, Data Assimilation and Predictability*, but it was always referred to as "the Data Assimilation book." The book was very well received by numerical weather prediction graduate students and researchers, with ~4,500 citations at this time. In the section that follows, "Reviews and Comments on the First Edition," we have included four book reviews as well as a sample of other reviews and endorsements of the first edition.

Since 2003, the ensemble-based assimilation methods (e.g., ensemble Kalman filter and particle filters) have evolved very fast, and the ensemble Kalman filter has been implemented in several operational centers. So, the most important chapter, 5 "Data Assimilation" (Chapter 5), has been completely rewritten. We have also introduced a new toy model (a stone in space) that makes the complex data assimilation equations easy to understand. Chapter 5 now discusses fundamental aspects of variational and ensemble methods and how to combine them (hybrid methods) to further improve the analysis. A short introduction to particle filters is also included.

The other chapters have also been updated. Chapter 1, "An Overview of Numerical Weather Prediction," which includes a history of numerical weather prediction and major developments, was rewritten to be accessible to an undergraduate-level reader. In Chapter 2, "The Continuous Equations," we have added an introduction to the governing equations of the oceans, Kelvin waves, and equatorially trapped waves. Chapter 3, "Numerical Discretization of the Equations of Motion" has also been updated. We have added sections on nonhydrostatic models and on the need to replace the long and successfully used spectral models because of the increased model resolution, and we discuss the next-generation global model at the National Centers for Environmental Prediction (NCEP). In Chapter 4, "Introduction to the Parameterization of Subgrid-Scale Physical Processes," we have added a section on the Simplified Parameterizations primitivE-Equation DYnamics (SPEEDY) model as an introduction to the dynamical core and the physical parameterizations of a full spectral Earth system model. A new section on cumulus parameterizations and super parameterization has also been included. Although we generally use the atmosphere as an example, the methodologies and the way of thinking presented in Chapters 2–6 are also applicable to ocean and land modeling data assimilation.

To accompany this book, we have provided additional learning materials (such as the computational project, as well as additional book sections) on the companion website. We will also continue to add new projects here.

During the past two decades, I have had the privilege of teaching at the University of Maryland (UMD) and the good fortune of being the advisor of about 40 students who have earned their PhDs, learning from them, and seeing many of them become recognized as leading experts in data assimilation. I have also developed a keen interest in the interaction of the Earth system with the human system, the overwhelming growth of the impact of the human system on the Earth system, and the impacts of inequality. I started working with Safa Mote (formerly Safa Motesharrei), a brilliant student who was equally interested in this problem. Jorge Rivas also had a keen interest in and deep knowledge of this subject and soon joined us. Together, we have written several papers, the first one being the highly cited human and nature dynamical (HANDY) model, a simple approach to modeling the interaction between the human and natural systems (Motesharrei et al., 2014). In 2016, we published another paper on "Modeling sustainability" (Motesharrei et al., 2016) with 20 coauthors, most of them leaders in their fields. We pointed out in this paper that since in the real world the human system has become the main driver of change in many of the physical subsystems of the Earth system, in order to understand the dynamics of *either* system, Earth system models *must* be coupled with human system models through bidirectional couplings representing the positive, negative, and delayed feedbacks that exist in the real systems. We also discuss applications of data assimilation methods to these coupled Earth–human system models. These include parameter estimation, sensitivity analysis, and ensemble runs to quantify uncertainty. We thus decided to change the title of the book to *Earth System Modeling, Data Assimilation, and Predictability*. Earth system models are generally considered to be the coupled modeling of the atmosphere, land, and oceans, as well as other subsystems, such as the biosphere and the cryosphere. However, the human system, which has come to completely dominate the evolution of the Earth system, has not yet been bidirectionally coupled to these natural systems (Motesharrei et al., 2016; Calvin and Bond-Lamberty, 2018). Our first HANDY model (Motesharrei et al., 2014), which models the dynamic interactions between the human and natural systems before the Industrial Revolution, is a minimal example of a coupled Earth–human system. In Chapter 6, "Atmospheric Predictability and Ensemble Forecasting," we now include several sections discussing the development of coupled Earth–human system modeling and how to use data assimilation to improve this modeling. In order to address the problems of climate change, it is both feasible and necessary to replace fossil fuels with renewables (e.g., Jacobson et al., 2015, 2018; IPCC, 2022). Modeling the Earth and human systems bidirectionally coupled together is necessary to inform policies that allow a timely transition to renewables.

In the last few years, artificial intelligence and machine learning, and particularly deep learning, have advanced extremely rapidly and offer the promise of being as accurate but much faster than standard numerical methods used in modeling and DA (e.g., Krasnopolsky, 2013). We emphasize in Chapter 3 the need to evaluate their performance in terms of not just comparing their RMS errors but, very importantly, whether the new methods have the essential ability to reproduce the atmospheric instabilities, measured, for example, using the breeding method.

I am very fortunate that Dr. Cheng Da and Dr. Safa Mote agreed to become my coauthors.

Cheng Da is an expert in (and has a passion for) data assimilation, something that he jokes is due to his last name. He earned his PhD degree under my supervision at the University of Maryland, focusing on precipitation assimilation to improve the prediction of tropical cyclones and a new multilayer observation localization method to accurately assimilate nonlocal observations in the local ensemble transform Kalman filter (LETKF). Before this, he earned his bachelor's and master's degree in meteorology at Florida State University. He is now working on coupled data assimilation as a postdoctoral research associate at the University of Maryland and the Global Modeling and Assimilation Office at NASA's Goddard Space Flight Center.

Safa Mote is an applied mathematician and Earth systems scientist who has expertise in dynamical systems and climate modeling, computational and data science, and data-driven prediction. He builds mathematical models to propose and assess holistic decisions and policies that lead to sustainability of the coupled climate–energy–water–food nexus. He also develops novel computational methods by combining dynamical systems, machine learning, and data assimilation to analyze large, diverse datasets of these complex systems and to create forecasts and projections. He is particularly interested in forecasting, and possibly mitigating, high-impact severe weather events such as wildfires, droughts, floods, and storms. I am deeply grateful to Professor Mote for presenting frequent guest lectures over the past decade in my graduate courses on applied statistics, Earth system modeling, and data assimilation.

Eugenia Kalnay

Reviews and Comments on the First Edition

Book Review by Dr. Andrew Lorenc

The original book review was published in the *Quarterly Journal of the Royal Meteorological Society*, 2003.

Eugenia Kalnay's enthusiasm for numerical weather prediction (NWP) shows through from the first chapter of her book. The introductory historical overview, from the first successes of Charney and others in the 1950s to modern ensemble forecasting systems, is designed to motivate graduate-level students by explaining how the rapid progress was achieved, and listing the outstanding challenges. Eugenia then sets out to give a solid grounding in all aspects of NWP, with chapters on the continuous equations, their numerical discretization, the parametrization of subgrid-scale physical processes, data assimilation and predictability, and ensemble forecasting. She uses simple examples and exercises to cover the ground from first principles to state-of-the-art NWP systems. Other exercises are more like essay topics, designed to make the reader think.

Chapter 2 is a concise but thorough exposition of the basic equations used for NWP, with discussion, through simple examples, of the types of waves which can occur, leading to understanding of the various filtering approximations. Chapter 3 is equally thorough in covering the numerical techniques used to construct forecast models. A student working through the examples and exercises should emerge confident to work on practical model development. Of course a book of this length cannot cover everything in depth; Chapter 4 is only a nine-page summary of parametrization methods, with references for more details but no exercises. In contrast, Chapter 5 on data assimilation is the longest in the book. Early empirical methods such as successive correction and nudging are described, before embarking, via simple examples, on the least-squares and Bayesian derivations of modern statistical methods. Advanced methods with evolving forecast error covariance are covered, with a mention of current research interest in the ensemble Kalman filter. Chapter 6 on predictability is another highlight, going from Lorenz's butterfly, through chaos, to modern ensemble forecasting systems.

Although she has succeeded in covering most aspects of NWP in a single book, Eugenia's background and particular enthusiasms show through in the emphasis given to each topic. For instance, 4D-Var is just described in the advanced data-assimilation methods section, without mentioning its success in operational use, and breeding is given more space than singular vectors in the section on ensembles. Yet her enthusiasm is a strength of the book. One is left with a feeling for the interesting problems that remain. As a broad introduction to the basis of modern NWP, this book has no equal.

Book Review by Dr. Peter Lynch

The original book review has been published in *Splanc*, Met Éireann Newsletter, 2003.

Computer modelling is now the primary means of forecasting the weather. The accuracy of Numerical Weather Prediction models has improved steadily over the half-century since the first tentative experiments. There are a huge number of technical papers and reports devoted to NWP, but very few books. So, a frisson of excitement accompanied the rumour that Eugenia Kalnay was writing a new book. Expectations were high, since she is a renowned expert in the field. She has not disappointed us.

Eugenia Kalnay got a PhD in MIT in 1971. Her advisor was Jule Charney (Ray Bates, a former Met Éireann Assistant Director, was another of his doctoral students). She was Director of the Environmental Modeling Center, U.S. National Weather Service, for ten years to 1997, so she knows a thing or two about NWP. Her book is called Atmospheric Modeling, Data Assimilation and Predictability, and covers all three topics in the title. It is a mathematical book, but don't stop reading, because it has a considerable amount of expository material, accessible to all. I am reviewing it here in the hope of persuading mathophobes (pace Mr. Gates) to look through the book. They will gain much by perusing the discursive sections even if they skip the sums.

The first of the six chapters is a historical survey, tracing the development of NWP from the ENIAC integrations up to the present, and ending with a peek into the future. It is almost completely non-mathematical and is eminently readable. The next two chapters deal with the equations of motion and the numerical methods of solving them. These are tough going, but are worth scanning, as there are some descriptive sections. In chapter 3.5, on regional models, Eugenia refers to a report by Aidan McDonald as an excellent review of the lateral boundary conditions used in operational regional NWP. Chapter 4 is an admirably succinct (i.e., mercifully short) introduction to the huge area of subgrid-scale physical processes. Again, much of it is accessible to the general reader.

Chapter 5, the longest in the book, is also the toughest, dealing with the vital subject of data assimilation. The material here is inherently difficult. However, a superficial scan will give you an overview of the subject, and many buzz-words to amaze your friends. The final chapter is on predictability and ensemble forecasting.

Forecasters should find this particularly useful and relevant. We have emerged from denial (or was it the Amazon?), and recognise the limitations on our predictive abilities. The chaotic nature of the atmosphere is unavoidable. We can forecast with probability but never with certainty, particularly at longer lead-times. Operational ensemble forecasting is reviewed in this chapter. The concluding sections are on the role of the oceans in long-range forecasting and on climate change. The bibliography is comprehensive, running to 44 pages. More than a dozen references are to work of current or former Met Éireann scientists. This is gratifying, indicating that a small, dedicated team can have an international impact. Some of the numerical techniques developed in Dublin, and now in widespread use, are described in this excellent book.

Book Review by Dr. Takemasa Miyoshi

The original book review in Japanese was published in the December 2003 issue of *Tenki*, the monthly bulletin of the Meteorological Society of Japan. ©Meteorological Society of Japan. Used with Permission. (Translated by Dr. Takemasa Miyoshi.)

Wide range of topics on numerical weather prediction (NWP) including its history, techniques in atmospheric modeling, data assimilation, and predictability are discussed in clear, self-contained, and precise way. The outstanding advantage is that basic ideas as well as details of techniques are included, which makes the discussion clearer. Not only readers new to NWP but also knowledgeable readers must be satisfied with the richness of the contents.

There has been a well-known textbook on NWP written by Haltiner and Williams (1980), but it is true that much progress including the realization of ensemble forecasting has been made thanks to the progress in the computational capability. Based on the latest situation, the present work describes the basics of atmospheric dynamics and dynamical equations in NWP, and in addition, the details of techniques in numerical integration methods. For instance, various numerical discretization schemes, Galerkin and spectral methods, semi-Lagrangian schemes, staggered grid structures, which are important in implementing NWP models, are precisely described. Furthermore, parameterization of subgrid-scale physical processes is described briefly. As for data assimilation, the discussion covers the basic idea of least square method, and the practical methods in realistic high-dimensional system including the optimum interpolation (OI), the three-dimensional variational method (3D-VAR), the four-dimensional variational method (4D-VAR), and the Kalman filtering, some of which are operationally used and some of which are latest research topics. There is no need to mention that initialization methods are included in the discussion. In the last chapter, predictability and ensemble forecasting, in which the author has been deeply involved, are discussed. The general discussion of chaos of nonlinear dynamical systems using a simple model is introduced, and the purpose of ensemble prediction is clearly explained. It follows that the basics of the typical error vectors such as singular vectors and Lyapunov vectors are explained, and then, the ensemble forecasting in the actual NWP cases is discussed. The discussion includes the Monte-Carlo method, the Lagged Averaged Forecasting (LAF) method, the breeding method, the singular vector method, and multimodel method. Climate change is also included, which is surprising because of the wide range of its topics.

This clear, precise, and comprehensive discussion in NWP was possible thanks to the author who had been involved in the development of NWP models at the National Centers for Environmental Prediction (NCEP) for many years. The author, Eugenia Kalnay, was a mentee of the famous Jule Charney, the "father" of NWP, and she is currently involved in the education and research in NWP at the University of Maryland,

College Park. The present work is based on her lecture notes, which is why the discussion is kind and precise. Moreover, representative papers in each field are included in the reference, which makes it best for beginners. For readers who have knowledge and experience in NWP, it helps to organize their knowledge, and the scales fall from their eyes because of the author's wisdom which has influenced places all over the work.

In conclusion, the present work crystallizes the author's wisdom, and a must in studying NWP.

Book Review by Dr. Christopher K. Wikle

The original book review was published in *Technometrics*, 2005.

With increasing concerns about potential climate change and its associated impacts, questions of uncertainty in climate prediction are becoming paramount. Naturally, such concerns about uncertainty suggest the need for statistical expertise. In addition, statisticians interested in problems in the environmental sciences (e.g., meteorology, oceanography, ecology) are incorporating deterministic dynamical models (e.g., partial differential equations) in their statistical models for spatiotemporal processes. Because it is relatively uncommon for statisticians to receive training in dynamical systems and partial differential equations (PDEs) in their formal academic courses, there is a need for a concise, yet thorough book that describes the critical aspects of these subjects, at least those related to environmental processes. This book does exactly that.

In essence, this book is an overview of modern numerical weather forecasting. After an accessible historical overview in Chapter 1, Chapter 2 provides a summary of the continuous system of equations that describe the state of the atmosphere. Although somewhat terse for a statistician with no previous exposure to the governing equations of the atmosphere, it is an excellent overview of atmospheric dynamics. Chapter 3 then describes the numerical solution of such systems of PDEs. This is an excellent chapter and would be a nice reference for statisticians interested in numerical solutions of PDEs, as well as those interested in the connection between PDEs and difference equations. Chapter 4 is a very short introduction to the parameterization of subgrid-scale processes that are necessary in atmospheric models as a result of the discretization limits imposed on the continuous system. Such parameterizations are critically important in weather forecasting and climate modeling, yet are rarely considered from a statistical perspective (either in terms of estimation of free model parameters or in terms of the uncertainty in the model specification).

Chapters 5 and 6 are more directly relevant to statistics. Chapter 5 gives a comprehensive overview of 'data assimilation," which involves combining uncertain observations with deterministic (or quasi-deterministic) dynamical models to obtain an estimate of the state of the system (and its uncertainty). For numerical weather forecasting, this is critical. To initialize the discretized deterministic weather forecasting model, one must have an estimate of the system state variables at each grid location (in three-dimensional space) Obviously, atmospheric observations do not occur at such resolution, so one must effectively interpolate the observations to obtain the initial state. The problem is complicated by the presence of large gaps of missing data and the fact that the initial state must be in some sort of dynamical balance (i.e., must be a physically realistic representation of the system state). This requires using a dynamical model to help fill in the missing information. Statisticians are familiar with these ideas in the context of spatial prediction (i.e., kriging). More generally, data assimilation can be posed from a Bayesian perspective in which one has a prior distribution for the state process that might be obtained from a numerical weather prediction model. The "likelihood" is then the distribution of the data conditional on this true process. One of the biggest challenges with this procedure is obtaining the "prior" covariance matrix. In the case of normal assumptions for both the prior and "likelihood," both the state vector and the associated covariance matrix can be obtained via the Kalman filter. However, with the high-dimensional state vectors common to atmospheric systems (on the order of millions!), and the nonlinearity of the dynamical evolution equations, standard Kalman filter methods are not practical. This book presents the outline of so-called "ensemble" Kalman filters, which overcome many of these problems and are currently the focus of intense research and development because of their outstanding potential for practical assimilation problems.

Chapter 6 is concerned with atmospheric predictability and ensemble forecasting. From a statistical perspective, this chapter provides a very nice and concise review of the basics of chaotic systems and such important concepts as singular vectors and Lyapunov vectors. The growth of errors in nonlinear dynamical systems is critical for prediction, and these concepts provide a fundamental way to evaluate predictability. There is also a discussion of the necessity of accounting for the uncertainty in the initial state for nonlinear prediction models due to chaos. This leads to the idea of ensemble forecasting, which is closely tied to the ensemble methods for data assimilation presented in Chapter 5.

In summary, this book is an excellent reference for statisticians interested in dynamical systems and/or spatiotemporal processes in the environment. The technical level is reasonable for statisticians with advanced degrees. Although a few token exercises are included, to use this book as a textbook the instructor would need to provide supplemental problems.

Short Reviews and Endorsements

...quite wonderful, achieving a tremendous balance between comprehensiveness and readability. I am especially pleased with the numerical analysis part, which is crystal clear and shows the benefits of classroom testing. I also like the tiny little touches, like the stepped-on butterfly story and the mention that Poincaré knew about chaos in celestial mechanics. Your book fills an enormous hole in the literature of NWP [numerical weather prediction].

Richard C. J. Somerville, Scripps Institution of Oceanography, San Diego

Reviews and Comments on the First Edition

Fantastic ... in content, format and practicability.
Kelvin K. Droegemeier, Regents' Professor of Meteorology, and Director, Center for Analysis and Prediction of Storms, University of Oklahoma

[I] admire the clarity and pedagogic superiority of [this] presentation.
Anders Persson, Swedish Meteorological and Hydrological Institute (SMHI)

... much better for learning about data assimilation than anything else currently available.
Richard Swinbank, United Kingdom Meteorological Office

... [the] presentation is impeccable and is very accessible to non-meteorologists like me.
Eric Kostelich, University of Arizona

... what a great wealth of historical information.
Lawrence Takacs, NASA, Data Assimilation Office

... [the] method in the [data] assimilation section of starting with 'baby' examples, and then working up through the full analysis, is great for understanding. On the predictability part, the history, and the explanations of how the unstable perturbations grow is the best I've seen.
Alexander E. MacDonald, Director, NOAA Forecast Systems Lab

... this book ... is extremely useful, informative, and well-written ... there are many instances where items that were only marginally familiar beforehand have now become very clear.
Brian O. Blanton, Senior Scientist/Oceanographer, University of North Carolina at Chapel Hill

Acknowledgments from the Second Edition

This book would not have been possible without tremendous support and helpful advice and feedback from numerous colleagues and students. We are grateful to numerous colleagues at the University of Maryland, Portland State University, the National Aeronautics and Space Administration (NASA), and many other institutions: Tony Busalacchi, Bob Cahalan, Mark Cane, Alberto Carrassi, Jim Carton, Rita Colwell, Dacian Daescu, Russ Dickerson, Paolo D'Odorico, Bill Dorland, Kelvin Droegemeier, Julien Emile-Geay, Clark Evans, Kuishuang Feng, Rachel Franklin, Inez Fung, Michael Ghil, Mitch Goldberg, Chris Jarzynski, Eric Hackert, Milt Halem, Pedram Hassanzadeh, Ross Hoffman, Klaus Hubacek, Brian Hunt, Eric Kostelich, Fred Kucharski, Doron Levy, Michel Loreau, Andrew Lorenc, Peter Lynch, Michael Mann, Fernando Miralles-Wilhelm, Takemasa Miyoshi, Sumant Nigam, Carlos Nobre, Ed Ott, Tim Palmer, George Philander, Drew Rice, Jorge Rivas, Raj Roy, Alfredo Ruiz-Barradas, Matthias Ruth, Roald Sagdeev, June Sherer, Adel Shirmohammadi, Jagadish Shukla, Jelena Srebric, Zoltan Toth, Mike Wallace, Christopher Wikle, Victor Yakovenko, Jim Yorke, Ning Zeng, and Aleksey Zimin.

We would like to thank many former students in our classes and/or research group members: Javier Amezcua, Eviatar Bach, Kriti Bhargava, George Britzolakis, Ming Cai, Chu-Chun Chang, Tse-Chun Chen, Matteo Corazza, Chris Danforth, Eli Dennis, Steve Greybush, David Groff, Matt Hoffman, Daisuke Hotta, Ji-Sun Kang, Maia Karpovich, Hong Li, Yan Li, Guo-Yuan Lien, Junjie Liu, Yun Liu, Erin Lynch, Adrienne Norwood, DJ Patil, Malaquias Peña, Steve Penny, Zhaoxia Pu, Juan Ruiz, Tamara Singleton, Travis Sluka, Qianqian Song, Luyu Sun, Shu-Chih Yang, and Takuma Yoshida. Many of these former research group members are currently professors at major research universities or lead scientists in meteorological agencies worldwide.

A complete list, even if possible, will inevitably become very long, so we would like to express our sincere appreciation to all colleagues whose names do not appear in the above lists due to space limitation.

We deeply appreciate continuous, impactful support for our scientific research from Eugenia and Michael Brin, as well as generous support from Rebecca Danesh, Peter Stansfield, and Fariborz Maseeh.

Our work over the past many years has been supported by scientific research grants from the National Science Foundation (NSF), NASA, the National Oceanic and Atmospheric Administration (NOAA), and the Indian Institute for Tropical Meteorology (IITM). Our current research grants include NASA grants IRET-QRS-22-0001, NNH23ZDA001N-ECIPES23-0020, NNH20ZDA001N-MAP20-0159, and 80NSSC23K0827, as well as NSF RTG grant DMS-2136228.

Acknowledgments from the First Edition

I drafted about two-thirds of this book while teaching the subject for the first time at Oklahoma University, during the fall of 1998. Oklahoma University provided me with a supportive environment that made it possible to write the first draft. I made major revisions and finished the book while teaching the course again in the fall in 1999 through 2001 at the University of Maryland. The students that took the course at University of Maryland and Oklahoma University gave me essential feedback and helped me find many (hopefully most) of the errors in the drafts.

In addition, several people helped to substantially revise one or more of the manuscript chapters, and their suggestions and corrections have been invaluable. Norm Phillips read an early draft of Chapter 1 and made important historical comments. Anders Persson wrote the notes on the early history of numerical weather prediction, especially in Europe, reproduced in an appendix. Alfredo Ruiz Barradas reviewed Chapter 2. Will Sawyer reviewed and made major suggestions for improvements for Chapter 3. Hua-lu Pan influenced Chapter 4. Jim Geiger reviewed Chapter 5 and pointed out sections that were obscure. Jim Purser also reviewed this chapter and not only made very helpful suggestions but also provided an elegant demonstration of the equivalence of the 3D-Var and OI formulations. Discussions with Peter Lyster on this chapter were also very helpful. D. J. Patil suggested many improvements to Chapter 6, and Bill Martin pointed out the story by Ray Bradbury concerning the "butterfly effect." Joaquim Ballabrera substantially improved the appendix on model output postprocessing. Shu-Chih Yang and Matteo Corazza carefully reviewed the complete book, including the appendices, and suggested many clarifications and corrections.

I am grateful to Malaquias Peña, who wrote the abbreviations list and helped with many figures and corrected references. Dick Wobus created the beautiful six-day ensemble forecast figure shown on the cover. Seon Ki Park provided the linear tangent and adjoint code in Appendix B. The help and guidance of Matt Lloyd and Susan Francis of Cambridge University Press, the editing of the text by Maureen Storey, and the kind foreword by Norm Phillips are also very gratefully acknowledged.

I began to learn numerical weather prediction (NWP) in the late 1960s from professors at the University of Buenos Aires, especially Rolando Garcia and Ruben Norscini, and from the inspiring book of P. D. Thompson. At the Massachusetts Institute of Technology (MIT), my thesis advisor, Jule Charney, and the lectures of Norm Phillips and Ed Lorenz, influenced me more than I can describe. The NWP class notes of Akio Arakawa at University of California, Los Angeles and the National Center for

Atmospheric Research text on numerical methods by John Gary helped me teach the subject at MIT. Over the last 30 years, I have continued learning from numerous colleagues at other institutions where I had the privilege of working. They include the University of Montevideo, MIT, NASA's Goddard Space Flight Center, Oklahoma University, and the University of Maryland. However, my most important experience came from a decade I spent as Director of the Environmental Modeling Center at the National Centers for Environmental Prediction (NCEP), where my extremely dedicated colleagues and I learned together how to best transition from research ideas to operational improvements.

Finally, I would like to express my gratitude for the tremendous support, patience, and encouragement that my husband, Malise Dick, my son, Jorge Rivas, and my sisters, Patricia and Susana Kalnay, have given me and for the love for education that my parents instilled in me.

Variables

a	radius of the Earth
$\mathbf{A}$	analysis error covariance matrix
$\mathbf{B}$	background error covariance matrix
$\mathbf{C}$	covariance matrix
C_p, C_v	specific heat at constant pressure, constant volume
$\mathbf{d}$	innovation or observational increments vector
D	fluid depth
$E()$	expected value
f	Coriolis parameter
g	gravitational constant
$\mathbf{H}$	linear observation operator matrix
H	observational operator, scale height of the atmosphere
$\mathbf{I}$	identity matrix
J	cost function
JM	maximum number of grid points j
$\mathbf{K}$	Kalman gain matrix
$L(t_0, t)$	resolvent or propagator of TLM
$\mathbf{M}$	TLM matrix
N	Brunt–Väisälä frequency
$\mathbf{P}$	projection matrix
p	pressure, probability, distribution function
q	mixing ratio of water vapor and dry air mass
$\mathbf{Q}$	forecast model error covariance
$\mathbf{r}$	position vector
$\mathbf{R}$	observations error covariance matrix
R	root mean square error, gas constant
R_d	Rossby radius of deformation
R_0	Rossby number
RE	relative error
T	temperature
TS	threat score
u, v	eastward and northward wind components
$\mathbf{W}$	weight matrix
W	vertical wind component, optimal weight

x,y	horizontal coordinates
δ_{ij}	Kronecker delta
ε_a	analysis error
ε_b	background error
η	absolute vorticity
Φ	geopotential height
φ	geopotential, latitude
λ	longitude
λ_i	global Lyapunov exponent
ρ_{ij}	element i,j of the correlation matrix C
σ	standard deviation
σ^2	variance
ψ	streamfunction
ω	vertical velocity in pressure coordinates, spectral frequency
ζ	relative vorticity

Abbreviations

3D-Var	Three-dimensional variational analysis
4D-Var	Four-dimensional variational analysis
4DDA	Four-dimensional data assimilation
AC	Anomaly correlation
ADI	Alternating direction implicit
AGCM	Atmospheric general circulation model
AI	Artificial intelligence
AMIP	Atmospheric Model Intercomparison Project (frequently refers to long model runs in which the observed SST is used instead of climatology)
AMSR2	Advanced Microwave Scanning Radiometer 2
AMV	Atmospheric motion vectors
AO	Arctic Oscillation
ARPS	Advanced Regional Prediction System
AVHRR	Advanced Very High Resolution Radiometer
AVN	NCEP's aviation (global) spectral model
BLUE	Best Linear Unbiased Estimation
BV	Bred vector
CAPS	Center for Analysis and Prediction of Storms
CFL	Courant–Friedrichs–Lewy
COAMPS	US Navy's Coupled Ocean/Atmosphere Mesoscale Prediction System
CONUS	Continental USA
CPC	Climate Prediction Center (NCEP)
CRCP	Cloud Resolving Convective Parameterization
CSE	Control Simulation Experiment
CSI	Critical success index (same as threat score)
CSRM	Cloud System Resolving Model
DL	Deep learning
DWD	German Weather Service
EAKF	Ensemble adjustment Kalman filter
ECMWF	European Centre for Medium-Range Weather Forecasts
EDA	Ensemble data assimilation
EFSO	Ensemble Forecast Sensitivity to Observations

List of Abbreviations

EKF	Extended Kalman filter
EMC	Environmental Modeling Center
ENIAC	Electronic numerical integrator and computer
ENSO	El Niño–Southern Oscillation
ETKF	Ensemble transform Kalman filter
EnKF	Ensemble Kalman filter
EnSRF	Ensemble square root filter
FASTEX	Fronts and Storm Track Experiment
FDE	Finite difference equation
FDR	Frequency dispersion relationship
FFSL	Flux-form semi-Lagrangian scheme
FSO	Forecast sensitivity to observations
FV3	Finite-volume, cubed-sphere
GCM	General circulation model
GEFS	Global Ensemble Forecast System
GFDL	Geophysical Fluid Dynamics Laboratory
GFS	Global Forecast System
GLE	Global Lyapunov exponents
GMI	Global Precipitation Measurement (GPM) Microwave Imager
GPS	Global Positioning System
GSI	Gridpoint Statistical Interpolation
hPa	Hectopascals (also known as millibars)
HFIP	Hurricane Forecast Improvement Program
HFSO	Hybrid forecast sensitivity to observations
HPC	Hydrometeorological Prediction Center (NCEP)
IAM	Integrated Assessment Model
ICTP	International Centre for Theoretical Physics
IFS	Integrated Forecasting System
IPCC	Intergovernmental Panel on Climate Change
JMA	Japan Meteorological Agency
JNWPU	Joint Numerical Weather Prediction Unit
KF	Kalman filter
LEKF	Local ensemble Kalman filter
LETKF	Local ensemble transform Kalman filter
LFM	Limited fine mesh
LLV	Local Lyapunov vectors
LPF	Local particle filter
MeteoFrance	National Meteorological Service for France
MCC	Mesoscale Compressible Community (model)
MJO	Madden–Julian Oscillation
ML	Machine learning
MM5	Fifth-Generation Penn State/NCAR Mesoscale Model
MMF	Multiscale modeling framework

MODIS	Moderate Resolution Imaging Spectroradiometer
MPAS	Model for Prediction Across Scales
MOS	Model Output Statistics
NAO	North Atlantic Oscillation
NASA	National Aeronautics and Space Administration
NCAR	National Center for Atmospheric Research
NCEP	National Centers for Environmental Prediction (US National Weather Service)
NCI	Nonlinear computational instability
NGGPS	Next Generation Global Prediction System
NGM	Nested Grid Model
NLNMI	Nonlinear normal mode initialization
NMC	National Meteorological Center
NOAA	National Oceanic and Atmospheric Administration
NORPEX	North Pacific Experiment
NWP	Numerical weather prediction
NWS	National Weather Service
OI	Optimal interpolation
OSSE	Observing System Simulation Experiments
PBL	Planetary boundary layer
PDE	Partial differential equation
PDO	Pacific Decadal Oscillation
PF	Particle filter
PIRCS	Project to Intercompare Regional Climate Systems
PQC	Proactive quality control
PQPF	Probabilistic Quantitative Precipitation Forecast
PSAS	Physical space analysis scheme
PVE	Potential vorticity equation
RAFS	Regional Analysis and Forecasting System
RAOB	Rawinsonde observation
RDAS	Regional Data Assimilation System
RF	Recursive filter
RIKEN	RIkagaku KENkyūjyo (Institute of Physical and Chemical Research in Japan)
RIP	Running in Place
RSM	NCEP's Regional Spectral Model
RUC	NCEP's Rapid Update Cycle
SAC	Standardized anomaly correction
SCM	Successive correction method
SLAF	Scaled Lagged Average Forecasting
SMAP	Soil Moisture Active Passive
SOR	Successive overrelaxation
SPEEDY	Simplified parameterizations, primitive-equation dynamics
SSI	Spectral Statistical Interpolation

SST	Sea surface temperature
SV	Singular vector
SWE	Shallow water equation
TAMC	Tangent and Adjoint Models Compiler
TOGA	Tropical Ocean, Global Atmosphere
TOVS	TIROS-N Operational Vertical Sounder
TS	Threat score
UKMO	United Kingdom Meteorological Office
UTC	Universal time or Greenwich time, e.g., 1200 UTC. Frequently abbreviated as 1200Z
WMO	World Meteorological Organization
WRF	Weather Research and Forecasting model

1 An Overview of Numerical Weather Prediction

1.1 Introduction

In this chapter, we give a historical overview of numerical weather prediction (NWP), which makes possible our daily and weekly forecasts.[*] Even seasonal forecasts become possible when ocean models are coupled to atmospheric models using similar methodologies to predict El Niño. In general, the public is not aware that our weather forecasts start out as initial-value problems on the supercomputers of the major national and international weather services or that the quality of the operational forecasts has undergone extraordinary improvements since their beginnings in the 1950s. These improvements are one of the most remarkable successes in the history of science, and the goal of this book is to clearly describe the major scientific developments that led to these improvements.

Numerical weather prediction provides the basic guidance for weather forecasting beyond the first few hours. For example, in the USA, computer weather forecasts issued by the National Centers for Environmental Prediction (NCEP) in College Park, MD, guide forecasts from the US National Weather Service (NWS). The NCEP forecasts are performed by "running" (i.e., integrating in time) computer models of the atmosphere that can simulate, given today's weather observations, the evolution of the atmosphere in the next few days. Because the time integration of an atmospheric model is an initial-value problem, in order to make a skillful forecast it is necessary that (a) *the computer model be an accurate representation of the atmosphere* and (b) *the initial conditions be also represented accurately.*

The NCEP (formerly the National Meteorological Center, or NMC) has performed operational computer weather forecasts since the 1950s. From 1955 to 1973, the forecasts included only the Northern Hemisphere (NH); they have been global since 1973. Over the years, the quality of the models and methods for using atmospheric observations has improved continuously, resulting in major forecast improvements. Weather centers have always kept track of the quality of the forecasts by comparing them with what actually happened, i e., by comparing the forecast maps with the verification maps.[1] Teweles and Wobus (1954) developed a measure of forecast skill known as the

---

[*] Note: Sections marked with an asterisk are written at an undergraduate level.

[1] Unfortunately, this is not the case in other sciences that issue periodic forecasts. For example, economic forecasts are not routinely verified, even though they are obviously wrong most of the time.

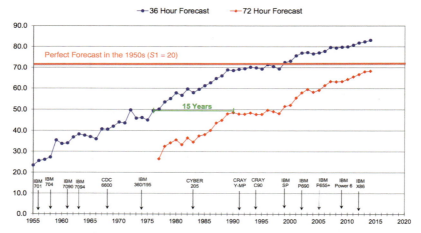

Figure 1.1.1 Historic evolution of the operational forecast skill of the NCEP (formerly NMC) models over North America. The score $100 * (1 - S1/70)$ is based on the $S1$ score that measures the relative error in the horizontal pressure gradient, averaged over the region of interest. The values $S1 = 70\%$ and $S1 = 20\%$ were empirically determined to correspond respectively to a "useless" and a "perfect" forecast when the score was designed. Note that the 72 hr forecasts are currently as skillful as the 36 hr forecasts were 10–20 years earlier (data courtesy S. Lilly, NCEP).

$S1$ score, which measures the relative error in the wind forecast through the estimation of the relative error in the forecast of the pressure gradient. Shuman (1989) pointed out that the $S1$ score was carefully calibrated to reflect the estimation of human forecasters of how useful the computer forecasts actually were. Forecasts with scores of $S1 = 70$ or larger were considered to be useless, and a score of $S1 = 20$ was considered a "perfect forecast," since that was the average score obtained when comparing hand-analyzed maps made by different expert weather analysts using the same observations over the data-rich area of North America. The NCEP still maintains a scaled version of the $S1$ score that reflects these limits: $100 \times (1 - S1/70)$, so that a "perfect score" of $S1 = 20$ is equal to about 71.4 in the scaled $S1$ score. Figure 1.1.1 shows the longest available record of the skill of NWP, measuring the scaled $S1$ score of 36 hr forecasts over North America at the constant pressure surface of 500 hPa (in the middle of the atmosphere, since the mean sea level pressure is about 1,000 hPa, or 1,000 mb). It is remarkable that, using the 1950s standard, the 36 hr forecast has been "perfect" since 1999. In the mid-1970s, NCEP started producing 72 hr forecasts, also shown in Figure 1.1.1, whose skill was comparable to that of the 36 hr forecasts made only 15–20 years earlier. The trend of the skill suggests that in a few years, the 500 hPa 72 hr forecast might also reach the "perfect" level, making it as close to the verifying analysis as two expert human analyses of the same abundant rawinsondes. This figure also includes information about the dates at which new supercomputers were installed at NCEP, since more powerful computers allowed the implementation of more accurate models and methods of data assimilation.

Other major operational centers such as the UK Meteorological Office (Met Office), the Japanese Meteorological Agency (JMA), and the Canadian Meteorological Centre (now Environment Canada, EC) made similar progress, and this shared experience led to the foundation, in 1975, of the European Centre for Medium-Range Weather Forecasts (ECMWF). As indicated by the name, its mission was to improve "medium-range weather forecasts" defined as 3 to 10 days. The ECMWF started issuing 10-day operational forecasts in 1979 and soon became the preeminent NWP operational center. It adopted a different skill measure, the anomaly correlation (AC), which computes the *pattern correlation* between the "anomalies" of a forecast map and of the corresponding verifying analysis (Miyakoda et al., 1976). Anomalies are defined as the difference between a field and its monthly climatology, so that the AC does not give credit to a forecast for just predicting climatology (e.g., predicting that winter is colder than summer, and higher latitudes colder than the tropics).

If the forecast is initialized from the analysis, the AC starts at 100%. As the forecast length increases, it becomes less skillful, and the AC becomes smaller. As they did with the $S1$ score, human forecasters calibrated the skill of the ECMWF forecasts by estimating that in order to provide useful forecast guidance, a forecast needs to have an AC larger than 60%. Figure 1.1.2 shows the historic ECMWF evolution of forecast skill in the NH extratropics measured by the time at which the AC reaches 60%. The figure shows that in 1980, on average, the forecasts "remained useful" until 5 to 6 days. The improvement in the quality of the forecasts is remarkable: Currently the ECMWF forecasts remain useful on average until 8 to 9 days and during some months until 10 days. This evolution in skill is paralleled by other centers (Figure 1.1.3).

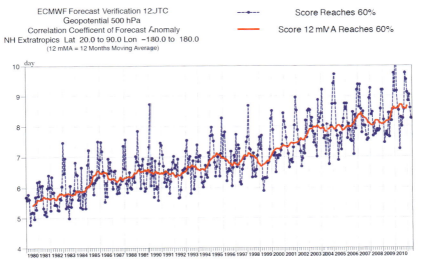

Figure 1.1.2 ECMWF forecast time at which the anomaly correlation (AC) reached the level of 60% (Figure courtesy of ECMWF, under the Licenses of CC-BY-4.0 and ECMWF Terms of Use).

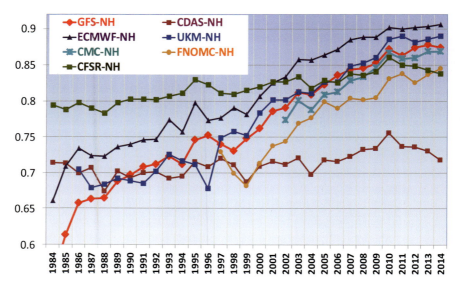

Figure 1.1.3 Evolution of the annual mean of the NH 500 hPa geopotential height 5-day anomaly correlation from 1984 to 2014. Included are the NCEP Global Forecasting System (GFS), the ECMWF, the UKMO, the Canadian Meteorological Centre (CMC), and the Fleet Numerical Meteorology and Oceanography Center (FNMOC, since 1997) operational model. Also included are two NCEP reanalyses run with frozen systems: the Climate Data Assimilation System (CDAS), also known as NCEP-NCAR Reanalysis, which is run with a system similar to the NCEP operational model circa 1995, and the Coupled Forecast System Reanalysis (CFSR), run with a Coupled Forecasting System circa 2005. Note that the two reanalyses, which use a "frozen" model and a "frozen" data assimilation system, show a skill that is almost constant, with only a slight improvement in time associated with the number of observations, and the improvement with time of their coverage and quality (figure courtesy of Fanglin Yang, NCEP).

These remarkable improvements in the skill of NWP are an extraordinary scientific achievement. They also demonstrate the benefits of national and international cooperation in sharing observations and new developments, as well as the impact of the friendly competition among scientists that try different methods in different research and operational centers, until eventually, the most successful two to three methods are selected.

In the USA, research on NWP takes place in the national laboratories of the National Oceanic and Atmospheric Administration (NOAA), the National Aeronautics and Space Administration (NASA), and the National Center for Atmospheric Research (NCAR), as well as in universities and centers such as the Center for Analysis and Prediction of Storms (CAPS) and the Weather and Chaos group at the University of Maryland. Internationally, major research takes place in large operational national and international centers (such as the ECMWF, NCEP, and the weather services of the UK, France, Germany, Scandinavian and other European countries, Canada, Japan, Australia, Korea, and others). In meteorology, there has been a long tradition of sharing both data and research improvements, with the result that progress in the science

1.1 Introduction

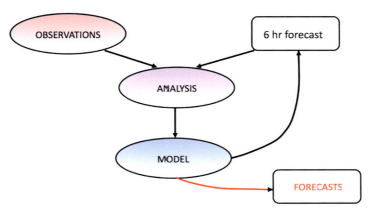

Figure 1.1.4 Schematic of the classic 6 hr analysis cycle that provides the initial conditions to the model. A 6 hr forecast is the first guess or background for the next analysis done 6 hr later and obtained by the optimal combination of the first guess and the new observations. The operational forecasts in red (which are two weeks long at NCEP) are used as guidance by the human forecasters.

of forecasting has taken place on many fronts, and all countries have benefited from this progress.

In order to understand how this scientific progress was achieved, let's consider the major components of NWP, namely the atmospheric *observations*, the *model*, and the *method used to create the initial conditions* for the model forecast. The initial conditions are known as the "analysis" because at the beginning of NWP, they were created by a human analyst who drew isolines of pressure, temperature, and wind, and this "analysis" was then interpolated to the model grid points. It soon became clear that rather than interpolating the analysis based on observations, it was better to interpolate the difference between the new observations obtained every 6 hr and the 6 hr forecast started from the previous analysis. This difference, which represents the new information brought by the observations, is also known as *analysis increment* or *innovation*. Thus was born the analysis cycle (Figure 1.1.4).

This type of "analysis cycle" where short forecasts are combined with new observations to obtain the best estimate of the state of the atmosphere (the analysis) came to be known as *data assimilation*, because the model is forced to "assimilate" observations of the atmosphere and thus to remain close to the real atmosphere. It is clear that in order to improve the forecasts, it is necessary to improve simultaneously the models, the observations, and the methods used to create the analysis (data assimilation). For example, improving the number and quality of the observations alone would not produce a good forecast if at the same time we use a mediocre model or initial conditions that come from an inaccurate analysis. In summary, the remarkable improvement in skill of NWP over the last half-century apparent in Figures 1.1.1–1.1.3 is due to four factors, the first two associated with model improvements, and the last two with improvements in data assimilation and in the quantity and quality of the observations:

- The increased power of supercomputers, allowing much *finer numerical resolution* and *fewer approximations* in the operational atmospheric models;
- The improved representation of *small-scale physical processes* (clouds, precipitation, turbulent transfers of heat, moisture, momentum, radiative transfer) within the models;
- The use of more accurate methods of data assimilation, which result in improved *initial conditions* for the models (analyses); and
- The increased availability and accuracy of observations, as well as advances in the use of satellite and aircraft data over oceans and the Southern Hemisphere.

This book is devoted to the exploration of NWP science (and similar approaches more recently used in ocean forecasting). This chapter contains a historical overview of NWP and an introduction to the main scientific advances that made possible the remarkable improvements we have seen and will continue to see in the future. Chapter 2 includes a derivation of the equations used in NWP and in ocean modeling, as well as the type of wave solutions that they allow. The presence of high-frequency gravity waves and sound waves in the equations of motion impose the use of very short time steps, which would be computationally extremely expensive. Chapter 2 also presents several approximations to the equations of motion that filter gravity and sound waves, such as the use of the hydrostatic or the anelastic approximations, and the limits in accuracy introduced by these approximations. Chapter 3 provides an introductory practical guide to the numerical methods used in atmospheric models, including finite differences, finite volume, and spectral schemes used in space discretization of the dynamical equations. Time schemes appropriate for different types of equations, as well as special methods such as semi-implicit and semi-Lagrangian schemes are also described. We also discuss the type of models (such as the Finite Volume, Cubed Sphere, or FV3) that are replacing the currently widely used global spectral models, since they become inefficient at horizontal grid scales smaller than about 10 km, which also require the use of nonhydrostatic equations. The textbook Durran (2010) covers this subject in more detail. Chapter 4 is a brief introduction to the problem of representing the impact (parameterizing) of those physical processes that take place at subgrid scales and that are not resolved by the model, such as radiation and turbulent transports of heat, moisture, and momentum. Since these subgrid-scale processes cannot be explicitly resolved by the numerical discretizations presented in Chapter 3, and they cannot be ignored beyond a few hours, they instead need to be parameterized. The textbook Stensrud (2007) is devoted to this subject. The new approach of "super-parameterizations" is briefly presented. Chapter 5, perhaps the most important in the book, is devoted to data assimilation, one of the pillars of NWP, where much progress has been made, after the early empirical methods to correct the short-range forecasts with observations were replaced with methods that accounted for the error statistics of the forecasts and observations. The first "statistical interpolation" methods (optimal interpolation and 3D-Var) assumed that the forecast error covariance is constant with time. More advanced methods (4D-Var and ensemble Kalman filter) can now account for the "errors of the day." All these methods require involved matrix algebra, so that they are introduced with "baby" scalar examples that make clear the interpretation of the otherwise very complex equations. We also review some recent developments in applications of data assimilation, where the pioneering NWP research is being explored in other sciences. Chapter 6 is devoted to chaos and predictability, ensemble forecasting, coupled modeling, and an introduction to climate change modeling. Here we respond to questions such as: "Can we take advantage of the variability of chaos and predict with skill some states of the atmosphere longer than others? How could we possibly predict climate change when we know that we cannot predict the weather more than two weeks ahead? Should we model and bidirectionally couple the human system with the Earth system models that we use to predict climate change?"

In this introductory chapter, we give an overview of the major components and milestones in numerical forecasting. They will be discussed in detail in the corresponding Chapters 2–6.

1.2 Early Developments

There are several outstanding scientists whose work was essential to the creation of NWP, but for brevity we just refer to the three most important giant advances:*

Vilhelm Bjerknes (1862–1951) was a Norwegian physicist who developed a "general circulation theorem" for the atmosphere published in 1900 in both *Monthly Weather Review and Meteorologische Zeitschrift*. He felt that empirical and statistical methods used at that time would never lead to a scientific approach for meteorology. Bjerknes set up the scientific basis for weather prediction in a subsequent paper entitled "The problem of weather forecasting as a problem in mechanics and physics," published in 1904 in *Meteorologische Zeitschrift* (Bjerknes, 1904). He did not use equations but posed the problem of NWP as integrating seven equations with seven unknowns using observations as initial data (Bjerknes, 1904; Gramelsberger, 2009). Since numerical solutions were at that time out of the question, he proposed instead the use of graphical methods of numerical integration. With his son Jacob Bjerknes (1897–1975) he established the famous Bergen School of Meteorology where the frontal theory for cyclone development was first developed and became the basis of synoptic meteorology. Jacob Bjerknes' amazing physical intuition made possible the prediction of El Niño: In 1969, he explained for the first time the phenomenon of El Niño as the ocean component of a coupled ocean–atmosphere oscillation, now referred to as the El Niño–Southern Oscillation, or ENSO.

L. F. Richardson (1881–1953) was also a physicist (and a pacifist) who refused to fight during World War I, instead driving a Quakers' ambulance. During the two years he spent at war, he developed the appropriate finite differences for a primitive equations model, solved many theoretical and practical problems in the physical and numerical formulations, and computed by hand the time derivative of the surface pressure for a single point in Germany, using maps created by Bjerknes (Richardson, 1922). Even though his methodology was correct, the result he obtained was catastrophically wrong: He predicted a huge, unrealistic change of 145 hPa in 6 hr, while the observed pressure hardly changed in those 6 hr. The reasons for this failure (the presence of high-frequency gravity waves that dominated the time derivative in the solution) are discussed in detail in Chapter 2. Despite this dramatic failure that kept other scientists from attempting further experiments in NWP, Richardson published an inspiring book (Richardson, 1922), describing in detail his methodology, which, as shown by Lynch (2006), was scientifically impeccable.

Charney, Fjørtoft, and von Neumann succeeded for the first time in predicting a moderately realistic 24 hr change in the atmosphere (Charney et al., 1950) by using, instead of the primitive equations as used by Richardson, the conservation of vorticity equation based on the quasi-geostrophic theory Charney had developed (Charney, 1948). This was a spectacular success (Figure 1.2.2) made possible by the access to a computer (the ENIAC) and especially by the choice of a simple but realistic equation to describe the atmospheric dynamics most relevant to weather prediction (the conservation of vorticity). Charney made this choice in order to *filter out the high-frequency gravity waves* that ruined Richardson's results (see Chapter 2).

Jule G. Charney (1917–1981) was one of the giants in the history of NWP. In his paper "Dynamic forecasting by numerical process" (Charney, 1951), he introduced the subject of this book (NWP) as well as it could be introduced today. We reproduce here parts of the paper (with emphasis added):

As meteorologists have long known, *the atmosphere exhibits no periodicities of the kind that enable one to predict the weather in the same way one predicts the tides.* No simple set of causal relationships can be found which relate the state of the atmosphere at one instant of time to its state at another. It was this realization that *led V. Bjerknes (1904) to define the problem of prognosis as nothing less than the integration of the equations of motion of the atmosphere* (Bjerknes, 1904).[2] But it remained for Richardson to suggest the practical means for the solution of this problem (Richardson, 1922). *He proposed to integrate the equations of motion numerically and showed exactly how this might be done. That the actual forecast used to test his method was unsuccessful was in no way a measure of the value of his work.* In retrospect it becomes obvious that the inadequacies of observation alone would have doomed any attempt, however well conceived, a circumstance of which Richardson was aware. The real value of his work lay in the fact that it crystallized once and for all the essential problems that would have to be faced by future workers in the field and it laid down a thorough groundwork for their solution.

[2] The importance of Bjerknes (1904) is clearly described by Thompson (1990), another pioneer of NWP, and the author of the first and still inspiring text on NWP (Thompson, 1961). His paper "Charney and the revival of numerical weather prediction" (Thompson, 1990) contains extremely interesting material on the history of NWP as well as on early computers: It was not until 1904 that Vilhelm Bjerknes – in a remarkable manifesto and testament of deterministic faith – stated the central problem of NWP. This was the first explicit, coherent recognition that the future state of the atmosphere is, *in principle*, completely determined by its detailed initial state and known boundary conditions, together with Newton's equations of motion, the Boyle–Charles–Dalton equation of state, the equation of mass continuity, and the thermodynamic energy equation. Bjerknes went further: He outlined an ambitious, but logical, program of observation, graphical analysis of meteorological data, and graphical solution of the governing equations. He succeeded in persuading the Norwegians to support an expanded network of surface observation stations, founded the famous Bergen School of synoptic and dynamic meteorology, and ushered in the famous polar front theory of cyclone formation. Beyond providing a clear goal and a sound physical approach to dynamical weather prediction, Bjerknes instilled his ideas in the minds of his students and their students in Bergen and in Oslo, three of whom were later to write important chapters in the development of NWP in the USA (Rossby, Eliassen, and Fjörtoft).

1.2 Early Developments

For a long time no one ventured to follow in Richardson's footsteps. The paucity of the observational network and the enormity of the computational task stood as apparently insurmountable barriers to the realization of his dream that one day it might be possible to advance the computations faster than the weather. But with the increase in the density and extent of the surface and upper-air observational network on the one hand, and the development of large-capacity high-speed computing machines on the other, interest has revived in Richardson's problem and attempts have been made to attack it anew.

These efforts have been characterized by a devotion to objectives more limited than Richardson's. Instead of attempting to deal with the atmosphere in all its complexity, one tries to be satisfied with *simplified models* approximating the actual motions to a greater or lesser degree. By *starting with models incorporating only what are thought to be the most important of the atmospheric influences*, and by gradually bringing in others, one is able to proceed inductively and thereby to avoid the pitfalls inevitably encountered when a great many poorly understood factors are introduced all at once.

A necessary condition for the success of this stepwise method is, of course, that the first approximations bear a recognizable resemblance to the actual motions. Fortunately, the science of meteorology has progressed to the point where one feels that at least the main factors governing the large-scale atmospheric motions are well known. *Thus integrations of even the linearized barotropic and thermally inactive baroclinic equations have yielded solutions bearing a marked resemblance to reality.* At any rate, it seems clear that the models embodying the collective experience and the positive skill of the forecast cannot fail utterly. This conviction has served as the guiding principle in the work of the meteorology project at The Institute for Advanced Study [in Princeton, New Jersey] with which the writer has been connected.

As indicated by Charney, Richardson performed a remarkably comprehensive numerical integration of the full primitive equations of motion (Chapter 2). He used a horizontal grid of about 200 km, and four vertical layers of approximately 200 hPa, centered over Germany. Using the observations at 7 UTC (Universal Coordinate Time) on May 20, 1910, he computed the time derivative of the pressure in central Germany between 4 and 10 UTC. *The predicted 6 hr change was 146 hPa, whereas in reality there was essentially no change observed in the surface pressure.* This huge error was discouraging, but it was due mostly to the fact that the initial conditions were *not quasi-geostrophically balanced*, and therefore included large amplitude, fast-moving gravity waves that masked *the initial rate of change* of the meteorological signal in the forecast (Figure 1.2.1). Moreover, if the integration had been continued, it would have suffered "computational blow-up" due to the violation of the Courant–Friedrichs–Lewy (CFL) condition (see Chapter 3), which requires that the time step should be smaller than the grid size divided by the speed of the fastest traveling signal (in this case, horizontally moving sound waves, traveling at about 300 m/s).

Charney (1948, 1949) and Eliassen (1949) solved both of these problems by deriving "filtered" equations of motion, based on quasi-geostrophic (slowly varying) balance, which filtered out (i.e., did not include) gravity and sound waves and were based on pressure fields alone. Charney points out that this approach was justified by the fact that forecasters' experience was that they were able to predict tomorrow's weather from pressure charts alone:

In the selection of a suitable first approximation, Richardson's discovery that the horizontal divergence was an immeasurable quantity had to be taken into account. Here a consideration of forecasting practice gave rise to the belief that this difficulty could be surmounted: forecasts

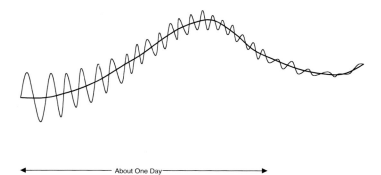

Figure 1.2.1 Schematic of a grid point forecast with slowly varying weather-related variations and superimposed high-frequency gravity waves that decay with time because they are dispersive, as in the shallow water experiment performed by Freeman described by Charney (1951). Note that even though the forecast of the slow waves is essentially unaffected by the presence of gravity waves, the initial time derivative is much larger in magnitude, as obtained by Richardson (1922) when he computed the initial time derivative at a grid point in a primitive equations model.

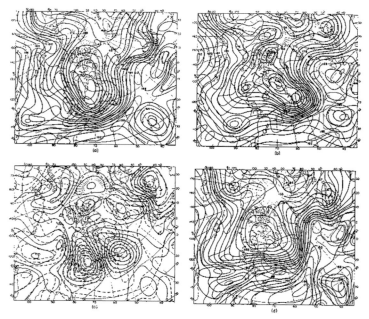

Figure 1.2.2 Forecast of January 30, 1949, 0300 GMT: (a) contours of observed z (height) and $\zeta + f$ (total vorticity) at $t = 0$; (b) observed z and $\zeta + f$ at $t = 24$ hr; (c) observed (continuous lines) and computed (broken lines) 24-hr height change; (d) computed z and $\zeta + f$ at $t = 24$ hr. The height unit is 100 ft and the unit of vorticity is $1/3 \times 10^{-4} s^{-1}$. Note that 1.2.2 (d) is the one day numerical forecast of 1.2.2 (b). ©American Meteorological Society. Used with permission

1.2 Early Developments 11

were made by means of geostrophic reasoning from the pressure field alone – forecasts in which the concept of horizontal divergence played no role.

In order to better understand Charney's comment, we quote an anecdote from Lorenz (1990) on his interactions with Jule Charney:

On another[3] occasion when our conversations had turned closer to scientific matters, Jule was talking again about the early days of NWP. For a proper perspective, we should recall that at the time when Charney was a student, pressure was king. The centers of weather activity were acknowledged to be the highs and lows. A good prognostic chart was one that had the isobars in the right locations. Naturally, then, the thing that was responsible for the weather changes was the thing that made the pressure change. This was readily shown to be the divergence of the wind field. The divergence could not be very accurately measured, and a corollary deduced by some meteorologists, including some of Charney's advisors, was that the dynamic equations could not be used to forecast the weather.

Such reasoning simply did not make sense to Jule. The idea that the wind field might serve instead of the pressure field as a basis for dynamical forecasting, proposed by Rossby, gave Jule a route to follow.[4] He told us, however, that what really inspired him to develop the equations that later became the basis for NWP was a determination to prove, to those who had assured him that the task was impossible, that they were wrong.

He proceeded to describe an unpublished study in which he and J. C. Freeman integrated barotropic primitive equations (i.e., shallow water equations, Chapter 2) that include not only the slowly varying quasi-geostrophic solution, but also fast gravity waves. They initialized the forecast assuming zero initial divergence, and compared the result with a barotropic forecast (with gravity waves filtered out). The results were similar to those shown schematically in Figure 1.2.1: they observed that over a day or so the gravity waves subsided (through a process that we call geostrophic adjustment) and did not otherwise affect the forecast of the slow waves. From this result Charney concluded that numerical forecasting could indeed use the full primitive equations (as eventually happened in operational practice). He listed in the paper the complete primitive equations in pressure coordinates, essentially as they are used in current operational weather prediction, but without heating (diabatic) and frictional terms, which he expected to have minor effects in one- or two-day forecasts. Charney concluded this visionary paper with the following discussion, which includes a list of the physical processes that take place at scales too small to be resolved,

---

[3] The previous occasion was a story about an invitation Charney received to appear on the *Today* show, to talk about how computers were going to forecast the weather. Since the show was at 7 am, Charney, a late riser, had never watched it. "He told us that he felt that he ought to see the show at least once before agreeing to appear on it, and so, one morning, he managed to pull himself out of bed and turn on the TV set, and the first person he saw was a chimpanzee. He decided he could never compete with a chimpanzee for the public's favor, and so he gracefully declined to appear, much to the dismay of the computer company that had engineered the invitation in the first place" (Lorenz, 1990).

[4] The development of the "Rossby waves" phase speed equation $c = U - \beta L^2/\pi^2$ based on the linearized, nondivergent vorticity equation (Rossby, 1939b,a), and its success in predicting the motion of the large-scale atmospheric waves, was an essential stimulus to Charney's development of the filtered equations (Phillips, 1990, 1998).

and are incorporated in present models through "parameterizations of the subgrid-scale physics" (condensation, radiation, and turbulent fluxes of heat, momentum and moisture, Chapter 4, see Stensrud (2007)):

Nonadiabatic and frictional terms have been ignored in the body of the discussion because it was thought that one should first seek to determine how much of the motion could be explained without them. Ultimately they will have to be taken into account, particularly if the forecast period is to be extended to three or more days.

Condensational phenomena appear to be the simplest to introduce: one has only to add the equation of continuity for water vapor and to replace the dry by the moist adiabatic equation. Long-wave radiation effects can also be provided for, since our knowledge of the absorptive properties of water vapor and carbon dioxide has progressed to a point where quantitative estimates of radiational cooling can be made, although the presence of clouds will complicate the problem considerably.

The most difficult phenomena to include have to do with the turbulent transfer of momentum and heat. A great deal of research remains to be done before enough is known about these effects to permit the assignment of even rough values to the eddy coefficients of viscosity and heat conduction. Owing to their statistically indeterminate nature, the turbulent properties of the atmosphere *place an upper limit to the accuracy obtainable by dynamical methods of forecasting*, beyond which we shall have to rely upon statistical methods. But it seems certain that much progress can be made before these limits can be reached.

This paper, which although written in 1951 has not become dated, predicted with almost supernatural vision the path that numerical weather forecasting was to follow over the next five decades. It described the need for objective analysis of meteorological data in order to replace the laborious hand analyses. We now refer to this process as data assimilation (Chapter 5), which uses both observations and short forecasts to estimate initial conditions. Note that at a time at which only one-day forecasts had ever been attempted, Charney already had the intuition that there was an *upper limit* to weather predictability, which Lorenz (1965) later estimated to be about two weeks. However, Charney attributed the expected limit to model deficiencies (such as the parameterization of turbulent processes), rather than to the chaotic nature of the atmosphere, which imposes a limit of predictability even if the model is perfect (Lorenz, 1963a) (see also Chapter 6 on Predictability). Charney was right in assuming that in practice model deficiencies, as well as errors in the initial conditions, would limit predictability. At the present time, however, the state of the art in numerical forecasting has advanced enough that, when the atmosphere is highly predictable, the theoretically estimated limit for weather forecasting (about two weeks) is occasionally reached and even exceeded through techniques such as ensemble forecasting (Chapter 6).

Following the success of Charney et al. (1950), Rossby moved back to Sweden, and was able to direct a group that reproduced similar experiments on a powerful Swedish computer known as BESK. As a result, the first operational (real time) numerical weather forecasts started in Sweden in 1954, with a 3-day forecast using a barotropic model, six months before the start-up of the US operational forecasts (Döös and Eaton, 1957; Wiin-Nielsen, 1991; Bolin, 1999). Japan started operational forecasts in 1959 with a barotropic system similar to Sweden. Anders Persson (2005a,b,c) has written a comprehensive, interesting historical description of how these developments took place in Sweden, in the UK Meteorological Office, and in other countries, where

1.3 Primitive Equations, Global and Regional Models, and Nonhydrostatic Models

operational NWP became operational only in 1965, with a baroclinic model that was delayed by problems associated with the use of too small of a domain (Persson, 2005a,b,c).

As predicted by Charney (1951, 1962), the filtered (quasi-geostrophic) equations, although very useful for understanding of the large-scale extratropical dynamics of the atmosphere, were not accurate enough to allow continued progress in NWP, and were eventually replaced by primitive equation models, as envisioned by Bjerknes (1904) and Richardson (1922).* The primitive equations (Chapter 2) are conservation laws applied to individual parcels of air: conservation of the three-dimensional momentum (equations of motion), conservation of energy (first law of thermodynamics), conservation of dry air mass (continuity equation), and equations for the conservation of moisture in all its phases, as well as the equation of state for perfect gases. Unlike the quasi-geostrophic equations, they include in their solution fast gravity and sound waves, and therefore in their space and time discretization they require the use of smaller time steps, or alternative techniques that slow them down (Chapter 3). For models with a horizontal grid size larger than 10 km, it is customary to replace the vertical component of the equation of motion with its hydrostatic approximation, in which the vertical acceleration is considered negligible compared with the gravitational acceleration (buoyancy). With this approximation, it is convenient to use atmospheric pressure, instead of height, as a vertical coordinate.

The continuous equations of motions are solved by discretization in space and in time using, for example, finite differences (Chapter 3). The accuracy of a model is very strongly influenced by the spatial resolution: in general, the higher the resolution, the more accurate the model. Increasing resolution, however, is extremely costly. For example, doubling the resolution in the three space dimensions also requires halving the time step in order to satisfy conditions for computational stability. Therefore, the computational cost of doubling the spatial resolution is a factor of 2^4 (three space and one time dimensions). Modern methods of discretization attempt to make the increase in accuracy less onerous by the use of semi-implicit and semi-Lagrangian time schemes. These schemes (pioneered by Canadian scientists under the leadership of André Robert) have less stringent stability conditions on the time step, and more accurate space discretization. Nevertheless, there is a constant need for higher resolution in order to improve forecasts, and as a result running atmospheric models has always been a major application of the fastest supercomputers available.

When the "conservation" equations are discretized over a given grid size (typically from a few to several hundred kilometers) it is necessary to add "sources and sinks" terms due to small-scale physical processes that occur at scales that cannot be explicitly resolved by the models. As an example, the equation for water vapor conservation in pressure coordinates is typically written as

$$\frac{\partial \bar{q}}{\partial t} + \bar{u}\frac{\partial \bar{q}}{\partial x} + \bar{v}\frac{\partial \bar{q}}{\partial y} + \bar{\omega}\frac{\partial \bar{q}}{\partial p} = \overline{E} - \overline{C} + \frac{\partial \overline{\omega' q'}}{\partial p}, \tag{1.3.1}$$

where q is the mixing ratio between water vapor and dry air mass, x and y are horizontal coordinates with appropriate map projections, p is pressure, t is time, u and v are the horizontal air velocity (wind) components, $\omega = dp/dt$ is the vertical velocity in pressure coordinates, and the product of primed variables represents turbulent transports of moisture on scales unresolved by the grid used in the discretization, with the overbar indicating a spatial average over the grid of the model. It is customary to call the left-hand side of the equation, the "dynamics" of the model, which is computed explicitly (Chapter 3).

The right-hand side represents the so-called "physics" of the model. For the moisture equation, it includes the effects of physical processes such as evaporation and condensation $\overline{E} - \overline{C}$, and turbulent transfers of moisture that take place at small scales that cannot be explicitly resolved by the "dynamics." These *subgrid-scale physical processes*, which are sources and sinks for the equations, are then "parameterized" in terms of the variables explicitly represented in the atmospheric dynamics (Chapter 4).

Two types of atmospheric models are in use for NWP: *global* and *regional* models (Chapter 3). Global models are generally used for guidance in medium-range forecasts (3 or more days), and for climate simulations. At NCEP, for example, the global models are run through 16 days every day. Because the horizontal domain of global models is the whole Earth, they usually cannot be run at high resolution. For more detailed forecasts it is necessary to increase the resolution, and this can only be done over limited regions of interest to the weather center that runs the models.

Regional models are used for shorter-range forecasts (typically 1–3 days), and are run with a resolution two or more times higher than global models. For example, the NCEP global model in 1997 was run with 28 vertical levels, and a horizontal resolution equivalent to 100 km for the first week, and 200 km for the second week. The regional model was run with a horizontal resolution of 29 km and 50 levels. Because of their higher resolution, regional models have the advantage of higher accuracy and the ability to reproduce smaller-scale phenomena such as fronts, squall lines, convection and much better orographic forcing than global models. On the other hand, regional models have the disadvantage that, unlike global models, they are not "self-contained" because they require *lateral boundary conditions* at the borders of the horizontal domain. These boundary conditions must be as accurate as possible, because otherwise the interior solution of the regional models quickly deteriorates. Therefore it is customary to "nest" the regional models within another model with coarser resolution, whose forecast provides the boundary conditions. For this reason, regional models are used only for short-range forecasts. After a certain period, which is proportional to the size of the model, the information contained in the high-resolution initial conditions is "swept away" by the influence of the boundary conditions, and the regional model becomes merely a "magnifying glass" for the coarser model forecast in the regional domain. This can still be useful, for example, in climate simulations performed for long periods (seasons to decades), and which therefore tend to be run at coarser resolution. A "regional climate model" can provide a more detailed version of the coarse climate simulation in a region of interest. Major NWP centers in Europe (United Kingdom,[5] France,[6] Germany[7]) and Japan,[8] Australia,[9] and Canada[10] have similar global and regional models, whose details can be obtained at their websites.

Most national centers that have a global model also use it to provide forecast boundary conditions to regional models with higher resolution. The regional models are now non-hydrostatic, even though the global model may still be hydrostatic (Chapter 3). There is a tendency toward the use of non-hydrostatic models that can be used globally as well. For example, in the United States, the regional non-hydrostatic Weather Research and Forecast (WRF) models are widely used. They have two alternative dynamical cores, the WRF-Advanced Research WRF developed at NCAR, and the WRF-Non-hydrostatic Mesoscale Model (WRF-NMM), developed at NCEP, as well as a version adapted for hurricane forecasting (HWRF). Details about these and other regional models can be obtained at the websites of the developers.

The hydrostatic approximation involves neglecting vertical accelerations in the equation of motion, except for gravitational acceleration. This is a very good approximation as long as horizontal scales of motion are larger than the vertical scales. The main advantage of the hydrostatic equation (Chapter 2) is that it filters sound waves (except those propagating horizontally, or Lamb waves). Because of the problem of computational instability, the absence of vertically propagating sound waves allows the use of larger time steps (the Lamb waves are generally handled with semi-implicit time schemes, discussed in Section 3.2.5).

In order to represent smaller-scale phenomena such as storms or convective clouds that have vertical accelerations not negligible compared to buoyancy forces, it is necessary to use the equations of motion without making the hydrostatic approximation, i.e., with non-hydrostatic models discussed in Section 3.6. Therefore, most regional models are non-hydrostatic and some global models are already non-hydrostatic as well. Since the widely used global models based on spectral methods become inefficient at grid sizes smaller than 10 km, the next generation of global models (such as the new FV3 model developed at GFDL (www.gfdl.noaa.gov/fv3/) and implemented at NCEP are non-hydrostatic, which will blur the separation between global and regional models.

1.4 Data Assimilation: Determination of the Initial Conditions for the Computer Forecasts

As indicated previously, NWP is an initial-value problem: given an estimate of the present state of the atmosphere based on a short forecast and the recent observations,

[5] www.metoffice.gov.uk/
[6] www.meteo.fr/
[7] www.dwd.de/
[8] www.jma.go.jp/
[9] www.bom.gov.au/
[10] www.canada.ca/en/environment-climate-change.html

the model simulates (forecasts) its evolution.* The problem of determination of the initial conditions for a forecast model is very important and complex, and has become a science in itself (Daley, 1991). In this section we briefly introduce methods that have been used for this purpose: the successive corrections method (SCM), optimal interpolation (OI), variational methods in three and four dimensions (3D-Var and 4D-Var), ensemble Kalman filtering (EnKF) and combinations of variational and EnKF known as "hybrids." We discuss this subject in detail in Chapter 5 and refer the reader to Daley (1991) for a comprehensive text on early atmospheric data analysis, and to Law et al. (2015) for a more mathematical introduction to modern data assimilation.

In their pioneering experiments, Richardson (1922) and Charney et al. (1950) used, as initial conditions, hand interpolations of the available observations to grid points, which were manually digitized, a very time-consuming procedure. The need for an automatic "objective analysis" quickly became apparent (Charney, 1951), and interpolation methods fitting data to grids were developed (e.g., Panofsky (1949); Gilchrist and Cressman (1954); Barnes (1964, 1978)). However, there is an even more important problem than spatially interpolating observations to gridded fields: the data available are not enough to initialize current models. Modern primitive equations models have a number of degrees of freedom of the order of 10^{7-9}. For example, a latitude-longitude model with a resolution of 1° in the horizontal and 20 vertical levels, as typically used in the 1990s, would have $360 \times 180 \times 20 = 1.3 \times 10^6$ grid points. At each grid point, we have to carry the values of at least four prognostic variables (two horizontal wind components, temperature, moisture), and the surface pressure for each column, giving over 5 million variables that need to be given an initial value. For any given time window of ± 3 hr, there were at that time typically $10-100$ thousand observations of the atmosphere, two orders of magnitude less than the number of degrees of freedom of the model. Moreover, their distribution in space and time is very non-uniform (Figure 1.4.1), with regions like North America and Eurasia that are relatively data-rich in conventional data such as rawinsondes, while others like the Southern Hemisphere and the oceans, much more poorly observed. We also show ocean observations in Figures 1.4.2 and 1.4.3.

For this reason, it became obvious rather early that it was necessary to use additional information (denoted *background, first guess* or *prior information*) to prepare initial conditions for the forecasts (Bergthorsson et al., 1955). Initially climatology was used as a first guess (e.g., Gandin (1963)), but as the forecasts became better, a short-range forecast was chosen as first guess or "prior" in the operational data assimilation systems or "analysis cycles." The intermittent data assimilation cycle shown schematically in Figure 1.1.4 is continued in present-day operational systems, which typically use a 6-hr cycle performed four times a day.

In the 6-hr data assimilation cycle for a global model, the background field is a model 6-hr forecast $\mathbf{x}^b$ (an array). To obtain the background or first guess "observations," the model forecast is interpolated to the observation location, and converted from model variables $\mathbf{x}^b$ (such as temperatures or moisture) to observation-like variables $\mathbf{y}^b$ (such as satellite radiances or radar reflectivities). The first guess (model forecast) of the observations is therefore $H(\mathbf{x}^b)$, where H is the *observation operator*

1.4 Data Assimilation

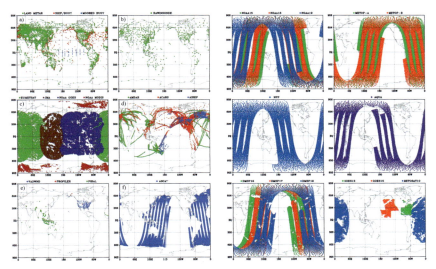

Figure 1.4.1 Typical distribution of observations in a +/−3 hr time window, courtesy of D. Kleist. On the top left are "conventional" (i.e., pre-satellite) observations, including a) surface land and ocean observations, and rawinsondes; b) additional vertical wind profiles; c) Atmospheric motion vectors (AMVs) from geostationary satellites, and MODIS winds. Geostationary satellites located at a fixed position at the Equator, at different longitudes, d) Aircraft Weather Reports; e) Wind sounders; f) Advanced Scatterometer (ASCAT). On the right side there are polar orbiter soundings: top left, NOAA-15/18/19, top right, MetOp-A/B, middle left, (Suomi) National Polar Partnership (NPP); middle right, NASA research Aqua; bottom left, Defense Meteorological Satellite Program (DMSP) satellites, bottom right, Geostationary satellite soundings, GOES-13/15 and Meteosat-10. Please see (Lord et al., 2016) for comprehensive discussion of the observing systems available around 2015 and their replacements.

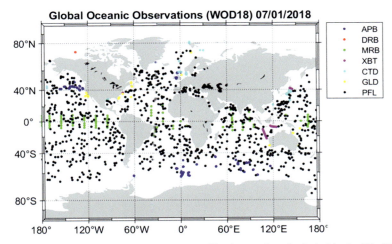

Figure 1.4.2 Distribution of the ocean profile observations included in the World Ocean Database 2018 (WOD18) on July 1 in 2018. More details can be found in Boyer et al. (2018).

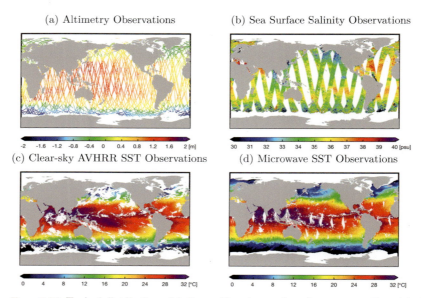

Figure 1.4.3 Typical distribution of daily satellite observations for oceans on June 1 in 2019. (a) Altimetry absolute dynamic topography retrievals from CryoSat-2, Jason-2/3, Sentinel-3A/3B and SARAL, (b) sea surface salinity retrievals from SMAP, (c) clear-sky AVHRR sea surface temperature from NOAA-18/19 and MetOp-A/B, and (d) microwave sea surface temperature from GMI, AMSR2, and WindSat.

that performs the necessary interpolation and transformation from model variables to observation space. The difference $\mathbf{y}^o - H(\mathbf{x}^b)$ between the real observations $\mathbf{y}^o$ and the model first guess is denoted "observational increments" or "innovations," since it represents the new information brought by the observations. The analysis $\mathbf{x}^a$ is obtained by adding the innovations to the model forecast (first guess) with weights $\mathbf{W}$ that are determined based on the estimated statistical error covariances of the forecast and the observations:

$$\mathbf{x}^a = \mathbf{x}^b + \mathbf{W}[\mathbf{y}^o - H(\mathbf{x}^b)] \qquad (1.4.1)$$

Different analysis schemes (SCM, OI, 3D-Var, 4D-Var, EnKF and hybrid methods combining Var and EnKF) are all based on (1.4.1) but differ by the approach taken to combine the background and the observations to produce the weights and the analysis. Earlier methods such as the SCM (Bergthorsson et al., 1955; Cressman, 1959; Barnes, 1964) had empirically estimated weights $\mathbf{W}$ in (1.4.1) as a function of the distance between the observation and the grid point, and the analysis scheme is then iterated several times. In OI (Gandin, 1963) the matrix of weights $\mathbf{W}$ was determined for the first time from the minimization of the statistical analysis errors at each grid point. Sasaki (1958, 1970) proposed to carry out numerical analyses by minimization of a cost function that measures the variance of the difference between observed and analyzed values, as well as how well the dynamic equations are satisfied. In 3D-Var, one defines a cost function proportional to the square of the distance between the

1.4 Data Assimilation   19

analysis and both the background and the observations, at the analysis time, and the analysis is the state of the model that minimizes the cost function. Lorenc (1986) proved that the OI and 3D-Var approaches are equivalent if the 3D-Var cost function is defined as:

$$J = \frac{1}{2}[y^o - H(\mathbf{x})]^T \mathbf{R}^{-1}[y^o - H(\mathbf{x})] + \frac{1}{2}[\mathbf{x} - \mathbf{x}^b]^T \mathbf{B}^{-1}[\mathbf{x} - \mathbf{x}^b] \qquad (1.4.2)$$

The cost function J in (1.4.2) measures the distance of a field x to the observations (the first term in the cost function) and the distance to the first guess or background (the second term in the cost function). These distances are scaled by the *observation error covariance* $\mathbf{R}$ and by the *background error covariance* $\mathbf{B}$ respectively. The minimum of the cost function is obtained for $\mathbf{x} = \mathbf{x}^a$, which is defined as the "analysis." The analysis obtained in (1.4.1) and (1.4.2) is the same if the weight matrix in (1.4.1) is given by one of these two equivalent formulations

$$\mathbf{W} = \mathbf{B}\mathbf{H}^T(\mathbf{H}\mathbf{B}\mathbf{H}^T + \mathbf{R})^{-1} = \mathbf{A}\mathbf{H}^T\mathbf{R}^{-1} \qquad (1.4.3)$$

where $\mathbf{A}$ is the *analysis error covariance*. Its inverse, the analysis "precision" $\mathbf{A}^{-1}$, is equal to the sum of the precision of the forecast $\mathbf{B}^{-1}$ plus the precision of the observations $\mathbf{R}^{-1}$ expressed in model space, implying that *the analysis combines the information of both the model forecast and the observations:*

$$\mathbf{A}^{-1} = \mathbf{B}^{-1} + \mathbf{H}^T\mathbf{R}^{-1}\mathbf{H} \qquad (1.4.4)$$

The reader should not worry about these apparently daunting data assimilation equations. In Chapter 5, they are introduced first with very intuitive "baby examples" that make their interpretation completely clear and understandable.

In summary, the initial conditions for model forecasts were first generated interpolating observations to the model grid, so that all the information needed to start a forecast came only from interpolating the observations. This was replaced by the *analysis cycle*, Figure 1.1.4, which combines a short forecast (typically 6 hr) that contains much information accumulated from past observations, and the new observations, and is always expressed as in Equation (1.4.1): The analysis is equal to the background plus the *statistically weighted* difference between the new observations and their value as forecasted by the model. also known as *innovation*.

Optimal interpolation (between the first guess and the new observations) was carried out for the first time by minimizing the analysis errors (Gandin, 1963) and it became adopted in several operational centers, but using a short model forecast as a first guess rather than using climatology, like Gandin did. 3D-Var (a variational analysis carried out in space at the analysis time) was first implemented at NCEP (Parrish and Derber, 1992). They addressed the difficult problem of creating an estimate of the model error covariance $\mathbf{B}$ by taking differences between one and two day forecasts verifying at the same time, and averaging over many days, as representative of the dynamical structure of forecast errors (an approach known as "the NMC method"). Because of its robustness and accuracy, 3D-Var became almost universally adopted during the 1990s, replacing OI.

Both OI and 3D-Var assume that the background error covariance $\mathbf{B}$ is constant, i.e., it is a climatological average of the estimation of many different forecast errors, obtained, for example, using the NMC method, and appropriately scaled to make them a realistic representation of the average 6 hr forecast error covariance. However, in the same way that weather is highly variable, weather forecast errors vary a lot from day to day (Chapters 5 and 6). Figure 5.5.1 shows very clearly that neglecting the variability of the forecast errors is not a good approximation.

Two advanced methods of data assimilation were successfully introduced in the last two decades to account for these "errors of the day": 4D-Var and ensemble Kalman filter (EnKF). Variational and EnKF methods are also being combined into "hybrids."

The first method to include "errors of the day" was 4D-Var, an extension of 3D-Var to 4 dimensions that allows the observations to be distributed over time in the assimilation window. A first version of 4D-Var was implemented at ECMWF at the end of 1997 (Rabier et al., 2000) and later was adopted by a number of centers, resulting in improved forecasts because in 4D-Var the forecast error covariance implicitly evolves within the data assimilation, and the observations are assimilated at their right time. Its disadvantages are that it is computationally much more expensive than 3D-Var, and that its method of solution requires the *adjoint model* (the transpose of *the linear tangent model*, Chapters 5 and 6, which requires complicated coding). In addition, although **B** implicitly evolves within the assimilation window, it still requires a constant first guess. At ECMWF the flow-dependence of the background error covariance is further estimated by performing an ensemble of lower resolution perturbed 4D-Var assimilations (Isaksen et al., 2010; Bonavita et al., 2015b). This Ensemble of 4D-Var Data Assimilations (EDA) has been very successful, but it is rather expensive.

Michael Ghil proposed using Kalman filter (KF) for advanced data assimilation, since in KF the background error covariance is forecasted at every assimilation step (Bengtsson et al., 1981). KF is thus an "ultimate" data assimilation system that forecasts the "errors of the day," but it is computationally completely unfeasible except for very small models.

Evensen (1994), proposed instead the second advanced method, ensemble Kalman filter (EnKF) where the background error covariance is estimated from the ensemble perturbations around the mean. Burgers et al. (1998), and, independently, Houtekamer and Mitchell (1998), formulated this type of EnKF where the Kalman filter is approximated with an ensemble of data assimilations, each of which uses observations perturbed with random errors. This type of filter is known as "perturbed observations" or "stochastic" EnKF. In the early 2000s a new class of EnKF methods where a single analysis ensemble is obtained using a square root approach (and the observations are not perturbed) were proposed. Tippett et al. (2003) discussed and compared three types of square root filters by Anderson (2001), Bishop et al. (2001), and Whitaker and Hamill (2002). Simultaneously, Ott et al. (2004) proposed a square root filter known as local ensemble Kalman filter (LEKF), with independent analyses performed at each grid point in local regions as in Kalnay and Toth (1994). Hunt et al. (2007) later combined the LEKF with the transform approach (Bishop et al., 2001) and created the local ensemble transform Kalman filter (LETKF). Both the perturbed observations or "stochastic" EnKFs, and the square root or "deterministic" EnKFs are widely used (see Houtekamer and Zhang (2016) for a thorough review of EnKF).

1.4 Data Assimilation

The advantages of EnKFs are their relative simplicity, including not requiring an adjoint model, and the fact that they account for the "errors of the day." Their main disadvantage is that the rank of the evolving forecast error covariance estimated from an ensemble of, let's say, $K = \mathcal{O}(100)$ members is $K-1$, much smaller than the rank of the forecast error covariance, which is of the order of the model size. This problem is partially addressed but not completely solved by the "localization" of the background error covariance (Whitaker and Hamill, 2002; Ott et al., 2004; Greybush et al., 2011).

Hamill and Snyder (2000) proposed a "hybrid" approach combining the variational and the ensemble estimation of the forecast error covariance. Several centers (JMA, UKMO, NCEP) have tested this "covariance hybrid" approach, implemented at NCEP in 2012 with a significant improvement of forecast scores (Kleist, 2012; Wang et al., 2013b; Kleist and Ide, 2015a,b). Penny (2014) proposed a different "gain hybrid" which is simpler to implement. A hybrid known as 4DEnVar became operational in Canada in 2015, replacing for the first time a previous 4D-Var system (Buehner et al., 2013). Bonavita et al. (2015a) and Hamrud et al. (2015), at ECMWF, compared the LETKF with a 4D-Var of similar resolution and found that they had similar forecast skill. The Gain hybrid of 4D-Var and LETKF, however, significantly outperformed both systems. Data assimilation is clearly an evolving science.

Finally, we note that in an analysis cycle, no matter which analysis scheme is employed, the use of the model forecast is essential in achieving "four-dimensional data assimilation" (4DDA). This means that the data assimilation cycle is like a long model integration, during which the model is "nudged" by the observational increments in such a way that it remains close to the real atmosphere. The importance of the model cannot be overemphasized: it accumulates information from previous observations, it transports information from data-rich to data-poor regions, and it provides a complete estimation of the four-dimensional state of the atmosphere. Figure 1.4.4 presents the rms difference between the 6-hr forecast (used as a first guess) and the

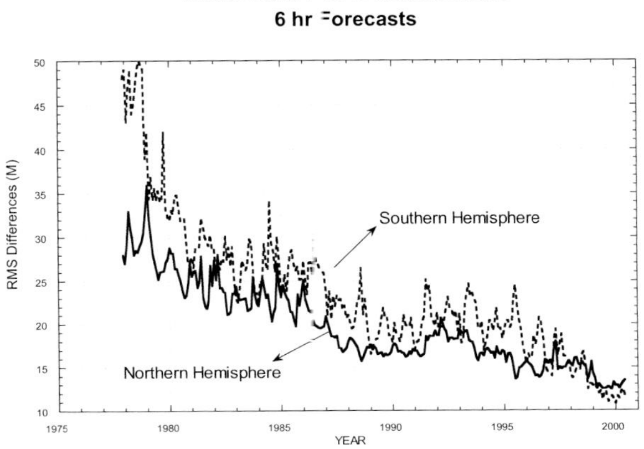

Figure 1.4.4 RMS observational increments (differences between 6-hr forecast and rawinsonde observations) for 500-hPa heights (data courtesy of Steve Lilly, NCEP).

rawinsonde observations from 1978 to 2000 (in other words, the rms of the observational increments for 500-hPa heights). It should be noted that the rms differences are not necessarily *forecast* errors, since the observations also contain errors. In the NH, the rms differences reduced from about 30 m in the late 1970s, to about 13 m in 2000, equivalent to a mean temperature error of about 0.65 K, similar to rawinsonde observational errors. In the Southern Hemisphere, the improvements are even larger, with the differences decreasing from about 47 m to about 12 m due to the introduction of satellite data starting in 1979 (the First GARP Global Experiment, or FGGE). The improvements in these short-range forecasts are a reflection of improvements in the model, the analysis scheme used to assimilate the data, and the quantity and quality of the observations (Chapter 5). Figure 1.4.5 compares the skill of the ECMWF forecasts in the Northern and the Southern extratropics. This record starts after the beginning of the era of satellite observations in 1979, and until 1995 the meteorological centers used temperature retrievals based on climatological first guess (an approach that is less than optimal). This made the NH, with many more rawinsondes than the SH, have much higher skill. If we compare these two figures, we can see that before 1979, when satellite retrievals of temperature began to be assimilated, the SH forecast errors were huge compared to the NH. After satellite retrievals were introduced, in the 1980s, the gap between the NH and SH decreased, but, because the methods of retrievals were not optimal, the NH, with many more rawinsondes, remained significantly more skillful. After the introduction of the assimilation of radiances, which is generally much

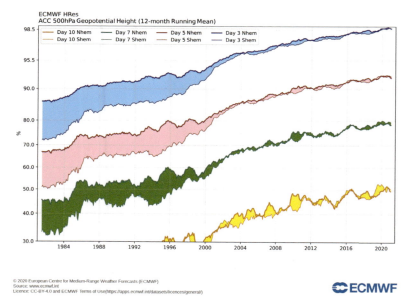

Figure 1.4.5 Comparison of the 12-month running mean 500hPa anomaly correlation of the ECMWF operational forecasts, for days 3, 5, 7, and 10, for the Northern and the Southern Hemisphere, from 1981 to 2021 (Figure courtesy of ECMWF, under the Licenses of CC-BY-4.0 and ECMWF Terms of Use).

1.5 Operational NWP and the Evolution of Forecast Skill

better than assimilating retrievals, increased satellite data, and improved methods of data assimilation (see Tables 1.5.1 and 1.5.2), the relative importance of the rawinsondes decreased, so that currently the skill of the SH forecasts is comparable to that of the NH, even though it has far less rawinsonde coverage.

1.5 Operational NWP and the Evolution of Forecast Skill

Major milestones of operational numerical weather forecasting include the paper by Charney et al. (1950) with the first successful forecast based on the quasi-geostrophic equations, and the first operational forecasts performed in Sweden in September 1954, followed 6 months later by the first operational (real time) forecasts in the USA.* We describe in what follows the evolution of NWP at NCEP, but as mentioned before, similar developments took place at several major operational NWP centers: in the UK, France, Germany, Japan, Australia, and Canada.

The early history of operational NWP at the NMC (now NCEP) has been reviewed by Shuman (1989) and Kalnay et al. (1998). It started with the organization of the Joint Numerical Weather Prediction Unit (JNWPU) on July 1, 1954, staffed by members of the US Weather Bureau (later the NWS), the Air Weather Service of the US Air Force, and the Naval Weather Service.[11] Shuman (1989) noted that in the first few years, numerical predictions could *not* compete with those produced manually. They had several serious flaws, among them overprediction of cyclone development. Far too many cyclones were predicted to deepen into storms. With time, and with the joint work of modelers and practising synopticians, major sources of model errors were identified, and operational NWP became the central guidance for operational weather forecasts.

In his paper, Shuman (1989) discusses some of the major system improvements that enabled NWP forecasts to overtake and surpass subjective forecasts. The first major improvement took place in 1958 with the implementation of a barotropic (one-level) model, which was actually a reduction from the three-level model first tried, but which included better finite differences and initial conditions derived from an objective analysis scheme (Bergthorsson et al., 1955; Cressman, 1959). It also extended the domain of the model to an octagonal grid covering most of the NH down to $9-15°N$. These changes resulted in numerical forecasts that for the first time were competitive with subjective forecasts, but in order to implement them JNWPU had to wait for the acquisition of a more powerful supercomputer, an IBM 704, to replace the previous IBM 701. This pattern of forecast improvements which depend on a combination of the better use of the data and better models, and would require more powerful supercomputers in order to be executed in a timely manner has been repeated throughout the history of operational NWP. Table 1.5.1 (adapted from Shuman (1989) and extended to about 2000, by P. Caplan, pers. comm.) summarizes the major improvements in the first

---

[11] In 1960 the JNWPU reverted to three separate organizations: the National Meteorological Center (NWS), the Global Weather Central (US Air Force) and the Fleet Numerical Oceanography Center (US Navy).

Table 1.5.1 Major changes in the NMC/NCEP global modeling and data assimilation from 1985 to 1998 (from compilations by Fedor Mesinger and Geoffrey DiMego, pers. comm.)

Year	Operational model	Computer
1955	Princeton three-level quasi-geostrophic model (Charney, 1954). Not used by the forecasters	IBM 701
1958	Barotropic model with improved numerics, objective analysis initial conditions, and an octagonal domain.	IBM 704
1962	Three-level quasi-geostrophic model with improved numerics	IBM 7090 (1960) IBM 7094 (1963)
1966	Six-layer primitive equations model (Shuman and Hovermale, 1968)	CDC 6600
1971	LFM model (Howcroft, 1971) (first regional model at NMC), used for 20 years but frozen in 1986.	
1978	Seven-layer primitive equation model (hemispheric)	
1978	OI (Bergman, 1979)	Cyber 205
1980	Global spectral model, R30/12 layers (Sela, 1980)	
1985	GFDL physics implemented on the global spectral model R40/18 layers	
1986	New OI code with new statistics	
1987		2nd Cyber 205
1987	T80/18 layers, Penman-Montieth evapotranspiration (Caplan and White, 1989; Pan, 1990)	
1988	Hydrostatic complex quality control (CQC) (Gandin, 1988)	
1990		Cray YMP
1991	T126/L18 and improved physics, mean orography (Kanamitsu et al., 1991)	
1991	New 3D-Var (Parrish and Derber, 1992)	
1992	First ensemble system based on breeding and Lagged Average Forecasting, providing 14 daily forecasts to 10 days. (Toth and Kalnay, 1993; Tracton and Kalnay, 1993)	
1993	T126/28 layers. Simplified Arakawa-Schubert cumulus convection (Pan and Wu, 1995).	
1994		Cray C90
1994	Second ensemble system: five pairs of bred forecasts at 00Z, two pairs at 12Z, extension of AVN, a total of 17 global forecasts started every day to 16 days	
1995	New soil hydrology (Pan and Mahrt, 1987), radiation, clouds, improved data assimilation. This was the model used in the NCEP/NCAR Reanalysis system, but at T62/L28.	
1995	Direct assimilation of TOVS cloud-cleared radiances (Derber and Wu, 1998). New planetary boundary layer (PBL) based on nonlocal diffusion (Hong and Pan, 1996).	
1997	New observational error statistics. Changes to assimilation of TOVS radiances and addition of other data sources	
1998	Assimilation of noncloud-cleared radiances (Derber et al. pers. comm.). Improved physics.	
1998	T170/40 layers (to 3.5 days). Improved physics. 3D ozone data assimilation and forecast. Nonlinear increments in 3D-Var. Resolution reduced to T62/28levels on October 1998 and upgraded back in January 2000	IBM SV2 256 processors
2000	Ensemble resolution increased to T126 for the first 60 hr	
2000	Tropical cyclones relocated to observed position every 6 hr	

1.5 Operational NWP and the Evolution of Forecast Skill

Table 1.5.2 Major changes in the ECMWF operational system (Magnusson and Källén, 2013). ©American Meteorological Society. Used with permission

Year	Change
1979	Start of operations (N48 gridpoint model)
1983	T63/L16 (spectral model)
1985	T106/L19
1991	T213/L31
1991	Semi-Lagrangian scheme
1996	3D-Var
1997	4D-Var
1998	T319/L31
1999	T319/L50
2000	T511/L60
2006	T799/L91
2007	New convection scheme
2010	T1279/L91
2011	Ensemble of data assimilations

40 years of operational numerical forecasts at the NWS. The first primitive equations model was implemented in 1966 (Shuman and Hovermale, 1968). The first regional, higher-resolution system (limited fine mesh or LFM model (Howcroft, 1971)) was implemented in 1971, and was the first to be upgraded by the use of fourth order differences (Gerrity et al., 1972). The LFM was remarkable because it remained in use for over 20 years, and it was the basis for the Model Output Statistics (MOS), but its development was frozen in 1986. A more advanced model and data assimilation system, the Regional Analysis and Forecasting System (RAFS) was implemented as the main guidance for North America in 1982. The RAFS was based on the two-way Nested Grid Model (NGM, Phillips, 1979) and on a regional OI scheme (DiMego, 1988). An 8 levels global model was developed in 1973, and in 1974 a 9 layers global model developed by Stackpole (1978) started to be used which included the Robert (1966) time filter for the leap-frog scheme and Fourier filtering of short zonal waves at high latitudes. It was initialized from an analysis of the Hough (tidal waves) spectral functions developed by (Flattery, 1971) which insured a balanced initialization. This finite difference global model was replaced in 1980 by the global spectral model (Sela, 1980) of 12 levels and rhomboidal truncation R30. Higher resolution and improved versions of this model remained in use at NCEP. At the time of this writing, the spectral model is being replaced by the Next Generation Global Prediction System (NGGPS), whose dynamical core will be the Finite Volume cubed sphere (FV3) selected among a number of non-hydrostatic advanced models.

What has been the impact of these many improvements?

Figure 1.5.1 shows threat scores for precipitation predictions made by expert forecasters from the NCEP Hydrometeorological Prediction Center (HPC, the Meteorological Operations Division of the former NMC) and annually averaged . The threat score (TS) is defined as the intersection of the predicted area of precipitation exceeding a particular threshold (P), in this case 0.5 inches in 24 hr, and the observed area

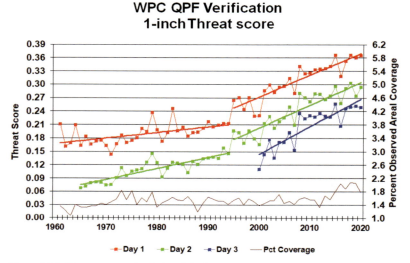

Figure 1.5.1 Threat scores for 3-day 1-inch quantitative precipitation forecast by human forecasters at NWS Weather Prediction Center (Figure courtesy of Louis Uccellini, NWS).

(O), divided by the union of the two areas: $TS = (P \cap O)/(P \cup O)$. The bias (not shown) is defined by P/O. The TS, also known as critical success index (CSI) is a particularly useful score for quantities that are relatively rare. Figure 1.5.1 indicates that the forecasters skill in predicting accumulated precipitation has been increasing with time, and that the average skill in the 1-day forecast of precipitation in the late 1960s was similar to the 2-day forecast skill in the mid 1990s, and to the 3-day forecast skill in the mid 2000s. The sharp skill improvement after the early 1990s is due to the introduction of global operational models and higher-resolution regional models in the mid 1980s, and their continuous improvements into the 1990s and beyond (i.e., better physics, higher model resolution, more observations, and improved data assimilation systems). Beyond the first 6–12 hr, the forecasts are based mostly on numerical guidance, so that the improvement reflects to a large extent improvements of the numerical forecasts, which the human forecasters in turn improve upon based on their knowledge and expertise. The forecasters also have access to several model forecasts, and they use their judgment in assessing which one is more accurate in each case. This constitutes a major source of the important "value-added" by the human forecasters.

The relationship between the evolution of human and numerical forecasts is clearly shown in a record compiled by the late L. Hughes (1987), reproduced in Figure 1.5.2. It is the first operational score maintained for the "medium-range" (beyond the first two days of the forecasts). The score used by Hughes was a standardized anomaly correlation (SAC), which accounted for the larger variability of sea level pressure at higher latitudes compared to lower latitudes. Unfortunately, the SAC is not directly comparable to other scores such as the AC (discussed in the next section). The fact that until 1976 the 3-day forecast scores from the model were essentially constant is an indication that their rather low skill was more based on synoptic experience than on

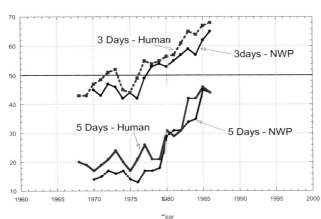

Figure 1.5.2 Hughes data: Comparison of the forecast skill in the medium-range from NWP guidance and from human forecasters.

model guidance. The forecast skill started to improve after 1977 for the 3-day forecast, and after 1980 for the 5-day forecast. Note that the human forecasts are on the average significantly more skillful than the numerical guidance, but it is the improvement in NWP forecasts that drives the improvements in the subjective forecasts.

Figure 1.5.3 shows the decadal improvements of hurricane predictions at the National Hurricane Center (NHC). There is substantial improvement in the track forecast skill in the 1990s, the first decade that the NHC forecasts were allowed as a formal policy to use models in their forecast out to day three. The track forecast skill improvements over decades were so tremendous that the official forecasts were extended to Day 5 in the 2000–2009 decade, and NHC are now experimenting with forecasts out to day 7. Note that now the average 7-day track forecast error is smaller than that during 1980–1989. Below is the chart from the National Hurricane Center tracking improvements in the official intensity forecasts. Dramatic improvements in the intensity forecasts have been achieved since 1990–1999. These remarkable achievements emanate from the improvements in the global models and the development of very high resolution hurricane models by the NOAA Environmental Modeling Center (EMC) through the Hurricane Forecast Improvement Program (HFIP) and the NOAA Geophysical Fluid Dynamics Laboratory (GFDL). The latter is probably the most crucial contributor to these improvements.

1.6 Weather Predictability, Ensemble Forecasting, and Seasonal to Interannual Prediction

In a series of remarkable papers, Lorenz (1963a,b, 1965, 1969) made the fundamental discovery that *even with perfect models and perfect observations, the chaotic nature*

An Overview of Numerical Weather Prediction

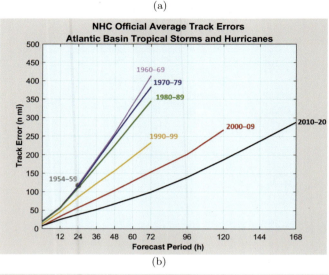

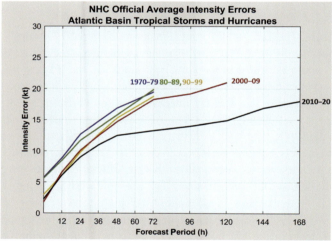

Figure 1.5.3 Decadal verification of the National Hurricane Center (NHC) official average (a) track and (b) intensity forecast error for the hurricane forecasts over the Atlantic Basin from NHC (Figure courtesy of Louis Uccellini, NWS).

of the atmosphere would impose a finite limit of about two weeks to the predictability of the weather.* He discovered this by running a very simple atmospheric model, introducing exceedingly small perturbations in the initial conditions due to computer truncation, and running the model again. With time, the small difference between the two forecasts became larger and larger, until after about two model weeks, the forecasts were as different as two randomly chosen states of the model. In the 1960s, Lorenz's discovery, which started the theory of chaos, was "of academic interest only" and not relevant to operational weather forecasting, since at that time the skill

1.6 Weather Predictability, Ensemble Forecasting, and Seasonal

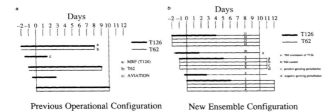

Figure 1.6.1 First NCEP ensemble forecasting system, implemented on December 7, 1992 from Toth and Kalnay (1993). In the new ensemble configuration (Tracton and Kalnay, 1993) the high resolution T126 model was extended at lower resolution (T62) to 12 days, and a T62 control forecast was combined with a pair of positive and negatively perturbed forecasts using the method of breeding (Toth and Kalnay, 1993) as explained in Chapter 6. Published (1993) by the American Meteorological Society.

of even two-day operational forecasts was very low. Since then, however, computer-based forecasts have improved so much that Lorenz's limit of predictability is starting to become attainable in practice, especially with ensemble forecasting. Furthermore, skillful prediction of longer lasting phenomena such as El Niño is becoming feasible (Chapter 6).

Because the skill of the forecasts decreases with their length, Epstein (1969) and (Leith, 1974a) suggested that instead of performing "deterministic" forecasts, stochastic forecasts providing an estimate of the skill of the prediction should be made. The only computationally feasible approach in order to achieve this goal is through "ensemble forecasting" in which several model forecasts are performed by introducing perturbations in the initial conditions or in the models themselves.

After considerable research on how to most effectively perturb the initial conditions, ensemble forecasting was implemented operationally in December 1992 at both NCEP and ECMWF (Tracton and Kalnay, 1993; Toth and Kalnay, 1993; Palmer et al., 1993; Molteni et al., 1996; Toth and Kalnay, 1997). NCEP combined pairs of bred vectors (Toth and Kalnay, 1993, 1997), which contain naturally growing dynamical perturbations in the atmosphere also present in the Analysis errors, with Lagged Averaged Forecasts, to create an ensemble of 14 global forecasts available daily (Figure 1.6.1). In 2005, the NCEP Global Ensemble Forecast System (GEFS) was extended to two weeks and run 4 times a day with five pairs of bred vector initial perturbations for a total of 40 ensemble members. An extended BV method with Ensemble Transform and Rescaling (BV-ETR) was implemented in 2006. The length of the forecasts allows the generation of "outlooks" for the second week. Starting in December 2015, the GEFS initial perturbations are obtained from the operational hybrid ensemble Kalman filter (Zhou et al., 2016). At ECMWF, the initial perturbation method is based on the use of *singular vectors*, which grow even faster than the *bred* or *Lyapunov* vector perturbations. The ECMWF ensemble contains 50 members (Chapter 6).

Ensemble forecasting has accomplished two main goals: the first one is to provide an ensemble average forecast that beyond the first few days is more accurate than individual forecasts, because *the components of the forecast that are most uncertain*

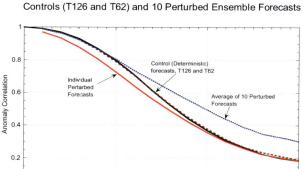

Figure 1.6.2 Anomaly correlation of the ensembles during the El Niño winter of 1997–8. The control forecasts T126 (solid black) and T62 (dashed black). The average of ten perturbed T62 ensemble forecasts is in red, showing the average skill of the perturbed T62 individual forecasts, and the skill of their ensemble average is in dotted blue (data courtesy of Jae Schemm, NCEP).

tend to be averaged out. The second and more important goal is to *provide forecasters with an estimation of the reliability of the forecast*, which because of changes in atmospheric predictability, varies from day to day and from region to region.

The first goal is illustrated in Figure 1.6.2, prepared at the Climate Prediction Center (CPC, the Climate Analysis Center of the former NMC) for the verification of the NCEP ensemble during the winter of 1997–8. This was an El Niño winter with major anomalies in the atmospheric circulation, and the operational forecasts had excellent skill. The control "deterministic forecasts" (T126 control in black full lines, and the T62 control in dashed lines) had an "anomaly correlation" (AC, pattern correlation between predicted and analyzed anomalies) in the 5-day forecast of slightly above and slightly below 80%, which was quite good. The ten perturbed ensemble members have individually a poorer verification with an average AC of about 73% at 5 days. This is because, in the initial conditions, the control starts from the best estimate of the state of the atmosphere (the analysis), but growing perturbations are added to this analysis for each additional ensemble member. However, the ensemble average forecast tends to average out uncertain components, and as a result, it has better skill than the control forecast starting at day 5. Note that the ensemble extends by one day the length of the useful forecast (defined as an AC greater than 60%) from about 7 days in the control to about 8 days in the ensemble average.

The second goal of the ensemble forecasting, to provide guidance on the uncertainty of each forecast, is accomplished best by the use of two types of plots. The "spaghetti" plots Figure 1.6.3 show a single contour line for all 17 forecasts, and the probabilistic plots show, for example, what percentage of the ensemble predicts 24-hr accumulated precipitation of more than 1 inch at each grid point (for probabilistic Quantitative Precipitation Forecasts or pQPF). Both of them provide guidance on the reliability of

1.6 Weather Predictability, Ensemble Forecasting, and Seasonal

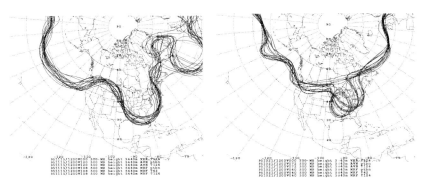

Figure 1.6.3 (a) "Spaghetti" plot for the 5-day forecast for November 15, 1995, a case of a very predictable storm over eastern USA. (Figure courtesy of R. Wobus, NCEP.) (b) Spaghetti plot for the 2.5-day forecast for October 21, 1995, the case of a very unpredictable storm over the USA (Courtesy of R. Wobus, NCEP). Dotted lines indicate the control forecast.

the forecasts in an easy-to-understand way. The use of the ensembles has provided the US NWS forecasters with the confidence to issue storm forecasts 5–7 days in advance when the spaghetti plots indicate good agreement in the ensemble, as in the first of the two plots. Conversely, the spaghetti plots also indicate when a short-range development may be particularly difficult to predict, so that the users should be made aware of the uncertainty of the forecast. Figure 1.6.3(a) shows an example of the 5-day forecast for November 15, 1995, the first East Coast winter storm of 1995–6: the fact that the ensemble showed good agreement provided the forecasters, for the first time, with the confidence to issue a storm forecast these many days in advance. By contrast, Figure 1.6.3(b) shows a 2.5-day forecast for a storm with verification time October 21, 1995, and it is clear that even at this shorter range, the atmosphere is much less predictable and there is much more uncertainty about the location of the storm.

It should be noted that in the United States, soon after NCEP introduced the use of ensemble forecasting, the public dissemination of the NWS forecasts was increased from 3 to 7 days. This is not because the quality of the forecast more than doubled, but because the confidence of the human forecasters in the quality of each individual forecast increased substantially.

The use of ensembles has also led to another major development, the possibility of an *adaptive* or *targeted* observing system. As an example, consider a case in which the lack of agreement among the ensemble members indicates that a 3-day forecast in a certain region is exceedingly uncertain, as in Figure 1.6.3(b). Several new techniques have been developed to trace such a region of uncertainty backward in time, for example 2 days. These techniques will point to a region or regions where additional observations would be especially useful. The additional observations could be dropwinsondes launched from a reconnaissance or a pilotless airplane, additional rawinsondes, or especially intensive use of satellite data such as a Doppler Wind Lidar. If additional observations are available 24 hr after the start of the originally critically uncertain 3-day forecast, they can increase substantially the usefulness of the 2-day forecast. Similarly, a few additional rawinsondes could be launched where short-range ensemble forecasts (12–24 hr) indicate that they are most needed. Preliminary tests of this approach of targeted observations have been successfully performed within an international Fronts and Storm Track Experiment (FASTEX) in the North Atlantic during January and February 1997, and in the North Pacific Experiment (NORPEX) in January and February 1998 (Szunyogh et al., 2000).

The cover of this book offers an interesting example of the representation of forecast uncertainties in the NCEP ensemble forecasts of the tropical cyclone Sandy that landed on 00Z October 30 in New Jersey, producing major devastation both there and in New York. The uncertainty of the forecasts, and the improvement brought by data assimilation are made clear by the fact that the earliest forecasts (in blue), started after the data assimilation of 12Z October 23, have the largest errors, with the majority of the forecasts having Sandy moving eastward with the large scale flow. The next assimilation, only 6 hr later, brings most of the forecasts (in green) westward, and after another assimilation, the forecasts started on 00Z October 24 (in red) represent much better the capture of Sandy by a deep trough that brought it westward, landing on the East Coast.

Ensemble forecasting also provides the basic tool to extend forecasts beyond Lorenz's 2-week limit of weather predictability (Chapter 6). Slowly varying surface forcing, especially from the tropical ocean and from land-surface anomalies, can produce atmospheric anomalies that are longer lasting and more predictable than individual weather patterns. The most notable of these is the El Niño–Southern Oscillation (ENSO) produced by unstable oscillations of the coupled ocean-atmosphere system, with a frequency of 3–7 years. Because of their long time scale, the ENSO oscillations should be predictable a year or more in advance (in agreement with the chaos theory). The first successful experiments in this area were made by Cane et al. (1986) with a simple coupled atmosphere-ocean model. The warm phases of ENSO (El Niño episodes) are associated with warm sea surface temperature (SST) anomalies in the equatorial central and eastern Pacific Ocean, and cold phases (La Niña episodes) with cold anomalies. NCEP started performing multi-seasonal predictions with coupled comprehensive atmosphere-ocean models in 1995, and ECMWF did so in 1997.

A single atmospheric forecast forced with the SST anomalies would not be useful beyond the first week or so, when unpredictable weather variability would mask the forced atmospheric anomalies. Ensemble averaging many forecasts made with atmospheric models forced by SST anomalies (and by other slowly varying anomalies over land such as soil moisture and snow cover) allows the filtering out of the unpredictable components of the forecast, and the retention of more of the forced predictable components. This filtering is reflected in the fact that the ensemble average for the second week of the forecasts for the winter of 1997–8 (Figure 1.6.1) had a high AC of 57%, much higher than previously obtained. Researchers at the JMA have performed forecasts for the 28-day average and also found that ensemble averaging substantially increased the information on the second week and the last 2 weeks of the forecast. The very successful operational forecasts of the ENSO episode of 1997–8 performed at both NCEP and ECMWF have been substantially based on the use of ensembles to extract the useful information on the impact of El Niño from the "weather noise."

1.7 The Future

The last decades have seen the expectations of Charney (1951) more than fulfilled, and an amazing improvement in the quality of the forecasts based on improved scientific NWP guidance.* The power of the new science of Data Assimilation is being recognized as being useful for any system where we can combine model forecasts and observations to estimate the evolution of a system.

From the active research that is taking place, one can envision that the next decades will continue to bring important improvements, including the following:

- detailed short-range forecasts, using ensembles of storm-scale models able to provide more skillful predictions of severe weather and extreme events;
- more sophisticated methods of data assimilation able to extract increased information from observing systems, and using ensemble forecast sensitivity to observations (EFSO) and proactive quality control (PQC) to identify and reject detrimental observations (Chen and Kalnay, 2020);
- Current data assimilation (DA) methods such as ensemble Kalman filters, 4D-Var, and hybrids used in operational weather prediction require using models that are linear, and errors that are Gaussian, which make it very difficult to accurately model non-linear systems. such as the human system which is increasingly dominating the Earth system. For many years modelers and tried to replace the EnKF approach with particle filters, since DA with particle filters is based on Bayes' Theorem, so that, unlike current DA techniques, it does not require using linear approximations for the models, and does not assume Gaussian background or observational errors. However, particle filters were confronted with a "curse of dimensionality" that made them unaffordable. In the last decade, building on significant progress in particle filters DA made by other modelers and geoscientists (see review by van Leeuwen et al. (2019)), Penny and Miyoshi (2016), and Poterjoy (2016) simultaneously developed localized particle filters that were competitive and more accurate than EnKF, so that particle filters may soon replace EnKF based DA systems;
- strongly coupled data assimilation of the Earth system components (atmosphere, ocean, land, ice, etc.), so that, for example, the ocean can also benefit from assimilating atmospheric observations and vice versa;
- applications of data assimilation to improve the models as well as the observations, including estimating optimal parameters and advanced quality control;
- adaptive observing systems, in which additional observations are placed where ensembles indicate that there is rapid error growth (low predictability);
- improvement in the usefulness of medium-range forecasts and the prediction of forecast skill, especially through the use of ensemble forecasting;
- fully coupled atmospheric-hydrological systems, where the atmospheric model precipitation is appropriately downscaled and used to extend and improve river flow prediction;
- more use of atmosphere-ocean-land coupled models, in which long-lasting coupled anomalies such as SST and soil moisture anomalies can lead to more skillful

predictions of anomalies in weather patterns beyond the limit of weather predictability;

- more guidance to governments and the public on subjects such as air pollution, ultraviolet radiation, transport of contaminants, and coupling modeling of evolution of pandemics and their prevention with atmospheric climatological and shorter range forecasts;
- a continued growth of commercial applications of NWP, from guidance on the state of highways to air pollution, flood prediction, guidance to agriculture, construction, etc.;
- the new science of data assimilation that has been so successful in improving weather prediction, will be adopted in other fields like medicine, agriculture, economics, etc., where models and observations can be combined to not only improve the forecasts, but also improve the models and observations;
- Develop a new science, "Modeling Sustainability" (Motesharrei et al., 2016) that will guide governments about policies needed to maintain a sustainable future for both the human system and the Earth system.

2 The Continuous Equations

2.1 Governing Equations

"If, as every scientifically inclined individual believes, atmospheric conditions develop according to natural laws from their precursors, it follows that the necessary and sufficient conditions for a rational solution of the problems of meteorological prediction are the following:

1. The condition of the atmosphere must be known at a specific time with sufficient accuracy;
2. The laws must be known, with sufficient accuracy, which determine the development of one weather condition from another" (Bjerknes, 1904).

In this "manifesto," Bjerknes (1904) pointed out for the first time that there is a complete set of seven equations with seven unknowns that governs the evolution of the atmosphere:

- Newton's second law or conservation of momentum (three equations for the three velocity components);
- the continuity equation or conservation of mass;
- the equation of state for ideal gases;
- the first law of thermodynamics or conservation of energy; and
- a conservation equation for water mass.

Bjerknes reasoned that if we know both the equations (with their appropriate boundary conditions) and the initial conditions with sufficient accuracy, we should be able to predict the weather.

In this section, we briefly derive the governing equations. The reader may refer to other texts, such as Haltiner and Williams (1980) or James (1994) for further details.

Newton's second law or conservation of momentum
On an inertial frame of reference (see Figure 2.1.1), the absolute acceleration of a parcel of air in three dimensions is given by the real forces $\mathbf{F}$ acting on a parcel of mass m:

$$\frac{d_a \mathbf{v}_a}{dt} = \mathbf{F}/m \qquad (2.1.1)$$

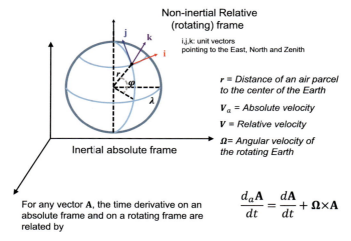

Figure 2.1.1 Schematic of an inertial or Newtonian reference frame, and a rotating reference frame where "apparent" forces like the Coriolis force and the centrifugal force appear.

The time derivative of any vector **A** is transformed from a non-rotating to a rotating frame of reference (Figure 2.1.1) by

$$\frac{d_a \mathbf{A}}{dt} = \frac{d\mathbf{A}}{dt} + \mathbf{\Omega} \times \mathbf{A}. \tag{2.1.2}$$

If we apply this transformation to a position vector **r**, we obtain $\frac{d_a \mathbf{r}}{dt} = \frac{d\mathbf{r}}{dt} + \mathbf{\Omega} \times \mathbf{r}$, which implies that the absolute velocity $\mathbf{v}_a$ is given by the sum of the relative velocity **v** on the rotating frame, plus the velocity due to the rotation with angular velocity $\Omega = 2\pi/24$ hrs:

$$\mathbf{v}_a = \mathbf{v} + \mathbf{\Omega} \times \mathbf{r} \tag{2.1.3}$$

We can apply again (2.1.2) to $\mathbf{v}_a$, which gives

$$\frac{d_a \mathbf{v}_a}{dt} = \frac{d\mathbf{v}_a}{dt} + \mathbf{\Omega} \times \mathbf{v}_a \tag{2.1.4}$$

Substituting (2.1.3) into (2.1.4) we obtain that the accelerations in an inertial (absolute) and a rotating frame of reference are related by

$$\frac{d_a \mathbf{v}_a}{dt} = \frac{d\mathbf{v}}{dt} + 2\mathbf{\Omega} \times \mathbf{v} + \mathbf{\Omega} \times (\mathbf{\Omega} \times \mathbf{r}) \tag{2.1.5}$$

This equation indicates that on a rotating frame of reference there are two *apparent* forces per unit mass: the Coriolis force (second term on the right-hand side) and the centrifugal force (third term).

The left-hand side of (2.1.5) represents the *real* forces acting on a parcel of air, i.e., the pressure gradient force $-\alpha \nabla p$, the gravitational acceleration $\mathbf{g}_e = -\nabla \phi_e$, and the frictional force **F**. Therefore in a rotating frame of reference moving with the Earth, the apparent acceleration is given by

$$\frac{d\mathbf{v}}{dt} = -\alpha\nabla p - \nabla\phi_e + \mathbf{F} - 2\mathbf{\Omega} \times \mathbf{v} - \mathbf{\Omega} \times (\mathbf{\Omega} \times \mathbf{r}) \tag{2.1.6}$$

Here $\alpha = 1/\rho$ is the specific volume (the inverse of the density ρ), p is the pressure, ϕ_e is the Newtonian gravitational potential of the Earth, and, as indicated before, the last two terms are the apparent accelerations, denoted the Coriolis force and centrifugal force respectively. We have not included the *tidal potential*, whose effects are negligible below about 100 km.

We can now combine the centrifugal force with the gravitational force, since $-\mathbf{\Omega} \times (\mathbf{\Omega} \times \mathbf{r}) = \Omega^2\mathbf{l} = \nabla(\Omega^2 l^2/2)$, where $\mathbf{l}$ is the position vector from the axis of rotation to the parcel. Therefore, we can define as the "geopotential" $\phi = \phi_e - \Omega^2 l^2/2$, and the apparent gravity is given by

$$-\nabla\phi = \mathbf{g} = \mathbf{g}_e + \Omega^2\mathbf{l} \tag{2.1.7}$$

We define the geographic latitude φ to be perpendicular to the geopotential ϕ. At the surface of the Earth, the geographic latitude and the geocentric latitude differ by less than 10 min of a degree of latitude. Therefore, Newton's law on the rotating frame of the Earth is written as

$$\frac{d\mathbf{v}}{dt} = -\alpha\nabla p - \nabla\phi + \mathbf{F} - 2\mathbf{\Omega} \times \mathbf{v} \tag{2.1.8}$$

Continuity equation or equation of conservation of mass
This can be derived as follows: Consider the mass of a small parcel of air of density ρ

$$M = \rho\Delta x\Delta y\Delta z \tag{2.1.9}$$

If we follow the parcel in time, it conserves its mass, i.e., the total time derivative (also called the substantial, individual or Lagrangian time derivative) is equal to zero: $dM/dt = 0$. If we take a logarithmic derivative of the mass

$$\frac{1}{M}\frac{dM}{dt} = 0$$

in (2.1.9) we obtain the continuity equation:

$$\frac{1}{\rho}\frac{d\rho}{dt} + \nabla_3 \cdot \mathbf{v} = 0 \tag{2.1.10}$$

since $\lim_{\Delta x \to 0} \frac{1}{\Delta x}\frac{d\Delta x}{dt} = \frac{\partial u}{\partial x}$ and similarly for the other directions y, z.

Now, the total derivative of any function $f(x,y,z,t)$, following a parcel, can be expanded as

$$\frac{df}{dt} = \frac{\partial f}{\partial t} + \frac{\partial f}{\partial x}\frac{dx}{dt} + \frac{\partial f}{\partial y}\frac{dy}{dt} + \frac{\partial f}{\partial z}\frac{dz}{dt} = \frac{\partial f}{\partial t} + \mathbf{v} \cdot \nabla f \tag{2.1.11}$$

Equation (2.1.11) indicates that the total (or Lagrangian or individual) time derivative of a property is given by the local (partial, Eulerian) time derivative (at a fixed point) plus the changes due to advection. If we expand $d\rho/dt$ in (2.1.10) using (2.1.11) we obtain an alternative form of the continuity equation, usually referred to as "in flux form":

$$\frac{\partial \rho}{\partial t} = -\nabla \cdot (\rho \mathbf{v}) \qquad (2.1.12)$$

Equation of state for perfect gases
The atmosphere can be assumed to be a perfect gas, for which the pressure p, specific volume α (or its inverse ρ, density), and temperature T are related by

$$p\alpha = RT \qquad (2.1.13)$$

where R is the gas constant for dry air. For moist air this has to take into account the partial pressure of moist air, usually approximated by defining the virtual temperature $T_v = T(1 + 0.6q)$, where q is the mixing ratio, i.e., the ratio of the mass of water vapor over the dry air mass. The virtual temperature allows to use the perfect gas law, but using the dry temperature that has the same density as moist air at the same pressure. The equation of state indicates that, given two thermodynamic variables, the others are determined.

Thermodynamic energy equation or conservation of energy equation
This equation expresses that if heat is applied to a parcel at a rate of Q per unit mass, this heat can be used to increase the internal energy $C_v T$ and/or to produce work of expansion:

$$Q = C_v \frac{dT}{dt} + p \frac{d\alpha}{dt} \qquad (2.1.14)$$

The coefficients of specific heat at constant volume C_v and at constant pressure C_p are related by $C_p = C_v + R$. We can use the equation of state (2.1.13) to derive another form of the thermodynamic equation:

$$Q = C_p \frac{dT}{dt} - \alpha \frac{dp}{dt} \qquad (2.1.15)$$

The rate of change of the specific entropy s of a parcel is given by $ds/dt = Q/T$, i.e., the diabatic heating divided by the absolute temperature. We now define potential temperature by $\theta = T (p_0/p)^{R/C_p}$, where p_0 is a reference pressure (1,000 hPa). With this definition, it is easy to show that the potential temperature and the specific entropy are related by

$$\frac{ds}{dt} = C_p \frac{1}{\theta} \frac{d\theta}{dt} = \frac{Q}{T} \qquad (2.1.16)$$

This shows that potential temperature is individually conserved in the absence of diabatic heating.

Equation for conservation of water vapor mixing ratio q
This equation simply indicates that the total amount of water vapor in a parcel is conserved as the parcel moves around, except when there are *sources* (evaporation E) and *sinks* (condensation C):

$$\frac{dq}{dt} = E - C \qquad (2.1.17)$$

2.2 Atmospheric Equations of Motion on Spherical Coordinates

Conservation equations for other atmospheric constituents can be similarly written in terms of their corresponding sources and sinks. If we multiply (2.1.17) by ρ, expand the total derivative $dq/dt = \partial q/\partial t + \mathbf{v} \cdot \nabla q$, and add the continuity equation (2.1.12) multiplied by q, we can write the conservation of water in an alternative "flux form":

$$\frac{\partial \rho q}{\partial t} = -\nabla \cdot (\rho \mathbf{v} q) + \rho(E - C) \tag{2.1.18}$$

The flux form of the time derivative is very useful in the construction of models. The first term of the right-hand side of (2.1.18) is the convergence of the flux of q. Note that we can include similar conservation equations for additional tracers such as liquid water and ozone, as long as we also include their corresponding sources and sinks.

We now have seven equations with seven unknowns: $\mathbf{v} = (u, v, w), T, p, \rho$ or α, and q. For convenience we repeat the governing equations, which (when written without friction $\mathbf{F}$) are sometimes referred to as "*the Euler equations*":

$$\frac{d\mathbf{v}}{dt} = -\alpha \nabla p - \nabla \phi + \mathbf{F} - 2\mathbf{\Omega} \times \mathbf{v} \tag{2.1.19}$$

$$\frac{\partial \rho}{\partial t} = -\nabla \cdot (\rho \mathbf{v}) \tag{2.1.20}$$

$$p\alpha = RT \tag{2.1.21}$$

$$Q = C_p \frac{dT}{dt} - \alpha \frac{dp}{dt} \tag{2.1.22}$$

$$\frac{\partial \rho q}{\partial t} = -\nabla \cdot (\rho \mathbf{v} q) + \rho(E - C) \tag{2.1.23}$$

2.2 Atmospheric Equations of Motion on Spherical Coordinates

Since the Earth is nearly spherical, it is natural to use spherical coordinates. Near the Earth, gravity is almost constant, and the ellipticity of the Earth is very small, so that one can accurately approximate scale factors by those appropriate for true spherical coordinates (Phillips, 1966, 1973, 1990). The three velocity components are then

$$\left. \begin{aligned} u &= \text{zonal (positive eastward)} = r \cos \varphi \frac{d\lambda}{dt} \\ v &= \text{meridional (positive northward)} = r \frac{d\varphi}{dt} \\ w &= \text{vertical (positive up)} = \frac{dr}{dt} \end{aligned} \right\}$$

Note that $\mathbf{v} = u\mathbf{i} + v\mathbf{j} + w\mathbf{k}$, where $\mathbf{i}, \mathbf{j}, \mathbf{k}$ are the unit vectors in the three orthogonal spherical coordinates. When the acceleration (total derivative of the velocity vector) is calculated, the rate of change of the unit vectors has to be included. For example, geometrical considerations show that

$$\frac{d\mathbf{k}}{dt} = \frac{u}{r \cos \varphi} \frac{\partial \mathbf{k}}{\partial \lambda} + \frac{v}{r} \frac{\partial \mathbf{k}}{\partial \varphi} = \frac{u\mathbf{i}}{r} + \frac{v\mathbf{j}}{r} \tag{2.2.1}$$

Exercise 2.2.1 *Use spherical geometry to show (2.2.1). Derive*

$$\frac{di}{dt} = \frac{u}{r\cos\varphi}(\boldsymbol{j}\sin\varphi - \boldsymbol{k}\cos\varphi) \quad and \quad \frac{dj}{dt} = \frac{1}{r\cos\varphi}(-u\boldsymbol{i}\sin\varphi - v\boldsymbol{k}\cos\varphi) \quad (2.2.2)$$

When we include these time derivatives, take into account that $\boldsymbol{\Omega} = \Omega\sin\varphi\boldsymbol{k} + \Omega\cos\varphi\boldsymbol{j}$, and expand the momentum equation (2.1.19) into its three components, we obtain

$$\left.\begin{array}{l}
\dfrac{du}{dt} = -\dfrac{\alpha}{r\cos\varphi}\dfrac{\partial p}{\partial\lambda} + F_\lambda + \left(2\Omega + \dfrac{u}{r\cos\varphi}\right)(v\sin\varphi - w\cos\varphi) \\[3mm]
\dfrac{dv}{dt} = -\dfrac{\alpha}{r}\dfrac{\partial p}{\partial\varphi} + F_\varphi - \left(2\Omega + \dfrac{u}{r\cos\varphi}\right)u\sin\varphi - \dfrac{vw}{r} \\[3mm]
\dfrac{dw}{dt} = -\alpha\dfrac{\partial p}{\partial r} - g + F_r + \left(2\Omega + \dfrac{u}{r\cos\varphi}\right)u\cos\varphi + \dfrac{v^2}{r}
\end{array}\right\} \quad (2.2.3)$$

The terms proportional to $u/r\cos\varphi$ are known as "metric terms."

A *"traditional approximation"* (Phillips, 1966) has been routinely made in NWP, since most of the atmospheric mass is confined to a few tens of kilometers. This suggests that in considering the distance of a point to the center of the Earth $r = a + z$, one can neglect z and replace r by the radius of the Earth $a = 6{,}371$ km, replace $\partial/\partial r$ by $\partial/\partial z$, and neglect the metric and Coriolis terms proportional to $\cos\varphi$. Then the equations of motion in spherical coordinates become

$$\left.\begin{array}{l}
\dfrac{du}{dt} = -\dfrac{\alpha}{a\cos\varphi}\dfrac{\partial p}{\partial\lambda} + F_\lambda + \left(2\Omega + \dfrac{u}{a\cos\varphi}\right)v\sin\varphi \\[3mm]
\dfrac{dv}{dt} = -\dfrac{\alpha}{a}\dfrac{\partial p}{\partial\varphi} + F_\varphi - \left(2\Omega + \dfrac{u}{a\cos\varphi}\right)u\sin\varphi \\[3mm]
\dfrac{dw}{dt} = -\alpha\dfrac{\partial p}{\partial z} - g + F_z
\end{array}\right\} \quad (2.2.4)$$

which possess the angular momentum conservation principle

$$\frac{d}{dt}[(u + \Omega a\cos\varphi)a\cos\varphi] = a\cos\varphi\left(-\frac{\alpha}{a\cos\varphi}\frac{\partial p}{\partial\lambda} + F_\lambda\right) \quad (2.2.5)$$

With the "traditional approximation" the total time derivative operator in spherical coordinates is given by

$$\frac{d(\ )}{dt} = \frac{\partial(\ )}{\partial t} + \frac{u}{a\cos\varphi}\frac{\partial(\ )}{\partial\lambda} + \frac{v}{a}\frac{\partial(\ )}{\partial\varphi} + w\frac{\partial(\ )}{\partial z} \quad (2.2.6)$$

and the three-dimensional divergence that appears in the continuity equation by

$$\boldsymbol{\nabla}_3\cdot\mathbf{v} = \frac{1}{a\cos\varphi}\left(\frac{\partial u}{\partial\lambda} + \frac{\partial v\cos\varphi}{\partial\varphi}\right) + \frac{\partial w}{\partial z} \quad (2.2.7)$$

2.3 Basic Wave Oscillations in the Atmosphere

In order to understand the problems in Richardson's result in 1922 (Figure 1.2.1) and the effect of the filtering approximations introduced by Charney et al. (1950), we need to have a basic understanding of the characteristics of the different types

2.3 Basic Wave Oscillations in the Atmosphere 41

of waves present in the atmosphere. The characteristics of these waves, (sound, gravity, and slower weather waves) have also profound implications for the present use of hydrostatic and nonhydrostatic models. The three types of waves are present in the solutions of the governing equations, and different approximations such as the hydrostatic, the quasi-geostrophic, and the anelastic approximations are designed to filter out some of them.

To simplify the analysis we make a tangent plane or "f-plane" approximation. We consider motions with horizontal scales L smaller than the radius of the Earth. On this tangent plane, we can approximate the spherical coordinates (Section 2.2) by

$$\frac{1}{a \cos \varphi_0} \frac{\partial}{\partial \lambda} \approx \frac{\partial}{\partial x} \qquad \frac{1}{a} \frac{\partial}{\partial \varphi} \approx \frac{\partial}{\partial y}, \qquad f \approx 2\Omega \sin \varphi_0$$

and ignore the metric terms, since $u/(a \tan \varphi)$ is small compared with Ω.

The governing equations on an f-plane (rotating with the local vertical component of the Earth rotation) are:

$$\frac{du}{dt} = +fv - \frac{1}{\rho} \frac{\partial p}{\partial x} \tag{2.3.1a}$$

$$\frac{dv}{dt} = -fu - \frac{1}{\rho} \frac{\partial p}{\partial y} \tag{2.3.1b}$$

$$\frac{dw}{dt} = -\frac{1}{\rho} \frac{\partial p}{\partial z} - g \tag{2.3.1c}$$

$$\frac{d\rho}{dt} = -\rho \left(\frac{\partial u}{\partial x} + \frac{\partial v}{\partial y} + \frac{\partial w}{\partial z} \right) \tag{2.3.1d}$$

$$\frac{ds}{dt} = \frac{Q}{T}; \qquad s = C_p \ln \theta \tag{2.3.1e}$$

$$p = \rho RT \tag{2.3.1f}$$

Consider a *basic state* at rest $u_0 = v_0 = w_0 = 0$. From (2.3.1a) and (2.3.1b), we see that p_0 does not depend on x, y, $p_0 = p_0(z)$. From (2.3.1c), ρ_0 and therefore the other basic state thermodynamic variables also depend on z only.

Assume that the motion is adiabatic and frictionless, $Q = 0$, $\mathbf{F} = 0$. Consider *small perturbations* $p = p_0 + p'$ etc. so that we can linearize the equations (neglect terms which are products of perturbations). For convenience, we define $u^* = \rho_0 u'$; $v^* = \rho_0 v'$; $w^* = \rho_0 w'$; $s^* = \rho_0 s'$. The perturbation equations are then

$$\frac{\partial u^*}{\partial t} = +fv^* - \frac{\partial p'}{\partial x} \tag{2.3.2a}$$

$$\frac{\partial v^*}{\partial t} = -fu^* - \frac{\partial p'}{\partial y} \tag{2.3.2b}$$

$$\frac{\partial w^*}{\partial t} = -\frac{\partial p'}{\partial z} - \rho' g \tag{2.3.2c}$$

$$\frac{\partial \rho'}{\partial t} = -\left(\frac{\partial u^*}{\partial x} + \frac{\partial v^*}{\partial y} + \frac{\partial w^*}{\partial z} \right) \tag{2.3.2d}$$

$$\frac{\partial s^*}{\partial t} = -w^* \frac{ds_0}{dz} \tag{2.3.2e}$$

$$\frac{p'}{p_0} = \frac{\rho'}{\rho_0} + \frac{T'}{T_0} \tag{2.3.2f}$$

The Continuous Equations

where

$$s^* = \rho_0 C_P \frac{\theta'}{\theta_0} = \rho_0 C_P \left(\frac{T'}{T_0} - \frac{R}{C_P} \frac{p'}{p_0} \right) = C_P \left(\frac{p'}{\gamma R T_0} - \rho' \right) \qquad (2.3.2g)$$

Exercise 2.3.1 *Derive (2.3.2a)–(2.3.2g), recalling that* $p = \rho R T, \theta = T(1000hP_a/p)^{R/C_P}, C_P = R + C_v, \gamma = C_P/C_v = 1.4,$ *and* $c_s^2 = \gamma R T \approx (320 \ m/s)^2$ *is the square of the speed of sound.*

2.3.1 Pure Types of Plane Wave Solutions

We first consider *special cases with pure wave type solutions*. They exist in their pure form only under very simplified assumptions. However, if we understand their basic characteristics, we will understand their role in the full nonlinear models and the methodology used for filtering some of the waves out. We will be assuming plane wave solutions aligning the x-axis along the horizontal direction of propagation:

$$(u^*, v^*, w^*, p') = (U, V, W, P)e^{i(kx + mz - vt)} \qquad (2.3.3)$$

Here $k = 2\pi/L_x$ and $m = 2\pi/L_z$ are horizontal and vertical wavenumbers, respectively, $v = 2\pi/T$ is the frequency, and $U, V, W,$ and P are constant amplitudes. We will aim to derive the *frequency dispersion relationship (FDR)* $v = f(k, m, parameters)$ for each type of wave by substituting the plane wave formulation (2.3.3) into the linear equation, and eliminating variables. The FDR gives us not only the frequency, but also the *phase speed* components $(v/k, v/m)$ as well as the *group velocity* components $(\partial v/\partial k, \partial v/\partial m)$. The phase speed is the speed of individual wave crests and valleys, and the group velocity is the speed at which wave energy propagates in the horizontal and vertical directions. A pure type of wave occurs under idealized conditions, such as no rotation, no stratification for sound waves, but its basic characteristics are retained even if the ideal conditions are not valid (sound waves are still present but slightly modified in the presence of rotation and stratification).

2.3.1.1 Pure Sound Waves

We neglect rotation, stratification and gravity: $f = 0, g = 0, ds_0/dz = 0$. From (2.3.2e), we have $s^* = 0$ (recall that s^* is a perturbation, and if it was constant, we would have included its value into the basic state s_0). Therefore $p' = c_s^2 \rho'$, and (2.3.2) reduce to

$$\frac{\partial u^*}{\partial t} = -\frac{\partial p'}{\partial x} \qquad (2.3.4a)$$

$$\frac{\partial v^*}{\partial t} = -\frac{\partial p'}{\partial y} \qquad (2.3.4b)$$

$$\frac{\partial w^*}{\partial t} = -\frac{\partial p'}{\partial z} \qquad (2.3.4c)$$

$$\frac{1}{c_s^2} \frac{\partial p'}{\partial t} = -\left(\frac{\partial u^*}{\partial x} + \frac{\partial v^*}{\partial y} + \frac{\partial w^*}{\partial z} \right) \qquad (2.3.4d)$$

2.3 Basic Wave Oscillations in the Atmosphere

These show that sound waves occur through adiabatic expansion and contraction (three-dimensional divergence), and that the pressure perturbation is proportional to the density perturbation.

Assuming plane wave solutions (2.3.3), with the x-axis along the horizontal direction of the waves, and substituting into (2.3.4), we get

$$-i\nu U = -ikP \tag{2.3.5a}$$

$$-i\nu V = 0 \tag{2.3.5b}$$

$$-i\nu W = -imP \tag{2.3.5c}$$

$$-i\nu P = -c_s^2(ikU + imW) \tag{2.3.5d}$$

From (2.3.5b) $V = 0$ and substituting U and W from (2.3.5a) and (2.3.5c) into (2.3.5d), we get the FDR:

$$\nu^2 = c_s^2(k^2 + m^2) \tag{2.3.6}$$

These are sound waves that propagate through air compression or three-dimensional divergence. The components of the phase velocity are $(\nu/k, \nu/m)$ and the total phase velocity is

$$\frac{\nu}{\sqrt{k^2 + m^2}} = \pm c_s$$

2.3.1.2 Lamb Waves (Horizontally Propagating Sound Waves)

We now neglect rotation and assume that there is only horizontal propagation (no vertical velocity), but we allow for the fluid to be gravitationally stratified. With $f = 0$ and $w^* = 0$, we again have $s^* = 0$, and from (2.3.2f) $p' = c_s^2 \rho'$, but from (2.3.2c) the flow is now hydrostatic: $\partial p'/\partial z = -\rho'g$. If we insert the same type of plane wave solutions (2.3.3) into (2.3.2), we find that $p' = Pe^{-(g/c_s^2)z}e^{i(kx-\nu t)}$, i.e., the vertical wavenumber is imaginary $m = ig/c_s^2$, and the phase speed is $\nu^2/k^2 = c_s^2$. Since the vertical wavenumber is imaginary, there is no vertical propagation, and the waves are *external*.

Therefore, a Lamb wave is a type of external horizontal sound wave, which is present in the solutions of models even when the hydrostatic approximation is made. This is very important because it means primitive equation models (which make the hydrostatic approximation) contain these fast moving horizontal sound waves. We will see that Lamb waves are also equivalent to the gravity waves in a shallow water model. Note also that the FDR is such that $\nu/k = \pm c_s$, so that the phase speed does not depend on the wavenumber. This implies that the group velocity $\partial \nu/\partial k = \pm c_s$. It is also independent of the wavenumber, and as a result Lamb waves without rotation are nondispersive, so that a package of waves will move together and not disperse.

2.3.1.3 Vertical Gravitational Oscillations

Now we neglect rotation and pressure perturbations, $f = p' = 0$, so that there is no horizontal motion, but allow for vertical stratification. Equations (2.3.2) become

$$\frac{\partial w^*}{\partial t} = -\rho' g \qquad (2.3.7a)$$

$$\frac{\partial \rho'}{\partial t} = \frac{w^*}{C_p}\frac{ds_0}{dz} = w^*\frac{d\ln\theta_0}{dz} \qquad (2.3.7b)$$

From these two equations, we get

$$\frac{\partial^2 w^*}{\partial t^2} + N^2 w^* = 0 \qquad (2.3.8)$$

and from the continuity equation, we obtain

$$\frac{\partial \rho'}{\partial t} = -\frac{\partial w^*}{\partial z} \qquad (2.3.9)$$

Substituting the plane wave solution (2.3.3) into (2.3.8), we obtain $\nu^2 = N^2$, where $N^2 = g\, d\ln\theta_0/dz$ is the square of the *Brunt–Väisälä frequency*. A typical value of N for the atmosphere is $N \sim 10^{-2}$ s^{-1}. A parcel displaced in a stable atmosphere will oscillate vertically with frequency N. Equations (2.3.7b) and (2.3.9) show that the amplitude of w^* will decrease with height as $e^{-(d\ln\theta/dz)z}$.

2.3.1.4 Inertia Oscillations

Inertia oscillations are horizontal and are due to the basic rotation. We now assume that $p' = 0$, $ds_0/dz = 0$, and there are no pressure perturbations and no stratification. Then $s^* = 0$, and, therefore, $\rho' = 0$ and the horizontal equations of motion become

$$\frac{\partial \mathbf{v}^*}{\partial t} = -f\mathbf{k}\times\mathbf{v}^* \qquad \text{or} \qquad \frac{\partial^2 \mathbf{v}^*}{\partial t^2} = f\mathbf{k}\times(f\mathbf{k}\times\mathbf{v}^*) = -f^2\mathbf{v}^* \qquad (2.3.10)$$

As indicated by (2.3.10), the frequency of inertia oscillations is $\nu = \pm f$, with the acceleration perpendicular to the wind, corresponding to a circular wind oscillation. In the presence of a basic flow, there is also a translation, and the trajectories look like Figure 2.3.1.

2.3.1.5 Lamb Waves in the Presence of Rotation and Geostrophic Modes

We now consider the same case as in Section 2.3.1.2 of horizontally propagating Lamb waves, but without neglecting rotation, i.e., $f \neq 0$, but the vertical velocity is still zero. From $w^* = 0$ and (2.3.2c), we have again $p' = c_s^2 \rho'$, and the hydrostatic balance in (2.3.2g) then implies $\partial p'/\partial z = -p'g/c_s^2$. Therefore, the three-dimensional perturbations can be written as $p'(x,y,z,t) = p'(x,y,0,t)e^{-z/\gamma H}$, where $\gamma H = c_s^2/g$.

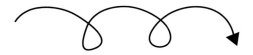

Figure 2.3.1 Schematic of an inertial oscillation in the presence of a basic flow to the right.

2.3 Basic Wave Oscillations in the Atmosphere   45

The system of equations (2.3.2) becomes

$$\left.\begin{aligned}
\frac{\partial \mathbf{v}^*}{\partial t} &= -f\mathbf{k} \times \mathbf{v}^* - \nabla p' \\
\frac{\partial p'}{\partial t} &= -c_s^2 \nabla \cdot \mathbf{v}^*
\end{aligned}\right\} \tag{2.3.11}$$

This system is completely analogous to the linearized *shallow water equations* (SWE) which are widely used in NWP as the simplest primitive equations model:

$$\left.\begin{aligned}
\frac{\partial \mathbf{v}}{\partial t} &= -f\mathbf{k} \times \mathbf{v} - \nabla \phi' \\
\frac{\partial \phi'}{\partial t} &= -\Phi \nabla \cdot \mathbf{v} \\
\text{where} \quad \phi &= \Phi + \phi'
\end{aligned}\right\} \tag{2.3.12}$$

If we assume plane wave solutions of the form $(u^*, v^*, p') = (U, V, P)e^{-i(kx - vt)}$ and substitute in (2.3.11), we obtain:

$$\left.\begin{aligned}
-ivU &= fV - ikP \\
-ivV &= -fU \\
-ivP &= -c_s^2 ikU
\end{aligned}\right\} \tag{2.3.13}$$

Therefore, the FDR is

$$v(v^2 - f^2 - c_s^2 k^2) = 0 \tag{2.3.14}$$

Note that this FDR contains *two* types of solution: one type is $v^2 = f^2 + c_s^2 k^2$, Lamb waves modified by inertia (rotation), or inertia Lamb waves. In the SWE analog, these are inertia-gravity waves (external gravity waves modified by inertia), $v^2 = f^2 + \Phi k^2$. Note that in the presence of rotation, the phase speed and group velocity depend on the wavenumber: rotation makes Lamb waves dispersive (and this helps with the problem of getting rid of noise in the initial conditions as in Figure 1.2.1).

The *second type of solution* (and for us the more important!) is *the steady state solution* $v = 0$. This means that $\partial()/\partial t = -iv() = 0$ for all variables. Without the presence of rotation, this steady state solution would be trivial: $u^* = v^* = w^* = p' = 0$. But *with rotation*, an examination of (2.3.13) or (2.3.12) shows that this is *the geostrophic mode*: $U = 0, \nabla \cdot \mathbf{v}^* = \partial U/\partial x = 0$, *but* $V = ikP/f$, i.e.,

$$v^* = \frac{1}{f}\frac{\partial p'}{\partial x}$$

This is a steady state, but nontrivial, geostrophic solution. If we add a dependence of f on latitude, the geostrophic solution becomes the Rossby waves solution, which is not steady state but still much *slower* than gravity waves or sound waves.

2.3.2   General Wave Solution of the Perturbation Equations in a Resting, Isothermal Atmosphere

So far we have been making drastic approximations to obtain "pure" elementary waves (sound, inertia, and gravity oscillations). We now consider a more general case,

46 The Continuous Equations

including all waves simultaneously. We consider again the equations for small perturbations (2.3.2), and assume a resting, isothermal basic state in the atmosphere: $T_0(z) = T_{00}$, a constant. Then

$$N^2 = g\frac{d\ln\theta}{dz} = -g\kappa\frac{d\ln p_0}{dz} \tag{2.3.15}$$

where $\kappa = R/C_p = 0.4$. Since the basic state is hydrostatic,

$$N^2 = g\kappa\frac{\rho_0 g}{p_0} = g\kappa\frac{g}{RT_0} = \frac{g\kappa}{H} \tag{2.3.16}$$

These equations show that for an isothermal atmosphere, both N^2 and the scale height $H = RT/g$ are constant.

We continue considering an f-plane, a reasonable approximation for horizontal scales L small compared to the radius of the Earth: $L \ll a$. If L is not small compared with the radius of the Earth, we would have to take into account the variation of the Coriolis parameter with latitude, and spherical geometry. With some manipulation, assuming that the waves propagate along the x-axis, and there is no y-dependence, the perturbation equations (2.3.2) become

$$\frac{\partial u^*}{\partial t} = +fv^* - \frac{\partial p'}{\partial x} \tag{2.3.17a}$$

$$\frac{\partial v^*}{\partial t} = -fu^* \tag{2.3.17b}$$

$$\alpha\frac{\partial w^*}{\partial t} = -\frac{\partial p'}{\partial z} - \rho'g \tag{2.3.17c}$$

$$\beta\frac{\partial\rho'}{\partial t} = -\left(\frac{\partial u^*}{\partial x} + \frac{\partial w^*}{\partial z}\right) \tag{2.3.17d}$$

$$\frac{g}{C_p}\frac{\partial s^*}{\partial t} = -w^*N^2 \tag{2.3.17e}$$

$$s^* = C_p\left(\frac{p'}{c_s^2} - \rho'\right) \tag{2.3.17f}$$

In these equations, we have introduced two constants α and β as *markers* for the hydrostatic and the quasi-Boussinesq approximations respectively. They can take the value 1 or 0. If we make $\alpha = 0$, it indicates that we are making the *hydrostatic* approximation, i.e., neglecting the vertical acceleration in (2.3.17c). If we make $\beta = 0$, it indicates that we are making the *anelastic or quasi-Boussinesq* approximation, i.e., assuming that the mass weighted three-dimensional divergence is zero. Otherwise the markers take the value 1. These markers will be used in the Section 2.4, where we discuss filtering approximations.

We now try plane wave solutions, where the basic state is a function of z of the form

$$(u^*, v^*, w^*, p', \rho') = (U(z), V(z), W(z), P(z), R(z))e^{i(kx-vt)} \tag{2.3.18}$$

Instead of assuming a z-dependence of the form $e^{i(mz)}$, we will determine it explicitly. If the horizontal scale is not small compared with the radius of the Earth,

2.3 Basic Wave Oscillations in the Atmosphere   47

$L \sim a$, then the solutions are of the form $(u^*, v^*, w^*, p', \rho') = [U(z), V(z), W(z), P(z), R(z)]A(\varphi)e^{i(s\lambda - \nu t)}$, and the equation obtained for $A(\varphi)$ is known as *the Laplace tidal equation.*

Substituting the assumed form of the solution (2.3.18) into (2.3.17), we get

$$-i\nu U = -ikP + fV \tag{2.3.19a}$$

$$-i\nu V = -fU \tag{2.3.19b}$$

$$-i\nu\alpha W = -Rg - \frac{dP}{dz} \tag{2.3.19c}$$

$$-i\nu\beta R = -ikU - \frac{dW}{dz} \tag{2.3.19d}$$

$$-i\nu\left(\frac{P}{c_s^2} - R\right) = -W\frac{N^2}{g} \tag{2.3.19e}$$

From (2.3.19a) and (2.3.19b),

$$U = \frac{k\nu}{\nu^2 - f^2}P \tag{2.3.19f}$$

From (2.3.19d) and (2.3.19f),

$$\beta R = \frac{k^2}{\nu^2 - f^2}P - \frac{i}{\nu}\frac{dW}{dz} \tag{2.3.19g}$$

From (2.3.19c) and (2.3.19e)

$$\frac{dP}{dz} + \frac{g}{c_s^2}P = \frac{i}{\nu}(\nu^2\alpha - N^2)W \tag{2.3.19h}$$

From (2.3.19e) and (2.3.19g)

$$\frac{dW}{dz} + \beta\frac{N^2}{g}W = \frac{i\nu}{c_s^2}\left[\frac{\beta(\nu^2 - f^2) - c_s^2 k^2}{\nu^2 - f^2}\right]P \tag{2.3.19i}$$

From (2.3.19h) and (2.3.19i)

$$\left(\frac{d}{dz} + \frac{g}{c_s^2}\right)\left(\frac{d}{dz} + \beta\frac{N^2}{g}\right)W = -\frac{1}{c_s^2}\left[\frac{(\beta(\nu^2 - f^2) - c_s^2 k^2)(\nu^2\alpha - N^2)}{\nu^2 - f^2}\right]W \tag{2.3.20}$$

or a similar equation for P. This last equation is of the form

$$\frac{d^2W}{dz^2} + A\frac{dW}{dz} + BW = 0$$

In order to eliminate the first derivative, we try a substitution of the form $W = e^{\delta z}\Omega$, and obtain $d^2\Omega/dz^2 + C\Omega = 0$. This requires that we choose $\delta = -A/2$, and in that case $C = B - A^2/4$.

From (2.3.20), the variable substitution, and additional sweat, we finally obtain

$$\frac{d^2\Omega}{dz^2} + n^2\Omega = 0 \tag{2.3.21}$$

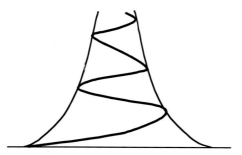

Figure 2.3.2 Schematic of density weighted internal (vertically propagating) waves.

where

$$n^2 = \frac{(\beta(v^2 - f^2) - c_s^2 k^2)(v^2 \alpha - N^2)}{c_s^2(v^2 - f^2)} - \frac{1}{4}\left(\beta\frac{N^2}{g} + \frac{g}{c_s^2}\right)^2 \quad (2.3.22)$$

This is the frequency dispersion relationship for waves in an atmosphere with an isothermal basic state. Given a horizontal structure of the wave (k), and its frequency (v), (2.3.22) determines the vertical structure (n) of Ω (and W) and vice versa. The same FDR would have been obtained making the substitution $Q = e^{-\delta z} P$, and solving for Q.

Equation (2.3.22) indicates that depending on the sign of n^2, we can have either external or internal wave solutions.

2.3.2.1 External Waves

If $n^2 < 0$, the vertical wavenumber n is imaginary, $n = im$. The solution of (2.3.21) is then $\Omega = Ae^{mz} + Be^{-mz}$, or, going back to the vertical velocity,

$$w^*(x,z,t) = e^{i(kx-vt)} e^{-\frac{1}{2}\left(\beta\frac{N^2}{g} + \frac{g}{c_s^2}\right)z} (Ae^{mz} + Be^{-mz}) \quad (2.3.23)$$

These are external waves (the waves do not oscillate in the vertical and, therefore, do not propagate vertically). If the boundary condition at the ground is that the vertical velocity is zero, then $\Omega = Ae^{mz} + Be^{-mz} = 0$ at $z = 0$, so that $A + B = 0$, and

$$w^*(x,z,t) = e^{i(kx-vt)} e^{-\frac{1}{2}\left(\beta\frac{N^2}{g} + \frac{g}{c_s^2}\right)z} 2A \sinh(mz)$$

which has an exponential behavior in z. Since $\sinh(mz)$ cannot be zero above the ground, an upper boundary condition of a rigid top can only be satisfied if $A = 0$. In other words, we cannot have external waves with rigid top and bottom boundary conditions: external waves require a free surface at the top (or at the bottom).

2.3.2.2 Internal Waves

If $n^2 > 0$, the vertical wavenumber n is real:

$$w^*(x,z,t) = e^{i(kx-vt)} (Ae^{inz} + Be^{-inz}) e^{-\frac{1}{2}\left(\beta\frac{N^2}{g} + \frac{g}{c_s^2}\right)z} \quad (2.3.24)$$

A, B are determined from the boundary conditions. Now there is both vertical and horizontal propagation. For example, if there is a rigid bottom, we have again $A + B = 0$, and the solution becomes

2.3 Basic Wave Oscillations in the Atmosphere

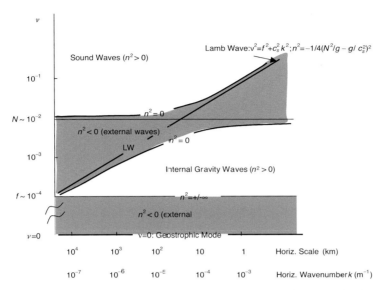

Figure 2.3.3 Schematic of the frequencies of small perturbations in an isothermal resting atmosphere as a function of k, the horizontal wavenumber (the horizontal scale is its inverse), and the vertical wavenumber n. Shaded regions represent $n^2 < 0$, external waves.

$$w^*(x,z,t) = A\left(e^{i(kx+nz-\nu t)} - e^{i(kx-nz-\nu t)}\right)e^{-\frac{1}{2}(\beta\frac{N^2}{g}+\frac{g}{c_s^2})z}$$

The shape of internal waves in the vertical is shown schematically in Figure 2.3.2.

2.3.3 Analysis of the FDR of Wave Solutions in a Resting, Isothermal Atmosphere

We will now plot the general FDR equation (2.3.22). We assume $T_{00} = 250$ K and $f = 2\Omega\sin 45° \approx 10^{-4}$ s^{-1}. Then, the speed of sound is $c_s^2 = \gamma RT \approx 10^5$ m^2/s^2, or $c_s \approx 320$ m/s, the scale height is $H = RT/g = 7.3$ km $= 7300$ m, and the Brunt–Väisälä frequency is $N^2 = gd(\ln\theta_0)/dz = g\kappa/H$ for the isothermal atmosphere, or about 4×10^{-4} s^{-2}. Note that the frequency associated with inertial oscillations is much lower than the frequency associated with gravitational oscillations.

$$f \sim 10^{-4} \text{ s}^{-1} \ll N \sim 10^{-2} \text{ s}^{-1} \tag{2.3.25}$$

We first plot in Figure 2.3.3 the FDR (2.3.22), with $\alpha = \beta = 1$, i.e., without making either the hydrostatic or the quasi-Boussinesq approximations. Note that this equation contains four solutions for the frequency ν, plus an additional solution $\nu = 0$, the geostrophic mode that satisfies nontrivially (2.3.19).

50   **The Continuous Equations**

2.4   Filtering Approximations

When we neglect the time derivative of one of the equations of motion, we convert it from a prognostic equation into a diagnostic equation, and eliminate with it one type of solution. Physically, we eliminate a restoring force that supports a certain type of wave. We call this a "filtering approximation." Use of the quasi-geostrophic filtering approximation that eliminates both sound and gravity waves made possible the successful forecast of (Charney et al., 1950). Until the 2010s, most global models and some regional models used the hydrostatic approximation, which filters sound waves. In this section we explore the effect of the filtering approximations.

2.4.1   Quasi-geostrophic Approximation

As we have already seen, without rotation, if we assume a steady state, the solution of (2.3.19) would be a trivial solution: all perturbations would be equal to zero. However, with rotation, if we assume steady state solutions, and neglect all time derivatives $v = 0$, we obtain from the perturbed equations (2.3.17), the geostrophic mode, a nontrivial solution:

$$\left. \begin{array}{l} V = ikP/f \\[4pt] U = 0 \\[4pt] \dfrac{dP}{dz} = -Rg \\[4pt] \dfrac{dW}{dz} = 0 \\[4pt] P = c_s^2 R \end{array} \right\} \tag{2.4.1}$$

For the continuous perturbation equations (2.3.17), this means:

$$\left. \begin{array}{ll} fv^* = -\dfrac{\partial p'}{\partial x} & \text{(geostrophically balanced flow)} \\[8pt] \dfrac{\partial v^*}{\partial t} = 0 & \text{(steady state flow)} \\[8pt] 0 = -\dfrac{\partial p'}{\partial z} - \rho'g & \text{(hydrostatically balanced flow)} \\[8pt] w^* = 0 & \\[8pt] \dfrac{\partial w^*}{\partial z} = \dfrac{\partial u^*}{\partial x} = 0 & \text{(horizontal, nondivergent flow)} \\[8pt] s^* = C_p \left(\dfrac{p'}{c_s^2} - \rho' \right) = 0 & \text{(pressure perturbations are proportional to density perturbations multiplied by the speed of sound squared, which is true whenever the hydrostatic equation is valid)} \end{array} \right\} \tag{2.4.2}$$

This is the "ultimate" filtering approximation: it filters out sound waves, inertia and gravity oscillations.

2.4 Filtering Approximations 51

For large horizontal scales, we have to include the effects of varying rotation, and the f-plane becomes a β-plane: $f = f_0 + \beta y$. When horizontal advection by the basic flow is included, the stationary geostrophic flow solution becomes quasi-stationary (slowly varying). The waves corresponding to the geostrophic mode are Rossby-type waves with a frequency small compared with the Coriolis or inertial frequency $v \approx Uk - \beta/k \sim 10^{-5} - 10^{-5}\ \text{s}^{-1}$. Rossby waves are quasi-geostrophic ($v^2 \ll f^2$), hydrostatically balanced, and the flow is quasi-horizontal ($w^*/H \ll U^*/L$), and therefore quasi-nondivergent ($\boldsymbol{\nabla} \cdot \mathbf{v}_h \approx 0$).

Note that this type of quasi-geostrophic solution, fundamental for NWP, is still present in the general equations of motion, and survives as a solution when we make either the anelastic or the hydrostatic approximation in order to filter out sound waves.

2.4.2 Quasi-Boussinesq or Anelastic Approximation

We now substitute $\beta = 0$ in (2.3.19d). This means that we neglected the time derivative $\partial \rho'/\partial t$ compared with $\boldsymbol{\nabla}_H \cdot v^*, \partial w^*/\partial z$ in the continuity equation. With this approximation, the equations become "anelastic," i.e., they do not allow the presence of sound waves, which require three-dimensional divergence and convergence for their propagation. Consider the terms that are neglected in the FDR (2.3.22):

1. $v^2 - f^2 \ll c_s^2 k^2$, i.e., the frequency of retained solutions is much smaller than that of sound waves; therefore, this also filters out the Lamb waves, i.e., horizontally propagating sound waves.
2. $N^2/g \ll g/c_s^2$. This approximation is justified if

$$\frac{N^2}{g} = \frac{1}{\theta_0}\frac{d\theta_0}{dz} \ll \frac{g}{\gamma RT_0}, \text{i.e.,}\ \frac{\gamma H}{\theta_0}\frac{d\theta_0}{dz} \ll 1.$$

In other words, the deep anelastic approximation is justified for a model for which the potential temperature does not change too much within the depth $\gamma RT/g \sim$ 10 km. This is a reasonable approximation for the standard troposphere (not for deep flow into the stratosphere), since for the troposphere: $\Delta\theta_0/\theta_0 \sim 30\ \text{K}/300\ \text{K} \sim 0.1$.

For models that are so shallow that not only $\Delta\theta_0/\theta_0 << 1$, but also $\Delta T_0/T_0 \ll 1$, we can also neglect $\partial\rho_0/\partial z$ in the continuity equation, and assume $\boldsymbol{\nabla}_3 \cdot \mathbf{v}' = 0$, not just $\boldsymbol{\nabla}_3 \cdot \mathbf{v}^* = 0$. In this case, we treat the atmosphere as if it was an incompressible fluid. This approximation is only accurate for very shallow atmospheric models (less than 1 km depth) but appropriate for ocean models, since water is well approximated as an incompressible fluid.

Figure 2.4.1 schematically shows the FDR when we make the anelastic approximation. From (2.3.22), and letting $\beta = 0$ (with $\alpha = 1$), we can derive the frequency of inertia-gravity waves with the anelastic approximation:

$$v^2 = f^2\frac{n^2 + p^2}{n^2 + p^2 + k^2} + N^2\frac{k^2}{n^2 + p^2 + k^2} \tag{2.4.3}$$

52 The Continuous Equations

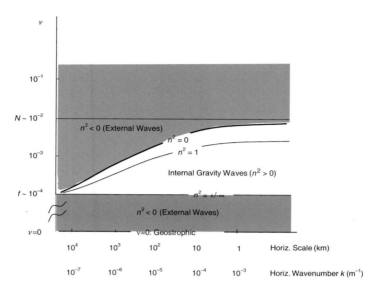

Figure 2.4.1 Schematic of the frequencies of small perturbations in an isothermal resting atmosphere when the quasi-Boussinesq or anelastic approximation is made ($\beta = 0$).

where p is like the inverse of a vertical wavelength

$$p = \frac{1}{2}\frac{g}{c_s^2} \sim \frac{1}{20}\text{ km}$$

From (2.4.3) we see that, for internal ($n^2 > 0$) inertia-gravity waves, $f^2 < \nu^2 < N^2$, the frequency ν is between the Coriolis and Brunt–Väisälä frequencies. Note from Figure 2.4.1 that for these waves, $\partial \nu^2/\partial k^2 > 0$, but $\partial \nu^2/\partial n^2 < 0$. This implies (since we can assume without loss of generality that $k > 0$) that the horizontal group velocity for gravity waves $\partial \nu/\partial k$ has the same sign as the phase velocity (the energy of gravity waves moves in the same direction as the phase speed in the horizontal). In the vertical the opposite is true: if the group velocity is upwards, which happens for example when gravity waves are generated by mountain forcing, the phase velocity is downwards.

Because the anelastic equation filters out acoustic internal waves (as well as the Lamb wave) it is widely used for problems in which the hydrostatic approximation cannot be made, as is the case for convection. For example, the ARPS model is based on deep anelastic equations. The FDR with the quasi-Boussinesq approximation is shown schematically in Figure 2.4.1.

2.4.3 Hydrostatic Approximation

If we neglect the vertical acceleration $\partial w^*/\partial t$ in the vertical momentum equation (2.3.17c), letting $\alpha = 0$ (with $\beta = 1$), we get the FDR

$$n^2 = -\frac{(\nu^2 - f^2 - c_s^2 k^2)N^2}{c_s^2(\nu^2 - f^2)} - \frac{1}{4}\left(\frac{N^2}{g} - \frac{g}{c_s^2}\right)^2 \tag{2.4.4}$$

2.4 Filtering Approximations

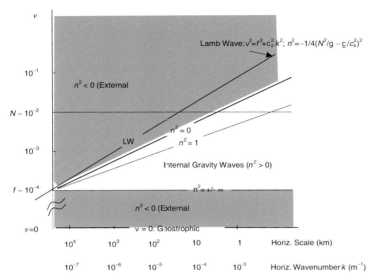

Figure 2.4.2 Schematic of the frequencies of small perturbations in an isothermal resting atmosphere when the hydrostatic approximation is made ($\alpha = 0$).

This FDR has two solutions: the horizontally propagating external sound wave (Lamb wave) solution, which unfortunately is retained:

$$v^2 = f^2 + c_s^2 k^2, \quad n^2 = -1/4(N^2/g - g/c_s^2)^2 \tag{2.4.5}$$

and inertia-gravity waves. From (2.4.4), we can derive the following relationship for inertia-gravity waves: using $N^2 = \kappa g/H$, $H = RT/g$, $c_s^2 = \gamma RT$

$$n^2 = \frac{N^2 k^2}{v^2 - f^2} - \frac{1}{4H^2}, \text{ or } v^2 = f^2 + \frac{N^2 k^2}{n^2 + \frac{1}{4H^2}} \tag{2.4.6}$$

Figure 2.4.2 shows the relationship between frequency and horizontal and vertical wavenumbers with the hydrostatic equation.

Exercise 2.4.1 *Derive (2.4.6) from (2.4.5).*

When are we justified in using the hydrostatic equation? By taking $\alpha = 0$, we neglected the time derivative of the vertical velocity compared to $\rho'/\rho_0 g$. Note that it is not enough to find $dw/dt \ll g$ to make the hydrostatic approximation: the vertical acceleration is small compared to gravity even for strong vertical motions, as in a cumulus cloud. The hydrostatic approximation requires that the vertical acceleration be small compared with the buoyancy $(\rho'/\rho_0)g$ or gravitational acceleration within the fluid. It can be shown by scale analysis that the hydrostatic approximation is valid as long as we are dealing with shallow flow ($H/L \ll 1$). For quasi-geostrophic flow, the condition for hydrostatic balance is valid even if $H/L \sim 1$. This implies that the hydrostatic approximation is very accurate for models with grid sizes of the order of 100 km or larger and still quite acceptable for quasi-geostrophic flow, even when the horizontal

54 The Continuous Equations

Table 2.4.1 Summary of wave characteristics and the filtering approximations (adapted from Zhang, personal communication, 1996)

Type of wave (typical amplitude)	Phase speed	Restoring force	Filtering approximations
Acoustic (less than 0.1 hPa, noise level)	$\sqrt{\gamma R T}$ (320 m/s)	Compression	Hydrostatic anelastic quasi-geostrophic
External gravity (if initial conditions are not balanced, 10 hPa)	$\sqrt{gH}$ (320 m/s for $H = 10$ km)	Gravity	No free surface at the top or the bottom, or no net column mass convergence
Internal gravity (0.1–1 hPa)	$\sim \frac{1}{k}\sqrt{\frac{N^2 k^2}{k^2 + n^2}}$ $\sim N/k$ (50 m/s for $L = 30$ km)	Buoyancy (gravitational acceleration within fluid)	Neutral stratification $(N = 0)$, or $\frac{\partial \nabla \cdot \mathbf{v}_H}{\partial t} = 0$
Inertia	f/k (15 m/s for $L = 1000$ km)	Coriolis force (f)	No rotation $(f = 0)$
Rossby (20 hPa)	$U - \beta/k^2$ (relative phase speed ~ 20–50 m/s depending on L)	Variation of f with latitude (β effect) $d\varsigma/dt = -\beta v$	Constant $f (\beta = 0)$

grid size of the model approaches 10 km. However, the hydrostatic equation is not valid for models with grid sizes of the order of 10 km that attempt to resolve explicitly cumulus convection. Figure 2.4.2 shows that for high frequencies $\nu \sim N$ or larger, or small horizontal scales the hydrostatic approximation distorts the original FDR (compare with Figure 2.3.3).

Exercise 2.4.2 *Show by scale analysis that as long as we are dealing with shallow flow ($H/L \ll 1$) the hydrostatic approximation is valid.*

Exercise 2.4.3 *Show that under the condition for quasi-geostrophic balance, the hydrostatic approximation is valid even if $H/L \sim 1$.*

We now summarize in Table 2.4.1 the characteristics of the different types of waves and the approximations that can be used to filter them out. For more details about Rossby waves and the filtering of inertia gravity waves, see Section 2.5, where these topics are discussed in the context of the SWEs.

Notes

(1) In normal mode analysis of large-scale (hydrostatic) motion or of atmospheric models, it is customary to find a horizontal structure equation and a vertical structure equation, associated by a separation constant h, where h is denoted as

"equivalent depth" (e.g., Temperton and Williamson, 1981). In our simple f-plane case, the horizontal structure equation for the inertia gravity waves (2.4.6) is

$$\frac{v^2 - f^2}{k^2} = gh \tag{2.4.7}$$

and the vertical structure equation

$$n^2 = \frac{1}{H^2}\left(\frac{\kappa H}{h} - \frac{1}{4}\right) \tag{2.4.8}$$

where we have used $N^2 = \kappa g/H$.

The reason h is called the equivalent depth is that internal modes are governed by equations similar to the SWEs with depth h. However, n is not a constant but a function of vertical wavenumber, and therefore the analogy is only approximate.

(2) With the hydrostatic approximation, the geopotential energy gz and the internal energy $C_v T$ of an air column are related to each other, since $\int_{z_s}^{\infty} \rho gz\, dz = \int_{z_s}^{\infty} -(\partial p/\partial z)z\, dz = [-pz]_{z_s}^{\infty} + \int_{z_s}^{\infty} p\, dz = p_s z_s + \int_{z_s}^{\infty} \rho RT\, dz$. Here the subscript s represents the Earth's surface, and $\lim_{z\to\infty} pz = 0$ is assumed. So, when $z_s = 0$, the ratio of the potential to the internal energy of a column is equal to $R/C_v = 0.4$. When z_s is not constant, the total potential energy (Lorenz, 1955) is given by

$$\int_{z_s}^{\infty} \rho(gz + C_v T)\,dz = p_s z_s + \int_{z_s}^{\infty} \rho C_p T\, dz.$$

2.5 Shallow Water Equations, Quasi-geostrophic Filtering, and Filtering of Inertia-Gravity Waves

Consider now the SWEs (Figure 2.5.1), valid for an incompressible hydrostatic motion of a fluid with a free surface $h(x, y, t)$. "Shallow" means that the vertical depth is much smaller than the typical horizontal scale which justifies the hydrostatic approximation. These equations are not only appropriate for representing a shallow mass of water (e.g., river flow, storm surges), but they are prototypical of the primitive equations based on the hydrostatic approximation and are frequently used to test numerical schemes. The shallow water horizontal momentum equations are

$$\frac{d\mathbf{v}}{dt} = -f\mathbf{k} \times \mathbf{v} - \boldsymbol{\nabla}\phi \tag{2.5.1}$$

Figure 2.5.1 Schematic of the shallow water model: a hydrostatic, incompressible fluid with a rigid bottom $h_s(x, y)$, a free surface $h(x, y, t)$, and horizontal scales L much larger than the mean vertical scale H.

The Continuous Equations

where

$$\frac{d}{dt} = \frac{\partial}{\partial t} + \mathbf{v} \cdot \boldsymbol{\nabla} \qquad \mathbf{v} = \mathbf{v}_H = u\mathbf{i} + v\mathbf{j} \qquad \phi = gh.$$

The continuity equation is

$$\frac{d(\phi - \phi_s)}{dt} = -(\phi - \phi_s)\boldsymbol{\nabla} \cdot \mathbf{v}, \tag{2.5.2a}$$

which can also be written as

$$\frac{\partial \phi}{\partial t} = -\boldsymbol{\nabla} \cdot [(\phi - \phi_s)\mathbf{v}]. \tag{2.5.2b}$$

Here $\phi_s = gh_s(x, y)$ and h_s is the bottom topography.

Exercise 2.5.1 *Derive the SWE from the primitive equations assuming hydrostatic, incompressible motion, and that the horizontal velocity is constant in height. Is the vertical velocity constant in height as well?*

We now derive the important conservation of potential vorticity equation making use of the useful equation,

$$\mathbf{v} \cdot \boldsymbol{\nabla}\mathbf{v} = \boldsymbol{\nabla}(v^2/2) + \varsigma\,\mathbf{k} \times \mathbf{v}$$

where $\varsigma = \mathbf{k} \cdot \boldsymbol{\nabla} \times \mathbf{v}$ is the vertical vorticity. We apply $\mathbf{k} \cdot \boldsymbol{\nabla} \times$ to the momentum equation (2.5.1) to derive the vorticity equation,

$$\frac{\partial \varsigma}{\partial t} + \mathbf{v} \cdot \boldsymbol{\nabla}\varsigma + \varsigma\boldsymbol{\nabla} \cdot \mathbf{v} = -f\boldsymbol{\nabla} \cdot \mathbf{v} - \mathbf{v} \cdot \boldsymbol{\nabla}f \tag{2.5.3}$$

or (since $df/dt = \mathbf{v} \cdot \boldsymbol{\nabla}f$)

$$\frac{d(f + \varsigma)}{dt} = -(f + \varsigma)\boldsymbol{\nabla} \cdot \mathbf{v} \tag{2.5.4}$$

which indicates that the absolute vorticity $(f + \varsigma)$ of a parcel of "water" increases with its convergence (or vertical stretching).

Eliminating the divergence between (2.5.4) and the first form of the continuity equation, we obtain the equation of conservation of potential vorticity:

$$\frac{d}{dt}\left(\frac{f + \varsigma}{\phi - \phi_s}\right) = 0 \tag{2.5.5}$$

where

$$q = \left(\frac{f + \varsigma}{\phi - \phi_s}\right) \tag{2.5.6}$$

is the potential vorticity: the absolute vorticity divided by the depth of the fluid.

Exercise 2.5.2 *Give a physical interpretation of the equation of conservation of potential vorticity.*

The conservation of potential vorticity is an extremely powerful dynamical constraint. In a multilevel primitive equation model, the *isentropic potential vorticity* (the absolute vorticity divided by the distance between two surfaces of constant potential temperature) is also individually conserved. If the initial potential vorticity distribution is accurately represented in a numerical model, and the model is able to transport potential vorticity accurately, then the forecast will also be accurate.

We now consider small perturbations on a flat bottom and a mean height $\Phi = gH = const.$ on a constant f-plane.

$$\frac{\partial \mathbf{v}'}{\partial t} = -f\mathbf{k} \times \mathbf{v}' - \boldsymbol{\nabla}\phi' \tag{2.5.7}$$

and

$$\frac{\partial \phi'}{\partial t} = -\Phi\boldsymbol{\nabla} \cdot \mathbf{v}' \tag{2.5.8}$$

(note that (2.5.7) and (2.5.8) are the same equations as in Section 2.3.1.5 on horizontal sound (Lamb) waves, with $gH = c_s^2 = \gamma RT_0, g\phi = p'/\rho_0$).

Assume solutions of the form $(u', v', \phi')e^{i(kx-\nu t)}$. Then the FDR is

$$\nu(\nu^2 - f^2 - \Phi k^2) = 0 \tag{2.5.9}$$

with three solutions for ν:

$$\nu^2 = f^2 + \Phi k^2 \tag{2.5.10}$$

the frequency of inertia-gravity waves, analogous to the inertia-Lamb wave, and $\nu = 0$, the geostrophic mode. As before, this is a geostrophic, nondivergent steady state solution $\frac{\partial}{\partial t}() = 0, \mathbf{v} = \frac{\mathbf{k} \times \boldsymbol{\nabla}\phi}{f}, \boldsymbol{\nabla} \cdot \mathbf{v} = 0$.

Following Arakawa (1997), we can now compare the FDR of inertia-gravity waves in the SWE with the FDR of a three-dimensional isothermal system using the hydrostatic approximation (2.4.4)–(2.4.6). We see that (2.5.10) is analogous to internal inertia-gravity waves for an isothermal hydrostatic atmosphere (2.4.6) if we define an equivalent depth such that $\Phi = gh_{eq}$:

$$h_{eq} = \frac{N^2/g}{n^2 + \dfrac{1}{4H_0^2}} = \frac{d\ln\theta_0/dz}{n^2 + \dfrac{1}{4H_0^2}} \tag{2.5.11}$$

and is analogous to the (external) inertia Lamb waves (2.4.5) if we define the equivalent depth as

$$h_{eq} = c_s^2/g = \gamma H_0. \tag{2.5.12}$$

2.5.1 Quasi-geostrophic Scaling for the SWE

If we want to filter the inertia-gravity waves, as Charney did in the first successful numerical weather forecasting experiment (Chapter 1), we can develop a quasi-geostrophic version of the SWE. We can do it first for an f-plane ($f = f_0$).

Assume that the atmosphere is in *quasi-geostrophic* balance: $\mathbf{v} = \mathbf{v}_g + \mathbf{v}_{ag} = \mathbf{v}_g + \varepsilon\mathbf{v}'$ where we assume that the typical size of the ageostrophic wind is much smaller (order $\varepsilon = U/fL$, the Rossby number) than the geostrophic wind $\varepsilon\mathbf{v}' \ll \mathbf{v}_g$, and that the same is true for their time derivatives $\varepsilon \partial \mathbf{v}'/\partial t \ll \partial \mathbf{v}_g/\partial t$. The geostrophic wind is given by

$$\mathbf{v}_g = \frac{1}{f}\mathbf{k} \times \boldsymbol{\nabla}\phi \tag{2.5.13}$$

Plugging these into the perturbation equations(2.5.7) and (2.5.8), we obtain

$$\frac{\partial \mathbf{v}_g}{\partial t} + \varepsilon \frac{\partial \mathbf{v}'}{\partial t} = -\boldsymbol{\nabla}\phi - f\mathbf{k} \times \mathbf{v}_g - \varepsilon f\mathbf{k} \times \mathbf{v}' = -\varepsilon f\mathbf{k} \times \mathbf{v}' \tag{2.5.14}$$

In this equation, the dominant terms (pressure gradient and Coriolis force on the geostrophic flow) cancel each other (geostrophic balance), so that the smaller effect of the Coriolis force acting on the ageostrophic flow is left to balance the time derivative. From (2.5.8),

$$\frac{\partial \phi}{\partial t} = -\Phi\boldsymbol{\nabla} \cdot \mathbf{v}_g - \varepsilon\Phi\boldsymbol{\nabla} \cdot \mathbf{v}' = -\varepsilon\Phi\boldsymbol{\nabla} \cdot \mathbf{v}' \tag{2.5.15}$$

Here the geostrophic wind is nondivergent, so that the time derivative of the pressure is given by the divergence of the smaller ageostrophic wind.

From (2.5.14) and (2.5.15), we can conclude that $\partial \mathbf{v}_g/\partial t$ and $\partial \phi/\partial t$ are of order ε, i.e., the geostrophic flow changes *slowly* (it is almost stationary compared with other types of motion), and that $\partial \mathbf{v}_{ag}/\partial t = \varepsilon \partial \mathbf{v}'/\partial t$, which is smaller than $\partial \mathbf{v}_g/\partial t$, is of order ε^2. With *quasi-geostrophic scaling*, we neglect terms of $O(\varepsilon^2)$, and we obtain the linearized quasi-geostrophic SWE:

$$\frac{\partial \mathbf{v}_g}{\partial t} = -\boldsymbol{\nabla}\phi - f\mathbf{k} \times \mathbf{v} = -f\mathbf{k} \times \mathbf{v}_{ag} \tag{2.5.16a}$$

$$\frac{\partial \phi}{\partial t} = -\Phi\boldsymbol{\nabla} \cdot \mathbf{v} = -\Phi\boldsymbol{\nabla} \cdot \mathbf{v}_{ag} \tag{2.5.16b}$$

$$\mathbf{v}_g = 1/f\,\mathbf{k} \times \boldsymbol{\nabla}\phi \tag{2.5.16c}$$

Note that in (2.5.16) there is only one independent time derivative because of the geostrophic relationship (we lost the other two time derivatives when we neglected the term $\partial \mathbf{v}_{ag}/\partial t$). Physically, this means that we only allow divergent motion to exist as required to maintain the quasi-geostrophic balance and eliminate the degrees of freedom necessary for the propagation of gravity waves.

We can rewrite (2.5.16) as

$$\frac{\partial u_g}{\partial t} = -\frac{\partial \phi}{\partial x} + fv = fv_{ag} \tag{2.5.17a}$$

$$\frac{\partial v_g}{\partial t} = -\frac{\partial \phi}{\partial y} - fu = -fu_{ag} \tag{2.5.17b}$$

$$\frac{\partial \phi}{\partial t} = -\Phi\left(\frac{\partial u}{\partial x} + \frac{\partial v}{\partial y}\right) = \Phi\left(\frac{\partial u_{ag}}{\partial x} + \frac{\partial v_{ag}}{\partial y}\right) \tag{2.5.17c}$$

$$u_g = -\frac{1}{f}\frac{\partial \phi}{\partial y}; v_g = \frac{1}{f}\frac{\partial \phi}{\partial x} \tag{2.5.17d}$$

We can compute the equation for the geostrophic vorticity evolution from (2.5.17) by taking the x-derivative of (2.5.17b) minus the y-derivative of (2.5.17a):

2.5 Shallow Water Equations 59

$$\frac{\partial \zeta}{\partial t} = -f_0 \left(\frac{\partial u}{\partial x} + \frac{\partial v}{\partial y} \right) - \beta v \tag{2.5.18}$$

where the last term in (2.5.18) appears if we are on a β-plane: $f = f_0 + \beta y$. Then we can eliminate the (ageostrophic) divergence between (2.5.18) and (2.5.17c) and obtain the linear quasi-geostrophic potential vorticity equation on a β-plane:

$$\frac{\partial}{\partial t} \left(\frac{\zeta}{f_0} - \frac{\phi}{\Phi} \right) = -\frac{\beta}{f_0^2} \frac{\partial \phi}{\partial x} \tag{2.5.19}$$

or, since $\zeta = \nabla^2 \phi / f_0$,

$$\frac{\partial}{\partial t} \left(\frac{\nabla^2 \phi}{f_0^2} - \frac{\phi}{\Phi} \right) = -\frac{\beta}{f_0^2} \frac{\partial \phi}{\partial x} \tag{2.5.20}$$

Note that there is a single independent variable (ϕ) so that there is a single solution for the frequency. If we neglect the β-term (i.e., assume an f-plane) and allow for plane-wave-type solutions $\phi = F e^{i(kx - vt)}$, the only solution of the FDR in (2.5.20) is $v = 0$, the geostrophic mode. This confirms that by eliminating *the time derivative* of the ageostrophic (divergent) wind $\mathbf{v}_{ag}$, we have eliminated the inertia-gravity wave solution. If we assume a β-plane, i.e., keep the β term in (2.5.20), the quasi-geostrophic FDR becomes

$$v = \frac{-\beta k}{k^2 + f_0^2 / \Phi}. \tag{2.5.21}$$

These Rossby waves are the essential "weather waves" and, as shown in Table 2.4.1, have rather large amplitudes (up to 50 hPa). The ageostrophic flow associated with these waves is responsible for the upward motion that produces precipitation ahead of the troughs.

In a multilevel model, the FDR (2.5.21) can be used with the equivalent depths (2.5.11), (2.5.12) applied to the baroclinic (internal) and barotropic Rossby waves, respectively. With these definitions, we can say that the waves in the atmosphere are analogous to the SWE waves. However, because h_{eq} appears as a separation constant in the definition of the normal modes of the atmosphere, the equivalent depth depends on the vertical wavenumber, and on the type of wave considered (Lamb or inertia-gravity waves).

Exercise 2.5.3 *Show that the quasi-geostrophic PVE (potential vorticity equation) for nonlinear SWE is*

$$\left(\frac{\partial}{\partial t} + u_g \frac{\partial}{\partial x} + v_g \frac{\partial}{\partial y} \right) \left(\frac{\nabla^2 \phi}{f_0^2} - \frac{\phi}{\Phi} \right) = -\frac{\beta}{f_0^2} \frac{\partial \phi}{\partial x} \tag{2.5.22}$$

using similar scaling arguments.

Exercise 2.5.4 *Allow for a basic flow $u_{g(total)} = U + u_g$; $v_{g(total)} = v_g$, in (2.5.22). How will this change the FDR (2.5.21)?*

Exercise 2.5.5 *Estimate the initial time derivative for typical values of the horizontal wavenumber ($L = 2000$ km, 8000 km), that Richardson would have observed, i.e.,*

60 The Continuous Equations

compare the frequency v for the external (barotropic, $H \sim 10$ km) mode for inertia–gravity waves and for Rossby waves.

Exercise 2.5.6 *Derive the formula for group velocity in the x-direction for Rossby waves.*

Exercise 2.5.7 *Using typical values of long and short synoptic waves (e.g., horizontal wavelengths of 8000 km and 2000 km, respectively), calculate the phase speed and the group velocity of Rossby waves for the barotropic mode and the first baroclinic mode ($H \sim 10$ km and 1 km, respectively).*

2.5.2 Inertia-Gravity Waves in the Presence of a Basic Flow

As we just saw, the SWEs are a simple version of the primitive equations and widely used to understand numerical and dynamical processes in primitive equations. As we noted in Chapter 1, filtered quasi-geostrophic models have been substituted by primitive equation models for NWP, because the quasi-geostrophic filtering is not an accurate approximation (it assumes that the Rossby number U/fL is much smaller than 1). Recall that quasi-geostrophic filtering was introduced by Charney et al. (1950) in order to eliminate the problem of gravity waves (which requires a small time step) whose high frequencies produced a huge time derivative in Richardson's computation, masking the time derivative of the actual weather signal.

An alternative way to deal with the presence of fast gravity waves without resorting to quasi-geostrophic filtering is the use of semi-implicit time schemes (to be discussed in Chapter 3). Consider small perturbations in the SWE including a basic flow U in the x-direction. Then the total linearized time derivative becomes

$$\frac{d}{dt} = \frac{\partial}{\partial t} + U\frac{\partial}{\partial x}$$

In that case, when we assume solutions of the form $Ae^{i(kx-vt)}$, $d/dt = i(-v + kU)$. Therefore the FDR remains the same except that v is replaced by $v - kU$. The FDR for small perturbations in the SWE with a basic flow U is therefore

$$(v - kU)[(v - kU)^2 - f^2 - \Phi k^2] = 0 \qquad (2.5.23)$$

As noted before, this has three solutions, quasi-geostrophic flow (which is steady state, except for the uniform translation with speed U) and two solutions for the inertia-gravity waves, modified by the basic flow translation:

$$(v_G - kU) = 0 \quad \text{(geostrophic mode)}$$

$$[(v_{IGW} - kU)^2 - f^2 - \Phi k^2] = 0 \quad \text{(inertia gravity waves, modified by the basic flow } U)$$

The phase speed of the inertia-gravity wave is given by

$$c_{IGW} = \frac{v_{IGW}}{k} = U \pm \sqrt{\frac{f^2}{k^2} + \Phi} \qquad (2.5.24)$$

Finally, we note that for the Lamb wave (as well as for the external gravity wave), the phase speed of the inertia-gravity wave is dominated by the term

$\sqrt{\Phi} \approx \sqrt{g \times 10 \text{ km}} \approx 300$ m/s. As we will see in Section 3.2.5. it is possible to avoid using costly small time steps by means of a *semi-implicit time scheme*. An implicit time scheme has no constraint on the time step. Therefore, in a semi-implicit scheme, the terms that give rise to the fast gravity waves, namely the horizontal divergence and the horizontal pressure gradient are written implicitly, while the rest of the SWE terms can be written explicitly. The terms generating the gravity wave are underlined in the following nonlinear SWE:

$$
\left.
\begin{aligned}
\frac{\partial u}{\partial t} + u\frac{\partial u}{\partial x} + v\frac{\partial u}{\partial y} &= \underline{-\frac{\partial \phi}{\partial x}} + fv \\
\frac{\partial v}{\partial t} + u\frac{\partial v}{\partial x} + v\frac{\partial v}{\partial y} &= \underline{-\frac{\partial \phi}{\partial y}} - fu \\
\frac{\partial \phi}{\partial t} + u\frac{\partial \phi}{\partial x} + v\frac{\partial \phi}{\partial y} &= \underline{-\Phi\left(\frac{\partial u}{\partial x} + \frac{\partial v}{\partial y}\right)} - (\phi - \Phi)\left(\frac{\partial u}{\partial x} - \frac{\partial v}{\partial y}\right)
\end{aligned}
\right\}.
\qquad (2.5.25)
$$

2.6 Primitive Equations and Vertical Coordinates

As Charney (1951) foresaw, most NWP modelers went back to using the primitive equations, with the hydrostatic approximation, but without quasi-geostrophic filtering. Quasi-geostrophic models are now reserved for simple problems where the main motivation is the understanding of atmospheric or ocean dynamics.

Exercise 2.6.1 *Give two or more reasons why using the primitive equations, with the hydrostatic approximation but without quasi-geostrophic filtering was a desirable goal.*

So far we have used z as the vertical coordinate. When we make the hydrostatic approximation, as in the primitive equations, the use of pressure vertical coordinates becomes very advantageous. We can also use any arbitrary variable $\zeta(x,y,z,t)$ as the vertical coordinate as long as it is a monotonic function of z (Kasahara, 1974). The most commonly used vertical coordinates are height z, pressure p, a normalized pressure σ (Phillips, 1957), potential temperature θ (Eliassen and Raustein, 1968), and several kinds of *hybrid* coordinates (e.g., Simmons and Burridge, 1981; Johnson et al. 1993; Purser, pers. comm.; and Bleck and Benjamin, 1993).

2.6.1 General Vertical Coordinates

When we transform the vertical coordinate, a variable $A(x,y,z,t)$ becomes $A(x,y,\zeta(x,y,z,t),t)$. The horizontal coordinates and time remain the same. Let s represent x, y, or t. Then, from Figure 2.6.1(a)

$$
\frac{D-B}{\Delta s} = \frac{C-B}{\Delta s} + \frac{D-C}{\Delta z} \cdot \frac{\Delta z}{\Delta s}
$$

so that

The Continuous Equations

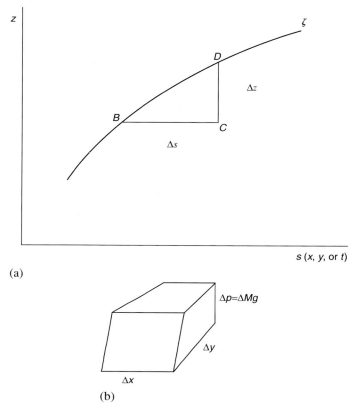

Figure 2.6.1 (a) Schematic showing the relationship between the derivatives of A at constant ζ and at constant Z. The points B and D represent values of A on a ζ-surface and B and C those on a constant z surface. (b) Schematic of a parcel of air in a hydrostatic system, where Δp is proportional to the change in mass per unit area ΔM.

$$\left(\frac{\partial A}{\partial s}\right)_{\zeta} = \left(\frac{\partial A}{\partial s}\right)_{z} + \left(\frac{\partial A}{\partial z}\right)_{s}\left(\frac{\partial z}{\partial s}\right)_{\zeta} \tag{2.6.1}$$

and

$$\frac{\partial A}{\partial \zeta} = \frac{\partial A}{\partial z}\frac{\partial z}{\partial \zeta}$$

or

$$\frac{\partial A}{\partial z} = \frac{\partial A}{\partial \zeta}\frac{\partial \zeta}{\partial z}. \tag{2.6.2}$$

Substituting (2.6.2) in (2.6.1), we get

$$\left(\frac{\partial A}{\partial s}\right)_{\zeta} = \left(\frac{\partial A}{\partial s}\right)_{z} + \left(\frac{\partial A}{\partial \zeta}\right)\left(\frac{\partial \zeta}{\partial z}\right)\left(\frac{\partial z}{\partial s}\right)_{\zeta}. \tag{2.6.3}$$

From this relationship (for $s = x, y$), we can get an expression for the horizontal gradient of a scalar A in ζ coordinates:

2.6 Primitive Equations and Vertical Coordinates 63

$$\nabla_\zeta A = \nabla_z A + \left(\frac{\partial A}{\partial \zeta}\right)\left(\frac{\partial \zeta}{\partial z}\right)\nabla_\zeta z \tag{2.6.4}$$

and for the horizontal divergence of a vector $\mathbf{B}$:

$$\nabla_\zeta \cdot \mathbf{B} = \nabla_z \cdot \mathbf{B} + \left(\frac{\partial \mathbf{B}}{\partial \zeta}\right)\cdot\left(\frac{\partial \zeta}{\partial z}\right)\nabla_\zeta z \tag{2.6.5}$$

The total derivative of $A(x,y,\zeta,t)$ is given by

$$\frac{dA}{dt} = \left(\frac{\partial A}{\partial t}\right)_\varsigma + \mathbf{v}\cdot\nabla_\varsigma A + \dot\varsigma\frac{\partial A}{\partial \varsigma} \tag{2.6.6}$$

The horizontal pressure gradient is therefore

$$\frac{1}{\rho}\nabla_z p = \frac{1}{\rho}\left[\nabla_\varsigma p - \left(\frac{\partial p}{\partial \varsigma}\right)\left(\frac{\partial \varsigma}{\partial z}\right)\nabla_\varsigma z\right] \tag{2.6.7}$$

which becomes, using the hydrostatic equation $\partial p/\partial\phi = -\rho$,

$$\frac{1}{\rho}\nabla_z p = \frac{1}{\rho}\nabla_\zeta p + \nabla_\zeta\phi. \tag{2.6.8}$$

In summary, the horizontal momentum equations become

$$\frac{d\mathbf{v}}{dt} = -\alpha\nabla_\varsigma p - \nabla_\varsigma\phi - f\mathbf{k}\times\mathbf{v} + F \tag{2.6.9}$$

and the hydrostatic equation $\partial p/\partial z = -\rho g$ becomes

$$\frac{\partial p}{\partial \varsigma}\frac{\partial \varsigma}{\partial z} = -\rho g$$

or

$$\frac{\partial p}{\partial \varsigma} = -\rho\frac{\partial\phi}{\partial \varsigma}. \tag{2.6.10}$$

The continuity equation can be derived from the conservation of mass for an infinitesimal parcel: The hydrostatic equation indicates that the mass of a parcel is proportional to the increase in pressure from the top to the bottom of the parcel (Figure 2.6.1(b)):

$$g\Delta M = \Delta x\Delta y\Delta p \tag{2.6.11}$$

Now, $\Delta p = (\partial p/\partial\varsigma)\Delta\varsigma$, so that taking a logarithmic total derivative, and noting that as $\Delta\varsigma \to 0$

$$\frac{1}{\Delta x}\frac{d\Delta x}{dt} \to \frac{\partial u}{\partial x}$$

and the same with the other space variables, we obtain

$$\frac{d}{dt}\left(\ln\frac{\partial p}{\partial \varsigma}\right) + \nabla\cdot\mathbf{v}_H + \frac{\partial\dot\varsigma}{\partial \varsigma} = 0. \tag{2.6.12}$$

The thermodynamic equation is as before

$$C_P\frac{T}{\theta}\frac{d\theta}{dt} = C_P\frac{dT}{dt} - \alpha\frac{dp}{dt} = Q. \tag{2.6.13}$$

The kinematic lower boundary condition is that the surface of the Earth is a material surface: the flow can only be parallel to it, not normal. This means that once a parcel touches the surface it is "stuck" to it, it cannot separate from the surface. This can be expressed as

$$\frac{d(\varsigma - \varsigma_s)}{dt} = 0 \quad at \quad \varsigma = \varsigma_s$$

or

$$\frac{d\varsigma}{dt} = \frac{\partial \varsigma_s}{\partial t} + \mathbf{v} \cdot \boldsymbol{\nabla} \varsigma_s \quad at \quad \varsigma = \varsigma_s \tag{2.6.14}$$

This kinematic boundary condition is well defined, although in practice it may not be accurate, for example, when there is subgrid-scale orography.

At the top, unfortunately, the boundary condition is not so well defined: As $z \to \infty$, $p \to 0$, but in general there is no satisfactory way to express this condition for a finite vertical resolution model. Most models assume a simple condition of a "rigid top" (i.e., making the top surface a material surface)

$$\frac{d\varsigma}{dt} = 0 \ at \ \varsigma = \varsigma_T, \tag{2.6.15}$$

but this is an artificial boundary condition that introduces spurious effects. For example, Kalnay and Toth (1996) showed that such a rigid top introduces artificial "upside-down" baroclinic instabilities in the NCEP global model, and similar observations were made by Hartmann et al. (1995) with the ECMWF model. If the top of the model is sufficiently high, and there is enough vertical resolution, the upward moving perturbations are damped in the model (as they are in nature), and the spurious interaction with the artificial top may remain small. Alternatively, radiation conditions enforcing the condition that energy can only propagate upwards can be used, but they are not simple to implement.

2.6.2   Pressure Coordinates

Pressure coordinates are a natural choice for a hydrostatic atmosphere (Eliassen, 1949). They greatly simplify the equations of motion: the horizontal pressure gradient becomes irrotational, and the continuity equation becomes simply zero three-dimensional divergence, a diagnostic linear equation.

As a result the geostrophic wind relationship is also simpler: $\mathbf{v}_g = (1/f)\mathbf{k} \times \boldsymbol{\nabla}\phi$. For this reason, rawinsonde measurements have been made in pressure coordinates since the early 1950s.

In pressure coordinates, $\partial p / \partial \varsigma \equiv 1$, the total derivative operator (2.6.6) is given by

$$\frac{d}{dt} = \frac{\partial}{\partial t} + \mathbf{v} \cdot \boldsymbol{\nabla} + \omega \frac{\partial}{\partial p}$$

where the vertical velocity in pressure coordinates is $\omega = dp/dt$. The primitive equations become:

2.6 Primitive Equations and Vertical Coordinates 65

$$\frac{d\mathbf{v}}{dt} = -\boldsymbol{\nabla}_p \phi - f\mathbf{k} \times \mathbf{v} + F \tag{2.6.16}$$

$$\frac{\partial \phi}{\partial p} = -\alpha \tag{2.6.17}$$

$$\boldsymbol{\nabla}_p \cdot \mathbf{v} + \frac{\partial \omega}{\partial p} = 0 \tag{2.6.18}$$

and the thermodynamic equation (2.6.13) is unchanged.

The geostrophic and thermal wind relationships are especially simple in pressure coordinates:

$$\mathbf{v}_g = \frac{1}{f}\mathbf{k} \times \boldsymbol{\nabla}\phi \qquad and \qquad \frac{\partial \mathbf{v}_g}{\partial p} = -\frac{R}{fp}\mathbf{k} \times \boldsymbol{\nabla}T \tag{2.6.19}$$

On the other hand, the bottom boundary condition is not simple in pressure coordinates because the pressure surfaces intersect the surface:

$$\omega = \frac{\partial p_s}{\partial t} + \mathbf{v} \cdot \boldsymbol{\nabla}p_s \qquad at \qquad p = p_s \tag{2.6.20}$$

This requires knowing the rate of change of p_s:

$$\frac{\partial p_s}{\partial t} + \mathbf{v} \cdot \boldsymbol{\nabla}p_s = -\int_0^\infty \boldsymbol{\nabla}_p \cdot \mathbf{v}dp \tag{2.6.21}$$

This complication of the surface boundary condition in pressure coordinates led Phillips (1957) to the invention of *sigma* coordinates (see next subsection).

Instead of the horizontal momentum equations, we can use the prognostic equations for the vorticity ζ and divergence δ, obtained by applying the operators $\mathbf{k} \cdot \boldsymbol{\nabla}x$ and $\boldsymbol{\nabla}\cdot$ to the momentum equations. In pressure coordinates, these equations are

$$\frac{\partial \varsigma}{\partial t} + \mathbf{v} \cdot \boldsymbol{\nabla}(f + \varsigma) + \omega\frac{\partial \varsigma}{\partial p} + (f + \varsigma)\boldsymbol{\nabla} \cdot \mathbf{v} + k \cdot \boldsymbol{\nabla}\omega \times \frac{\partial \mathbf{v}}{\partial p} = 0, \tag{2.6.22}$$

$$\frac{\partial \delta}{\partial t} + \boldsymbol{\nabla} \cdot (\mathbf{v} \cdot \boldsymbol{\nabla}\mathbf{v}) + \omega\frac{\partial \delta}{\partial p} + \boldsymbol{\nabla}\omega \cdot \frac{\partial \mathbf{v}}{\partial p} + \boldsymbol{\nabla} \cdot (f\mathbf{k} \times \mathbf{v}) + \nabla^2\phi = 0. \tag{2.6.23}$$

2.6.3   Sigma, Eta, and Hybrid Coordinates

Because of the complication of the bottom boundary conditions, Phillips (1957) introduced "normalized pressure" or "sigma" coordinates, where $\sigma = p/p_s$ and $p_s(x, y, t)$ is the surface pressure. These are by far the most widely used vertical coordinates. At the surface, $\sigma = 1$, and at $p = 0$, $\sigma = 0$, so that the top and bottom boundary conditions are $\dot{\sigma} = 0$. More generally, allowing for a rigid top at a finite pressure $p_T = const.$,

$$\sigma = \frac{p - p_T}{p_S - p_T} = \frac{p - p_T}{\pi} \tag{2.6.24}$$

with $\dot{\sigma} = 0$ at $\sigma = 0, 1$.

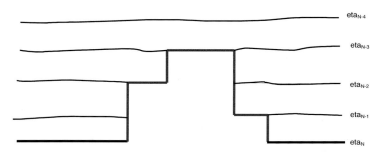

Figure 2.6.2 Schematic of the eta coordinate.

The continuity equation is

$$\frac{\partial \pi}{\partial t} = -\nabla \cdot (\pi \mathbf{v}) - \frac{\partial \pi \dot{\sigma}}{\partial \sigma}. \qquad (2.6.25)$$

The surface pressure tendency equation is:

$$\frac{\partial \pi}{\partial t} = \frac{\partial p_S}{\partial t} = -\nabla \cdot \int_0^1 (\pi v) d\sigma. \qquad (2.6.26)$$

Substituting back into the continuity equation, one can determine $\dot{\sigma}$ diagnostically from the horizontal wind field **v**.

Exercise 2.6.2 *Derive (2.6.25) and (2.6.26)*

Despite their popularity, sigma coordinates have a serious disadvantage: The pressure gradient becomes the difference between two terms:

$$\frac{d\mathbf{v}}{dt} = -\alpha \nabla_\sigma p - \nabla_\sigma \phi - f \mathbf{k} \times \mathbf{v} + \mathbf{F} \qquad (2.6.27)$$

where the first term, if sigma surfaces are steep, may not have the information that went into the finite difference calculation of the second. To avoid the resulting errors, Mesinger (1984) introduced a step-mountain coordinate denoted "eta" (used in the Eta model at NCEP, e.g., Mesinger et al., 1988; Janjić, 1990; Black, 1994):

$$\eta = \frac{p}{p_S} \frac{p_o(z)}{1000 \text{ mb}} \qquad (2.6.28)$$

The first factor is the standard sigma coordinate, the second is a scaling factor, with $p_o(z)$ the pressure in the standard atmosphere. Mountains are defined as boxes, whose tops have to coincide with a model eta level (Figure 2.6.2). As a result of the scaling, the eta surfaces are almost horizontal, and the pressure gradient is computed accurately.

Another disadvantage of the terrain-following coordinate is that the effect of the mountains on the pressure gradient is felt all the way to the top of the model. Simmons and Burridge (1981) introduced a "hybrid" vertical coordinate that tends uniformly to pressure at upper levels and thus reduces the errors of the pressure gradients over steep

2.6 Primitive Equations and Vertical Coordinates   67

orography. Other hybrid coordinates have been used for isentropic coordinates (e.g., Bleck and Benjamin, 1993).

2.6.4   Isentropic Coordinates

The fact that under adiabatic motion, *potential temperature is individually conserved* suggested long ago that it could be used as a vertical coordinate. The main advantage, which makes it an almost ideal coordinate, is that "vertical" motion $\dot{\theta}$ is approximately zero in these coordinates (except for diabatic heating). This reduces finite difference errors in areas such as fronts, where pressure or z-coordinates tend to have large errors associated with poorly resolved vertical motion.

Hydrostatic equation: from the definition of potential temperature, and using the hydrostatic and state equations, we get

$$\frac{d\theta}{\theta} = \frac{dT}{T} - \frac{R}{C_p}\frac{dp}{p} = \frac{dT}{T} - \frac{1}{C_p}\frac{d\phi}{T} \tag{2.6.29}$$

If we define the Exner function $\Pi = C_p T/\theta = C_p (p/p_0)^{R/C_p}$ and the Montgomery potential $M = C_p T + \phi$, we see from the previous equation that

$$\frac{\partial M}{\partial \theta} = \Pi \tag{2.6.30}$$

The horizontal pressure gradient becomes very simple, so that for $\zeta = \theta$ the momentum equation is

$$\frac{d\mathbf{v}}{dt} = -\boldsymbol{\nabla}_\theta M - f\mathbf{k} \times \mathbf{v} + \mathbf{F}. \tag{2.6.31}$$

The continuity equation is

$$\frac{d}{dt} \ln \frac{\partial p}{\partial \theta} + \boldsymbol{\nabla}_\theta \cdot \mathbf{v} + \frac{\partial \dot{\theta}}{\partial \theta} = 0. \tag{2.6.32}$$

The potential vorticity is conserved for adiabatic, frictionless flow (Ertel's theorem). This general property can be posed in its simplest formulation in isentropic coordinates:

$$\frac{dq}{dt} = 0 \tag{2.6.33}$$

where $q = (f + \mathbf{k} \cdot \boldsymbol{\nabla}_\theta \times \mathbf{v})\, \partial\theta/\partial p$ and integrating between two isentropic surfaces, the potential vorticity is

$$q = \frac{(f + \mathbf{k} \cdot \boldsymbol{\nabla}_\theta \times \mathbf{v})}{\Delta p} \tag{2.6.34}$$

which is similar to the SWE potential vorticity.

Although the isentropic coordinates have many advantages, they have also two main disadvantages: The first is that isentropic surfaces intersect the ground (as do other vertical coordinates except for sigma-type coordinates). In practice, this implies that it is difficult to enforce strict conservation of mass, and this is important for long (climate) integrations. For this reason, hybrid sigma–theta coordinates have been used

68   **The Continuous Equations**

(e.g., Johnson et al., 1993). Bleck and Benjamin (1993) designed an isentropic model with a hybrid coordinate approach adopted for the operational RUC/MAPS model, and for the GFDL Cubed Sphere FV3 model, implemented at NCEP, and Arakawa and Konor (1996) designed another hybrid isentropic model. The second disadvantage is that only statically stable solutions are allowed, since the vertical coordinate has to vary monotonically with height. There are situations, e.g., over hot surfaces, where this is not true even at a grid scale. Moreover, in regions of low static stability, the vertical resolution of isentropic coordinates can be poor.

Exercise 2.6.3 *Derive (2.6.31) from (2.6.9) for $\zeta = \theta$, and (2.6.32) from the logarithmic derivative of*

$$\Delta M = \frac{\partial p}{\partial \theta} \cdot \Delta\theta \cdot \Delta x \cdot \Delta y \qquad (2.6.35)$$

where ΔM is proportional to the mass of a parcel in isentropic coordinates

2.7   Introduction to the Equations for Ocean Models

2.7.1   Primitive Equations for the Oceans

In this introduction to the dynamic equations used in classic ocean circulation modeling, we follow Haidvogel and Bryan (1992). Chapter 3 briefly discusses numerical approaches used in current, eddy-resolving ocean models.

The dynamics of the atmosphere and the ocean share important similarities. For example, the large-scale flow in both the ocean and the atmosphere are quasi-geostrophic and in hydrostatic equilibrium. They also have important differences: unlike air, sea water is nearly incompressible, and salinity plays a very important role. Thermal forcings (both solar and long-wave radiation) take place throughout the atmosphere, whereas the ocean is both thermally and mechanically forced primarily by the atmosphere at the surface. The equations of motion for the ocean can be derived in a similar way as the atmospheric equations (Section 2.1). Since in the oceans typical horizontal motions take place on scales of the order of 10–100 km, whereas vertical motions have typical scales of only 10–100 m, the hydrostatic approximation (2.7.4) is quite accurate. In addition, the density of the ocean water has variations in space and time that are much smaller than the mean value:

$$\rho_{tot}(x,y,z,t) = \rho_0 + \rho(x,y,z,t) \quad \text{with } |\rho| \ll \rho_0 \qquad (2.7.1)$$

Therefore, we can make the Boussinesq approximation and neglect the variations in density, except in the buoyancy term, which has the advantage of filtering out sound waves. With these approximations (Haidvogel and Bryan, 1992), the governing equations for the ocean in Cartesian coordinates are the 3 equations of motion,[1]

---

[1] For a more comprehensive and updated review of the equations used in advanced ocean models, see Griffies and Adcroft (2008) and McWilliams (1996).

2.7 Introduction to the Equations for Ocean Models 69

$$\frac{du}{dt} = fv - \frac{\partial \Phi}{\partial x}, \tag{2.7.2}$$

$$\frac{dv}{dt} = -fu - \frac{\partial \Phi}{\partial y}, \tag{2.7.3}$$

$$\frac{\partial \Phi}{\partial z} = -\frac{\rho}{\rho_0} g. \tag{2.7.4}$$

Here u, v, w are the currents in the x, y, z directions, respectively, $\Phi = p/\rho_0$ is a scaled pressure, and, as in atmospheric models, $g \approx 9.8$ m/s^2 includes the effect of the centrifugal force and is assumed to be approximately constant.

The equations of conservation of internal energy (proportional to the temperature T) and of salinity S, in the absence of sources and sinks, are

$$\frac{dT}{dt} = 0, \tag{2.7.5}$$

$$\frac{dS}{dt} = 0, \tag{2.7.6}$$

and the continuity equation

$$\frac{\partial u}{\partial x} + \frac{\partial v}{\partial y} + \frac{\partial w}{\partial z} = 0. \tag{2.7.7}$$

Salinity is measured in Practical Salinity Units (psu), which indicate the number of grams of salt per kilogram of water. The average salinity is 35 psu, which indicates a mean water density of 1.035 in units of 1,000 Kg/m^3.

The system of equations is closed with an equation of state for sea water, expressing the density of sea water as a function of temperature T (in °C), salinity S (in psu), and ocean pressure p (in bars):

$$\rho(S,T,p) = \rho(S,T,0)/[1 - \rho(S,T,0)/K(S,T,p)]. \tag{2.7.8}$$

Here $\rho(S,T,0)$ and the coefficient $K(S,T,p)$ are empirical polynomials of (S,T) and (S,T,p), respectively (e.g., Millero and Poisson, 1981). The ocean pressure p is measured relative to the surface, neglecting the atmospheric pressure. More updated formulations for the empirical equation of state are available (e.g., Jackett et al., 2006).

As is the case for the atmosphere (Chapter 4), it is necessary to represent the transports of heat, salinity and momentum by the subgrid scale turbulent motions, many orders of magnitude larger than the corresponding molecular transports. For this purpose, the prognostic variables are also separated into the large scale resolved by the numerical model and the subgrid scale perturbations, e.g., $u = \bar{u} + u'$. Here the overbar represents the average over a grid box, and the prime the subgrid scale values. Performing "Reynolds averaging" as in Chapter 4 and dropping the overbars, we obtain the equations for the large-scale flow:

$$\frac{du}{dt} = fv - \frac{1}{\rho_0}\frac{\partial p}{\partial x} + \frac{1}{\rho}\left[\frac{\partial}{\partial x}\overline{(u'u')} + \frac{\partial}{\partial y}\overline{(v'u')} + \frac{\partial}{\partial z}\overline{(w'u')}\right], \tag{2.7.9}$$

where, for example, $\tau_{xz} = \overline{(w'u')} \approx \nu_V \frac{\partial u}{\partial z}$ are the vertical Reynolds stresses, and ν_V is the "turbulent vertical eddy viscosity" coefficient. The resulting zonal momentum

70 The Continuous Equations

equation (dropping the overbar) then becomes

$$\frac{\partial u}{\partial t} + \mathbf{v} \cdot \nabla u - fv = -\frac{\partial \Phi}{\partial x} + v_h \nabla_h^2 u + v_v \frac{\partial^2 u}{\partial z^2}. \tag{2.7.10}$$

Exercise 2.7.1 *Derive (2.7.9) from (2.7.2). Hint: First split $u = \bar{u} + u'$, and we have $\overline{u'} = 0, \overline{\bar{u}u'} = 0$. In the derivation, we will also utilize the continuity equation (2.7.7) that $\nabla \cdot \mathbf{v} = 0$.*

Similar equations are obtained for the v-momentum, temperature and salinity equations. As for the atmosphere (Chapter 4), a crude "K-theory" is frequently applied assuming that the turbulent transports are proportional to the gradient of the large-scale variable and to a horizontal or vertical "turbulent eddy viscosity (or diffusivity) coefficient." Unfortunately, the large-scale circulation is obtained with this or other approximations is quite sensitive to this arbitrary formulation, especially of the horizontal turbulent transports, and the solution of this closure problem may have to wait until ocean models with enough resolution are feasible (Chassignet and Garraffo, 2001). At that time, it was estimated that $0.25°$ resolution was required to permit the impact of ocean eddies. By the 2020s, some models are going to $0.1°$ and even higher resolution.

2.7.2 Ocean Boundary Conditions and Coupled Atmosphere–Ocean Models

Lateral and bottom boundary conditions usually assume no fluxes of temperature and salinity, which imply that the normal gradient of T and S is zero. For the momentum equations, the normal fluxes of u and v are prescribed in terms of a bottom drag law. At the lateral boundaries, the boundary conditions are either free-slip (the normal stress is assumed to vanish) or no-slip (with both the normal and tangential velocities vanishing). If there are "open" lateral boundary conditions, the lateral boundary conditions are similar to boundary conditions of a nested atmospheric model (Chapter 3).

For the coupling of ocean and atmosphere models, in principle, the fluxes from the atmosphere to the ocean should be equal and opposite to the fluxes from the ocean to the atmosphere. As graphically described by (Neelin et al., 1994): "In a typical coupling scheme for an ocean–atmosphere model, the ocean passes the SST to the atmosphere, while the atmosphere passes back heat flux components, freshwater flux and horizontal momentum fluxes (i.e., surface stress)." The SST allows the atmospheric model parameterizations to estimate the surface heat flux. The freshwater flux is calculated within the atmospheric model by evaporation, precipitation, river inflow and ice melting.

In most modern coupled models, more generally known as Earth system models, there is a "model coupler" that ensures the fluxes and matter are exchanged without loss between the coupled sub-models. For example, the GFDL coupler, known as Flexible Modeling System, allows for an atmospheric model, an ocean model, an ocean surface including sea ice, a land surface and an ocean that exchange matter and heat at each coupling time step, and they do so in a conserving fashion, so that what one model component loses, the other gains. The conservation property is maintained by ensuring the overlap areas are properly accounted for in the flux coupler. The GFDL coupler ensures conservation when coupling between any ocean grid and any atmospheric grid. However, if the coupling is done at longer intervals, like every few days, it is not clear that the impact of the fluxes will be the same as what would be obtained coupling the models at every time step.

Valcke et al. (2012) give an overview of different coupling systems for Earth system models, including the characteristics of the GFDL FMS, of the Earth System Modeling Framework (designed and used by a consortium of US agencies), the CPL7 of NCAR designed for the Community Climate System models, the widely used OASIS3 coupler of CERFACS, the Model Computing Toolkit that can be used for any multiphysics coupled model, also for more general applications.

2.8 Kelvin Waves and Equatorially Trapped Waves

2.8.1 Kelvin Waves

Kelvin waves are a special kind of waves that can appear in both the atmosphere and the ocean in the presence of a barrier such as a coast or a mountain chain like the Rockies (see schematic Figure 2.8.1). For simplicity, let's consider the SWE on an f-plane and a North-South barrier to the right of the domain:

$$\frac{\partial u}{\partial t} = fv - g\frac{\partial h}{\partial x}; \quad \frac{\partial v}{\partial t} = -fu - g\frac{\partial h}{\partial y}; \quad \frac{\partial h}{\partial t} = -\frac{\partial hu}{\partial x} - \frac{\partial hv}{\partial y} \qquad (2.8.1)$$

where $h = H + h'$, and H is the mean height of the barrier (much larger than the perturbations h').

If we linearize the SWE and assume the flow is parallel to the barrier ($u = 0$), we get

$$fv = g\frac{\partial h'}{\partial x}; \quad \frac{\partial v}{\partial t} = -g\frac{\partial h'}{\partial y}; \quad \frac{\partial h'}{\partial t} = -H\frac{\partial v}{\partial y}. \qquad (2.8.2)$$

Eliminating h' from the second and third equations, we obtain

$$\frac{\partial^2 v}{\partial t^2} + gH\frac{\partial^2 v}{\partial y^2} = 0, \qquad (2.8.3a)$$

and from the first equation in (2.8.1)

$$fv = g\frac{\partial h'}{\partial x}. \qquad (2.8.3b)$$

Equation (2.8.3a) indicates that waves can propagate in the y direction (parallel to the barrier) with the speed of external gravity waves (or in the atmosphere as the analogous external Lamb waves). However, equation (2.8.3b) indicates that in x, the direction perpendicular to the wall, there is geostrophic balance! At a Kelvin wave crest (Figure 2.8.1), the Coriolis force acts to push the water (or air) toward the barrier at the right, and the pressure gradient pushes it to the left, balancing each other and conversely at the trough. Because of the convergence and divergence $\dfrac{\partial v}{\partial y}$ that this produces, the

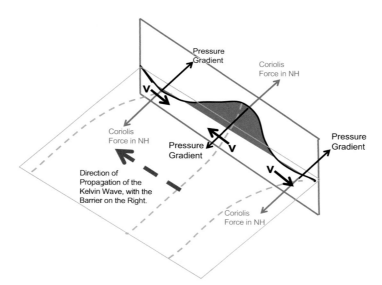

Figure 2.8.1 Schematic of a Kelvin wave, trapped against a barrier in the NH. The Coriolis force is to the right of the flow in the NH, always balancing the pressure gradient, so that the flow **v** satisfies the geostrophic balance. The convergence and divergence of the **v** velocity shows that in the NH, the Kelvin wave will propagate leaving the barrier to the right. In the SH, the Coriolis force is to the left of the flow, so the Kelvin propagates leaving the barrier to the left. At the Equator, two Kelvin waves can act as mutual barriers and move eastward together as an Equatorially trapped Kelvin wave.

Kelvin waves move always leaving the barrier on the right in the Northern Hemisphere and on the left in the Southern Hemisphere.

2.8.2 Equatorially Trapped Waves

What happens at the Equator, where the Coriolis parameter is zero? Close to the Equator, the Coriolis parameter f can be approximated by $f \approx \beta y$, where $\beta = \dfrac{df}{d\phi}$. This allows to have a single Kelvin wave symmetric across the Equator, where the Southern half acts as a barrier for the Northern half Kelvin wave, and the Northern half acts as a barrier for the Southern half Kelvin wave, with both halves moving together to the East. These equatorial Kelvin waves not only actually exist in reality, but they play an important role in triggering El Niño. The amplitudes of these equatorial Kelvin waves vanish exponentially at higher latitudes, hence they are a type of "equatorially trapped waves."

Matsuno (1966b) published in the February issue of the Journal of the Meteorological Society of Japan,[2] a mathematically elegant paper on the complete wave

---

[2] Remarkably, Matsuno (1966a) also published in the same issue of the JMSJ a second extremely famous paper on his widely used "Matsuno time scheme," an Euler backward time scheme whose property of damping high frequencies such as the Lamb waves that ruined the results of Richardson (1922) made

2.8 Kelvin Waves and Equatorially Trapped Waves

dynamics at the Equator. He discovered that there are several kinds of equatorially trapped waves, as shown in the Figure 2.8.2, adapted from Wheeler (2003), who used and further clarified Matsuno's derivation and illustrated it with very clear examples.

We follow Wheeler and Matsuno in using the shallow water equations linearized about a zero motion basic flow, in order to identify pure types of equations (as we did in Section 2.3):

$$\frac{\partial u_l'}{\partial t} = \beta y v_l' - \frac{\partial \phi_l'}{\partial x}; \quad \frac{\partial v_l'}{\partial t} = -\beta y u_l' - \frac{\partial \phi_l'}{\partial y}; \quad \frac{\partial \phi_l'}{\partial t} = -g h_l \left[\frac{\partial u_l'}{\partial x} + \frac{\partial v_l'}{\partial y} \right]. \quad (2.8.4)$$

The subscript l indicates that these equations govern a particular type of vertical normal mode, and h_l represents the equivalent depth for that mode. We seek solutions in the form of zonally propagating wave perturbations with amplitudes that are a function of y:

$$(u_l', v_l', \phi_l') = \text{Re}\left\{ (\hat{u}_l(y), \hat{v}_l(y), \hat{\phi}_l(y)) \exp[i(kx - vt)] \right\}. \quad (2.8.5)$$

Replacing (2.8.5) with (2.8.4) gives ordinary differential equations for the y-structure of the amplitudes (denoted with a hat):

$$-iv\hat{u} - \beta y \hat{v} = -ik\hat{\phi}; \quad -iv\hat{v} + \beta y \hat{v} = -\frac{d\hat{\phi}}{dy}; \quad -iv\hat{\phi} + g h_l \left(ik\hat{u} + \frac{d\hat{v}}{dy} \right) = 0 \quad (2.8.6)$$

Eliminating from equations (2.8.6) $\hat{u}$ and $\hat{\phi}$, we obtain a second order differential equation in $\hat{v}(y)$:

$$\frac{d^2\hat{v}}{dy^2} + \left(\frac{v^2}{g h_l} - k^2 - \frac{k}{v}\beta - \frac{\beta^2 y^2}{g h_l} \right)\hat{v} = 0. \quad (2.8.7)$$

As suggested by Matsuno (1966b) and Wheeler (2003), it is convenient to make time and distance in y non-dimensional by multiplying them by $(\beta \sqrt{g h_l})^{1/2}$ and by $(\beta / \sqrt{g h_l})^{1/2}$ respectively. The non-dimensional frequency v^* and non-dimensional zonal wavenumber k^* are obtained by multiplying them by the inverse of these scales: $v^* = v/(\beta \sqrt{g h_l})^{1/2}$ and $k^* = k(\sqrt{g h_l}/\beta)^{1/2}$.

Equation (2.8.7) in non-dimensional form becomes simpler

$$\frac{d^2\hat{v}^*}{dy^{*2}} + \left(v^{*2} - k^{*2} - \frac{k^*}{v^*} - y^{*2} \right)\hat{v}^* = 0. \quad (2.8.8)$$

As pointed out by Matsuno (1966b), if we require that the amplitude of the waves decay away from the Equator, the boundary condition for (2.8.8) is $\hat{v}^*(y) \to 0$ when $|y| \to 0$. With this boundary condition, equation (2.8.7) is equivalent to Schrödinger's equation for a harmonic oscillator. In order to satisfy the boundary condition, the constant terms in the parenthesis have to be equal to an odd integer:

$$v^{*2} - k^{*2} - \frac{k^*}{v^*} = 2n + 1. \quad (2.8.9)$$

it very useful for data assimilation applications (Chapter 3, Table 3.2.1). Both papers were part of his doctoral thesis under Prof. S. Syono at Tokyo University.

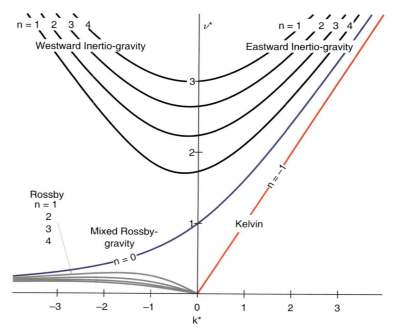

Figure 2.8.2 Dispersion curves for equatorial waves, n is the meridional mode number (plotted up to $n = 4$) as a function of the nondimensional frequency ν^*, and zonal wavenumber k^*. (Adapted from Wheeler, 2003 (©Elsevier. Used with permission.) and Matsuno (1966b)).

where n is a non-dimensional wavenumber associated with the meridional direction. The remarkable different sets of equatorially trapped waves is shown in Figure 2.8.2, adapted from Wheeler (2003) and Matsuno (1966b).

For low frequencies ν^* one can neglect the first term in (2.8.9) and obtain $\nu^* = -\dfrac{k^*}{k^{*2} + (2n + 1)}$, which in dimensional form clearly represents the frequency dispersion relationship of equatorial Rossby waves:

$$\nu_R = -\frac{\beta k}{k^2 + (2n + 1)\beta / \sqrt{gh_l}}. \tag{2.8.10}$$

For high frequencies ν^*, one can neglect the third term in (2.8.9) and obtain $\nu^{*2} = (2n + 1) + k^{*2}$, which in dimensional form clearly represents the frequency dispersion relationship of equatorial inertia-gravity waves:

$$\nu_{IG} = \pm \left[(2n + 1)\beta \sqrt{gh_l} + k^2 g h_l\right]^{1/2}. \tag{2.8.11}$$

Curiously, for $n = 0$, the behavior of (2.8.9) is similar to that of an inertia-gravity wave for $k^* > 0$, and to the Rossby waves for $k^* < 0$, and in the limit $|k^*| \to 0$ both solutions coincide. Not surprisingly, this solution is known as the Mixed Rossby-gravity equatorial wave.

2.8 Kelvin Waves and Equatorially Trapped Waves

Another surprising solution are the Kelvin waves that, as Matsuno pointed out, formally correspond to $n = -1$. Kelvin waves have to be symmetrical about the Equator moving eastward, so the meridional velocity is zero, $v' = 0$.

Setting $v' = 0$, the first and the third equations in (2.8.4) correspond to a gravity wave moving in the zonal direction, and the second equation establishes that u is in geostrophic balance, and is positive (eastward) in order for the geopotential to decay away from the Equator. The first relationship imposes the frequency dispersion relationship for Kelvin waves $v_K = \sqrt{gh_l}\,k$, and the second establishes that $u' > 0$. Note that although (2.8.7) is not valid for Kelvin waves (since $v' = 0$), equation (2.8.9) is still satisfied by the FDR of Kelvin waves (which in non-dimensional form is $v^* = k^*$) for $n = -1$.

3 Numerical Discretization of the Equations of Motion

3.1 Classification of Partial Differential Equations

3.1.1 Reminder about PDEs

Second order linear PDE

$$\alpha \frac{\partial^2 u}{\partial x^2} + 2\beta \frac{\partial^2 u}{\partial x \partial y} + \gamma \frac{\partial^2 u}{\partial y^2} + \delta \frac{\partial u}{\partial x} + \varepsilon \frac{\partial u}{\partial y} + \eta = 0.$$

Second order linear partial differential equations are classified into three types depending on the sign of $\beta^2 - \alpha\gamma$ (e.g., Courant and Hilbert, 1962).* Equations are hyperbolic, parabolic, or elliptic if the sign is positive, zero, or negative, respectively. The simplest (canonical) examples of these equations are

(a) $\dfrac{\partial^2 u}{\partial t^2} = c^2 \dfrac{\partial^2 u}{\partial x^2}$ Wave equation (hyperbolic). Examples: vibrating string, water waves.

(b) $\dfrac{\partial u}{\partial t} = \sigma \dfrac{\partial^2 u}{\partial x^2}$ Diffusion equation (parabolic). Examples: heated rod, viscous damping.

(c) $\dfrac{\partial^2 u}{\partial x^2} + \dfrac{\partial^2 u}{\partial y^2} = 0$ (*or* $f(x,y)$)   Laplace's or Poisson's equations (elliptic). Examples: steady state temperature of a plate, streamfunction/vorticity relationship.

The behavior of the solutions, the proper initial and/or boundary conditions, and the numerical methods that can be used to find the solutions *depend essentially on the type of PDE* that we are dealing with. Although nonlinear multidimensional PDEs cannot in general be reduced to these canonical forms, we need to study these prototypes of the PDEs to develop an understanding of their properties and then apply similar methods to the more complicated NWP equations.

(d) $\dfrac{\partial u}{\partial t} = -c \dfrac{\partial u}{\partial x}$ Advection equation, with solution $u(x,t) = u(x - ct, 0)$.

The *advection* equation, very important in atmospheric sciences, is a first order PDE, but it can also be classified as hyperbolic, since its solutions satisfy the wave equation (a), and the latter can be written as the system

$$\frac{\partial \mathbf{u}}{\partial t} = \mathbf{A} \frac{\partial \mathbf{u}}{\partial x},$$

3.1 Classification of Partial Differential Equations   77

where

$$\mathbf{u} = \begin{pmatrix} \dfrac{\partial u}{\partial t} \\ c\dfrac{\partial u}{\partial x} \end{pmatrix},$$

and

$$\mathbf{A} = \begin{bmatrix} 0 & c \\ c & 0 \end{bmatrix} \quad \text{or an equivalent transformation.}$$

3.1.2 Well-posedness, Initial and Boundary Conditions

A *well-posed* initial/boundary condition problem has a unique solution that depends continuously on the initial/boundary conditions.* Clearly, the specification of proper initial conditions and boundary conditions for a PDE is essential in order to have a well-posed problem:

- If too many initial/boundary conditions are specified, there will be no solution.
- If too few are specified, the solution will not be unique.
- If the number of initial/boundary conditions is right, but they are specified at the wrong place or time, the solution will be unique, but it will not depend smoothly on initial/boundary conditions, i.e., small errors in the initial/boundary conditions will produce huge errors in the solution.

In any of these cases, we have an *ill-posed problem* . And we can *never* find a numerical solution of a problem that is ill posed: The computer will show its disgust by "blowing up."

We briefly discuss well-posed initial/boundary conditions. The number of initial conditions required is equal to the order of the time derivative, and the number of boundary conditions required is equal to the number of space derivatives, so that

- Linear parabolic equations, e.g., $\dfrac{\partial u}{\partial t} = \sigma \dfrac{\partial^2 u}{\partial x^2}$ require one initial condition at the initial time and one boundary condition at each point of the spatial boundaries (if they exist).
- Linear hyperbolic equations, e.g., $\dfrac{\partial^2 u}{\partial t^2} = c^2 \dfrac{\partial^2 u}{\partial x^2}$ require as many initial conditions as the number of characteristics that come out of every point in the surface $t = 0$, and as many boundary conditions as the number of characteristics that cross a point in the (space) boundary pointing *inwards* (into the spatial domain). Example: to solve $\partial/\partial t = -c\partial u/\partial x$ for $x > 0$, $t > 0$; characteristics: solutions of $dx/dt = c$; space boundary: $x = 0$ (see Figure 3.1.1(a),(b)). If $c > 0$, we need the initial condition $u(x,0) = f(x)$ and the boundary condition $u(0,t) = g(t)$. If $c < 0$, we need the initial condition $u(x,0) = f(x)$ but *no boundary conditions*.
- Second order elliptic equations, e.g., $\dfrac{\partial^2 u}{\partial^2 x^2} + \dfrac{\partial^2 u}{\partial^2 y} = f(x,y)$, require one boundary condition on each point of the spatial boundary. These are "boundary value,"

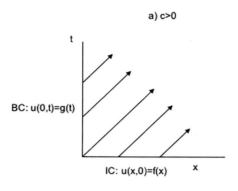

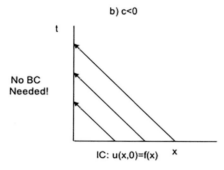

Figure 3.1.1 Schematic of the characteristics of the advection equation $\partial u/\partial t = -c\partial u/\partial x$ for (a) positive and (b) negative velocity c and the corresponding well-posed initial/boundary conditions (IC/BC).

time-independent problems, and the methods used to solve them are introduced in Section 3.4. The boundary conditions may be given on the value of the function (Dirichlet problem), as when we specify the temperature in the borders of a plate, or on its normal derivative (Neumann problem), as when we specify the heat flux. We could also have a mixed "Robin" boundary condition, involving a linear combination of the function and its derivative.

For nonlinear equations, no general statements can be made, but physical insight and local linearization can help to determine proper initial/boundary conditions. For example, in the nonlinear advection equation $\partial u/\partial t = -u\partial u/\partial x$, the characteristics are $dx/dt = u$, and since we don't know *a priori* the sign of u at the boundary, and whether the characteristics will point inwards or outwards, we have to estimate the value of u from the nearby solution, and define the boundary condition accordingly.

One method of solving simple PDEs is the method of separation of variables, but unfortunately in most cases, it is not possible to use it (hence the need for numerical models!). Nevertheless, it is useful to try to solve some simple PDE's analytically.

3.1 Classification of Partial Differential Equations 79

Exercise 3.1.1 Solve by the method of separation of variables these prototype PDEs:

(1) $\dfrac{\partial^2 u}{\partial x^2} + \dfrac{\partial^2 u}{\partial y^2} = 0 \quad 0 \le x \le 1 \quad 0 \le y \le 1.$

Boundary conditions (one boundary condition at each boundary point):

$u(x,0) = f(x), u(x,1) = u(0,y) = u(1,y) = 0.$

Assume

$$f(x) = \sum_{k=1}^{\infty} a_k \sin k\pi x \text{ with } \sum_{k=1}^{\infty} k^2 |a_k| < \infty.$$

Find the solution

$$u(x,y) = \sum_{k=1}^{\infty} a_k \frac{\sinh k\pi(1-y)}{\sinh k\pi} \sin k\pi x.$$

(2) $\dfrac{\partial u}{\partial t} = \sigma \dfrac{\partial^2 u}{\partial x^2} \quad 0 \le x \le 1 \quad t \ge 0$

Boundary conditions (one at each boundary): $u(0,t) = u(1,t) = 0$;

Initial condition: $u(x,0) = f(x) = \sum_{k=1}^{\infty} a_k \sin k\pi x.$

Find the solution

$$u(x,t) = \sum_{k=1}^{\infty} a_k e^{-\sigma k^2 \pi^2 t} \sin k\pi x$$

Note that the higher the wavenumber, the faster it goes to zero, i.e., the solution is smoothed as time goes on.

(3) $\dfrac{\partial^2 u}{\partial t^2} = c^2 \dfrac{\partial^2 u}{\partial x^2} \quad 0 \le x \le 1 \quad 0 \le t \le 1$

Boundary conditions (one at each boundary): $u(0,t) = u(1,t) = 0$;

Initial conditions (we need two): $u(x,0) = f(x) = \sum_{k=1}^{\infty} a_k \sin k\pi x.$

Find the solution

$$\frac{\partial u}{\partial t}(x,0) = g(x) = \sum_{k=1}^{\infty} b_k \sin k\pi x.$$

(4) Same as (3), but now, instead of two initial conditions, we give an initial and a "final" condition:

Boundary conditions: $u(0,t) = u(1,t) = 0$;

Initial condition: $u(x,0) = f(x)$; "final condition": $u(x,1)=g(x)$.

In other words, we try to solve a hyperbolic (wave) equation as if it was a boundary value problem. Show that the solution is unique, but it does not depend continuously on the boundary conditions, and therefore it is not a well-posed problem.

Conclusion: Before trying to solve a problem numerically, make sure that it is *well posed*: it has a unique solution that depends continuously on the data that define the problem.

Exercise 3.1.2 Lorenz showed that the atmosphere has a finite limit of predictability: even if the models and the observations were perfect, "the flapping of a butterfly in Brazil (not taken into account in the model) will result in a completely different forecast over the US after a couple of weeks." Does this mean that the problem of NWP is not well posed?

3.2 Initial Value Problems: Numerical Solution

Hyperbolic and parabolic PDEs are *initial value or marching* problems: The solution is obtained by using the known initial values and marching or advancing in time. If boundary values are necessary, they are called "mixed initial-boundary value problems." Again, the simplest prototypes of these initial value problems are the wave or advection equation:

$$\frac{\partial u}{\partial t} = -c\frac{\partial u}{\partial x}, \tag{3.2.1}$$

with solution $u(x,t) = u(x - ct,0)$, a *hyperbolic* equation, and the diffusion equation,

$$\frac{\partial u}{\partial t} = \sigma\frac{\partial^2 u}{\partial x^2}, \tag{3.2.2}$$

a *parabolic* equation.

3.2.1 Finite Differences Method

Figure 3.2.1 shows that we can estimate the slope of a function by taking centered differences, which are more accurate (closer to the exact slope) than forward or backward differences.

We now take discrete values for x and t: $x_j = j\Delta x$, $t_n = n\Delta t$ (as in schematic Figure 3.2.2). The solution of the finite difference equation is then defined at those discrete points: $U_j^n = U(j\Delta x, n\Delta t)$. We will use a small u to denote the solution of the PDE (continuous) and capital U to denote the solution of the finite difference equation (FDE), a discrete solution.

3.2 Initial Value Problems: Numerical Solution

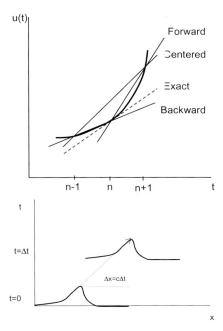

Figure 3.2.1 (a) Schematic of finite difference estimating the time derivative $\partial u/\partial t$ at time $t_n = n\Delta t$: centered $(\partial u/\partial t)_n \approx (u_{n+1} - u_{n-1})/(2\Delta t)$, forward $(\partial u/\partial t)_n \approx (u_{n+1} - u_n)/\Delta t$, and backward $(\partial u/\partial t)_n \approx (u_n - u_{n-1})/\Delta t$. The three estimates are *consistent* with $\partial u/\partial t$ since they all converge to $\partial u/\partial t$ as $\Delta t \to 0$. However, the slope calculated from centered differences is much closer to the exact derivative because its truncation errors are second order. (b) Schematic of the solution of the wave equation.

Consider again the advection equation (3.2.1). Suppose that we choose to approximate this PDE with the following FDE (called an "upstream scheme," Figure 3.2.2a):

$$\frac{U_j^{n+1} - U_j^n}{\Delta t} + c\frac{U_j^n - U_{j-1}^n}{\Delta x} = 0 \qquad (3.2.3a)$$

Note that both the space and time differences are non-centered with respect to the point $(j\Delta x, n\Delta t)$, which makes them less accurate (Figure 3.2.1).

Alternatively, we could use the "leap-frog" scheme:

$$\frac{U_j^{n+1} - U_j^{n-1}}{2\Delta t} + c\frac{U_{j+1}^n - U_{j-1}^n}{2\Delta x} = 0 \qquad (3.2.3b)$$

We should now ask two fundamental questions:

1. Are these FDEs *consistent* with the PDEs?
2. For any given time $t > 0$, will the solution U of the FDE *converge* to u as $\Delta x \to 0$, $\Delta t \to 0$?

Let us now clarify these questions.

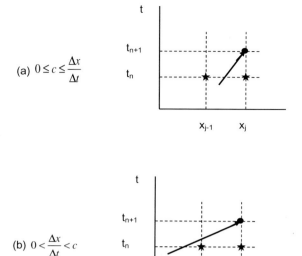

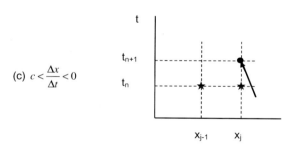

Figure 3.2.2 Upstream equation: Schematic of the relationship between Δx, Δt, and c leading to interpolation of the solution at time level $n + 1$ (case a), or to extrapolation (cases b and c) depending on the value of the Courant number $\mu = \dfrac{c \Delta t}{\Delta x}$.

3.2.2 Truncation Errors and Consistency

We say that the FDE is *consistent* with the PDE if, in the limit $\Delta x \to 0, \Delta t \to 0$ the FDE coincides with the PDE. Obviously, this is a first requirement that the FDE should fulfill if its solutions are going to be good approximations of the solutions of the PDE. The difference between the PDE and the FDE is the discretization error or *local* (in space and time) *truncation error*. Consistency is rather simple to verify: Substitute U by u in the FDE, evaluate all terms using a Taylor series expansion centered on the point (j, n), and then subtract the PDE from the FDE. If the difference (or local truncation error τ) goes to zero as $\Delta x \to 0, \Delta t \to 0$, then the FDE is consistent with the PDE.

3.2 Initial Value Problems: Numerical Solution 83

Example 3.2.1 We verify the consistency of (3.2.3a) with (3.2.1) by a Taylor series expansion:

$$\left.\begin{aligned} u_j^{n+1} &= \left(u + u_t \Delta t + u_{tt} \frac{\Delta t^2}{2} + \cdots \right)_j^n \\ u_{j-1}^n &= \left(u - u_x \Delta x + u_{xx} \frac{\Delta x^2}{2} + \cdots \right)_j^n \end{aligned}\right\} \tag{3.2.4}$$

Substitute the series (3.2.4) in the FDE (3.2.3a)

$$\left(u_t + u_{tt} \frac{\Delta t}{2} + \cdots + c u_x - c u_{xx} \frac{\Delta x}{2} + \cdots \right)_j^n \approx 0 \tag{3.2.5}$$

and when we subtract the PDE (3.2.1) we get the (local) *truncation error*

$$\tau = u_{tt} \frac{\Delta t}{2} - c u_{xx} \frac{\Delta x}{2} + \text{higher order terms} = 0(\Delta t) + 0(\Delta x) \tag{3.2.6}$$

so that $\lim_{\Delta t \to 0, \Delta x \to 0} \tau \to 0$. Therefore *the FDE is consistent*. Note that both the *time and the space truncation errors are of first order*, because the finite differences are uncentered in both space and time. Truncation errors for *centered differences are second order*, and therefore centered differences are more accurate than uncentered differences (see Figure 3.2.1(a) and the leapfrog scheme, based on centered differences in space and in time).

Exercise 3.2.1 Show that the leap-frog scheme (3.2.3b) is consistent with the advection equation (3.2.1) and that it has second order local truncation errors.

3.2.3 Convergence and Criteria for Computational Stability

The second question posed in Section 3.2.1 was whether the solution of the FDE converges to the PDE solution, i.e., whether $U(j\Delta x, n\Delta t) \to u(x,t)$ when $j\Delta x \to x$, $n\Delta t \to t$, $\Delta x \to 0$, $\Delta t \to 0$. This is of evident practical importance but can only be answered after considering another problem, that of *computational stability*. Consider again the PDE (3.2.1), which has the solution $u(x,t) = u(x - ct, 0)$, i.e., the initial shape of u translates with velocity c.

The upstream FDE (3.2.3a) can be written as

$$U_j^{n+1} = (1 - \mu)U_j^n + \mu U_{j-1}^n \tag{3.2.7}$$

where $\mu = c\Delta t/\Delta x$ is the *Courant number*. Assume that $0 \le \mu = c\Delta t/\Delta x \le 1$, as in Figure 3.2.2(a). Then the FDE solution at the new time level U_j^{n+1} is *interpolated* between the values U_j^n and U_{j-1}^n. In this case the advection scheme works the way it should, because we know the true solution is in between those values. However, if this condition is not satisfied, and $\mu = c\Delta t/\Delta x > 1$ (as in Figure 3 2.2(b)) or $\mu = c\Delta t/\Delta x < 0$ (as in Figure 3.2.2(c)), then the value of U_j^{n+1} is *extrapolated* from the values U_j^n and U_j^{n-1}. The problem with extrapolation is that the maximum absolute value of the solution U_j^n increases with each time step, leading to computational instability. For the upstream scheme, we can use a simple criterion to derive a condition for no extrapolation, and hence, computational stability.

3.2.3.1 Criterion of the Maximum

Taking absolute values of the upstream FDE (3.2.7) and letting $U^n = \max_j |U_j^n|$, we obtain

$$|U_j^{n+1}| \leq |U_j^n| |1 - \mu| + |U_{j-1}^n| |\mu|$$

so that

$$U^{n+1} \leq \left\{ |1 - \mu| + |\mu| \right\} U^n.$$

Then $U^{n+1} \leq U^n$ if and only if the Courant number satisfies the condition $0 \leq \mu \leq 1$.

If the condition $0 \leq \mu \leq 1$ is *not* satisfied, then *the solution is not bounded* and it grows with n. If we let $\Delta t, \Delta x \to 0$ with $\mu = const.$, it only makes things worse, because then $n \to \infty$. In practice, if the stability condition $0 \leq \mu \leq 1$ is not satisfied, the FDE "blows up" in a few time steps and even faster for nonlinear problems. We define now computational stability: We say that an FDE is *computationally stable* if the solution of the FDE at a fixed time $t = n\Delta t$ remains bounded as $\Delta t \to 0$. The condition on the Courant number of being less than 1 in absolute value is usually known as the Courant–Friedrichs–Lewy or CFL condition.

We can now state the fundamental Lax–Richtmyer theorem: Given a *properly posed linear* initial value problem, and a finite difference scheme that satisfies the *consistency* condition, then the *stability* of the FDE is the necessary and sufficient condition for *convergence*.

The theorem is useful because it allows us to establish convergence by examining separately the easier questions of consistency and stability. We are interested in convergence not because we want to let $\Delta t, \Delta x \to 0$, but because we want to make sure that if Δt and Δx are small, then the accumulated or *global truncation errors* $u(j\Delta x, n\Delta t) - U_j^n$ at a finite time $t_n = n\Delta t$ and a location $x_j = j\Delta x$ are acceptably small.

We can also use the criterion of the maximum to study the stability condition of the following FDE, which approximates the parabolic diffusion equation $\partial u/\partial t = \sigma \partial^2 u \partial x^2$:

$$\frac{U_j^{n+1} - U_j^n}{\Delta t} = \sigma \frac{U_{j+1}^n - 2U_j^n + U_{j-1}^n}{\Delta x^2} \tag{3.2.8}$$

The verification of consistency is immediate. Note that, because the differences are centered in space but forward in time, the truncation error is first order in time and second order in space $O(\Delta t) + O(\Delta x)^2$.

We can write (3.2.8) as

$$U_j^{n+1} = \mu U_{j+1}^n + (1 - 2\mu)U_j^n + \mu U_{j-1}^n$$

where $\mu = \sigma \Delta t / \Delta x^2$. If we take absolute values, and let $U^n = \max_j |U_j^n|$, we get

$$U^{n+1} \leq \{|\mu| + |1 - 2\mu| + |\mu|\} U^n \tag{3.2.9}$$

So we obtain the condition $0 \leq \mu \leq 1/2$ to insure that the solution remains bounded as $n \to \infty$, i.e., as the necessary condition for stability of the FDE.

3.2 Initial Value Problems: Numerical Solution 85

Exercise 3.2.2 The condition on the wave equation $0 \leq \mu \leq 1$ for the upstream FDE is interpreted as "the time step should be chosen so that a signal cannot travel more than one grid size in one time step." Give a physical interpretation of the stability condition and the equivalent "Courant number" $\mu = \sigma(\Delta t/\Delta x^2) < 1/2$ for the diffusion equation.

Unfortunately, the criterion of the maximum, which is intuitively very clear, can only be applied in very few cases. In most FDEs, some coefficients of the equations analogous to (3.2.9) are negative, and therefore the criterion cannot be applied.

3.2.3.2 Von Neumann Stability Criterion

Another stability criterion that has much wider application than the criterion of the maximum is *the von Neumann stability criterion*. Assume that the boundary conditions allow expansion of the solution of the FDE in an appropriate set of space eigenfunctions. For simplicity, we will assume an expansion into Fourier series (e.g., periodic boundary conditions):

$$U(x,t) = \sum_{\mathbf{k}} Z_{\mathbf{k}} e^{i\mathbf{k}\cdot\mathbf{x}} \tag{3.2.10}$$

The space variable, $\mathbf{x}$, and the wavenumber $\mathbf{k}$ can be multidimensional, e.g., $\mathbf{x} = (x_1, x_2, x_3)$, $\mathbf{k} = (k_1, k_2, k_3)$. For a system of equations, the dependent variable U would be a vector, not a scalar.

Let $x_j = j\Delta x$ (or $\mathbf{x}_j = (j_1\Delta x_1, j_2\Delta x_2, j_3\Delta x_3)$). We define p as the wavenumber for the finite Fourier series: $p = k\Delta x$ or $\mathbf{p} = (k_1\Delta x_1, k_2\Delta x_2, k_3\Delta x_3)$. Let $t_n = n\Delta t$. Then the Fourier expansion is

$$U_j^n = \sum_p Z_p^n e^{ipj} \tag{3.2.11}$$

(where for multiple dimensions $\mathbf{p} \cdot \mathbf{j} = p_1 j_1 + p_2 j_2 + p_3 j_3$).

When we substitute this Fourier expansion into a linear FDE, we obtain a system of equations

$$Z_p^{n+1} = G_p Z_p^n$$

G is an "amplification" matrix that, when applied to the pth Fourier component of the solution at time $n\Delta t$ "advances" it to the time $(n + 1)\Delta t$; G depends on p, Δt and Δx. If we know the initial conditions

$$U_j^0 = \sum_p Z_p^0 e^{ipj} \tag{3.2.12}$$

then the solution of the FDE in (3.2.11) is

$$Z_p^n = G_p^n Z_p^0. \tag{3.2.13}$$

Therefore, stability, i.e., boundedness of the solution for any permissible initial condition at any fixed time is guaranteed if the matrix G^n *is bounded for all* p when $\Delta t \to 0$ and $n \to \infty$. So, we must have $\|G_n\| < M$ for all p, as $n \to \infty$. Here $\|\mathbf{A}\|$ is a norm or measure of the "size" of a matrix $\mathbf{A}$. If $\sigma(G)$ is the *spectral radius* of G, i.e., $\sigma(G) = \max_i |\lambda_i|$, where λ_i, are the eigenvalues of G, then it can be shown that for any norm,

$$[\sigma(G)]^n \leq \|G^n\| \leq \|G\|^n \tag{3.2.14}$$

The equal sign is valid if G is normal, i.e., if $GG^* = G^*G$, where G^* is the transpose-conjugate of G, but in general the amplification matrices arising from FDEs are not normal.

Thus a necessary condition for stability of an FDE, and therefore a necessary condition for convergence, is that

$$\lim_{\Delta t \to 0, n\Delta t \to t} [\sigma(G)]^n = \text{finite} = e^{const.} \tag{3.2.15}$$

Then

$$\sigma(G) \leq [\sigma(G)^n]^{1/n} \leq e^{const./n} = e^{const.\Delta t/t} \approx 1 + \frac{const.\Delta t}{t},$$

or

$$\sigma(G) \leq 1 + O(\Delta t). \tag{3.2.16}$$

This is the von Neumann necessary condition for computational stability. The term $O(\Delta t)$ allows bounded growth with time if this growth is "legitimate," i.e., if it arises from a physical instability present in the PDE. If the exact solution grows with time, then the FDE cannot both satisfy $\sigma(G) \leq 1$ and be consistent with the PDE. Sufficient conditions are very complicated, and are known only for special cases. In practice, it is generally observed that eliminating the equal sign in equation (3.2.16) is enough to ensure computational stability.

In principle, this method can also be used to study the stability of the boundary conditions, if they are appropriately included in the amplification matrix. In practice this is complicated, and computational stability of the boundary conditions is usually obtained by ensuring well posedness, and testing the stability experimentally. For simple equations, and without considering the effect of boundary conditions, the von Neumann criterion can be simplified by assuming solutions with an amplification factor ρ rather than a matrix. The solution for the amplification factor ρ then coincides with the eigenvalues of the amplification matrix, and the von Neumann stability criterion is $\rho \leq 1 + O(\Delta t)$.

Example 3.2.2 Application of the Von Neumann criterion to the upstream scheme.

$$\text{PDE: } \frac{\partial u}{\partial t} + c\frac{\partial u}{\partial x} = 0$$

$$\text{FDE: } \frac{U_j^{n+1} - U_j^n}{\Delta t} + c\frac{U_j^n - U_{j-1}^n}{\Delta x} = 0 \quad \text{(upstream scheme)} \tag{3.2.17}$$

We have already studied consistency, and used the criterion of the maximum to get a sufficient condition for stability. Let us now apply the von Neumann criterion:

3.2 Initial Value Problems: Numerical Solution 87

Assume

$$U_j^n = \sum_p Z_p^n e^{ipj} = \sum_p A\rho_p^n e^{ipj}$$

We substitute in equation (3.2.17) and eliminate Ae^{ipj} and obtain

$$\frac{\rho_p^{n+1} - \rho_p^n}{\Delta t} + \frac{\rho_p^n (1 - e^{-ip})}{\Delta x} = 0 \quad \text{for all } p. \tag{3.2.18}$$

The amplification factor is the same as the 1×1 amplification matrix G for each wavenumber p, and thus the same as its spectral radius $\sigma(G)$. The stability condition is therefore $|\rho_p| \le 1$ for all wavenumbers p. We need to estimate the maximum value of the spectral radius (or amplification factor in this case):

$$\rho_p = 1 - \mu(1 - e^{-ip}) = 1 - \mu(1 - \cos p + i \sin p) \tag{3.2.19}$$

$$|\rho_p|^2 = (1 - \mu(1 - \cos p))^2 + \mu^2 \sin^2 p \tag{3.2.20}$$

We make use of the trigonometrical relationships

$$\cos p = \cos^2 \frac{p}{2} - \sin^2 \frac{p}{2} \qquad \sin p = 2 \sin \frac{p}{2} \cos \frac{p}{2},$$

and obtain

$$|\rho_p|^2 = 1 - 4\mu(1 - \mu) \sin^2 \frac{p}{2}. \tag{3.2.21}$$

Now consider the $\sin^2(p/2)$ term: *The shortest wave that can be present in the finite difference solution is $L = 2\Delta x$, therefore the maximum value that $p = k\Delta x = 2\pi\Delta x/L$ can take is $p = \pi$, so that the maximum value that $\sin^2(p/2)$ can attain is 1. The other factor, $\mu(1 - \mu)$, is a parabola whose maximum value is 0.25 when $\mu = 0.5$. So the amplification factor squared will remain less than or equal to 1 as long as $0 \le \mu \le 1$. This coincides with the condition we obtained from the criterion of the maximum (and also with the notion that we should not extrapolate but interpolate the new values at time level $t = (n + 1)\Delta t$, as in Figure 3.2.2).

It is important to note that the *amplification factor* ρ_p indicates how much the amplitude of *each* wavenumber p will decrease or increase with each time step. The upstream scheme decreases the amplitude of all Fourier wave components of the solution, since, for $0 < \mu < 1$, $\rho_p < 1$ for all wavenumbers. This is therefore a very dissipative FDE: It has strong "numerical diffusion." Figure 3.2.3 shows the decrease in amplitude when using the upstream scheme after one time step and after 100 time steps for each wavenumber p, using a Courant number $\mu = 0.1$, a typical value for advection given the presence of fast gravity waves. Since the upstream scheme truncation errors are large (of first order), in general, it is not recommended except for special situations (e.g., for outflow boundary conditions, where damping is beneficial). An alternative, less damping scheme known as Matsuno or Euler-backward scheme (see Table 3.2.1), frequently used in combination with the leapfrog scheme, is also shown in this figure. Note that a "downstream" scheme (Figure 3.2.2(c)) would be unstable.

Numerical Discretization of the Equations of Motion

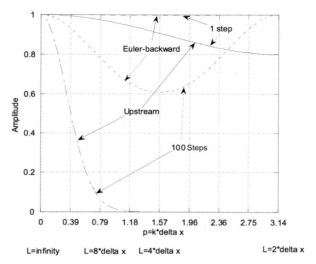

Figure 3.2.3 Amplification factor of wave components of the wave equation using either the "upstream," or Euler-backward (Matsuno) schemes with $\mu = 0.1$. L is the wavelength in units of Δx. After 100 time steps the strong damping of the upstream scheme is very apparent.
$\rho_{Upstream} = [1 - 4\mu(1 - \mu)\sin^2 p]^{1/2}$,
$\rho_{Euler-Backward} = [1 - \mu^2 \sin^2 p + \mu^4 \sin^4 p]^{1/2}$.

Example 3.2.3 Application of the Von Neumann criterion to the leapfrog scheme

$$\text{PDE:} \quad \frac{\partial u}{\partial t} + c\frac{\partial u}{\partial x} = 0$$

$$\text{FDE:} \quad \frac{U_j^{n+1} - U_j^{n-1}}{2\Delta t} + c\frac{U_{j+1}^n - U_{j-1}^n}{2\Delta x} = 0 \quad (3.2.22)$$

This is the most popular of the schemes used for hyperbolic equations. In Exercise 3.2.2, we found that the leapfrog scheme is consistent with the PDE and that because the space and time finite differences are centered, the local truncation error is of second order in both space and time.

Stability: Assume $U_j^n = \sum_p Z_p^n e^{ipj} = \sum_p A\rho_p^n e^{ipj}$.
Substitute in the FDE (and dropping the subindex p)

$$\frac{\rho^{n+1} - \rho^{n-1}}{2\Delta t} + c\frac{\rho^n(e^{ip} - e^{-ip})}{2\Delta x} = 0, \quad (3.2.23)$$

Therefore

$$\rho^2 + 2i\mu \sin p\, \rho - 1 = 0. \quad (3.2.24)$$

Because we have three, not two, time levels ρ^{n+1}, ρ^n, and ρ^{n-1}, we have a quadratic equation and (unfortunately) *two* solutions for the amplification factor ρ:

$$\rho = (-i\mu \sin p) \pm \sqrt{(-\mu^2 \sin^2 p + 1)}. \quad (3.2.25)$$

Since the last term in the quadratic equation (3.2.24) is -1, and this is the product of the roots, the term inside the root $(-\mu^2 \sin^2 p + 1)$ must be positive, since otherwise

3.2 Initial Value Problems: Numerical Solution   89

the roots would be purely imaginary, and one of them would be larger than 1, which violates the stability criterion. In order for $\sqrt{(-\mu^2 \sin^2 p + 1)}$ to be real for all p, we must have $\mu^2 \leq 1$. Therefore, the stability condition for the leapfrog scheme becomes

$$-1 \leq c\Delta t/\Delta x \leq 1. \tag{3.2.26}$$

Exercise 3.2.3 Create a schematic figure equivalent to the upstream scheme Figure 3.2.2 but for the Leapfrog scheme to explain why the sign of the Courant number does not matter for the Leapfrog stability criterion.

Exercise 3.2.4 Create a similar figure for the forward scheme,

$$\frac{U_j^{n+1} - U_j^n}{\Delta t} + c\frac{U_{j+1}^n - U_{j-1}^n}{2\Delta x} = 0.$$

Show that it is computationally absolutely unstable, even though the necessary but not sufficient "no extrapolation" criterion of the maximum would suggest that it has the same stability criterion as the leapfrog scheme.

In order to explore the properties of the Leap-Frog scheme, it is useful to look at the exact solution of the leapfrog FDE (3.2.22), as well as of the original PDE. Recall that the exact solution of the PDE $\partial u/\partial t + c\partial u/\partial x = 0$ is simply a translation with velocity c: $u(x,t) = u(x - ct, 0)$. Therefore, it has plane wave solutions of the form $A_k e^{ik(x-ct)} = A_k e^{i(kx-\omega t)}$, and the Frequency Dispersion Relationship (FDR) $\omega = kc$ gives the exact PDE relationship between frequency ω and wavenumber k. By analogy, we try to find solutions of the FDE of the form $A_p e^{i(pj-\theta n)}$, where $\theta = \nu\Delta t$ represents the computational frequency ν multiplied by Δt (the computational frequency ν is in general different than the exact frequency ω). Substituting in the FDE and dividing by $e^{i(pj-\theta n)}$, we get

$$(e^{-i\theta} - e^{i\theta}) + \mu(e^{ip} - e^{-ip}) = 0$$

or

$$\sin \theta = \mu \sin p \tag{3.2.27}$$

the FDR for the leapfrog scheme. Since $\sin \theta = \sin(\pi - \theta)$, the two solutions for the finite difference FDR are

$$\left.\begin{aligned}\theta_1 &= \arcsin(\mu \sin p) \\ \theta_2 &= \pi - \arccos(\mu \sin p)\end{aligned}\right\} \tag{3.2.28}$$

Substituting into the FDR, and assuming that the initial amplitude for the wavenumber p is 1, we obtain that the solution of the FDE is a sum of two terms corresponding to θ_1 and θ_2 respectively:

$$U_j^n = A_p e^{i(pj-\theta n)} + (1 - A_p)e^{i(pj+\theta n)}(-1)^n \tag{3.2.29}$$

where $\theta = \arcsin(\mu \sin p)$, and $e^{i\pi} = -1$. Of the two terms in the solution (3.2.29), the first one is the "legitimate" solution, which approximates the PDE solution $A_k e^{ik(x-ct)}$.

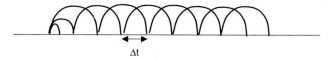

Figure 3.2.4 Schematic of the Leap-Frog scheme with a half time initial step.

The second term in (3.2.29) changes its sign every time step, and it moves in the wrong direction. This unphysical term is called "computational mode." It arises because the leapfrog scheme has three time levels, rather than two, giving rise to an additional spurious solution in the amplification factor. Although the leapfrog scheme is simple and accurate, its three-time level character gives rise to two problems that have to be dealt with: The need for a two step initialization and the need to deal with the growing amplitude of the computational mode.

3.2.3.3 Leapfrog Scheme Initialization

The first problem arising from the fact that leapfrog is a 3-level scheme is that it needs a special initial step to get to the first-time level U^1 from the initial conditions U^0, before it can be started (Figure 3.2.4). This can be done in several simple ways:

(a) Simply set $U^1 = U^0$. Since $u^1 = u^0 + u_t \Delta t + \ldots$, this introduces errors of order $O(\Delta t)$ and is not recommended.
(b) Use for the first time step a forward time scheme. The forward scheme has truncation errors of order $O(\Delta t)$, but since the time step is only used once, its contribution to the global error is multiplied by Δt, so that the total error is still of $O(\Delta t)^2$. The forward scheme is unstable for hyperbolic equations, but if it is executed only once, the computational instability is not a significant problem. A safer approach is to use an Euler-backwards (Matsuno) scheme for the first time step (see Table 3.2.1).
(c) Use half (or a quarter, eighth, etc.) of the initial time step for the forward time step (Figure 3.2.4), followed by leapfrog time steps. This will halve (or reduce by a quarter, eighth, etc.) the error introduced in the unstable first step.

Exercise 3.2.5 Show that a forward time step is unstable for hyperbolic equations (like the wave equation).

3.2.3.4 Robert–Asselin and Williams Time Filters for Leapfrog

In addition to the need to initialize the first step, the 3-level leapfrog scheme has a second, more important problem: For nonlinear models, it has a tendency to increase the amplitude of the computational mode with time. This also separates the space dependence in a checkerboard fashion between the even and odd time steps, until the model integration blows up. One solution to this problem is to restart leapfrog every 50 steps or so, before the separation becomes noticeable, but a more popular approach is to apply the Robert–Asselin time filter.

After the leapfrog scheme is used to obtain the solution at $t = (n + 1)\Delta t$, Robert (1969) suggested adding a slight "time smoothing" to the solution at time $n\Delta t$.

3.2 Initial Value Problems: Numerical Solution

$$\bar{U}^n = U^n + \frac{\nu}{2}(U^{n+1} - 2U^n + \bar{U}^{n-1}) \tag{3.2.30}$$

The filtered value $\bar{U}^n$ immediately replaces the previous value U^n, so that there is no memory overhead. ν is a small number of the order of 0.01. The Robert–Asselin filter is widely used with the leapfrog scheme because it efficiently damps the computational mode. However, since it uses $\bar{U}^{n-1}$ rather than U^{n-1}, (3.2.30) introduces a slight distortion of a centered time filter, and it not only strongly damps the computational mode, as desired, but it also has some damping effect on the physical mode (Asselin, 1972).

Williams (2009) introduced a significant improvement upon the Robert filter, with corrections made on both the solutions at time n and $n + 1$ (Figure 3.2.5). The Robert–Asselin–Williams (RAW) filter starts with the standard leapfrog scheme with $\bar{\bar{U}}_{n-1}$ having been modified twice in the previous two-time steps, and $\bar{U}_n$ modified once in the previous time step:

$$U_{n+1} = \bar{\bar{U}}_{n-1} + 2\Delta t \cdot F(\bar{U}_n). \tag{3.2.31}$$

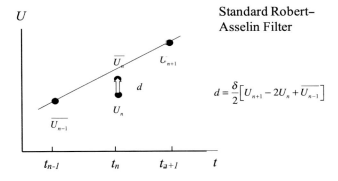

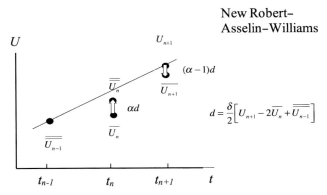

Figure 3.2.5 Schematic of the standard Robert–Asselin filter and the new Robert–Asselin–Williams filter. Adapted from Williams (2009). ©American Meteorological Society. Used with permission.

Numerical Discretization of the Equations of Motion

The leapfrog step is followed by two filtering steps:

$$\bar{\bar{U}}_n = \bar{U}_n + \frac{\nu\alpha}{2}\left[\bar{\bar{U}}_{n-1} - 2\bar{U}_n + U_{n+1}\right]$$

$$\bar{U}_{n+1} = U_{n+1} - \frac{\nu(\alpha-1)}{2}\left[\bar{\bar{U}}_{n-1} - 2\bar{U}_n + U_{n+1}\right]. \qquad (3.2.32)$$

The filtered values immediately replace the previous values so that, as in the Robert filter, there is no memory overhead. α is a coefficient which is equal to 1 for the Robert filter, and for the RAW filter is close to $1/2$. Williams showed that the unfiltered, single filtered, and doubly filtered values share a common complex amplification factor,

$$A = \frac{U_{n+1}}{U_n} = \frac{\bar{U}_{n+1}}{\bar{U}_n} = \frac{\bar{\bar{U}}_{n+1}}{\bar{\bar{U}}_n}. \qquad (3.2.33)$$

The amplitude of a coupled inertial wave equation $\frac{du}{dt} = fv$, $\frac{dv}{dt} = -fu$ decreases substantially in time with the Robert–Asselin (RA) Filter ($\alpha = 1$), whereas the choice $\alpha = 1/2$ maintains its correct amplitude and phase almost exactly. Williams recommends using $\alpha = 0.53$ that has the effect of keeping the amplification factor smaller but very close to 1, while still damping the computational mode as well as the RA filter. The new RAW filter, was tested and compared with its standard RA filter on 6-day forecasts with the SPEEDY model initialized and verified with the NCEP-NCAR Reanalysis (Amezcua et al., 2011). They found that using the RAW filter did not produce significant changes on the model climatology, which avoids having to retune the model physical parameterizations. However, the forecast verifications were improved with the RAW filter and that this improvement increased with forecast length.

We now present in Table 3.2.1 some of the most used time schemes.

In Table 3.2.1, the Lorenz N-cycle is an economical time integration scheme that requires only one function evaluation per time step and a minimal memory footprint, but yet possesses a high order of accuracy every N steps. Despite these advantages, it has remained less commonly used in meteorological applications, partly because of its lack of semi-implicit formulation. Hotta et al. (2016) developed a new semi-implicit modification of the Lorenz N-cycle scheme that does not require the RA filter and maintains second order accuracy, as tested in the SPEEDY model.

Exercise 3.2.6 Discuss the results that you obtain with these numerical experiments.

1) Program in MATLAB or FORTRAN the coupled inertia wave equations, with $f = 10^{-4}\sec^{-1}$ being the Coriolis parameter:

$$\begin{cases} \dfrac{du}{dt} = fv \\ \dfrac{dv}{dt} = -fu \end{cases}$$

using leap-frog and computing the RHS exactly at time n. Explore the stability for different values of $f\Delta t$ (the "Courant number"). Plot the results (u, v) versus time.

3.2 Initial Value Problems: Numerical Solution   93

Table 3.2.1 Time schemes for initial value problems $dU/dt = F(U)$ for schemes (a)–(i); $dU/dt = F_1(U) + F_2(U)$ for schemes (j) and (k)

(a) $\dfrac{U^{n+1} - U^{n-1}}{2\Delta t} = F(U^n)$

Leapfrog (good for hyperbolic equations, unstable for parabolic equations)

(a') $\dfrac{U^{n+1} - \bar{U}^{n-1}}{2\Delta t} = F(U^n)$

$\bar{U}^n = U^n + \dfrac{\nu}{2}(U^{n+1} - 2U^n + \bar{U}^{n-1})$

Leapfrog smoothed with the Robert–Asselin time filter; $\nu/2 \sim 1\%$(see also the Robert–Asselin–Williams filter)

(b) $\dfrac{U^{n+1} - U^n}{\Delta t} = F(U^n)$

Euler (forward, good for diffusive terms, unstable for hyperbolic equations)

(c) $\dfrac{U^{n+1} - U^n}{\Delta t} = F\left(\dfrac{U^n + U^{n+1}}{2}\right)$

Crank–Nicholson or centered implicit

(c') $\dfrac{U^{n+1} - U^n}{\Delta t} = F(\beta U^n + (1 - \beta)U^{n+1});$

$\beta < 0.5$

Implicit, slightly damping

(d) $\dfrac{U^{n+1} - U^n}{\Delta t} = F(U^{n+1})$

Fully implicit or backward

(e) $\dfrac{U^* - U^n}{\Delta t} = F(U^n); \quad \dfrac{U^{n+1} - U^n}{\Delta t} = F(U^*)$

Euler-backward or Matsuno: good for damping high frequency waves

(f) $\dfrac{U^* - U^n}{\Delta t} = F(U^n); \quad \dfrac{U^{n+1} - U^n}{\Delta t} = F\left(\dfrac{U^n + U^*}{2}\right)$

Another predictor–corrector scheme (Heun)

(g) $\dfrac{U^{n+1} - U^n}{\Delta t} = F\left(\dfrac{3}{2}U^n - \dfrac{1}{2}U^{n-1}\right)$

Adams–Bashford (second order in time).

(h) $\dfrac{U^{n+1/2^*} - U^n}{\Delta t/2} = F(U^n);$

Runge–Kutta (fourth order)

$\dfrac{U^{n+1/2^{**}} - U^n}{\Delta t/2} = F(U^{n+1/2^*});$

$\dfrac{U^{n+1^*} - U^n}{\Delta t} = F(U^{n+1/2^{**}});$

$\dfrac{U^{n+1} - U^n}{\Delta t} = \dfrac{1}{6}\Big[F(U^n) - 2F(U^{n+1/2^*})$
$\qquad\qquad + 2F(U^{n+1/2^{**}}) + F(U^{n+1^*})\Big]$

(i) $a = 0; \; b = 1/\Delta t$

$\quad U^* \leftarrow (aU^* + F(U^n))/b$

$\quad U^* \leftarrow U^n + U^*$

$\quad a \leftarrow a - 1/(N\Delta t); \; b \leftarrow b - 1/(N\Delta t)$

$\Big\}$ N-times

Lorenz's N-cycle, Nth order accuracy every Nth step. $N = 4$ is usually chosen.

(j) $\dfrac{U^{n+1} - U^{n-1}}{2\Delta t} = F_1(U^n) + F_2\left(\dfrac{U^{n+1} + U^{n-1}}{2}\right)$

Semi-implicit

(k) $\dfrac{U^* - U^n}{\Delta t} = F_1(U^n); \quad \dfrac{U^{n-1} - U^*}{\Delta t} = F_2(U^*)$

Fractional steps

94 Numerical Discretization of the Equations of Motion

2) Implement the Robert filter and the RAW filter (as in Williams, 2009, Figure 2). Compare the results with $\delta = 0.2$, $\alpha = 1$ and $\alpha = 0.53$. What happens with $\alpha = 0.5$?

3) Program a wave/dissipation equation

$$\begin{cases} \dfrac{du}{dt} = fv - \sigma u \\[2mm] \dfrac{dv}{dt} = -fu - \sigma v, \end{cases}$$

explore the results using a leapfrog for the first (wave) terms, and forward (with two-time steps) for the dissipation terms.

4) Same as 3) but using the implicit backward scheme for the wave terms.

Exercise 3.2.7 Examine the FORTRAN code of the SPEEDY model and show that it uses a leapfrog scheme with a Robert filter for the "wave-like" terms and a two-time-step forward scheme for the parameterization schemes.

Example 3.2.4

$$\frac{\partial u}{\partial t} = \sigma \frac{\partial^2 u}{\partial x^2} + bu \tag{3.2.34}$$

This is the heat or diffusion equation with a "source of growth" bu.

$$\text{FDE:} \quad \frac{U_j^{n+1} - U_j^n}{\Delta} = \sigma \frac{U_{j+1}^n - 2U_j^n + U_{j-1}^n}{\Delta x^2} + bU_j^n \tag{3.2.35}$$

Exercise 3.2.8 Show that the amplification factor is

$$\rho = 1 - \frac{4\sigma \Delta t}{(\Delta x)^2} \sin^2 \frac{p}{2} + b\Delta t \leq 1 + O(\Delta t). \tag{3.2.36}$$

Therefore, the stability criterion is still $\sigma \Delta t / \Delta x^2 \leq 1/2$, as we obtained with the criterion of the maximum.

Exercise 3.2.9 Explain physically why the term $b\Delta t$ does not influence the stability criterion.

3.2.4 Implicit Time Schemes

In these schemes, the advection or diffusion terms are written in terms of not only the old-time level variables, but the new-time level variables as well.

Example 3.2.5

$$\text{PDE:} \quad \frac{\partial u}{\partial t} + c \frac{\partial u}{\partial x} = 0$$

$$\text{FDE:} \quad \frac{U_j^{n+1} - U_j^n + U_{j+1}^{n+1} - U_{j+1}^n}{2\Delta t}$$

$$+ c \frac{\alpha \left(U_{j+1}^n - U_j^n \right) + (1 - \alpha) \left(U_{j+1}^{n+1} - U_j^{n+1} \right)}{\Delta x} = 0. \tag{3.2.37}$$

3.2 Initial Value Problems: Numerical Solution

The factor α determines the weight of the "old" time values compared with the "new" time values in the right-hand side of the FDE. Using the von Neumann stability criterion method, we substitute $U_j^n = A\rho^n e^{ipj} = Ae^{i(pj-\theta n)}$ into (3.2.37).

Note that the scheme is centered in time (if $\alpha = 1/2$) at the point $U_{j+1/2}^{n+1/2}$. For this reason, we multiply by $e^{-ip/2}$, and obtain the amplification factor:

$$\rho = \frac{\cos\frac{p}{2} - i2\mu\alpha\sin\frac{p}{2}}{\cos\frac{p}{2} + i2\mu(1-\alpha)\sin\frac{p}{2}} \quad (3.2.38)$$

or

$$|\rho|^2 = \frac{1 + 4\mu^2\alpha^2 \tan^2\frac{p}{2}}{1 + 4\mu^2(1-\alpha)^2 \tan^2\frac{p}{2}}. \quad (3.2.39)$$

This implies that $\rho \leq 1$ if $\alpha \leq 0.5$, i.e., if the new values are given at least as much weight as the old values in computing the RHS. In this case, *there is no restriction on the size that Δt can take!* This result (absolute stability, independent of the Courant number) is typical of implicit time schemes. Figure 3.2.6 shows that in an implicit scheme, a point at the new time level is influenced by all the values at the new level, which avoids extrapolation and therefore is *absolutely stable*. Note also that if $\alpha < 0.5$, the implicit time scheme reduces the amplitude of the solution: It is an example of a *damping scheme*. This property is useful for solving some problems such as spuriously growing mountain waves in semi-Lagrangian schemes.

In summary, if we consider a marching equation

$$\frac{dU}{dt} = F(U) \quad (3.2.40)$$

explicit methods such as the forward scheme

$$\frac{U^{n+1} - U^n}{\Delta t} = F(U^n) \quad (3.2.41)$$

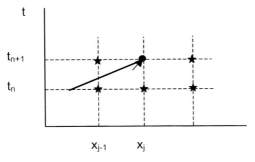

Figure 3.2.6 Schematic of an implicit scheme. The dot represents the value being updated and the stars the values that influence it. Note that with the implicit scheme there is no extrapolation, and therefore no limit to the size of Δt.

Numerical Discretization of the Equations of Motion

or the leapfrog scheme

$$\frac{U^{n+1} - U^{n-1}}{2\Delta t} = F(U^n) \tag{3.2.42}$$

are either *conditionally stable* (when there is a condition on the Courant number or the equivalent stability number for parabolic equations) or *absolutely unstable*.

A fully implicit scheme

$$\frac{U^{n+1} - U^n}{\Delta t} = F(U^{n+1}) \tag{3.2.43}$$

and a centered implicit scheme (Crank–Nicholson)

$$\frac{U^{n+1} - U^n}{\Delta t} = F\left(\frac{U^n + U^{n+1}}{2}\right) \tag{3.2.44}$$

are *absolutely stable*. The latter scheme is attractive because it is centered in time (around $t_{n+1/2}$), and it can be written with centered space differences, which makes it second order in space and in time. Also, it only has two time levels so it does not have a computational mode. But, like all implicit schemes, it also has a great disadvantage. Since U^{n+1} appears on the left- and on the right-hand sides, the solution for U^{n+1}, unlike explicit schemes, in general requires the solution of a system of equations. If it involves only tridiagonal systems, this is not an obstacle, because there are fast methods to solve them (Section 3.4.2). There are also methods, such as fractional steps (with each spatial direction solved successively), where one space dimension is considered at a time, that allow taking advantage of the large time steps permitted by implicit schemes without paying a large additional computational cost. Moreover, we will see in the next section that the possibility of using a time step with a Courant number much larger than 1 in an implicit scheme does not imply that we will obtain accurate results economically. The implicit scheme maintains stability by *slowing down* the solutions, so that the slower waves do satisfy the CFL condition. For this reason, implicit schemes are only useful for those modes (such as the Lamb wave or vertical sound waves) that are very fast but of little meteorological importance (semi-implicit schemes, see Section 3.2.5).

A Simple Check of the Stability of Hyperbolic or Parabolic Equations
(1) It is easy to check the properties of the time schemes in Table 3.2.1 when applied to hyperbolic equations by testing them with a simple harmonic equation:

$$\frac{\partial U}{\partial t} = -i\nu U \tag{3.2.45}$$

with solution $U(t) = U(0)e^{-i\nu t}$. After one time step, the exact solution is

$$U((n+1)\Delta t) = U(n\Delta t)e^{-i\nu \Delta t} \tag{3.2.46}$$

which indicates that the exact magnification factor is $e^{-i\nu \Delta t}$.

In (3.2.45), ν is the computational frequency for a wave equation for a given space discretization. For example, if we were using second order centered differences in space, $v = (\sin k\Delta x / \Delta x) c$, for a spectral scheme, $v = kc$. For the fully implicit time scheme (d), the amplification factor is

$$\frac{1}{1 + iv\Delta t} = \frac{1 - iv\Delta t}{1 + (v\Delta t)^2}$$

Since the exact amplification factor has an amplitude equal to 1, this shows that the implicit scheme is dissipative; similarly, comparing the imaginary components of the exact and approximate amplification factors, it is clear that the implicit solution is slowed down by a factor of about $1/[1 + (v\Delta t)^2]$.

Exercise 3.2.10 Show that the Crank–Nicholson scheme significantly slows down the angular speed of the solution by deriving the magnification factor for this scheme and comparing it with the exact magnification factor $e^{-iv\Delta t}$. Determine the limit of the Crank–Nicholson amplification factor for the Courant number $v\Delta t \to \infty$.

$$U^{n+1} = U^n \rho_{CN} \tag{3.2.47}$$

(2) Equations with damping terms (such as the parabolic equation) can also be simply represented by the equation:

$$\frac{\partial U}{\partial t} = -\mu U \tag{3.2.48}$$

In (3.2.48), μ can be considered as the computational rate of damping. For example, for the diffusion equation, using centered differences in space,

$$\mu = \frac{4\sigma}{(\Delta x)^2} \sin^2 \frac{k\Delta x}{2}$$

Exercise 3.2.11 Show that the leapfrog scheme is unstable for a damping term.

Exercise 3.2.12 Write a numerically stable scheme for the equation with both wave-like and damping terms $\partial U / \partial t = -(iv + \mu)U$ using a three-time level scheme.

Exercise 3.2.13 Show that for a wave equation the forward time scheme with centered differences in space is absolutely unstable. Note that this scheme shows that the "no extrapolation" rule is a necessary but not a sufficient condition for stability of wave equations.

3.2.5 Semi-implicit Schemes

Consider the SWEs that we discussed in Section 2.4.1:

$$\left. \begin{aligned} \frac{\partial u}{\partial t} + u\frac{\partial u}{\partial x} + v\frac{\partial u}{\partial y} &= -\frac{\partial \phi}{\partial x} + fv \\ \frac{\partial v}{\partial t} + u\frac{\partial v}{\partial x} + v\frac{\partial v}{\partial y} &= -\frac{\partial \phi}{\partial y} - fu \\ \frac{\partial \phi}{\partial t} + u\frac{\partial \phi}{\partial x} + v\frac{\partial \phi}{\partial y} &= -\Phi\left(\frac{\partial u}{\partial x} + \frac{\partial v}{\partial y}\right) - (\phi - \Phi)\left(\frac{\partial u}{\partial x} + \frac{\partial v}{\partial y}\right) \end{aligned} \right\} \tag{3.2.49}$$

As indicated in that section, the phase speed of the inertia-gravity wave is given by

$$c_{IGW} = \frac{\nu_{IGW}}{k} = U \pm \sqrt{\frac{f^2}{k^2} + \Phi} \approx U \pm 300 \,\text{m/s}$$

and the terms that give rise to the fast gravity waves (pressure gradient and divergence) are underlined. This means that the Courant number $\mu = c_{IGW} \Delta t / \Delta x$ is dominated by the speed of external inertia-gravity waves (equivalent to the Lamb waves, horizontal sound waves), and an explicit scheme would therefore require a time step an order of magnitude smaller than that required for advection. For this reason, Robert (1969) introduced the use of semi-implicit schemes to slow down the gravity waves. We write such a scheme using the compact finite difference notation for differences and averages:

$$\left. \begin{aligned} \delta_x f &= \frac{f_{j+1/2} - f_{j-1/2}}{\Delta x} \\ \overline{f}^x &= (f_{j+1/2} + f_{j-1/2})/2 \end{aligned} \right\} \tag{3.2.50}$$

and similarly for differences in y or t. With this notation, assuming uniform resolution,

$$\left. \begin{aligned} \delta_{2x} f &= \delta_x \overline{f}^x = \frac{f_{i+1} - f_{i-1}}{2\Delta x} \\ \overline{f}^{2x} &= (f_{i+1} + f_{i-1})/2 \end{aligned} \right\} \tag{3.2.51}$$

Using this compact finite difference notation, we can write the leapfrog semi-implicit SWE as

$$\left. \begin{aligned} \delta_{2t} u + u \delta_{2x} u + v \delta_{2y} u &= -\delta_{2x} \overline{\phi}^{2t} + fv \\ \delta_{2t} v + u \delta_{2x} v + v \delta_{2y} v &= -\delta_{2y} \overline{\phi}^{2t} - fu \\ \delta_{2t} \phi + u \delta_{2x} \phi + v \delta_{2y} \phi &= -\Phi \overline{(\delta_{2x} u + \delta_{2y} v)}^{2t} - (\phi - \Phi)(\delta_{2x} u + \delta_{2y} v) \end{aligned} \right\} \tag{3.2.52}$$

Everything that does not have a time average involves only terms evaluated explicitly at the nth time step. We can rewrite the FDEs (3.2.52) as

$$\left. \begin{aligned} \frac{u^{n+1} - u^{n-1}}{2\Delta t} &= -\delta_{2x}(\phi^{n+1} + \phi^{n-1})/2 + R_u \\ \frac{v^{n+1} - v^{n-1}}{2\Delta t} &= -\delta_{2y}(\phi^{n+1} + \phi^{n-1})/2 + R_v \\ \frac{\phi^{n+1} - \phi^{n-1}}{2\Delta t} &= -\Phi[\delta_{2x}(u^{n+1} + u^{n-1})/2 + \delta_{2y}(v^{n+1} + v^{n-1})/2] + R_\phi \end{aligned} \right\} \tag{3.2.53}$$

where the "R" terms are the "rest" of the terms evaluated at the center time $n\Delta t$. For example, $R_u = fv - u\delta_{2x}u - v\delta_{2y}u$, and similarly for R_v and R_ϕ.

From these three equations, we can eliminate u^{n+1}, v^{n+1} and obtain an elliptic equation for ϕ^{n+1}:

$$\begin{aligned} \left(\delta_{2x}^2 + \delta_{2y}^2 - \frac{1}{\Phi \Delta t^2} \right) \phi^{n+1} &= -\left(\delta_{2x}^2 + \delta_{2y}^2 + \frac{1}{\Phi \Delta t^2} \right) \phi^{n-1} \\ &\quad + 2(\delta_{2x} R_u + \delta_{2y} R_v) + \frac{2}{\Delta t}(\delta_{2x} u^{n-1} + \delta_{2y} v^{n-1}) - \frac{2}{\Phi \Delta t} R_\phi = F_{i,j}^n \end{aligned} \tag{3.2.54}$$

3.2 Initial Value Problems: Numerical Solution 99

Note that the right-hand side of this elliptic equation is evaluated at $t = n\Delta t$ or $(n - 1)\Delta t$, so that it is known. Solving this elliptic equation provides ϕ^{n+1}, and once this is known, it can be plugged back into the first two equations of (3.2.53), and thus (u^{n+1}, v^{n+1}) can be obtained.

The elliptic operator in brackets in the left-hand side of (3.2.54), is a finite difference equivalent to $(\nabla^2 - \lambda^2)$,

$$\left(\delta_{2x}^2 + \delta_{2y}^2 - \frac{1}{\Phi \Delta t^2}\right) \phi = \frac{\phi_{i+2,j} + \phi_{i-2,j} + \phi_{i,j+2} + \phi_{i,j-2} - \left(4 + \frac{1}{\mu^2}\right) \phi_{i,j}}{4\Delta^2}$$

$$(3.2.55)$$

where we have assumed for simplicity that $\Delta x = \Delta y = \Delta$, and $\mu^2 = \Phi \Delta t^2 / \Delta^2$ is the square of the Courant number for gravity waves. Since $\mu^2 \gg 1$, the semi-implicit scheme distorts the gravity waves solution, slowing the gravity wave down until they satisfy the von Neumann criterion. This is an acceptable distortion since we are interested in the slower "weather-like" processes, and since the slower modes satisfy the CFL (von Neumann) stability criterion, and they are written explicitly, they are not slowed down or distorted in a significant way.

In the same way that the terms giving rise to gravity waves can be written semi-implicitly, the terms giving rise to sound waves can also be written semi-implicitly (Robert, 1982). They are the three-dimensional divergence in the continuity equation (Sections 2.3.2 and 2.3.3). This has allowed the use of nonhydrostatic models without the use of the anelastic approximation or the hydrostatic approximation. Robert (1982) created a model that could be considered an "ultimate" atmospheric model. It treats the terms generating sound waves (anelastic terms, i.e., three-dimensional divergence), and the terms generating gravity waves (pressure gradient and horizontal divergence) semi-implicitly, and it uses a three-dimensional semi-Lagrangian scheme for all advection terms. This model, denoted the "Mesoscale Compressible Community" (MCC) model, is a "universal" model designed so that it can tackle accurately atmospheric problems from the planetary scale through mesoscale, convective and smaller (Laprise et al., 1997). There is another approach followed by several major nonhydrostatic models (e.g., MM5 and ARPS): the use of fractional steps (see Table 3.2.1, scheme (k)), with the sound-wave terms integrated with small time steps. In addition, the ARPS model uses a semi-implicit scheme for vertically propagating sound waves (Xue et al., 1995).

Non-hydrostatic models are becoming more widely used as the resolution of even global models becomes closer to the "grey zone" of about 5km. They are discussed in more detail in the new Section 3.6.

Exercise 3.2.14 Consider the diffusion equation $\partial u / \partial t = \sigma \partial^2 u / \partial x^2$ with initial conditions $u = x$ for $x \leq 0.5$ and $u = 1 - x$ for $x \geq 0.5$. Compute the first two time steps using an explicit scheme (forward in time, centered in space) with five points between $x = 0$ and $x = 1$, and a time step such that $r = \sigma \Delta t / (\Delta x)^2$ is equal to $r = 0.1, 0.5, 1.0$. Repeat using Crank–Nicholson's scheme. Discuss the results.

3.3 Space Discretization Methods

3.3.1 Space Truncation Errors, Computational Phase Speed, Second- and Fourth-Order Schemes

It is convenient to separate the truncation errors in a discretized model into space truncation errors and time truncation errors. For explicit finite difference models, the errors introduced by space truncation tend to dominate the total forecast errors because, in the presence of Lamb waves, the time step and the Courant number used are much smaller than what would be required to resolve the slow "weather waves." Let's neglect for the moment time truncation errors and consider the wave equation $\partial U/\partial t = -c\partial U/\partial x$ discretized only in space.

If we approximate $\partial U/\partial x$ using space centered differences, we get

$$\delta_{2x}U_j = \frac{U_{j+1} - U_{j-1}}{2\Delta x} = U_x + \frac{\Delta x^2}{6}U_{xxx} + \frac{\Delta x^4}{120}U_{xxxxx} + HOT$$
$$= U_x + A\Delta x^2 + B\Delta x^4 + \cdots \tag{3.3.1}$$

If instead of the closest neighboring points $j+1, j-1$, we use the points $j+2, j-2$, we get

$$\delta_{4x}U_j = \frac{U_{j+2} - U_{j-2}}{4\Delta x} = U_x + 4A\Delta x^2 + 16B\Delta x^4 + \cdots \tag{3.3.2}$$

This is also a second order scheme, but the truncation errors are four times larger. We can now eliminate from (3.3.1) and (3.3.2) the term $A\Delta x^2$ and obtain

$$\frac{4}{3}\delta_{2x}U_j - \frac{1}{3}\delta_{4x}U_j = U_x - 4B\Delta x^4 + \cdots \tag{3.3.3}$$

Now (3.3.3) is a *fourth order* approximation of the space derivative. So

$$\frac{dU_j}{dt} = -c\delta_{2x}U_j \tag{3.3.4}$$

is a second order FDE and

$$\frac{dU_j}{dt} = -c\left(\frac{4}{3}\delta_{2x}U_j - \frac{1}{3}\delta_{4x}U_j\right) \tag{3.3.5}$$

is a fourth order FDE.

Assume solutions of the form

$$U_j(t) = Ae^{ik(x_j - c't)} = Ae^{i(kx_j - v't)} \tag{3.3.6}$$

where c' is the computational phase speed, and v' is the computational frequency, so that $dU_j/dt = -iv'U_j$. Making use of $\delta_{2x}U_j = i\left(\sin k\Delta x/\Delta x\right)U_j$ and replacing in (3.3.4), and (3.3.5) we find that for second order differences,

$$c'_2 = \frac{\sin k\Delta x}{k\Delta x}c \tag{3.3.7}$$

and for fourth order differences,

$$c'_4 = \left(\frac{4}{3}\frac{\sin k\Delta x}{k\Delta x} - \frac{1}{3}\frac{\sin 2k\Delta x}{2k\Delta x}\right)c \tag{3.3.8}$$

3.3 Space Discretization Methods

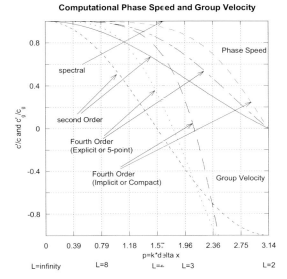

Figure 3.3.1 Ratio of the computational to the physical phase speed and group velocity for a simple wave equation, neglecting time truncation errors, for second order, fourth order explicit and implicit and spectral schemes.

Note that (3.3.7) and (3.3.8) imply that the phase speed is always underestimated by space finite differences. For the smallest possible wavelength, $L = 2\Delta x$, $k\Delta x = \pi$, the computational phase speed is zero for both second and fourth order differences: The shortest waves don't move at all (as shown in Figure 3.3.1). For $L = 4\Delta x, k\Delta x = \pi/2$, a much more accurate approximation is obtained with fourth order than with second order differences: $c'_2 = 0.64c$, $c'_4 = 0.85c$, and the fourth order advantage becomes even better for longer waves: for $L = 8\Delta x, c'_2 = 0.90c, c'_4 = 0.99c$.

We can also compute the *computational group velocity* $\partial v'/\partial k$, where

$$v' = c'k = c\frac{\sin k\Delta x}{\Delta x} \tag{3.3.9}$$

for second order differences. Then,

$$\frac{\partial v'}{\partial k} = c \cos k\Delta x \tag{3.3.10}$$

for second order differences. Therefore, for the shortest waves, $L = 2\Delta x, k\Delta x = \pi$, with both second and fourth order differences, the energy moves in the opposite direction to the real group velocity (equal to the phase speed c): $c'_g = -c_g$. Figure 3.3.1 shows the computational phase speed and group velocity for second and fourth order differences. As a result of the negative group velocity, space centered FDEs of the wave equation tend to leave a trail of short-wave computational noise upstream of where the real perturbation should be. This problem is greatly reduced using more advanced schemes such as those of Takacs (1985) and Smolarkiewicz and Grabowski (1990).

102 Numerical Discretization of the Equations of Motion

This problem of the appearance of trail of positive and negative values when using second or fourth order schemes for the advection equation, is quite serious for models where the quantity being transported can only be positive or zero. Several approaches greatly reduced using more advanced schemes such as those of Takacs (1985) and Flux Corrected Transport (FCT, e.g., Boris and Book, 1973; Zalesak, 1979). Smolarkiewicz (1983, 1984) introduced a simple modification of the upstream scheme by taking advantage of the fact that this scheme, being an interpolation from the previous time step values, does not create negative values of the tracer in the new time step (Figure 3.2.2a). The upstream scheme (3.2.3a), however, introduces strong numerical diffusion (Figure 3.3.1), similar to the diffusion equation (3.2.8). Smolarkiewicz (1983) combines the upstream equation as a predictor, with a corrector negative diffusion equation. This approach has been widely used. The problem of avoiding spurious negative values is also addressed with finite volume methods (Section 3.3.6).

A second type of fourth order finite difference scheme, known as the *compact* or *implicit* fourth order scheme, can be obtained by again making use of (3.3.1) but replacing the third derivative in the truncation error for the centered differences by its finite difference approximation $U_{xxxj} \approx (U_{xj+1} - 2U_{xj} + U_{xj-1})/(\Delta x)^2 + O(\Delta x)^2$. The new fourth order scheme then becomes

$$U_{xj+1} + 4U_{xj} + U_{xj-1} = 6\frac{U_{j+1} - U_{j-1}}{2\Delta x} \tag{3.3.11}$$

It is called "compact" because it involves only the point j and its closest neighbors, and "implicit" because (3.3.11) results in a system of (tridiagonal) equations for the x-derivative, rather than an explicit estimate such as (3.3.4) or (3.3.5).

With this scheme, the finite difference space derivative for a given wavenumber is given by

$$U_x \simeq U\frac{i \sin k\Delta x}{\Delta x}\frac{6}{4 + 2\cos k\Delta x}$$

so that

$$\frac{dU_j}{dt} = -cU_j\frac{i \sin k\Delta x}{\Delta x}\frac{6}{4 + 2\cos k\Delta x} = -iv_{4I}'U_j$$

and the computational phase speed becomes

$$c_{4I}' = \frac{\sin k\Delta x}{k\Delta x}\frac{6}{4 + 2\cos k\Delta x}c, \tag{3.3.12}$$

and for $L = 4\Delta x, k\Delta x = \pi/2$, the phase speed is $c_{14}' = 0.955c$, which is considerably better than even the regular fourth order differences phase speed.

The group velocity for this scheme,

$$\frac{\partial v_{4I}'}{\partial k} = \left[\frac{6\cos k\Delta x}{4 + 2\cos k\Delta x} + \frac{2\sin^2 k\Delta x}{(4 + 2\cos k\Delta x)^2}\right]c \tag{3.3.13}$$

is already positive for $L = 3\Delta x$ (Figure 3.3.1). For implicit schemes where one is already solving a tri-diagonal equation (see Section 3.4.2), this compact fourth order scheme, which has an accuracy equivalent to linear finite elements, is very accurate

3.3 Space Discretization Methods   103

and involves little additional computational cost. The compact scheme is similar to Galerkin finite element approximation to space derivatives (Durran, 1999).

3.3.2   Galerkin and Spectral Space Representation

The use of spatial finite differences, as we saw in the previous section, introduces errors in the space derivatives, resulting in a computational phase speed slower than the true phase speed, especially for short waves.

The Galerkin approach to ameliorate this problem is to perform the space discretization using a sum of basis functions $U(x,t) = \sum_{k=1}^{K} A_k(t)\varphi_k(x)$. Then, the *residual* (error) $R(U) = \partial U/\partial t + F U)$ of the original PDE $\partial u/\partial t + F(u) = 0$ is required to be orthogonal to the basis functions $\varphi(x)$. The space derivatives are computed directly from the known $d\varphi(x)/dx$. This procedure leads to a set of ordinary differential equations for the coefficients $A_k(t)$. If the basis functions chosen for the discretization are orthogonal and satisfy the boundary conditions, the derivation becomes simpler. The use of *local* basis functions (e.g., if $\varphi_i(x)$ is a piecewise linear function equal to 1 at a grid point i and zero at the neighboring points) gives rise to the *finite element method*, with accuracy similar to that of the compact fourth order scheme. Another popular type of Galerkin approach is the use of a global spectral expansion for the space discretization, which allows the space derivatives to be computed analytically rather than numerically. In one dimension, periodic boundary conditions suggest the use of complex Fourier series as a basis.

Consider a periodic domain of length L, with a number of grid points $J_{max} = JM$, and scale x by $2\pi/L$. If we use discrete complex Fourier series truncated to include wavenumbers up to K, the spectral representation is:

$$U(x_j,t) = \sum_{k=-K}^{K} A_k(t)e^{ikx_j} \tag{3.3.14}$$

where $A_{-k}(t) = A_k^*(t)$, and the star represents the complex conjugate.

Alternatively, (3.3.14) can be written using real Fourier series as

$$U(x_j,t) = a_0 + \sum_{k=1}^{K} a_k \cos(kx) + \sum_{k=1}^{K} b_k \sin(kx) \tag{3.3.15}$$

where

$$A_k(t) = \frac{a_k}{2} - i\frac{b_k}{2} \quad k > 0 \quad A_0 = a_0.$$

There are $2K + 1$ distinct real coefficients that are determined by

$$A_k(t) = \frac{1}{JM} \sum_{j=0}^{JM-1} U(x_j,t)e^{-ikx_j}. \tag{3.3.16}$$

Here we have used the orthogonality property

$$\frac{1}{JM} \sum_{j=0}^{JM-1} e^{-ikx_j} e^{ilx_j} = \delta_{kl} = \begin{cases} 1 & \text{if } k = l \\ 0 & \text{otherwise} \end{cases}. \tag{3.3.17}$$

If $JM = 2K + 1$, the grid representation (left-hand side of (3.3.14)) and the spectral representation (right-hand side of (3.3.14)) contain the same number of degrees of freedom, and the same information.

Then, in the wave equation $\partial U/\partial t = -c\partial U/\partial x$, we can discretize U in space as in (3.3.14) and compute the space derivative analytically:

$$\frac{\partial U(x,t)}{\partial x} = \sum_{k=-K}^{K} ik A_k(t) e^{ikx} \tag{3.3.18}$$

If we neglect the time discretization errors, as before, and assume solutions of the form $U(x,t) = Ae^{ik(x-c't)}$, we find that $c' = c$, i.e., the computational phase speed is *equal to the true speed* (Figure 3.3.1). *The space discretization based on a spectral representation is extremely accurate* (the space truncation errors are of "infinite" order). This is because the space derivatives are computed analytically, not numerically, as done in finite differences.

If the PDE is nonlinear, for example $\partial U/\partial t = -U\partial U/\partial x$, then both the grid-point ("physical space") representation and the spectral representation are very useful: Derivatives are computed efficiently and accurately in spectral space, whereas nonlinear products are computed efficiently in physical space. This leads to the so-called *transform method* used for spectral models: The space derivatives are computed in spectral space, then U is transformed back into grid space, and the product $U_j (\partial U/\partial x)_j$ is computed locally in grid space. We will see later that in order to avoid *nonlinear computational instability* introduced by aliasing of wavenumbers beyond K that appear in quadratic terms, the grid representation requires about 3/2 as many points as the minimum number of points required for a linear transform ($JM = 2K+1$). For this reason the new values of U at time $(n+1)\Delta t$ are usually stored in their spectral representation, which is more compact.

We can use von Neumann's criterion to determine the maximum time step allowed for stability using, for example, the leapfrog time scheme. The FDE is

$$\frac{U^{n+1} - U^{n-1}}{2\Delta t} = -ikcU^n. \tag{3.3.19}$$

Assuming solutions for the wave equation of the form $U^n = \rho^n e^{ikx}$, we obtain that the amplification factor is $\rho = -ikc\Delta t \pm \sqrt{1 - k^2 c^2 \Delta t^2}$, and in order to have $|\rho| \leq 1$, we need to satisfy the stability condition

$$(kc\Delta t)^2 \leq 1 \tag{3.3.20}$$

Since the highest wavenumber present corresponds to $L = 2\Delta x$, the stability criterion for spectral models is therefore $c\Delta t/\Delta x \leq 1/\pi$. So, the stability criterion is more restrictive (by a factor of π) for spectral models than for finite difference models, but this is compensated by the fact that the accuracy, especially for shorter waves, is much higher, and therefore fewer short waves need to be included (Figure 3.3.1).

The basis functions used in spectral methods are usually the eigensolutions of the Laplace equation. In a rectangular domain, they are sines and cosines (e.g., the Regional Spectral Model (RSM), Juang et al., 1997). On a circular plate, one would instead use Bessel functions.

Global atmospheric models use as basis functions spherical harmonics, which are the eigenfunctions of the Laplace equation on the sphere:

$$
\begin{aligned}
\nabla^2 Y_n^m &= \frac{1}{a^2}\left[\frac{1}{\cos^2\varphi}\frac{\partial^2 Y_n^m}{\partial\lambda^2} + \frac{1}{\cos\varphi}\frac{\partial}{\partial\varphi}\left(\cos\varphi\frac{\partial Y_n^m}{\partial\varphi}\right)\right]\\
&= \frac{-r(n+1)}{a^2}Y_n^m
\end{aligned}
\tag{3.3.21}
$$

The spherical harmonics are products of Fourier series in longitude and associated Legendre polynomials in latitude:

$$
Y_n^m(\lambda,\varphi) = P_n^m(\mu)e^{im\lambda}
\tag{3.3.22}
$$

where $\mu = \sin\varphi$, m is the zonal wavenumber and n is the "total" wavenumber in spherical coordinates (as suggested by the Laplace equation (3.3.21)). P_n^m are the associated Legendre polynomials in $\mu = \sin\varphi = \cos\theta$, where $\theta = \pi - \varphi$ is the colatitude. For example, the $P_0^0 = 1$; $P_1^0 = \cos\theta$; $P_1^1 = \sin\theta$; $P_2^0 = 1/2\,(3\cos^2\theta - 1)$; $P_2^1 = 3\sin\theta\cos\theta$; $P_2^1 = 3\sin^2\theta$; ...

Using "triangular" truncation

$$
U(\lambda,\varphi,t) = \sum_{n=0}^{N}\sum_{m=-n}^{n}U_n^m(t)Y_n^m(\lambda,\varphi)
\tag{3.3.23}
$$

the spatial resolution is uniform throughout the sphere. This is a major advantage over finite differences based on a latitude–longitude grid, where the convergence of the meridians at the poles requires very small time steps. Although there are solutions for this "pole problem" for finite differences, the natural approach to solve the pole problem for global models is the use of spherical harmonics. Williamson and Laprise (2000) provide a comprehensive description of numerical methods for global models.

Figure 3.3.2(a) shows the shape of three spherical harmonics with total wavenumber $n = 6$, and zonal wavenumber $m = 0$, 3 and 6. Note that the distance between neighboring maxima and minima is similar for the three harmonics, and is associated with the "total" (two-dimensional) wavenumber n. Figures 3.3.2(b) and (c) show that the amplitude of the Legendre polynomials for high zonal wavenumbers are indeed suppressed near the poles. This suppression eliminates the need for small time steps due to the convergence of the meridians in the poles, which are not singular points spectral models.

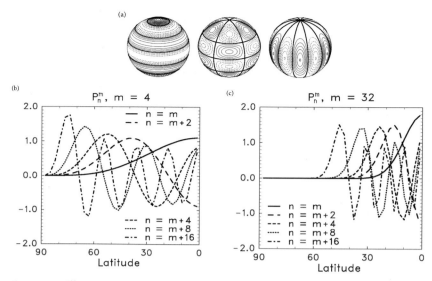

Figure 3.3.2 Illustration of the characteristics of spherical harmonics: (a) Depiction of three spherical harmonics with total wavenumber $n = 6$. Left, zonal wavenumber $m = 0$; center, $m = 3$; right, $m = 6$. Note that n is associated with the total wavelength (twice the distance between a maximum and a minimum), which is the same for the three figures. (b) and (c) Amplitude of Legendre polynomials for different combinations of m and n showing how high zonal wavenumbers are suppressed near the poles, so that the horizontal resolution is uniform when using a spectral representation with triangular truncation. Adapted from Williamson and Laprise (2000). ©SNCSC. Used with permission.

3.3.3 Semi-Lagrangian Schemes

Another numerical method that has become very popular in NWP models is the *semi-Lagrangian* scheme. The equations of motion, as we have seen, can in general be written as conservation equations

$$\frac{du}{dt} = S(u) \qquad (3.3.24)$$

where the left-hand side of the equation represents a *total time derivative* (following an individual parcel) of the vector of dependent variables u. The total time derivative (also known as individual, substantial or Lagrangian time derivative) is zero, i.e., the parcel conserves its total u, except for the changes introduced by the source or sink S.

In a truly Lagrangian scheme, one would follow individual parcels (transporting them with the three-dimensional fluid velocity), and then add the source term at the right time. This is not practical in general because one has to keep track of many individual parcels, and with time they may "bunch up" in certain areas of the fluid, and leave others without parcels to track.

The semi-Lagrangian scheme avoids this problem by using a regular grid as in the previous schemes discussed (which are denoted Eulerian, because the partial derivative $\partial u/\partial t$ is estimated instead of the total derivative). At every new time step we find

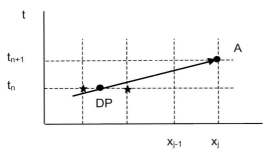

Figure 3.3.3 Schematic of the semi-Lagrangian scheme. The circles represent the arrival point (AP) at the new time level. The thick arrow represents the advection from the departure point (DP). The value of the variables at AP, the arrival point, is equal to their value at departure point, which is obtained by interpolation between neighboring points. Because there is no extrapolation, the semi-Lagrangian schemes are absolutely stable.

out where the parcel arriving at a grid point (denoted arrival point or AP) *came from* in the previous time step (denoted departure point or DP). The value of u at the DP is obtained by interpolating the values of the grid points surrounding the departure point. Figure 3.3.3 suggests that, because there is no extrapolation, the semi-Lagrangian scheme is absolutely stable with respect to advection, which can be confirmed by doing a von Neumann criterion check (Bates and McDonald, 1982).

The semi-Lagrangian scheme can be written using two- or three-time levels. In a three-level time scheme, for example, if MP is the middle point between the DP and AP, the scheme can be written as

$$(U_j^{n+1})_{AP} = (U^{n-1})_{DP} + 2\Delta t S(U^n)_{MP} \qquad (3.3.25)$$

In a two-time level scheme it could be written as

$$(U_j^{n+1})_{AP} = (U^n)_{DP} + \frac{\Delta t}{2}[S(U^n)_{DP} + S(U_j^{n+1})_{AP}]$$

In general, for nonlinear equations $\partial q/\partial t = -u\,\partial q/\partial x + S(q)$, so the semi-Lagrangian scheme for the quantity q can be written as

$$q_{AP} = q_{DP} + \Delta t S(q_{MP}) \qquad (3.3.26)$$

However, the DP has to be determined from the trajectory $dx/dt = u$ integrated between the DP and AP, for example as

$$x_{DP} = x_{AP} - \frac{\Delta t}{2}(U_{DP} + U_{AP}) \qquad (3.3.27)$$

Since u evolves with time, U_{AP} and U_{DP} are not known until the DP has been determined, this is an implicit equation that needs to be solved iteratively. For three-level semi-Lagrangian schemes, the approximation

$$x_{DP} = x_{AP} - 2\Delta t U_{MP} \qquad (3.3.28)$$

also has to be solved iteratively for U_{MP}, but this is simpler than for the two-level time scheme.

The accuracy of the semi-Lagrangian scheme depends on the accuracy of the determination of the DP, on the determination of the value of U_{DP}, and the other conserved quantities q by interpolation from the neighboring points. A linear interpolation between neighboring points results in excessive smoothing, especially for the shortest waves. For this reason cubic interpolation is preferred (Williamson and Laprise, 2000). This is a costly overhead of semi-Lagrangian schemes. Despite the additional costs, in practice this scheme has been found to be accurate and efficient (see the general review of semi-Lagrangian methods by Staniforth and Côté (1990). A "cascade" method has been proposed that results in a very efficient high order interpolation between the distorted Lagrangian grid and the regular Eulerian grid (Purser and Leslie, 1991; Leslie and Purser, 1995). This allowed Purser and Leslie to suggest a forward trajectory semi-Lagrangian approach instead of the conventional backward trajectory that we have so far described, which has additional advantages. (See Staniforth and Côté (1990); Bates et al. (1995); Purser and Leslie (1996); Williamson and Laprise (2000) for further details.) Combining the semi-Lagrangian approach with a semi-implicit treatment of gravity waves (Section 3.2.5), as first suggested by Robert (1982) and Robert et al. (1985), increases its efficiency. Laprise et al. (1997) have documented a "mesoscale compressible community model," which is nonhydrostatic, three-level semi-Lagrangian, and uses the semi-implicit approach for both the elastic terms (three-dimensional divergence) and the gravity wave terms. As such, it is a flexible and accurate model that has been used for a wide range of scales.

3.3.4 Nonlinear Computational Instability, Quadratically Conservative Schemes, and the Arakawa Jacobian

In 1956, Phillips published the first "climate" or "general circulation" simulation ever made with a numerical model of the atmosphere. He started with a baroclinically unstable zonal flow using a two-level quasi-geostrophic model, added small random perturbations, and was able to follow the baroclinic growth of the perturbations, and their nonlinear evolution. He obtained very realistic solutions that contributed significantly to the understanding of the atmospheric circulation in mid-latitudes.

However, his climate simulation only lasted for about 16 days: the model "blew up" despite the fact that care had been taken to satisfy the von Neumann criterion for linear computational instability. In 1959, Phillips pointed out that this instability, which he named nonlinear computational instability (NCI), was associated with nonlinear terms in the quasi-geostrophic equations, in which products of short waves create new waves shorter than $2\Delta x$. Since these waves cannot be represented in the grid, they are "aliased" into longer waves. The shortest wave that can be represented with a grid (with a wavelength $2\Delta x$) corresponds to the maximum computational wavenumber $p_{max} = 2\pi/L_{min} = \pi$. However, quadratic terms with Fourier components will generate higher wavenumbers: $e^{\pm i p_1} e^{\pm i p_2} = e^{\pm i(p_1 \pm p_2)}$, doubling the maximum wavenumber. The new shorter waves, with wavenumbers $p = \pi + \delta$, cannot be represented in the grid, and become folded back (aliased) into $p' = \pi - \delta$, leading to a spurious accumulation of energy at the shortest range (Figure 3.3.4).

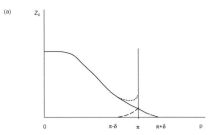

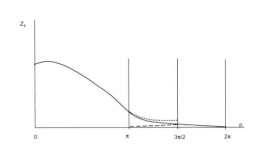

Figure 3.3.4 (a) Schematic of the effect of aliasing: the waves above $p = \pi$ (solid line) get folded back (dashed line) and are added to the original spectrum, producing a spurious maximum in the energy spectrum at the cut-off wavelength (dotted line). (b) Schematic showing that if we allow 3/2 of the total spectrum in the Fourier transform of a quadratic product, aliasing within the original spectrum $(0 - \pi)$ is avoided.

The effect of NCI can be seen clearly in the following simple example: consider the nonlinear (quasi-linear) PDE $\partial u/\partial t = -u\partial u/\partial x$ and the corresponding FDE $\partial U_j/\partial t = -U_j(U_{j+1} - U_{j-1})/2\Delta x$. Suppose that at a given time t, we have $U_1 = 0$, $U_2 > 0$, $U_3 < 0$, $U_4 = 0$. Then $\partial U_1/\partial t = 0$, $\partial U_2/\partial t > 0$, $\partial U_3/\partial t < 0$, $\partial U_4/\partial t = 0$, i.e., U_2 and U_3 will grow without bound and the FDE will blow up. In fact, this will happen even for a linear model $\partial u/\partial t = -a(x)\partial u/\partial x$ if $a_1 = 0$, $a_2 > 0$, $a_3 < 0$, $a_4 = 0$. On the other hand, if $a(x)$ is always of the same sign, and we use the same FDE

$$\frac{\partial U_j}{\partial t} = -a_j \frac{U_{j+1} - U_{j-1}}{2\Delta x} \qquad (3.3.29)$$

we can show that

$$\frac{\partial}{\partial t} \sum_j \frac{U_j^2}{a_j} = 0$$

i.e., that the solution will remain bounded. Numerical experiments show that nonlinear computational instability arises only when there are changes in sign in the velocity.

Exercise 3.3.1 Prove that the above solution will remain bounded.

There are basically two approaches for dealing with the problem of nonlinear computational instability.

Numerical Discretization of the Equations of Motion

Filtering out High Wavenumbers

Phillips (1959) proposed transforming the grid-space solution into Fourier series (with sine and cosine wavenumbers from 0 to π), and chopping out the upper half of the spectrum (wavenumbers above $\pi/2$). Since the maximum wavenumber generated in a quadratic term is twice the original wavenumber, this avoids spurious aliasing, and, indeed, Phillips found that the model could then be run indefinitely. However, the procedure is rather inefficient, since half of the spectrum is not used.

For grid-point models, complete Fourier filtering of the high wavenumbers has been found to be an unnecessarily strong measure to avoid nonlinear computational instability. Some models filter high wavenumbers but only enough to maintain computational stability. Experience shows that as long as the amplitude of the highest wavenumbers is not allowed to acquire finite amplitude, nonlinear computational stability can be avoided. For example, Kalnay-Rivas (1977) combined the use of an energy-conserving fourth order model with a sixteenth order filter (similar to the eighth power of the horizontal Laplacian (Shapiro, 1970)). This efficiently filtered out the shortest waves (mostly between $2\Delta x$ and $3\Delta x$) without affecting much waves of wavelength $4\Delta x$ or longer, and resulted in an accurate and economic model that, not formally, but in practice, is enstrophy (vorticity squared) conserving and therefore not affected by nonlinear computational instability. The Shapiro filter of order n of a field U_i is a simple and efficient operator given by $\overline{U_j}^{2n} = [1 - (-D)^n]U_j$, where the "diffusion" operator $DU_j = (U_{j+1} - 2U_j + U_{j-1})/4$ is applied to the original field n times. For a Fourier component e^{ip} with wavenumber $p = 2\pi\Delta x/L$, the response of the operator is $De^{ipj} = -(\sin^2 p/2)e^{ipj}$ so that the second order Shapiro filter $\overline{U_j}^2 = [1 - (-D)^2]U_j = \frac{1}{4}(U_{j+1} + 2U_j + U_{j-1})$ has a response $\overline{U_j}^2 = (1 - \sin^2 p/2)U_j$. This is a strong filter that zeroes out the highest wavenumber ($L = 2\Delta x$), but also reduces the amplitude of even the longest waves, which is not desirable. But a higher order filter, for example $2n = 16$, however, has the following desirable response: $\overline{U_j}^{16} = (1 - \sin^{16} p/2)U_j$, which still filters out $2\Delta x$ waves, dampens waves shorter than $4\Delta x$, and essentially leaves longer waves unaffected.

Spectral models with a wavenumber cut-off of M (i.e., with $2M + 1$ degrees of freedom) require at least $2M + 1$ grid points to be transformed into equivalent solutions in grid space. Orszag (1971) showed that if they are transformed into $3M + 1$ grid points before a quadratic term is computed in physical space, then aliasing is avoided. In other words, it is not necessary to perform the space transform into $4M$ points. The reason for this is shown schematically in Figure 3.3.4(b): Even if there is aliasing, it only occurs on the part of the spectrum (above $p = \pi$) that is eliminated anyway on the back transformation into spectral space. For this reason, in two horizontal dimensions, spectral models use "quadratic grids" with about $(3/2)^2$ as many grid points as spectral degrees of freedom, and therefore spectral models are "alias-free" for quadratic computations. Triple products in spectral models still suffer from aliasing, but this is generally not a serious problem.

Using Quadratically Conserving Schemes

Lilly (1965) showed that it is possible to create a spatial finite difference scheme that conserves both the mean value and the mean square value when integrated over a closed domain. Quadratic conservation will generally ensure that catastrophic NCI does not take place. Arakawa (1966) created a numerical scheme for the vorticity equation that conserves the mean vorticity, the mean square vorticity (enstrophy), and the kinetic energy. This ensures that the mean wavenumber is also conserved (as it is in the continuous equation), and therefore that even in the absence of diffusion the solution remains realistic. Arakawa and Lamb (1977) showed how an equivalent "Arakawa Jacobian" can be written for primitive equation models.

Consider first a conservation equation for the SWE written in advective form (as an example relevant to primitive equations):

$$\left. \begin{aligned} \frac{\partial \alpha}{\partial t} + \mathbf{v} \cdot \nabla \alpha &= 0 \\[2ex] \frac{\partial h}{\partial t} + \mathbf{v} \cdot \nabla h + h \nabla \cdot \mathbf{v} &= \frac{\partial h}{\partial t} + \nabla \cdot h \mathbf{v} = 0 \end{aligned} \right\}. \tag{3.3.30}$$

If we multiply the first equation by h and the second by α, and add them, we can write the conservation equations in *flux* form:

$$\left. \begin{aligned} \frac{\partial h \alpha}{\partial t} + \nabla \cdot h \mathbf{v} \alpha &= 0 \\[2ex] \frac{\partial h}{\partial t} + \nabla \cdot h \mathbf{v} &= 0 \end{aligned} \right\}. \tag{3.3.31}$$

Note that from the continuity equation written in flux form (3.3.31), the total mass is conserved in time.

Exercise 3.3.2 Show that from the continuity equation written in flux form, the total mass is conserved in time:

$$\frac{\partial}{\partial t} \int\int h \, dx \, dy = 0$$

Now, consider any function of $G(\alpha)$. Multiply the conservation equation by $g(\alpha) = dG/d\alpha$ and integrate over a closed domain (i.e., a domain bounded by walls with zero normal velocities or by periodic boundary conditions). It is easy to show that the mean value of $G(\alpha)$ will be conserved in time:

$$\frac{\partial}{\partial t} \int\int h G(\alpha) \, dx \, dy = 0 \tag{3.3.32}$$

Therefore, the mass weighted mean and the mean squared value of α (as well as all its higher moments) will be conserved. With finite differences, we can only enforce two independent conservation properties (Arakawa and Lamb, 1977). We discuss now how to enforce mean and quadratic conservation, as suggested by Lilly (1965). The simplest approach is to write first the FDE continuity equation in flux form. This constitutes the backbone of a quadratically conservative scheme, and it is also similar to the simplest finite volume schemes (Section 3.3.6).

Exercise 3.3.3 Consider any function $G(\alpha)$ and multiply the conservation equation by $g(\alpha) = dG/d\alpha$ and integrate over a closed domain. Show that the mean value of $G(\alpha)$ will be conserved as in (3.3.32).

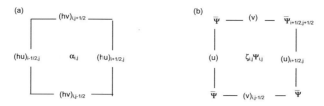

Figure 3.3.5 (a) A grid element with the value of α defined in the center, and estimates of the normal fluxes at its boundaries. These estimates are used for casting the continuity FDE and for constructing a quadratically conservative FDE for α (primitive equation model). (b) Grid for a simple enstrophy conserving FDE (quasi-geostrophic model).

Consider Figure 3.3.5, which shows a typical grid element with the value of α defined in the center and estimates of the normal mass fluxes at its boundaries (e.g., $(hu)_{i+1/2,j}$ at the right wall). These estimates are used for casting the continuity FDE and for constructing a quadratically conservative FDE for α. The continuity FDE in flux form is

$$\frac{\partial h_{ij}}{\partial t} + \frac{(hu)_{i+1/2,j} - (hu)_{i-1/2,j}}{\Delta x} + \frac{(hv)_{i,j+1/2} - (hv)_{i,j-1/2}}{\Delta y} = 0 \qquad (3.3.33)$$

It is easy to check that this FDE will conserve total mass $(\partial/\partial t) \sum_{i,j} h_{i,j} \Delta x_{i,j} \Delta y_{i,j} = 0$ since the mass flux going out from one grid box is equal to the mass flux going into the neighboring box. We now write the FDE for α using *any* consistent estimate of α at the normal walls of the grid box:

$$\frac{\partial h_{ij}\alpha_{i,j}}{\partial t} + \frac{(hu)_{i+1/2,j}(\alpha)_{i+1/2,j} - (hu)_{i-1/2,j}(\alpha)_{i-1/2,j}}{\Delta x}$$
$$+ \frac{(hv)_{i,j+1/2}(\alpha)_{i,j+1/2} - (hv)_{i,j-1/2}(\alpha)_{i,j-1/2}}{\Delta y} = 0 \qquad (3.3.34)$$

Again it is easy to check that this FDE will conserve the total (mass weighted) value of α: $(\partial/\partial t) \sum_{i,j} h_{i,j} \alpha_{i,j} \Delta x_{i,j} \Delta y_{i,j} = 0$. *This is a general property of FDEs written in flux form.*

Finally, we choose to estimate the value of α at the walls of the grid-cells as an average between the two contiguous cells $\alpha_{i+1/2,j} = (\alpha_{i,j} + \alpha_{i+1,j})/2$ and similarly for the other walls. With this particular choice, we obtain:

$$\frac{\partial h_{i,j}\alpha_{i,j}}{\partial t} + \frac{(hu)_{i+1/2,j}(\alpha_{i,j} + \alpha_{i+1,j}) - (hu)_{i-1/2,j}(\alpha_{i,j} + \alpha_{i-1,j})}{2\Delta x}$$
$$+ \frac{(hv)_{i,j+1/2}(\alpha_{i,j} + \alpha_{i,j+1}) - (hv)_{i,j-1/2}(\alpha_{i,j} + \alpha_{i,j-1})}{2\Delta y} = 0. \qquad (3.3.35)$$

We can show that this scheme is quadratically conservative. First note that we can construct a mass weighted quadratic conservation equation for α from either the advection of the flux form prognostic equation for α and the continuity equation (prognostic equation for h):

3.3 Space Discretization Methods 113

$$\frac{\partial \left(h\frac{\alpha^2}{2}\right)}{\partial t} = -\frac{\alpha^2}{2}\frac{\partial h}{\partial t} + \alpha\frac{\partial h\alpha}{\partial t}. \tag{3.3.36}$$

This equality suggests how to test quadratic conservation of α: Multiply the FDE continuity equation (3.3.33) by $\alpha_{i,j}^2/2$ and subtract it from the flux form prognostic equation (3.3.35) multiplied by $\alpha_{i,j}$. If we do this, we find that (because of cancellations of mass weighted fluxes of $\alpha_{i,j}$ on the grid-box walls), there is indeed quadratic conservation:

$$\frac{\partial}{\partial t}\sum_{i,j} h_{i,j}\frac{1}{2}\alpha_{i,j}^2\Delta x_{i,j}\Delta y_{i,j} = 0. \tag{3.3.37}$$

Note that this is true no matter how the FDE for the continuity equation is written. We could choose several finite difference formulations, and as long as the flux form of the FDE for $h\alpha$ is consistent with the continuity equation, and as long as we estimate α at the walls by a simple average, we have quadratic conservation and the danger of NCI is small.

Exercise 3.3.4 Show that the FDE (3.3.33) will conserve total mass.

Exercise 3.3.5 Show that the FDE (3.3.34) will conserve the mass weighted total value value of α.

Exercise 3.3.6 Prove from (3.3.36) that there is quadratic conservation.

Exercise 3.3.7 Write two different FDEs for the continuity equation, i.e., two different estimates of the normal mass fluxes at the walls, $(hu)_{i+1/2,j}$, etc.

Finally, we consider the vorticity equation, which represents much of the dynamics of the real atmosphere:

$$\frac{\partial \zeta}{\partial t} = -\mathbf{v}\cdot\boldsymbol{\nabla}\zeta = -\boldsymbol{\nabla}\cdot(\mathbf{v}\zeta) = -J(\Psi,\zeta) \tag{3.3.38}$$

where $\zeta = \mathbf{k}\cdot\boldsymbol{\nabla}\times\mathbf{v} = \boldsymbol{\nabla}^2\Psi$, $\mathbf{v} = \mathbf{k}\times\boldsymbol{\nabla}\Psi$. The flow is nondivergent, so that the continuity equation is simply $\boldsymbol{\nabla}\cdot\mathbf{v} = 0$.

In this case, a simple scheme that conserves the mean vorticity and its mean square (i.e., an enstrophy conserving scheme) can be written following the recipe given earlier (Kalnay-Rivas and Merkine, 1981). The continuity equation is (cf. Figure 3.3.5(b))

$$0 = -\frac{(u)_{i-1/2,j} - (u)_{i-1/2,j}}{\Delta x} - \frac{(v)_{i,j+1/2} - (v)_{i,j-1/2}}{\Delta y} \tag{3.3.39}$$

where the normal velocity estimates are obtained from

$$u_{i+1/2,j} = -\frac{\partial \overline{\Psi}_{i+1/2,j}}{\partial y} \approx -\frac{\overline{\Psi}_{i+1/2,j+1/2} - \overline{\Psi}_{i+1/2,j-1/2}}{\Delta y} \tag{3.3.40}$$

and similarly for the other velocities. Note that this satisfies the continuity equation automatically. Then we write the forecast equation for the vorticity in a way consistent

Numerical Discretization of the Equations of Motion

with the continuity equation, thus ensuring conservation of the mean vorticity and enstrophy (mean square vorticity).

$$\frac{\partial \zeta_{i,j}}{\partial t} = -\frac{(u)_{i+1/2,j}(\zeta_{i,j} + \zeta_{i+1,j}) - (u)_{i-1/2,j}(\zeta_{i,j} + \zeta_{i-1,j})}{2\Delta x}$$
$$- \frac{(v)_{i,j+1/2}(\zeta_{i,j} + \zeta_{i,j+1}) - (v)_{i,j-1/2}(\zeta_{i,j} + \zeta_{i,j-1})}{2\Delta y}. \tag{3.3.41}$$

After a new vorticity field is obtained at $t = (n + 1)\Delta t$ using, for example, leapfrog, we have to determine the new streamfunction ψ. This is done by solving the elliptic equation $\zeta = \nabla^2 \Psi$, which in finite differences can be written as

$$\frac{\Psi_{i+1,j} - 2\Psi_{i,j} + \Psi_{i-1,j}}{\Delta x^2} + \frac{\Psi_{i,j+1} - 2\Psi_{i,j} + \Psi_{i,j-1}}{\Delta y^2} = \zeta_{i,j}. \tag{3.3.42}$$

In Section 3.4, we will discuss how to solve this *boundary value problem*.

Once we obtain $\psi_{i,j}$ we can obtain $\overline{\Psi}_{i+1/2,j+1/2}$ by averaging the corresponding four surrounding values of $\psi_{i,j}$. This is probably the simplest FDE model of the barotropic atmosphere devoid of nonlinear computational instability that we can construct.

Before we discuss the Arakawa Jacobian, let's note that the continuous vorticity equation conserves total (kinetic) energy as well as enstrophy. Multiply the vorticity equation by the streamfunction:

$$\Psi \frac{\partial \zeta}{\partial t} = -\Psi \nabla \cdot (\mathbf{v}\zeta) = -\nabla \cdot (\mathbf{v}\zeta \Psi) + \zeta \mathbf{v} \cdot \nabla \Psi \tag{3.3.43}$$

The last term on the right-hand side vanishes because $\mathbf{v}$ is perpendicular to $\nabla \Psi$. The left-hand side can be shown to be the time derivative of the kinetic energy:

$$\Psi \frac{\partial \zeta}{\partial t} = \Psi \frac{\partial \nabla \cdot \nabla \Psi}{\partial t} = \frac{\partial \nabla \Psi \cdot \nabla \Psi}{\partial t} - \nabla \Psi \cdot \frac{\partial \nabla \Psi}{\partial t}$$
$$= \frac{\partial \mathbf{v} \cdot \mathbf{v}}{\partial t} - \frac{\partial \mathbf{v} \cdot \mathbf{v}/2}{\partial t} = \frac{\partial |\mathbf{v}|^2/2}{\partial t} = \frac{\partial KE}{\partial t}. \tag{3.3.44}$$

Therefore, integrating (3.3.43) over the domain, the mean kinetic energy is conserved. The simple scheme described above conserves vorticity and squared vorticity but not kinetic energy.

Arakawa (1966) introduced a Jacobian that conserves all three properties: it is based on the FDE corresponding to these three equivalent formulations of the Jacobian:

$$J(\Psi, \zeta) = J_1 = J_2 = J_3 \tag{3.3.45}$$

where

$$\left. \begin{aligned} J_1 &= \frac{\partial \Psi}{\partial x}\frac{\partial \zeta}{\partial y} - \frac{\partial \Psi}{\partial y}\frac{\partial \zeta}{\partial x} \\ J_2 &= \frac{\partial}{\partial x}\left(\Psi\frac{\partial \zeta}{\partial y}\right) - \frac{\partial}{\partial y}\left(\Psi\frac{\partial \zeta}{\partial x}\right) \\ J_3 &= \frac{\partial}{\partial y}\left(\zeta\frac{\partial \Psi}{\partial x}\right) - \frac{\partial}{\partial x}\left(\zeta\frac{\partial \Psi}{\partial y}\right) \end{aligned} \right\} \tag{3.3.46}$$

The Arakawa Jacobian is the finite difference Jacobian corresponding to $J_A = (J_1 + J_2 + J_3)/3$ and it conserves kinetic energy and enstrophy. Arakawa and Lamb (1977) showed how the Arakawa Jacobian could also be approximately constructed for primitive equation models.

Exercise 3.3.8 Write the finite difference equivalent of J_1, J_2 and J_3

The ratio of enstrophy to kinetic energy is proportional to the mean square of the wavenumber, and this quantity is conserved by the continuous frictionless vorticity equation. Therefore, the Arakawa conservation ensures that a long model run will conserve the mean square of the wavenumber, and generally look realistic (i.e., not become dominated by small scale noise) even without horizontal diffusion. In the real atmosphere, however, turbulent dissipation acts as a control on the amplitude of the smallest waves, and leaks their energy out of the system, so that strict conservation is not truly relevant. For this reason, there is no consensus among the community of modelers on whether the use of strictly conserving FDEs is an essential requirement. On the one hand, some models are based on schemes that are as conservative as possible (remember that the continuous equations without friction conserve all moments of the quantities being advected, whereas the FDEs can only conserve one or two moments). Other modelers prefer to use less conservative but more accurate and simpler schemes. They include dissipation acting at the highest wavenumbers that mimics the leakage of energy that takes place in reality. Experience shows that an energy-conserving scheme, for example, combined with a small amount of high-order horizontal diffusion, in practice also behaves very realistically, approximately conserving enstrophy. This is because a catastrophic loss of enstrophy occurs only when energy is allowed to accumulate in the shortest waves because of aliasing, and they acquire large amplitudes. The dispute as to whether it is more important to have conservative FDEs or accurate (higher order or semi-Lagrangian) FDEs that are not conservative but still avoid NCI has thus not been resolved.

3.3.5 Staggered Grids

So far all the variables we have used (e.g., h, u, v) have been defined at the same location in a grid cell. This means that in order to compute centered space differences at a point j, for example, we need to go to $j + 1$, and $j - 1$, and the differences are computed over a distance of $2\Delta x$ (Figure 3.3.6(a)). If we use instead a staggered grid,

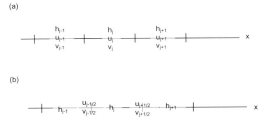

Figure 3.3.6 Staggered grids: (a) Example of unstaggered grid in one dimension; (b) example of a staggered grid in one dimension.

Numerical Discretization of the Equations of Motion

certain differences (such as the pressure gradient for the u equation and the horizontal convergence term for the h equation) can be computed over just $1\Delta x$, and, for those terms, it is equivalent to doubling the horizontal resolution. (Figure 3.3.6(b)). However, the advection terms still have to be computed over $2\Delta x$ (or $2d$, where d is the distance between closest grid points of the same class).

Let's consider again the SWE in two dimensions:

$$\left.\begin{array}{l}
\dfrac{\partial h}{\partial t} = -\left[h\left(\dfrac{\partial u}{\partial x} + \dfrac{\partial v}{\partial y}\right)\right] - u\dfrac{\partial h}{\partial x} - v\dfrac{\partial h}{\partial y} \\[3mm]
\dfrac{\partial u}{\partial t} = -\left[g\dfrac{\partial h}{\partial x} + fv\right] - u\dfrac{\partial u}{\partial x} - v\dfrac{\partial u}{\partial y} \\[3mm]
\dfrac{\partial v}{\partial t} = -\left[g\dfrac{\partial h}{\partial y} - fu\right] - u\dfrac{\partial v}{\partial x} - v\dfrac{\partial v}{\partial y}
\end{array}\right\} \qquad (3.3.47)$$

The terms in square brackets in (3.3.47) are the dominant terms for the geostrophic and the inertia-gravity wave dynamics. They are computed in different ways depending on the type of grid we use. The advective terms are less affected by the choice of alternative (staggered) grids.

In two dimensions, there are several possibilities for staggered grids (Arakawa and Lamb, 1977), which are shown schematically in Figure 3.3.7. Grid A (unstaggered) has several advantages and disadvantages. The advantages are its simplicity, and, because all variables are available at all the grid points, it is easy to construct a higher order accuracy scheme (Kalnay, 1983). Grid A tends to be favored by proponents of the philosophy "accuracy is more important than conservation." Its main disadvantage is that all differences occur on distances $2d$, and that neighboring points are not coupled for the pressure and convergence terms. This can give rise in time to a horizontal uncoupling (checkerboard pattern), which needs to be controlled by using a high order diffusion (e.g., Janjić, 1974; Kalnay-Rivas, 1977).

Grid C has the advantage that the convergence and pressure terms in square brackets in (3.3.47) are computed over a distance of only $1d$, which is equivalent *to doubling the resolution of grid A*. For this reason, geostrophic adjustment (the dispersion of gravity waves generated when the fields are not in geostrophic balance) is computed much more accurately (Arakawa, 1997). The Coriolis acceleration terms, on the other hand, require horizontal averaging, making the inertia-gravity waves less accurate. This makes grid C less attractive for situations in which the length of the Rossby radius of deformation $R_d = \sqrt{gH}/f$ is not large compared to the grid size d. The equivalent depth of the atmosphere, H, is about 10 km for the external mode, so that R_d is about 3000 km , but H is an order of magnitude smaller for the second vertical mode, and it becomes much smaller for higher vertical modes. Therefore, some atmospheric models use grid B, where the minimum distance for horizontal differences is $\sqrt{2}d$, rather than $1d$ as in grid C, but where u and v are available at the same locations. The NCEP Eta model is defined on a grid B rotated by 45°, denoted grid E by Arakawa and Lamb (1977), see Figure 3.3.7.

3.3 Space Discretization Methods

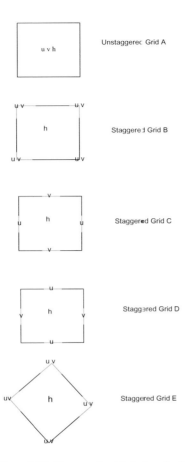

Figure 3.3.7 Staggered grids in two horizontal dimensions: Arakawa and Lamb (1977) classification.

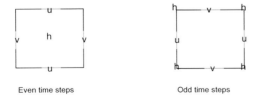

Figure 3.3.8 The Eliassen grid, staggered in both space and time.

The disadvantages of staggered grids are: (a) The terms in square brackets are hard to implement in higher-order schemes, and (b) The staggering introduces considerable complexity in, for example, diagnostic studies and graphical output.

Grid D seems to have no particular merit, but, if also staggered in time (as suggested by Eliassen), it becomes ideal for atmospheric flow using the leapfrog scheme (see Figure 3.3.8). In the Eliassen grid, *all* differences are computed on a distance d (the advection also requires a horizontal average over one grid length, but this is a small

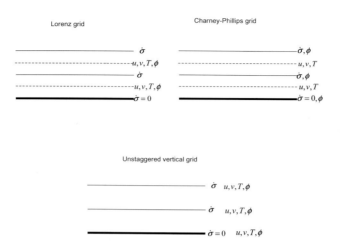

Figure 3.3.9 Staggering in vertical grids (Arakawa and Konor, 1996). ©American Meteorological Society. Used with permission.

drawback). Despite its apparent optimality, this grid has not been adopted in any major model. Lin and Rood (1997) have also adopted a combination of the C and the D grids for the global atmospheric model known as Finite-Volume Cubed-Sphere (FV3) because it is optimal for the transport of vorticity.

In the vertical direction, most models have adopted a staggered grid, with the vertical velocity defined at the boundary of layers and the prognostic variables in the center of the layer (Figure 3.3.9). This type of grid, introduced by Lorenz in 1960, allows simple quadratic conservation, and the boundary conditions of no flux at the top and the bottom are easily fulfilled. However, as pointed out by Arakawa and Moorthi (1988), the Lorenz grid allows the development of a spurious computational mode, since the geopotential in the hydrostatic equation (and therefore the acceleration of the wind components) is insensitive to temperature oscillations of $2\Delta\sigma$ wavelength. The Lorenz grid is being replaced in some newer models by a vertical grid similar to the one introduced by Charney and Phillips (1953) for a two-level model. In the Charney–Phillips grid, the vertical staggering is more consistent with the hydrostatic equation and therefore it does not have the additional computational mode (Arakawa, 1997). A nonstaggered vertical grid, allowing a simple implementation of higher order differences in the vertical, would also be possible, but it would also have more computational modes present in the solution.

3.3.6 Finite Volume Methods

We present here a brief introduction to the finite volume approach, which is discussed in more detailed in texts such as Durran (1999), Fletcher (1988) and Gustafsson et al. (1995). The basic idea of this method is that the governing equations are first written in an integral form for a finite volume, and only then are they discretized. This is in contrast to the methods we have seen so far, in which the equations in differential form

3.3 Space Discretization Methods

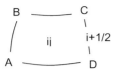

Figure 3.3.10 Schematic of the two-dimensional volume centered at the point i,j and with walls at which fluxes are computed in a finite volume method.

are discretized using finite differences or spectral methods. The two approaches may or may not lead to similar discretized schemes.

Consider for example the continuity equation and a conservation equation for a shallow water model written in flux form, as in (3.3.31), and integrate them over a volume limited by walls $\overline{AB}, \overline{BC}, \overline{CD}$, and $\overline{DA}$ (in this two-dimensional case, the volume of integration is the horizontal area, Figure 3.3.10).

If we integrate (3.3.31) within the volume $\overline{ABCD}$, and apply Green's theorem, we obtain

$$\left. \begin{array}{l} \dfrac{d}{dt} \int h\,dx\,dy + \oint \mathbf{H} \cdot \mathbf{n}\,ds = 0 \\[1ex] \dfrac{d}{dt} \int h\alpha\,dx\,dy + \oint (\mathbf{H}\alpha) \cdot \mathbf{n}\,ds = 0 \end{array} \right\} \quad (3.3.48)$$

where $\mathbf{H}$ is the normal flux of h across the walls, and $\mathbf{n}$ is the normal vector to the wall. These equations can be discretized, for example, as

$$\left. \begin{array}{l} \dfrac{d}{dt}\left(\overline{h}^{ij}\Delta x_{ij}\Delta y_{ij}\right) = -(\overline{hu}^{i+1/2\,j})\Delta y_{i+1/2\,j} + (\overline{hu}^{i-1/2\,j})\Delta y_{i-1/2\,j} \\[1ex] \qquad\qquad\qquad\qquad -(\overline{hv}^{ij+1/2})\Delta x_{ij+1/2} + (\overline{hv}^{ij-1/2})\Delta x_{ij-1/2} \\[1ex] \dfrac{d}{dt}\left(\overline{\alpha h}^{ij}\Delta x_{ij}\Delta y_{ij}\right) = -(\overline{hu\alpha}^{i+1/2\,j})\Delta y_{i+1/2\,j} + (\overline{hu\alpha}^{i-1/2\,j})\Delta y_{i-1/2\,j} \\[1ex] \qquad\qquad\qquad\qquad -(\overline{hv\alpha}^{ij+1/2})\Delta x_{ij+1/2} + (\overline{hv\alpha}^{ij-1/2})\Delta x_{ij-1/2} \end{array} \right\} \quad (3.3.49)$$

Here, the overbar indicates a suitable average over the grid volume or area. It is evident that any scheme based on these finite volume equations will conserve the average mass and average mass weighted α. There are a number of choices of how this average can be carried out over this *subgrid* domain of each grid volume: One can assume that h and α are constant within the volume, or that they vary linearly, and so on. A simple choice for the estimates of the average values at the center and at the walls leads naturally to the quadratically conservative differences presented earlier in (3.3.34) and (3.3.35):

$$\left. \begin{array}{l} \overline{h}^{ij} = h_{ij} \quad \overline{hu}^{i+1/2\,j} = (h_{ij} + h_{i+1\,j})(u_{ij} + v_{i+1\,j})/4 \\[1ex] \overline{hu\alpha}^{i+1/2\,j} = (\overline{hu}^{i+1/2\,j})(\alpha_{ij} + \alpha_{i+1\,j})/2 \end{array} \right\} \quad (3.3.50)$$

Although in this case both methods lead to the same discretization, the finite volume approach allows additional flexibility in the choice of discretization. For example,

Numerical Discretization of the Equations of Motion

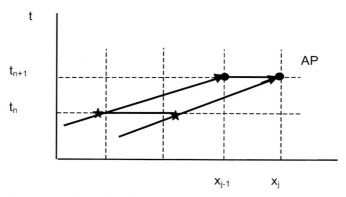

Figure 3.3.11 Schematic of the flux-form semi-Lagrangian scheme (Lin and Rood, 1996). It differs from the regular semi-Lagrangian scheme (Figure 3.3.3) in that the walls of the volume are transported to the "arrival walls." The mass weighted average of the variables at the arrival volume is equal to the value at the departure volume (indicated by the thick segments), ensuring mass conservation. Because there is no extrapolation, the flux-form semi-Lagrangian scheme is still absolutely stable. Published (1996) by the American Meteorological Society.

Lin and Rood (1996) developed a combination of semi-Lagrangian and finite volume methods, in which the boundaries of the grid volume are transported to the new time step, rather than the centers of the volume as is done in the conventional semi-Lagrangian schemes (Figure 3.3.11). Although the order of the scheme is formally low, the method seems very promising, but it requires considerable care in the detailed formulation in order both to remain conservative and to maintain the shape of the transported tracers. Lin (1997) also developed a remarkably simple finite volume expression to compute the horizontal pressure gradient force that can be applied to any hydrostatic vertical coordinate system. It avoids the problem of having two large terms in the pressure gradient computation that almost cancel each other, which is characteristic of the sigma and other vertical coordinate systems. It also ensures that the pressure force exerted by a finite volume on a neighboring volume cancels exactly with the pressure force exerted back by the neighbor, satisfying Newton's action and reaction principle.

3.4 Boundary Value Problems

3.4.1 Introduction

Elliptic equations are boundary value problems, with either a fixed time, or a steady state solution at long times. Two examples of such problems arising in NWP are:

(a) Finding the new streamfunction from the vorticity after the latter has been updated to time $(n + 1)\Delta t$. For example, in Section 3.3.4, we introduced the following enstrophy-conserving numerical scheme:

$$\frac{\zeta_{i,j}^{n+1} - \zeta_{i,j}^{n-1}}{2\Delta t} = -\frac{(u)_{i+1/2,j}(\zeta_{i,j} + \zeta_{i+1,j}) - (u)_{i-1/2,j}(\zeta_{i,j} + \zeta_{i-1,j})}{2\Delta x}$$
$$-\frac{(v)_{i,j+1/2}(\zeta_{i,j} + \zeta_{i,j+1}) - (v)_{i,j-1/2}(\zeta_{i,j} + \zeta_{i,j-1})}{2\Delta y} \qquad (3.4.1)$$

where we used the leapfrog scheme, and the right-hand side is evaluated at time $t = n\Delta t$. After solving for $\zeta_{i,j}^{n+1}$, we can obtain the streamfunction by solving the elliptic equation (Laplace) valid at $t = (n+1)\Delta t$:

$$\frac{\Psi_{i+1,j} - 2\Psi_{i,j} + \Psi_{i+1,j}}{\Delta x^2} + \frac{\Psi_{i,j+1} - 2\Psi_{i,j} + \Psi_{i,j-1}}{\Delta y^2} = \zeta_{i,j} \qquad (3.4.2)$$

For this particular scheme, after solving for $\Psi_{i,j}^{n+1}$, we obtain $\overline{\Psi}_{i+1/2,j+1/2}^{n+1}$ by averaging from the four surrounding corners.

(b) Solving a semi-implicit elliptic equation for the heights also at $(n+1)\Delta t$ (Section 3.2.5):

$$\left(\delta_{2x}^2 + \delta_{2y}^2 - \frac{1}{\Phi\Delta t^2}\right)\phi^{n+1} = -\left(\delta_{2x}^2 + \delta_{2y}^2 + \frac{1}{\Phi\Delta t^2}\right)\phi^{n-1}$$
$$+ 2(\delta_{2x}R_u + \delta_{2y}R_v)$$
$$+ \frac{1}{\Delta t}(\delta_{2x}u^{n-1} + \delta_{2y}v^{n-1})$$
$$- \frac{2}{\Phi\Delta t}R_\phi = F_{i,j}^n \qquad (3.4.3)$$

These linear elliptic equations are easily solved with spectral methods in which the basis functions are eigenfunctions of the Laplace equation. For example, if we use spherical harmonics on the globe and make use of

$$\nabla^2 Y_n^m = \frac{1}{a^2}\left[\frac{1}{\cos^2\varphi}\frac{\partial^2 Y_n^m}{\partial\lambda^2} + \frac{1}{\cos\varphi}\frac{\partial}{\partial\varphi}\left(\cos\varphi\frac{\partial Y_n^m}{\partial\varphi}\right)\right]$$
$$= \frac{-n(n+1)}{a^2}Y_n^m \qquad (3.4.4)$$

we can solve the semi-implicit equation for $\phi(\lambda,\varphi,t) = \sum_{n=0}^{N}\sum_{m=-n}^{n}\phi_n^m(t)Y_n^m(\lambda,\varphi)$ simply by writing the Helmholtz linear equation $\nabla^2\phi^{n+1} - (1/\Phi\Delta t^2)\phi^{n+1} = F$ corresponding to (3.4.3) component by component

$$\nabla^2\phi_m^n(t_{p+1})Y_n^m(\lambda,\varphi) - \frac{1}{\Phi\Delta t^2}\phi_m^n(t_{p+1})Y_n^m(\lambda,\varphi) = F_m^n(t_{p+1})Y_n^m(\lambda,\varphi) \qquad (3.4.5)$$

so that the solution for each spherical harmonic coefficient is given by

$$\phi_m^n(t_{p+1}) = -\frac{1}{\left[\dfrac{n(n+1)}{a^2} + \dfrac{1}{\Phi\Delta t^2}\right]}F_m^n(t_{p+1}) \qquad (3.4.6)$$

(Note that in (3.4.5) and (3.4.6), we have used p instead of n for the time step to avoid confusion with the total wavenumber n.) The simplicity with which the semi-implicit scheme can be computed is a major advantage of spectral models. For finite differences, the solution is much more involved.

122   Numerical Discretization of the Equations of Motion

The methods of solution for elliptic equations (discretized in space) are basically of two types: *direct* and *iterative*. Here, we only present some simple examples of both types of methods and refer the reader to texts such as Golub and Van Loan (1996), Ferziger and Perić (2002), Dahlquist and Björck (1974), and Gustafsson et al. (1995) for more complete discussions of direct and iterative schemes. In the last decade, considerable work has also been done on the solution of nonsymmetric systems. Books on computational methods for these types of problems include Barrett et al. (1994), Bruaset (1995), Greenbaum (1997), and Meurant (1999).

3.4.2   Direct Methods for Linear Systems

We saw that for spectral models, the direct solution of the linear elliptic equation arising from the semi-implicit method is trivial. For finite differences, however, direct methods involve solving equations like (3.4.2) or (3.4.3), which can be written in matrix form as

$$\mathbf{A}\phi = F \tag{3.4.7}$$

using any direct solver. They are related to Gaussian elimination. If the matrix $\mathbf{A}$ is fixed (e.g., independent of the time step), the LU decomposition of $\mathbf{A} = LU$, where the diagonal of L are $l_{ii} = 1$, allows us to perform the decomposition once and then solve $LX = F$, followed by $U\Phi = X$. Here L and U are lower and upper triangular matrices.

If the matrix is tridiagonal, the direct problem is particularly easy to solve. A tridiagonal problem can be written as:

$$a_j U_{j-1} + b_j U_j + c_j U_{j+1} = d_j \tag{3.4.8}$$

with general boundary conditions

$$U_0 = A_1 U_1 + A_2 \quad U_J = B_1 U_{J-1} + B_2 \tag{3.4.9}$$

An algorithm based on Gaussian elimination is the "double sweep" method: Assume that

$$U_j = E_j U_{j+1} + F_j \tag{3.4.10}$$

Then $U_{j-1} = E_{j-1} U_j + F_{j-1}$ which can be substituted into the tridiagonal equation (3.4.8) to obtain:

$$(a_j E_{j-1} + b_j) U_j + c_j U_{j+1} = d_j - a_j F_{j-1} \tag{3.4.11}$$

From this we deduce that

$$\left. \begin{aligned} E_j &= \frac{-c_j}{a_j E_{j-1} + b_j} \\ F_j &= \frac{d_j - a_j F_{j-1}}{a_j E_{j-1} + b_j} \end{aligned} \right\} \tag{3.4.12}$$

So the method of solution is:

(a) use the lower boundary condition $U_0 = A_1 U_1 + A_2$ to determine
$E_0 = A_1$, $F_0 = A_2$;
(b) sweep forward using (3.4.12) to obtain E_j, F_j, $\quad j = 1, \dots, J - 1$;
(c) determine U_J, U_{J-1} from $U_{J-1} = E_{J-1} U_J + F_{J-1}$ and the upper boundary
condition $U_J = B_1 U_{J-1} + B_2$;
(d) determine $U_j, j = J - 2, \dots, 1$ using (3.4.10).

Tridiagonal matrices can thus be solved very efficiently, although problems arise when the denominator in (3.4.12) is close to zero.

3.4.3 Iterative Methods for Solving Elliptic Equations

The system $\mathbf{A}\phi = F$ can be solved iteratively by transforming it into another system,

$$\phi = (I - A)\phi + F$$

or

$$\phi = M\phi + F \tag{3.4.13}$$

choosing an initial guess ϕ^0 and then iterating (3.4.13): $\phi^{\nu+1} = M\phi^{\nu} + F$. The method converges if the spectral radius $\sigma(M) = \max |\lambda_i| < 1$, where λ_i are the eigenvalues of M. The asymptotic convergence rate is defined as

$$R = -\log_{10}[\sigma(M)] \tag{3.4.14}$$

We now give an example for a simple elliptic equation to provide an idea of how to attack the problem. The reader is referred to the references cited in Subsection 3.4.1 for a more comprehensive discussion.

For a uniform grid with $\Delta x = \Delta y = \Delta$, an elliptic equation like (3.4.3) can be written as

$$\delta^2 \phi_{i,j} - \alpha \phi_{i,j} = g_{i,j} \tag{3.4.15}$$

where the finite difference Laplace operator is

$$\delta^2 \phi_{i,j} = (\phi_{i+1,j} + \phi_{i-1,j} + \phi_{i,j+1} + \phi_{i,j-1} - 4\phi_{i,j}) \tag{3.4.16}$$

Suppose we are in iteration ν. Then

$$\delta^2 \phi^{\nu}_{i,j} - \alpha \phi^{\nu}_{i,j} = g_{i,j} + \epsilon^{\nu}_{i,j} \tag{3.4.17}$$

where $\epsilon^{\nu}_{i,j}$ is the error in iteration ν. If we assume at the point i, j

$$\phi^{\nu+1}_{i,j} = \phi^{\nu}_{i,j} + \delta \phi^{\nu}_{i,j} \tag{3.4.18}$$

and choose $\delta \phi^{\nu}_{i,j}$ to make $\epsilon^{\nu+1}_{i,j} = 0$, we get

$$\phi^{\nu+1}_{i,j} = \phi^{\nu}_{i,j} + \frac{\delta^2 \phi^{\nu}_{i,j} - \alpha \phi^{\nu}_{i,j} - g_{i,j}}{4 + \alpha} \tag{3.4.19}$$

124   Numerical Discretization of the Equations of Motion

This is the *Jacobi simultaneous relaxation method*. If we start at the southwest corner, and sweep to the right and up, by the time we reach the point i,j we have already updated the neighboring points to the west and the south, so we can use these updated values:

$$\phi_{i,j}^{v+1} = \phi_{i,j}^v + \frac{\phi_{i-1,j}^{v+1} + \phi_{i+1,j}^v + \phi_{i,j-1}^{v+1} + \phi_{i,j+1}^v - \alpha\phi_{i,j}^v - g_{i,j}}{4 + \alpha} \qquad (3.4.20)$$

This is the *Gauss–Seidel* or *successive relaxation method*.

If instead, we *overcorrect* by changing the sign of $\epsilon_{i,j}^{v+1}$ rather than making it equal to zero, i.e.,

$$\phi_{i,j}^{v+1} = \phi_{i,j}^v + \omega\frac{\phi_{i-1,j}^{v+1} + \phi_{i+1,j}^v + \phi_{i,j-1}^{v+1} + \phi_{i,j+1}^v - \alpha\phi_{i,j}^v - g_{i,j}}{4 + \alpha}$$

$$\text{with } 1 < \omega < 2 \qquad (3.4.21)$$

the rate of convergence is further increased. This is the *successive overrelaxation* (SOR) method. Optimal values for ω can be obtained analytically for simple geometries such as a rectangular domain. For the equation above, the spectral radius of the Jacobi matrix $\mathbf{M}$ is

$$\lambda_1 = 1 - \varepsilon$$

where

$$\varepsilon = \sin^2 \frac{\pi}{2(JM + 1)} + \sin^2 \frac{\pi}{2(KM + 1)}$$

and *JM, KM* are the number of intervals in the x and y directions of the problem. Then the optimum value of the overrelaxation coefficient is

$$\omega_{opt} = \frac{2}{1 + \sqrt{1 - \lambda_1^2}}$$

Since the maximum error is reduced after each Jacobi iteration by the spectral radius $\lambda_1 = (1 - \varepsilon)$, we can define the rate of convergence as ε.

The rates of convergence of the three methods are then:

ε = rate of convergence of the Jacobi iteration;
2ε = rate of convergence of the Gauss–Seidel iteration;
$2\sqrt{2\varepsilon}$ = rate of convergence of the SOR iteration with optimum overrelaxation.

3.4.4   Other Iterative Methods

We give only a simple introduction to other methods and refer the reader for further details to the references cited in Section 3.4.1. A short but comprehensive summary of these methods is given in Appendix C of Buckeridge et al. (2011) in the context of inverting the potential vorticity equation in order to use it as a control variable in variational data assimilation.

3.4 Boundary Value Problems 125

Alternating Direction Implicit (ADI)

An efficient fractional time steps time scheme (Table 3.2.1) is used to obtain the solution of the elliptic equation as a steady state solution. For example, to solve the Laplace equation, we write the parabolic equation

$$\frac{\partial u}{\partial t} = \sigma \left(\frac{\partial^2 u}{\partial x^2} + \frac{\partial^2 u}{\partial y^2} \right) \tag{3.4.22}$$

The asymptotic long-time solution of (3.4.22) is the solution to the Laplace equation. Equation (3.4.22) is integrated numerically by separating it into two fractional steps (similar to the time scheme (k) in Table 3.2.1)

$$\left. \begin{array}{l} \dfrac{u^* - u^n}{\Delta t} = \sigma \delta_x{}^2 u^* \\[2mm] \dfrac{u^{n+1} - u^*}{\Delta t} = \sigma \delta_y{}^2 u^{n+1} \end{array} \right\} \tag{3.4.23}$$

Since each fractional step is implicit, large time steps can be used. And since the solution of each fractional step involves only inverting tridiagonal matrices, it can be performed very efficiently (see, e.g., Hageman and Young (1981)).

Multigrid Methods

The speed of convergence for iterative schemes depends on the number of grid points and is much faster for coarser grids (see expression for λ_1 above). Moreover, the errors take longest to converge correspond to long waves (i.e., they are smooth), whereas the shortest waves are damped fastest. Multigrid methods take advantage of this and use both coarse and fine grids (see Briggs (1987), Hackbusch (1985), Barrett et al. (1994)). The procedure is as follows: Several steps of a basic method on the full grid are performed in order to smooth out the error (pre-smoothing). A coarse grid is selected from a subset of the grid points, and the iterative method is used to solve the problem on this coarse grid. The coarse grid solution is then interpolated back to the original grid, and the original method applied again for a few iterations (post-smoothing). In carrying out the solution in the second step, the method can be applied recursively to coarser grids, until the number of grid points is small enough that a direct solution can be obtained.

The method of descending through a sequence of coarser grids and then ascending back to the full grid is known as a V-cycle. A W-cycle results from visiting the coarse grid twice, with some smoothing steps in between. Some multigrid methods have an (almost) optimal number of operations, i.e., almost proportional to the number of variables.

Krylov Subspace Methods

There are a number of iterative algorithms for solving the linear problem of (3.4.7) in the Krylov subspace, defined by

$$K_m(\mathbf{A}, r_0) = \mathrm{span} \left\{ r_0, \mathbf{A} r_0, \mathbf{A}^2 r_0, \ldots, \mathbf{A}^{m-1} r_0 \right\} \tag{3.4.24}$$

where $r_0 = F - \mathbf{A}\phi_0$ is the *residual* for an arbitrary initial error ϕ_0. The approximate solution ϕ_m lies in the space $\phi_0 + K_m(\mathbf{A}, r_0)$. The residual after m steps has to satisfy certain conditions, and the choice of the condition gives rise to different types of iterative methods (e.g., Sameh and Sarin, 1999). The requirement that the residual be orthogonal to the Krylov subspace, $F - \mathbf{A}\phi_m \perp K_m(\mathbf{A}, r_0)$ leads to the *conjugate gradient* and the *Lanczos* methods. Methods like GMRES, MINRES and ORTHODIR are obtained by requiring that the residual be minimized over the Krylov subspace. The bi-conjugate gradient and QMR methods are derived requiring the residual to be orthogonal to $K_m(\mathbf{A}^T, r_0)$. The discussion of these methods applicable to nonsymmetric systems is beyond the scope of this book but given in the texts referred to Section 3.4.1.

3.5 Lateral Boundary Conditions for Regional Models

3.5.1 Introduction

The use of regional models for weather prediction has arisen from the desire to reduce the model errors through an increase in horizontal resolution that cannot be afforded in a global model. Operational regional models have been embedded or "nested" into coarser resolution hemispheric or global models since the 1970s. In the USA, the first regional model was the LFM model (Chapter 1). The nesting of regional models requires the use of updated lateral boundary conditions obtained from the global model.

We have seen that for pure hyperbolic equations, there should be as many boundary conditions imposed at a given boundary as the number of characteristics moving *into* the domain. Parabolic equations with second order diffusion require one boundary condition at every point in the boundary for each prognostic equation. Second order elliptic equations (such as Laplace, Poisson, and Helmholtz equations) also require one boundary condition. The first forecast experiment of Charney et al. (1950) used the barotropic vorticity equation (conservation of absolute vorticity) and already had to deal with boundary conditions. They solved the hyperbolic equation $\partial \zeta / \partial t = -\mathbf{v} \cdot \nabla(\zeta + f)$ followed by the Poisson (elliptic) equation $\nabla^2 \partial \Psi / \partial t = \partial \zeta / \partial t$. Therefore, Charney et al. (1950) had to impose a boundary condition on the streamfunction at all the boundary points (needed to solve the Poisson equation) and a boundary condition for the vorticity at the inflow points. They used persistence in both cases: For the elliptic equation, they used as boundary condition $\partial \Psi / \partial t = 0$ (i.e., the normal wind remains constant), and then specified that the vorticity also remained constant ($\partial \nabla^2 \Psi / \partial t = 0$) at the inflow points and extrapolated the vorticity using upstream differences at the outflow points.

For the SWEs, there are three characteristics, one corresponding to a geostrophic solution, moving with the speed of the flow U, and the other two corresponding to inertia-gravity waves, moving with speed $U \pm \sqrt{f^2 k^2 + \Phi}$. At the boundaries, if the speed of inertia-gravity waves is larger than U and the flow is inward, we have to specify two boundary conditions. If the flow is outward, we have to specify one

3.5 Lateral Boundary Conditions for Regional Models 127

boundary condition (corresponding to the inertia-gravity waves moving in). If U is greater than the speed of the inertia-gravity waves, we have to specify all three boundary conditions at the inflow points and none at the outflow points. For parabolic equations (with horizontal diffusion), each predicted variable has to be specified as well at all lateral boundaries.

Oliger and Sundström (1978) showed that the hydrostatic primitive equations are not purely hyperbolic (because of the loss of the time derivative of the vertical velocity), and that they do not have a well-posed set of boundary conditions. In an excellent review of the lateral boundary condition used in operational regional NWP models, McDonald (1997) pointed out that with the presence of horizontal diffusion in models, there is a feeling that we can "over-specify slightly the lateral boundary conditions and not do very much damage."

In practice, boundary conditions are chosen pragmatically and tested numerically to check their appropriateness. Several methods have been tried over the years, but the most widely used is the boundary relaxation scheme introduced by Davies (1976). Davies (1983) has a very illuminating analysis of the impact of the different types of boundary conditions and their generation of spurious reflection using simple examples of wave equations and SWEs. He points out that an overspecifying boundary condition scheme is satisfactory if: (a) It transmits incoming waves from the "host" model providing boundary information without appreciable change of phase or amplitude, and (b) at the outflow boundaries, reflected waves do not reenter the domain of interest with appreciable amplitude. We follow the Davies (1983) analysis and the review by McDonald (1997) in the rest of this section. Durran (1999), Chapter 8, is also devoted to this subject.

3.5.2 Lateral Boundary Conditions for One-Way Nested Models

The majority of regional models have "one-way" lateral boundary conditions, i.e., the host model, with coarser resolution, provides information about the boundary values to the nested regional model, but it is not affected by the regional model solution. This approach has some advantages: (a) It allows for independent development of the regional model, and (b) the host model can be run for long integrations without being "tainted" by problems associated with nonuniform resolution or from the regional model. Overall, the regional one-way nesting can be considered to have been successful, in the sense that the boundary information from the host model is able to penetrate the regional model, and the regional model solution is able to leave the domain without appreciable deterioration of the solutions. The success can also be measured by the fact that there have been several attempts to perform long-term integrations of nested regional models. In these long-term integrations, the initial regional information is swept out of the domain in the first day or two, and all the additional information comes from the global model integration. This approach is denoted "regional climate modeling." Takle et al. (1999) discuss the Project to Intercompare Regional Climate Systems (PIRCS). In these extended integrations, the regional model acts as a "magnifying glass" for the global solution, allowing the large-scale flow to interact with

Numerical Discretization of the Equations of Motion

smaller scale forcing such as orography, variations in soil moisture and land–sea contrast, and as a result tend to give a more realistic solution. The "added value" over the global solutions empirically indicates the overall success of the one-way boundary conditions used in different models.

There are four types of "pragmatic" boundary conditions that have been formulated for one-way lateral boundary conditions:

Pseudo-radiation Boundary Conditions
Orlanski (1976) proposed a finite difference approximation of the "radiation condition," i.e., specifying well-posed boundary conditions for pure hyperbolic equations. One assumes that the prognostic equations locally satisfy $\partial u/\partial t + c\partial u/\partial x = 0$ and then estimates the phase speed c using a finite difference equivalent of

$$c = -\frac{\partial u}{\partial t} \bigg/ \frac{\partial u}{\partial x} \tag{3.5.1}$$

at the points immediately inside the boundary (denoted by $b - 1$). Miller and Thorpe (1981) used first order upstream approximation

$$c' = -\frac{u_{b-1}^n - u_{b-1}^{n-1}}{\Delta t} \bigg/ \frac{u_{b-1}^{n-1} - u_{b-2}^{n-1}}{\Delta x} \tag{3.5.2}$$

as well as higher-order approximations. After estimating c', if it points into the domain, u_b^{n+1} is specified. If it points out, the upstream scheme is used: $u_b^{n+1} = u_b^n - c'\Delta t/\Delta x(u_b^n - u_{b-1}^n)$. If $c'\Delta t/\Delta x > 1$ because the space derivative of u is small, Orlanski (1976) suggested limiting the value of c' to $c' = \Delta x/\Delta t$. Klemp and Lilly (1978) pointed out reasons why the approximate "radiation schemes" are not completely successful in avoiding spurious reflection: there can be overspecification at the boundaries, specification of the right number of boundary conditions but not their correct values, and errors in the estimation of c'. The radiation condition has been used for research models (e.g., Durran et al., 1993). Klemp and Durran (1983) and Bougeault (1983) used radiation boundary conditions at the top of the model. Operational models generally do not use radiation boundary conditions and instead impose the condition that the vertical velocity be zero at the top (e.g., $\dot\sigma = 0$ at $\sigma = \sigma_T$ for sigma coordinates). As a result, the presence of this artificial "rigid top" leads to spurious wave reflections and even generates instabilities near the top, analogous to baroclinic waves, that remain attached to the top (e.g., Kalnay and Toth, 1996; Hartmann et al., 1995).

Diffusive Damping in a Boundary Zone or "Sponge Layer" (Burridge, 1975; Mesinger, 1977)
In this method, the global (or host) model boundary conditions are specified for all variables, and horizontal diffusion is added over a boundary zone to dissipate the noisy waves generated by the boundary conditions:

$$\frac{\partial u}{\partial t} + c\frac{\partial u}{\partial x} = \left(\frac{\partial}{\partial x} v \frac{\partial u}{\partial x}\right)_{BZ} \tag{3.5.3}$$

3.5 Lateral Boundary Conditions for Regional Models 129

This would seem to be a natural choice for regional model boundary conditions since by increasing the order of the equation to make it parabolic within a limited boundary zone, it is possible to specify all variables at the boundary without overspecifying. However, this approach also has clear disadvantages: it damps the incoming waves from the global model (unless they are long compared to the width of the damping zone). It also produces spurious reflections of outgoing waves if v increases abruptly, and if it increases slowly, it may not be enough to damp the reflected waves. As a result, this method is not very much in use at this time.

Tendency Modification Scheme (Perkey and Kreitzberg, 1976)
The wave equation is replaced by

$$\frac{\partial u}{\partial t} + c\frac{\partial u}{\partial x} = -\gamma\frac{\partial(u - \overline{u})}{\partial t} \tag{3.5.4}$$

where $\overline{u}$ is prescribed from the host model (which is assumed to be correct near the boundary), and γ is zero in the interior and increases to large values at the boundaries. Since the host model follows the wave equation

$$\frac{\partial \overline{u}}{\partial t} + c\frac{\partial \overline{u}}{\partial x} = 0 \tag{3.5.5}$$

we can write an "error" equation for the difference u' between the regional and the host model:

$$\frac{\partial u'}{\partial t} + c^*\frac{\partial u'}{\partial x} = 0 \tag{3.5.6}$$

where $c^* = c/(1+\gamma)$. Therefore the time tendency scheme advects the error and slows it down to almost zero at the boundaries, thus avoiding overspecification. In practice, this scheme is also found to produce spurious reflections.

Flow Relaxation Scheme (Davies, 1976, 1983)
As indicated before, this is the most widely used scheme. The forecast equations are modified by adding a Newtonian relaxation term over a boundary zone:

$$\frac{\partial u}{\partial t} + c\frac{\partial u}{\partial x} = -K(u - \overline{u}) \tag{3.5.7}$$

The "error" equation is now

$$\frac{\partial u'}{\partial t} + c\frac{\partial u'}{\partial x} = -Ku' \tag{3.5.8}$$

indicating that the error is advected to or from the boundary and damped. At the inflow boundaries only the differences between the regional and the host model are damped. Therefore, this scheme mitigates the effects of overspecification at the outflow boundaries without introducing deleterious effects in the inflow boundaries.

If K increases abruptly, it can also introduce some spurious reflection. For this reason, Kallberg (1977) proposed the use of a smoothly growing function for K. Let's consider a complete prognostic equation for the regional model near the boundaries

Numerical Discretization of the Equations of Motion

$$\frac{\partial u}{\partial t} = F - K(u - \overline{u}) \tag{3.5.9}$$

In (3.5.9), F includes all the regular "forcing terms" in the interior time derivative (e.g., advection, sources/sinks, etc.). We can discretize it in time, using, for example, the leapfrog scheme for the regular terms and backward implicit scheme for the boundary relaxation term, as

$$\frac{u^{n+1} - u^{n-1}}{2\Delta t} = F^n - K(u_i^{n+1} - \overline{u}^{n+1}) \tag{3.5.10}$$

Here the overbar represents the host model, u^{n+1} is the updated regional model, and the subscript i indicates the regional model (internal) solution obtained *before* relaxing toward the host model values $\overline{u}^{n+1}$:

$$u_i^{n+1} = u^{n-1} + 2\Delta t F^n \tag{3.5.11}$$

From (3.5.10) and (3.5.11) we can now write

$$u^{n+1} = u_i^{n+1} - K2\Delta t u_i^{n+1} + K2\Delta t \overline{u}^{n+1} = (1 - \alpha)u_i^{n+1} + \alpha \overline{u}^{n+1} \tag{3.5.12}$$

Here $\alpha = 2\Delta t K$ varies from 0 in the interior ($K = 0$), to 1 at the boundary, where the regional model solution is specified to coincide with the host model solution. McDonald (1997) mentioned three functions that have been proposed for $\alpha(j)$, where we define $j = 0, \alpha(0) = 1$ at the boundary, and assume that the boundary zone has n points so that for $j \geq n, \alpha(j) = 0$. The first function, found to be optimal in minimizing false reflection of both Rossby and gravity waves by Kallberg (1977), starts gently in the interior and has the steepest slope at the boundary: $\alpha = 1 - \tanh(j/2)$. Jones et al. (1995) used a linear profile $\alpha = 1 - j/n$, and McDonald and Haugen (1992) proposed a cosine profile $\alpha = [1 + \cos(j\pi/n)]/2$, which has the steepest slope at the center of the boundary zone. Benoit et al. (1997) in the MC2 model used $\alpha(j) = \cos^2(j\pi/2n)$ and reported good results.

3.5.3 Other Examples of Lateral Boundary Conditions

Tatsumi (1983, 1986) suggested adding an "error diffusion" at the boundaries as well, which can also help to reduce the boundary errors without affecting the incoming wave:

$$\frac{\partial u}{\partial t} + c\frac{\partial u}{\partial x} = -K(u - \overline{u}) + \frac{\partial}{\partial x}\left(v\frac{\partial u - \overline{u}}{\partial x}\right) \tag{3.5.13}$$

This is used in the regional spectral model of the Japan Weather Service (Tatsumi, 1986).

Juang and Kanamitsu (1994) and Juang et al. (1997) also developed a RSM nested in the NCEP global spectral model, but they cast it as *a perturbation* model, so that the full RSM solution includes the global model solutions plus the regional perturbations. They use an "implicit" variation of the tendency modification approach with

$$\frac{\partial u}{\partial t} = F - \mu(u^{n+1} - \overline{u}^{n+1}) \tag{3.5.14}$$

3.5 Lateral Boundary Conditions for Regional Models 131

where $\mu = \alpha/T, T$ is an e-folding time (3 hr), and

$$u^{n+1} = u^{n-1} + 2\Delta t \frac{\partial u}{\partial t} \quad \overline{u}^{n+1} = \overline{u}^{n-1} + 2\Delta t \frac{\partial \overline{u}}{\partial t} \tag{3.5.15}$$

so that for the perturbation $u' = u - \overline{u}$, the implicit relaxation is given by

$$\frac{\partial u'}{\partial t} = \frac{F - \dfrac{\partial \overline{u}}{\partial t} - \mu u'^{n-1}}{1 + 2\mu\Delta t} \tag{3.5.16}$$

They found that the orography of the regional model also has to be blended with the global orography in the boundary zone in order to avoid spurious noise.

The Eta model at NCEP (Mesinger et al., 1988; Janjić, 1994; Black, 1994) uses an "almost well-posed" approach. It uses boundary values from the NCEP global model only at the outermost row. When the flow is inwards, all the prognostic variables are prescribed from the global model. At the outflow points, the tangential velocities are extrapolated from the interior of the integration domain. The variables in the second row are a blend from the outermost and the third row. The "interior" is defined as the third row inwards, but the Eta model uses an upstream advection scheme for the five outer rows of the domain in order to minimize possible reflections at the boundary.

3.5.4 Two-Way Interactive Boundary Conditions

Finally, we note that some regional models have been developed using two-way interaction in the boundary conditions, i.e., the (presumably more accurate) regional solution, in turn, also affects the global solution. Although in principle, this would seem a more accurate approach than the one-way boundary condition, care has to be taken that the high-resolution information does not become distorted in the coarser resolution regions, which can result in worse overall results, especially at longer time scales. There are basically two types of two-way boundary condition approaches.

The first approach corresponds to a truly nested model, with abrupt changes in the resolution, but with the inner or nested solution also used to modify the global or outer model solution. The first operational example of this type of two-way interaction was the Nested Grid Model (NGM) developed by Phillips (1979). Zhang et al. (1986) implemented two-way boundary conditions for the nesting in the MM5 model. See also Kurihara and Bender (1980), and Skamrock (1989).

The second approach is simpler, and it involves the use of continuously stretched horizontal coordinates so that only the region of interest is solved with high resolution. It is evident that with this approach, the equations in the regional high-resolution areas do not require special boundary conditions, and that they do influence the solutions in the regions more coarsely resolved, so that they can be considered as two-way interactive nesting. There have been a few methods used to obtain regional high resolution using stretched global coordinates:

(a) Uniform latitude–longitude stretching (Staniforth and Daley, 1977; Benoit et al., 1989; Fox-Rabinovitz et al., 1997). This method was used by the Canadian regional operational system.

(b) Stretched spherical harmonics (Courtier and Geleyn, 1988). This method is used in the French regional operational system.

(c) A regular volume (such as a cube or an icosahedron) projected on the sphere and then stretched (Rančić et al., 1996; Taylor et al., 1997; Putman and Lin, 2007; Satoh et al., 2014). This approach was explored somewhat unsuccessfully during the 1960s (Williamson, 1968; Sadourny et al., 1968), and is now again in vogue. The use of a regular volume to generate the grid avoids the pole problem of the convergence of the meridians in the latitude-longitude grid.

3.6 Nonhydrostatic Models

We saw in Section 2.4 that the hydrostatic approximation filters out vertically propagating sound waves, but not horizontally propagating sound waves (Lamb waves). Moreover, the hydrostatic approximation is accurate if $H/L \ll 1$, where H and L are the vertical and horizontal scales of motion. For quasi-geostrophic flow the hydrostatic approximation is accurate even for $H/L \leq 1$, since in quasi-geostrophic flow the vertical velocity is constrained to be small, $W/H \ll U/L$. As computer power increases and the model resolution increases, it becomes possible to resolve cumulus convection explicitly, rather than by parameterization. This has the potential of significantly improving the quality of the forecast, but the hydrostatic approximation cannot be applied anymore.

For non-hydrostatic models, the new challenge is to develop an approach that is computationally stable for both the vertical and the horizontal propagation of sound waves without having to use unfeasibly small time steps. Since the vertical grid sizes are much smaller than horizontal grid sizes, the CFL criterion is much more stringent for the vertically propagating sound waves ($c_s \Delta t / \Delta z < 1$) than for the horizontally propagating sound (Lamb) waves ($c_s \Delta t / \Delta x < 1$).

In hydrostatic models horizontal sound waves (and internal gravity waves) can be handled semi-implicitly (Section 3.2.5), but in non-hydrostatic models the treatments of both vertically and horizontally propagating sound waves are critical, so that they form the basis of Saito et al. (2007)'s classification of non-hydrostatic models, reproduced below in Table 3.6.1.

Klemp and Wilhelmson (1978), in a non-hydrostatic, quasi-compressible cloud model, proposed to solve the terms associated with just the vertical sound waves semi-implicitly, and to integrate horizontal sound waves explicitly but with smaller time steps than those used for the meteorologically meaningful processes. This approach (similar to scheme (k) in Table 3.2.1) is known as *split explicit* (for the horizontal waves). Saito et al. (2007) in an excellent summary of the methods used at that time in non-hydrostatic models, refer to the Klemp and Wilhelmson (1978) split explicit approach to integrating sound waves without violating the CFL condition as *"Horizontally Explicit-Vertically Implicit"* (HE-VI). Since only the vertical sound waves are integrated implicitly, the necessary elliptic equation that has to be solved for the pressure is a one-dimensional Helmholtz equation, which can be computed efficiently.

3.6 Nonhydrostatic Models   133

Table 3.6.1 Some of the non-hydrostatic, fully compressible models, and their approach to the horizontal and vertical sound waves. Adapted from Saito et al. (2007). The original version was published in Journal of Meteorological Society of Japan

Model	Country/Center	Method	References
UM	UK	SI-SL	Cullen (1993), Staniforth et al. (2002), Davies et al. (2005)
JMA-NHM	Japan (JMA, MRI)	HE-VI (HI-VI option)	Ikawa (1991), Saito et al. (2001, 2006)
WRF-NMM	USA (NCEP)	HE-VI	Janjic et al. (2001); Janjic (2003)
WRF-ARW	USA (NCAR)	HE-VI	Skamarock et al. (2005), Klemp et al. (2007), Skamarock and Klemp (2008)
RSM	USA (NCEP)	HI-VI (Spectral)	Juang (2000)
GEM	Canada (MSC)	SI-SL	Côté et al. (1998a,b), Yeh et al. (2002)
ALADIN-NH	France etc.	HI-VI	Bubnová et al. (1995)
AROME	France (Meteo-France)	SI-SL	Bouttier (2003), Bénard et al. (2005)
NH HIRLAM	Denmark etc.	HE-VI or SI-SL	Männik and Room (2001), Room et al. (2006)

Alternatively, it is possible to write the terms associated with both horizontally and vertically propagating sound waves implicitly, with other slower frequency terms (like advection) written explicitly. This approach (usually known as semi-implicit) is referred to by Saito et al. (2007) as *"Horizontally Implicit-Vertically Implicit"* (HI-VI). Because the semi-implicit scheme is 3-dimensional, each time step requires solving a 3-dimensional Helmholtz pressure equation.

The non-hydrostatic models can further relax the CFL condition on the time step by including, in addition to the (3-dimensional) semi-implicit sound waves approach (HI-VI), the use a semi-Lagrangian for the advection terms, which eliminates the constraint associated with the *wind* Courant number (Section 3.3.3). This approach has been adopted by a number of models and is known as *"Semi-Implicit, Semi-Lagrangian"* (SI-SL).

Finally, let's recall that it is possible to filter both horizontal and vertical sound waves by applying the incompressible or Boussinesq approximation, which requires that the 3-dimensional mass divergence $\frac{d\rho}{dt} = -\nabla \cdot \mathbf{v}_3 = 0$ be zero, and hence filters out all sound waves. This approximation is reasonable to filter sound waves in the ocean, since water is essentially incompressible, but not for the atmosphere. A less stringent approximation to filter the sound waves is the anelastic or quasi-Boussinesq approximation (Ogura and Phillips, 1962), which separates the solution into a reference state that is hydrostatic and a perturbation state, and requires that the 3-dimensional mass divergence be zero: $\nabla \cdot (\overline{\rho} \mathbf{v}_3) = 0$. This approximation filters all sound waves (Section 2.4.2) but is not accurate enough for deep models. As a result, anelastic models (e.g., Sommeria, 1976. Soong and Ogura, 1980) are not widely used. Another approximation, the *quasi-compressible* approximation, allows for air compressibility and therefore, includes sound waves, but uses the reference density in the momentum equations. This type of approximation has been used in Klemp and Wilhelmson

Numerical Discretization of the Equations of Motion

(1978), Xue et al. (1995, 2003), Pielke et al. (1992), but it is not as accurate as the fully compressible models that do not linearize with respect to the reference atmosphere. The first non-hydrostatic *fully compressible model* was implemented at the UKMO (Tapp and White, 1976). Now, the fully compressible models constitute the majority of operational and research regional non-hydrostatic models, and it is starting to be used in the next generation global models.

For hydrostatic models, the natural vertical coordinate is (hydrostatic) pressure as introduced by Eliassen (1949), and essentially all hydrostatic models have used either pressure or coordinates related to pressure. Examples are the *sigma* coordinates (Phillips, 1957) that elegantly deals with surface topography, *isentropic* coordinates (Eliassen and Raustein, 1968, Section 2.6.4), that are very accurate by making the flow essentially horizontal, since the vertical velocity $\dot{\theta} \approx 0$, and *hybrids* that are linear combinations of terrain following coordinates near the surface, and pressure or isentropic coordinates in the free atmosphere. Hybrid vertical coordinates are terrain following near the surface and become closer to pressure coordinates at higher levels.

For non-hydrostatic models the natural vertical coordinates is height, and several models have been based on height (z). However, Laprise (1992) developed a vertical coordinate system based on *hydrostatic pressure* π (i.e., mass-based) that has become popular. It is formally similar to the hydrostatic pressure coordinates and it allows for a simple transition from hydrostatic to non-hydrostatic, and vice-versa. The non-hydrostatic prognostic equations using hydrostatic pressure vertical coordinates are:

$$\frac{d}{dt}\mathbf{v} + \alpha \nabla_\pi p + \frac{\partial p}{\partial \pi} \nabla_\pi \phi = \mathbf{F} \tag{3.6.1a}$$

$$\gamma \frac{dw}{dt} + g\left(1 - \frac{\partial p}{\partial \pi}\right) = \gamma F_z, \quad \gamma = \begin{cases} 0, \text{hydrostatic} \\ 1, \text{non-hydrostatic} \end{cases} \tag{3.6.1b}$$

$$\frac{dT}{dt} - \frac{\alpha}{C_p}\frac{dp}{dt} = \frac{Q}{C_p} \tag{3.6.1c}$$

$$\frac{dp}{dt} + \frac{C_p p}{C_v} D_3 = \frac{Qp}{C_v T} \tag{3.6.1d}$$

(one prognostic equation for each of the dependent variables $(\mathbf{v}, w, T, p)$, so that the equations (3.6.1) form a complete set) with additional diagnostic relations and operators defined as

$$\nabla_\pi \cdot \mathbf{v} + \frac{\partial \dot{\pi}}{\partial \pi} = 0, \quad \dot{\pi} \equiv \frac{d\pi}{dt} \tag{3.6.2a}$$

$$\alpha = \frac{RT}{p} \tag{3.6.2b}$$

$$\phi = \phi_0 - \int_{\pi_0}^{\pi} \alpha \, d\pi' \tag{3.6.2c}$$

$$(\cdot) \equiv \frac{d}{dt} \equiv \left(\frac{\partial}{\partial t}\right)_\pi + \mathbf{v} \cdot \nabla_\pi + \dot{\pi}\frac{\partial}{\partial \pi} \tag{3.6.2d}$$

$$D_3 \equiv \nabla_\pi \cdot \mathbf{v} + \rho(\nabla_\pi \phi) \cdot \left(\frac{\partial \mathbf{v}}{\partial \pi}\right) - \rho g \frac{\partial w}{\partial \pi} \tag{3.6.2e}$$

As an example, the NCEP version of the Weather Research and Forecasting model, known as the Nonhydrostatic Mesoscale Model (WRF-NMM, Janjic et al., 2001; Janjic, 2003) is based on hydrostatic pressure, hybrid terrain following vertical coordinates (Simmons and Burridge, 1981), explicit horizontal advection (modified Adams-Bashford for u, v, T and Coriolis force), and an efficient Forward-Backward scheme (Gadd, 1978) for fast waves. In the vertical u, v, T are advected with Crank-Nicholson, and sound waves are solved implicitly.

3.7 Need to Replace Spectral Models: Experiments to Choose the Next Generation Global Model at NCEP

Spectral models have been extremely effective for operational global forecasting, with many advantages over finite differences, including uniform resolution on the sphere (no pole singularities), and especially very high accuracy due to the analytical (exact) calculation of the space derivatives from the spherical harmonics.

However, terms like advection terms, that contain products of variables, cannot be computed efficiently in spectral space, so spectral models need to compute them in physical (grid point) space. The physical parameterizations are also computed in physical space. So, at every time step, both the physical and spectral representations of the state of the model are needed, and a fast transformation between them is necessary. The transformation in longitude is just a Fourier transform, and so it can be efficiently done with the Fast Fourier Transform algorithm, which reduces the cost from $O(m^2)$ to $O(m\log(m))$, with m being the maximum zonal wave number. This is a very large speedup. Unfortunately, in the meridional direction, there is no equivalent of a Fast Transform for the associated Legendre polynomials and their cost increases with n^2, the square of the total wavenumber. There have been moderately successful attempts to solve this problem by using Fourier Transforms in both latitude and longitude, but they have not been operationally implemented.

As a result, the computational cost of spectral models increases fast with resolution, and cannot be used beyond the hydrostatic range. For example, the last global spectral model implemented at NCEP in 2015, with semi-Lagrangian dynamics, has a resolution of 13km, near the limit of the hydrostatic range, as well as the limit for an effective computational use of spectral dynamics. It is also in the "grey zone," between about 5 and 15 km, where the resolution not high enough to resolve the cumulus convection explicitly, and not low enough to use cumulus parameterizations (Chapter 4). In addition, the next generation models will have to scale well in supercomputers with over 100K processors, and spectral, non-hydrostatic models do not scale well at this range. In order to address this major change, NOAA/NCEP has decided to replace the Global Forecasting System by a "Next Generation Global Prediction System (NGGPS)," and for this purpose organized an intercomparison of a number of promising, advanced non-hydrostatic global models with different types of dynamical cores. We discuss this intercomparison here because it shows the kind of models that, at the time of this writing, are being considered for future high-resolution global models. The last two

Table 3.7.1 Non-hydrostatic global models that participated in the NGGPS intercomparison

Model	Organization	Numerical method	Grid
NIM (Non-hydrostatic Icosahedral Model)	NOAA/ESRL	Finite Volume	Icosahedral
MPAS (Model for Prediction Across Scales)	NCAR	Finite Volume	Icosahedral/ Unstructured
NEPTUNE (Navy Environmental Prediction sysTem Using the NUMA corE)	Navy/NRL	Spectral Element	Cubed-Sphere with Adaptive Mesh Refinement (AMR)
FV3 (Finite Volume Cubed Sphere)	NOAA/GFDL	Finite Volume	Cubed-Sphere, nested
NMM-UJ (Non-Hydrostatic Multiscale Model, Uniform Jacobian)	NOAA/EMC	Finite Differences	Cubed-Sphere
GFS-Hydrostatic	NOAA/EMC	Semi-Lagrangian/ Spectral	Reduced Gaussian
IFS(RAPS13) Hydrostatic	ECMWF	Semi-Lagrangian/ Spectral	Reduced Gaussian

models (GFS and IFS) are the current hydrostatic spectral operational models at NCEP and ECMWF.

The new types of models included in this intercomparison were NIM (NOAA/ ESRL), MPAS (NCAR), NEPTUNE (Navy/NRL), FV3 (NOAA/GFDL), NMM-UJ (NOAA/NCEP), and the hydrostatic operational models GFS (NOAA/NCEP), IFS (ECMWF) used for benchmarking. There were two phases in the intercomparison. In Phase 1, the ability of the models to integrate for 2 hr within 21 sec at a nominal resolution of 13 km was tested. The NMM-UJ was the fastest, followed by the FV3 model running single-precision. NIM and the double-precision FV3 were tied. MPAS and NEPTUNE required more cores to meet the operational time requirement. There were a number of comparisons testing different properties of the models. In these tests, the FV3 and MPAS produced the most realistic solutions, much more similar to each other than to the rest of the models. It was decided that MPAS and FV3 were ready to advance to Phase 2.

The full forecast experiment comparisons (with GFS initial conditions and physics) showed FV3 with similar skill as the GFS, and MPAS with some instability problems that resulted into significantly lower anomaly correlation. In the high resolution, real data runs, MPAS used a non-uniform mesh, and FV3 used a combination of a global stretched grid and a nest. The effective resolution, measured by the KE spectrum, indicated that for total wavenumbers larger than about 80, the MPAS and the FV3 were similar and had much better resolution than the GFS. In a comparison of Data Assimilation Cycling, using the same initial conditions as the GFS, the FV3 was slightly better than the GFS, and the MPAS was significantly worse. Overall, at this point in time, the assessment of the risk associated with the FV3 was considered to be low, and for the MPAS it was considered to be higher. The FV3 was found to meet all the technical needs, and to be less expensive to implement, with higher readiness for implementation, and with significantly better technical and computational performance. So the recommendation was to select the FV3 and proceed to NGGPS Phase 3 dynamic core integration and implementation.

The finite-volume, cubed-sphere (FV3) model, developed at NASA Goddard, and later at GFDL by S. J. Lin and collaborators, is a combination of remarkable advances in global modeling. It includes monotonicity preserving finite-volume schemes (Lin and Rood, 1996), a shallow-water formulation without false vorticity generation (Lin and Rood, 1997), a formulation of the pressure gradient that ensures Newton's action and reaction principle is satisfied between cells (Lin, 1997), and a vertically Lagrangian scheme (Lin, 2004). It is documented at the GFDL web page FV3: Finite Volume Cubed-Sphere Dynamical Core (www.gfdl.noaa.gov/fv3/).

3.8 How to Validate NWP Models That Are Based on Machine Learning and Artificial Intelligence?

In the last decade or two, applications of machine learning (ML) and artificial intelligence (AI) have exploded. Although the centers for modeling and data assimilation of the Earth system models were not leaders in this field, their activities have now rapidly increased, as shown by two papers just published in the Bulletin of the American Meteorological Society. The first paper summarizes the workshop on "Leveraging AI in the Exploitation of Satellite Earth Observations and Numerical Weather Prediction" held by the National Oceanic and Atmospheric Administration (NOAA) in April 2019 (Boukabara et al., 2021). The second paper Bonavita et al. (2021), discusses the First ECMWF-ESA Workshop on Machine Learning for Earth System Observation and Prediction that took place online at ECMWF on October 5–8, 2020. Both meetings attracted a very large (more than 400 each) number of participants. The presentations and discussions were recorded at ECMWF (https://events.ecmwf.int/event/172/). These two papers are a good introduction and overview to several current and planned applications of AI, ML and deep learning (DL) to Earth system modeling and data assimilation.

For the NWP model development, different ML/AI or hybrid methods have been explored to accelerate the calculation of the dynamic core (Lu et al., 2018; Arcomano et al., 2020; Kochkov et al., 2021) and physical parameterizations (Krasnopolsky, 2013). One inevitable and interesting question is about how should we validate those ML/AI-based prediction models? Besides the usual root-mean-square-error calculation with some reference data, it is necessary to apply the same dynamic-core tests and climatology checks to those models, as we did for the traditional numerical models. Perhaps it is even more crucial to ensure those ML/AI-based models have similar error growth rates as the traditional NWP models, which describe the more fundamental behavior of the atmosphere as a dynamical system. Atmospheric models exhibit several types of instabilities, each of which has a unique error growth rate. A valid ML/AI-based model should be able to capture the error growth across all scales. Chapter 6 introduces several tools to quantify these growth errors. For example, bred vectors (Section 6.5.1) with different scaling can be used to capture different types of instabilities for the atmospheric model (Norwood et al., 2013). It is essential to verify those bred vectors generated by the ML/AI-based models against those from the traditional NWP models during the prediction phase, or further incorporate the information of these bred vectors during the training phase for those ML/AI-based models.

4 Introduction to the Parameterization of Subgrid-Scale Physical Processes

4.1 Introduction

In Chapter 2, we derived the equations that govern the evolution of the atmosphere, and in Chapter 3, we discussed the numerical discretizations that allow the numerical integration of those equations on a computer.* The discretization of the continuous governing equation is limited by the model resolution, i.e., by the size of the smallest resolvable scale. We have seen that in a finite difference scheme, the smallest scales of motion that can be (poorly) resolved are those which have a wavelength of two grid sizes. In spectral models, the motion of the smallest wave present in the solution is more accurately computed, but for these and for any type of numerical discretization, there is always a minimum resolvable scale. At the time of this writing ($\sim$2015), *climate models* typically have a horizontal resolution of the order of 50–100 km, *global weather forecast models* have resolutions of 10–50 km, and regional *mesoscale models* of 1–10 km. *Storm-scale models* have even higher resolution, with grid sizes of the order of 0.1–5 km. In the vertical direction, model resolution and vertical extent have also been increased substantially, with current models having typically between 20 and 100 vertical levels, and extending from the surface to the stratosphere or even the mesosphere. As computer power continues to increase, so does the resolution of atmospheric models.

Despite the continued increase of horizontal and vertical resolution, it is obvious that there are many important processes and scales of motion in the atmosphere that cannot be explicitly resolved with present or future models. They include turbulent motions with scales ranging from a few centimeters to the size of the model grid, as well as processes that occur at a molecular scale, like condensation, evaporation, friction, and radiation. We refer to all the processes that cannot be resolved explicitly as "subgrid-scale processes." An example of an important process that takes place at a subgrid scale is the turbulent mixing in the planetary boundary layer. During the daytime, the solar heating at the Earth's surface not only warms the soil but also causes the plants to transpire and soil moisture to evaporate, thus transporting moisture into the atmosphere. Surface heating leads to turbulent motion that is on the scale of a few meters to a few hundred meters. Models with a horizontal grid size of 10–100 km cannot resolve these motions. Yet the transport of the heat and moisture into the boundary layer is very crucial to the development of afternoon thunderstorms and a host of other phenomena that are important to the resolvable atmospheric fields.

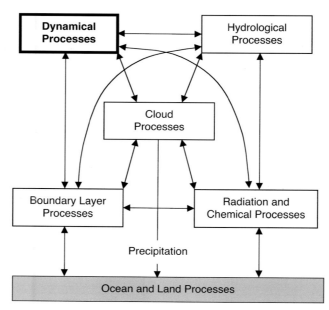

Figure 4.1.1 Physical processes in the atmosphere and their interactions. The dynamical processes for resolvable scales, in bold, are explicitly computed by the model "dynamics" (discussed in Chapters 2 and 3). The other subgrid-scale processes are parameterized in terms of the resolved-scale fields. Adapted from Arakawa (1997, 2004). ©American Meteorological Society. Used with permission.

Another notable example is tropical cumulus convection. The cumulus clouds in the tropics are known to be extremely important to the global energy balance, yet each cloud typically occupies only a few kilometers of space horizontally and vertically.

Although these processes occur at small scales, they depend on, and in turn, affect the larger-scale fields and processes that are explicitly resolved by a numerical model. For example, condensation of water vapor on a subgrid scale occurs if the resolved-scale humidity field is sufficiently high, and, in turn, condensation releases latent heat that warms the grid-scale temperature field. For this reason, it is not possible to ignore the effect of the subgrid processes on the resolvable-scale fields without degrading the quality of the forecast. To reproduce the interaction of the grid and subgrid-scale processes, the subgrid-scale phenomena are "parameterized," i.e., their effect is formulated in terms of the resolved fields. Figure 4.1.1, adapted from Arakawa (1997, 2004), indicates schematically the resolved processes (usually referred to as the "dynamics of the model"), and the processes that must be parameterized ("the model physics"), and their interactions. Arakawa (1997) points out that some subgrid-scale processes can be interpreted as *adjustment* processes. For example, the atmosphere adjusts to the surface conditions through boundary layer adjustment processes, which are very efficient if the planetary boundary layer is unstable. Radiative fluxes occur because temperature tends to adjust toward radiative equilibrium. Convective processes occur in the presence of an unstable stratification and adjust the field toward a more neutrally stable state. Because radiative equilibrium is convectively unstable for the lower troposphere,

4.2 Subgrid-Scale Processes and Reynolds Averaging

radiative–convective adjustment is a dominant process controlling the vertical thermal structure of the troposphere.

The details of the parameterizations have a profound effect on the model forecast, especially at longer time scales, and are the subject of very intense research. In this chapter, we provide only a very elementary introduction to model parameterizations. A short but inspiring introduction is presented in Arakawa (1997) and updated with a strong vision of history and the future in Arakawa (2004). An overview of different subgrid processes, and their parameterizations in atmospheric models appears in Haltiner and Williams (1980), and a review, including ocean and land models, is available in *Climate system modeling* edited by Trenberth (1992). Stull (1988) and Garratt (1994) are texts on the atmospheric boundary layer processes. Emanuel and Raymond (1993) edited a volume including detailed discussions of a number of cumulus parameterizations. Randall (2000) has edited a book honoring Akio Arakawa on the occasion of his retirement, which includes many review papers on areas related to physical parameterizations (as well as numerical modeling). Stensrud (2007) is a textbook on parameterization schemes.

4.2 Subgrid-Scale Processes and Reynolds Averaging

Consider the prognostic equation for water vapor written in flux form in z-coordinates (Section 2.1):

$$\frac{\partial \rho q}{\partial t} = -\frac{\partial \rho u q}{\partial x} - \frac{\partial \rho v q}{\partial y} - \frac{\partial \rho w q}{\partial z} + \rho E - \rho C \qquad (4.2.1)$$

In the real atmosphere, both u and q contain scales that are resolved by the grid of the model, and smaller, subgrid scales. We write then

$$\left.\begin{array}{l} u = \bar{u} + u' \\ q = \bar{q} + q' \end{array}\right\} \qquad (4.2.2)$$

where the overbar represents the spatial average over a grid, and the primes, the subgrid-scale perturbation. We can neglect the subgrid-scale variations of ρ. By definition, the grid-box average of all quantities linear in the perturbations is zero, e.g., $\overline{q'} = 0, \overline{u'\bar{q}} = 0$. Also, averaging a grid-average quantity does not change it, e.g., $\overline{\bar{u}\bar{q}} = \bar{u}\bar{q}$. These are the rules for *Reynolds averaging*, a method originally developed by Reynolds in 1895 for use in time averages, but that we apply to grid-box averages. We can substitute (4.2.2) in the moisture equation (4.2.1), take a grid average, and obtain:

$$\frac{\partial \rho \bar{q}}{\partial t} = -\frac{\partial \rho \bar{u}\bar{q}}{\partial x} - \frac{\partial \rho \bar{v}\bar{q}}{\partial y} - \frac{\partial \rho \bar{w}\bar{q}}{\partial z} - \frac{\partial \rho \overline{u'q'}}{\partial x}$$

$$- \frac{\partial \rho \overline{v'q'}}{\partial y} - \frac{\partial \rho \overline{w'q'}}{\partial z} + \rho E - \rho C \qquad (4.2.3)$$

The first three terms of the right-hand side are the grid-scale (resolved) advection terms, whose numerical discretization we have studied in Chapter 3. They are

The Parameterization of Subgrid-Scale Physical Processes

included in the "dynamical processes" box of Figure 4.1.1. The next three terms are *the divergences of the eddy fluxes of moisture* or turbulent moisture transports. The last two terms (evaporation and condensation) are subgrid-scale processes that occur at a molecular scale and that we still need to parameterize. Both the molecular-scale processes and eddy fluxes that occur at scales much larger than molecular, but smaller than the grid resolution, are denoted collectively as "subgrid-scale processes." As indicated in the introduction, the impact of at least some of these physical processes on the larger scales explicitly represented in the model must be included. Without the parameterization of at least the most important subgrid-scale processes, the model integrations cease to be realistic in a very short period, from a day or two for large-scale flow, to less than an hour for storm-scale simulations.

There are several choices for the parameterization of the effect of turbulent transport terms in terms of the resolved scales. Consider, for example, the vertical turbulent flux of moisture (which, because of the strong vertical gradients, especially in the planetary boundary layer, is by far the dominant component of the eddy fluxes). We can choose to:

(a) Neglect the vertical turbulent flux, assuming that, in the boundary layer, the grid-scale field is well mixed:

$$-\rho \overline{w'q'} = 0 \tag{4.2.4}$$

This is known as a "zeroth order" closure, in which only the average properties are sought. An example is the bulk parameterization of the mixed boundary layer (Deardorff, 1972), in which the potential temperature, water vapor, and wind are assumed to be well mixed, and only the depth of the layer is forecast.

(b) Parameterize the vertical flux as a "turbulent diffusion process" in terms of $\overline{q}$ and the other grid-scale variables (this is a first order closure and is the most commonly used):

$$-\rho \overline{w'q'} = K \frac{\partial \overline{q}}{\partial z} \tag{4.2.5}$$

This represents the effect of turbulent mixing due to parcels moving up or down, bringing with them the moisture from their original level, and mixing with the environment at the new level. The main problem in "K-theory," as this approach is also known, is to find a suitable formulation of the eddy diffusivity K, which also depends on the grid-average fields and the stability of the flow.

(c) Obtain a prognostic equation for $\overline{w'q'}$ by multiplying the vertical equation of motion by ρq and adding it to (4.2.1) multiplied by w. We obtain an equation with many terms like

$$\frac{\partial \rho wq}{\partial t} = -\frac{\partial \rho uwq}{\partial x} - \cdots \tag{4.2.6}$$

We can then take its Reynolds average and subtract it from (4.2.6) and derive a prognostic equation for the turbulent fluxes $\partial \rho \overline{w'q'}/\partial t = \cdots - \partial \rho \overline{w'w'q'}/\partial z \cdots$. This equation can be included as an additional model equation. Since it contains

4.2 Subgrid-Scale Processes and Reynolds Averaging 143

triple products of turbulent terms, these terms, in turn, have to be parameterized in terms of the double products:

$$-\rho\overline{w'w'q'} = K'\frac{\partial\overline{\rho w'q'}}{\partial z} \qquad (4.2.7)$$

This is a second-order closure. Second-order closure models have many additional prognostic equations (for all the products of turbulent variables) but are an alternative to high-resolution models to obtain an estimate of turbulent transports (e.g., Moeng and Wyngaard, 1988). Mellor and Yamada (1974, 1982) show how to construct a hierarchy of closures for vertical fluxes and provide simplifying assumptions.

If an important physical process that occurs in the real atmosphere on a scale unresolved by the model is not parameterized, it may still appear in the model integration "aliased" into the resolved scales. For example, primitive equation model integrations will be ruined (become too noisy) by dry convective instability if it is not parameterized. In the real atmosphere, if the potential temperature decreases with height in some area, the unstable convective circulation that will take place occurs at very small horizontal scales, of the same order as the depth of the unstable layer, typically 1 km or less. Since this cannot be resolved with horizontal grids of the order of 10–100 km, models with unstable layers develop an unrealistic appearance of "vertical noodles," with narrow columns moving up and down side by side. In order to handle this problem, Manabe et al. (1965) developed the *dry convective adjustment*, a simple parameterization of dry convection still used in many present-day models. In this parameterization, when the grid-scale atmosphere lapse rate exceeds the dry adiabatic lapse rate $\Gamma_d = g/C_p \approx 10$ K/km, the unstable atmospheric column is instantaneously adjusted to an adiabatic or very slightly stable profile, while keeping constant the layer total heat content (enthalpy). Moist (cumulus) convection that occurs when there is grid-scale saturation and the temperature gradient exceeds the moist adiabatic lapse rate also results in a "wet noodles" circulation. This led to the *moist convective adjustment*, the first parameterization of cumulus convection (Manabe et al., 1965), adjusting to a moist adiabatic profile. The moist convective adjustment was not found to be a sufficiently realistic cumulus parameterization and has since been replaced by other convective parameterizations by Kuo (1974), Arakawa and Schubert (1974), Betts and Miller (1986), and Kain and Fritsch (1990). See the volume edited by Emanuel and Raymond (1993) for a detailed review of cumulus parameterizations and an inspiring paper by Arakawa (2004) that reviews "the past, present and future of cumulus parameterization." Cumulus convection is one of the most important parameterizations in determining the characteristics of the model climatology (e.g., Miyakoda and Sirutis, 1977) and is reviewed in Arakawa (2004).

When a process occurs at scales not much smaller than the grid size, it presents an additional difficulty: The resolved scales and the unresolved scales to be parameterized are not well separated. An example of a process only marginally resolved in present-day models, which therefore appears aliased into the shortest waves present in the solution, is the sea-breeze circulation. A model with a grid size

144 The Parameterization of Subgrid-Scale Physical Processes

of 50–100 km (or more) cannot resolve the real sea-breeze circulation that takes place, for example, over a distance of the order of 1–20 km in the Florida peninsula on summer days. Therefore, in the model, the sea-breeze coastal circulation becomes distorted into a $2\Delta x$ circulation, and because the scales are not well separated, its effects on the large scales are difficult to parameterize. Similar effects are observed near heated mountain slopes when they are not properly resolved. The same problem of lack of scale separation complicates cumulus convection parameterization in models with a resolution of the order of 5–10 km, which is close to the horizontal scale of the convection, but not high enough to resolve convection explicitly. This resolution is called "the grey zone" and is the subject of much research.

4.3 Overview of Model Parameterizations

In a typical hydrostatic model on pressure coordinates, the governing equations (Chapter 2), including parameterized subgrid-scale processes, denoted with a tilde, are written as:

$$\frac{d\overline{\mathbf{v}}}{dt} = -\boldsymbol{\nabla}_p\overline{\phi} - f\,k \times \overline{\mathbf{v}} - g\frac{\partial\tilde{\tau}}{\partial p} \tag{4.3.1}$$

for the two horizontal equations of motion, including the effect of the dominant vertical eddy fluxes of momentum,

$$\frac{\partial\overline{\phi}}{\partial p} = -\overline{\alpha} \tag{4.3.2}$$

the hydrostatic equation,

$$\boldsymbol{\nabla}_p \cdot \overline{\mathbf{v}} + \frac{\partial\overline{\omega}}{\partial p} = 0 \tag{4.3.3}$$

the continuity equation,

$$\frac{\partial\overline{p}_s}{\partial t} + \overline{\mathbf{v}} \cdot \nabla\overline{p}_s = -\int_0^\infty \boldsymbol{\nabla}_p \cdot \overline{\mathbf{v}}dp \tag{4.3.4}$$

the rate of change of the surface pressure,

$$\overline{p}\,\overline{\alpha} = R\overline{T} \tag{4.3.5}$$

the equation of state,

$$C_P\frac{\overline{T}}{\overline{\theta}}\frac{d\overline{\theta}}{dt} = \tilde{Q} = \tilde{Q}_{rad} - g\frac{\partial\tilde{F}_\theta}{\partial p} + L(\tilde{C} - \tilde{E}) \tag{4.3.6}$$

and the first law of thermodynamics, which includes radiative heating and cooling, sensible heat fluxes and condensation and evaporation, and

$$\frac{d\overline{q}}{dt} = \tilde{E} - \tilde{C} - g\frac{\partial\tilde{F}_q}{\partial p} \tag{4.3.7}$$

the conservation equation for water vapor. Condensation takes place when the grid average value oversaturates (stable or grid-scale condensation), or when there is moist

4.3 Overview of Model Parameterizations

convective instability and cumulus convection. The condensed water falls as precipitation, and may evaporate if the layers below are not saturated. Additional conservation equations can be written for cloud and rain water in models with prognostic (rather than diagnostic) clouds, and for other substances such as ozone.

In these equations the quantities with an overbar are the grid-averaged quantities computed by the model dynamics, and the terms with the tilde represent terms that are parameterized. In a typical model, the vertical eddy flux of momentum $\tilde{\tau} = \rho\overline{w'u'}\,\mathbf{i} + \rho\overline{w'v'}\,\mathbf{j}$ (also known as eddy stress), of sensible heat $\tilde{F}_T = \rho C_p \overline{w'T'}$ and of moisture $\tilde{F}_q = \rho\overline{w'q'}$, may be represented using K-theory in the boundary layer and neglected in the free atmosphere above the boundary layer (using $K = 0$ or a very small value). The vertical derivatives of the turbulent fluxes that appear in the right-hand sides of (4.3.1), (4.3.6), and (4.3.7) introduce a requirement for lower boundary conditions for the surface fluxes of heat, moisture, and momentum. These surface fluxes are computed using a bulk parameterization based on the Monin and Obukhov (1954) similarity theory. This theory concludes that the profiles of wind and temperature in the turbulent *surface layer* can be described by a set of equations that depends only on a few parameters, including the surface roughness length z_0. The hypothesis of similarity, based on many observational studies, suggests that the fluxes of momentum and heat are nearly constant with height in the *surface layer* (of depth 10–100 m, which is much thinner than the planetary boundary layer). The fluxes in the surface or *constant flux layer* are usually represented with bulk aerodynamic formulas:

$$\left.\begin{aligned}
\tau &= -\rho C_D |\mathbf{v}|\mathbf{v} \\
F_\theta &= -\rho C_H |\mathbf{v}|C_p(\theta - \theta_S) \\
F_q &= -\rho C_E |\mathbf{v}|\beta(q - q_S)
\end{aligned}\right\} \tag{4.3.8}$$

Here, $\mathbf{v}, \theta, q$ are the velocity, potential temperature, and mixing ratio in the surface layer, respectively, and the variables with an S subscript are the corresponding values at the underlying ocean or land surface ($\mathbf{v}_S = 0$). C_D, C_H, and C_E are transfer coefficients (C_D is known as "drag coefficient") and they depend on the stability of the surface layer (measured by the bulk Richardson number $Ri_B = gz[(\theta - \theta_S)/\overline{\theta}]/\mathbf{v}^2$, the height z and the surface roughness length). They are nondimensional and have typical values of the order of 10^{-3} for stable conditions and 10^{-2} for unstable conditions (Louis, 1979). β is a coefficient representing the degree of saturation of the underlying surface (1 for oceans, 0–1 for land depending on the degree of saturation in the soil moisture content). The surface layer values are either obtained through the use of a thin (order 10 m) prognostic layer or diagnosed.

The radiative heating in (4.3.6) is determined from the vertical divergence of the upward and downward fluxes of *short- and long-wave radiation*, obtained using the radiative transfer equation. See Kiehl (1992) for a review of the parameterization of radiation. The interaction between *clouds and radiation* is very complex, and is a major area of research. Early models specified clouds climatologically (Manabe et al., 1965). In the 1980s the cloud cover was specified diagnostically, based on relative humidity (Slingo, 1987; Campana, 1994). More recently, cloud and rain water were predicted using budget equations and cloud cover was deduced from the amount of cloud water (e.g., Zhao and Carr, 1997). The cloud properties are also important: rather than plane slabs, as generally assumed, clouds have a fractal structure, which effectively reduces their albedo and increases atmospheric absorption of solar radiation (Cahalan et al., 1994).

An important area of research is the effect of subgrid-scale mountains. Wallace et al. (1983) proposed representing the blocking effect of subgrid-scale hills and valleys by increasing the effective height of mountains above its grid average by a factor of order one, times the standard deviation of the subgrid-scale orography. This approach has been denoted "envelope orography." Similarly, Mesinger et al. (1988) chose a method that essentially defines the grid mountain height by the tallest peaks ("silhouette orography"). Lott and Miller (1997) formulated a new parameterization using developments in the nonlinear theory of stratified flows around obstacles, paying special attention to the parameterization of the blocked flow when the effective height of the subgrid-scale orography is high enough. They showed that this method can duplicate the results using envelope orography. In addition to its blocking effect, under stable conditions, small-scale orography generates internal gravity waves that propagate upwards, increase their amplitude, and eventually break at upper levels, depositing their low-level momentum Lilly and Kennedy (1973). The net result is a deceleration due to surface orography at upper levels. Modelers have introduced a *gravity-wave parameterization* following Palmer et al. (1986), McFarlane (1987), and Lindzen (1988). Kim and Arakawa (1995) developed a parameterization of the drag due to gravity waves.

Other areas of research in the parameterization of subgrid processes are related to the fact that the underlying surfaces (ocean and land) have their own evolution and therefore provide a "longer memory" to the forecast model which cannot be represented diagnostically. Equation (4.3.8) indicates that over ocean it is necessary to know the surface stress τ and the sea surface temperature (SST). Short-range forecasts have been performed with observed SSTs, under the assumption that they do not change significantly with time, but this is clearly not a reasonable assumption for medium-range or longer forecasts. There are two major reasons why this is not a reasonable assumption. First, the assumption of constant temperature is not valid if the wind is strong. For example, if a hurricane goes through a region in the ocean, the enhanced vertical mixing will cool the path of the hurricane and make it harder for another hurricane to follow the same path. In addition, forcing the ocean SSTs to remain constant, or even to be as observed, is not realistic because this coupling does not allow feedbacks from the ocean to the atmosphere (e.g., Peña et al., 2003), and the ocean is thus forcing the atmosphere everywhere (Ruiz-Barradas et al., 2017).

For *seasonal and interannual predictions*, the SST is predicted using an ocean model coupled to the atmospheric model (Ji et al., 1994; Trenberth, 1992). In addition, the surface fluxes over the ocean depend on the surface waves, which are driven by the wind. Currently most models use the Charnock (1955) parameterization relating an effective roughness length to the surface stress. In an iterative procedure the stress and the roughness length are obtained and bulk-aerodynamical formulas used to deduce the sensible and latent heat fluxes. However, in reality, ocean waves have a memory of their previous interactions with the atmosphere: swell ("old sea") is smoother than "new sea" where waves are driven by sudden changes in the wind, and, in turn, this affects the surface stress and the fluxes of heat and moisture. To take this effect into account, it is necessary to couple atmospheric models with ocean wave models (e.g., Tolman, 2009).

Over land, similarly, the surface fluxes of heat and moisture are strongly dependent on the vegetation and soil moisture. Older models followed Manabe et al. (1965) by representing the effect of available soil moisture with a simple 15-cm "bucket" model, whose content was reduced by evaporation and increased by precipitation, with overflow representing river runoff. The surface temperature was obtained diagnostically assuming zero heat capacity for the land. Current models include coupling the atmospheric model with multilevel soil models with prognostic equations for the soil temperature and moisture, and include the very important controlling effect of plants on evapotranspiration (see reviews by Sellers, 1992; Dickinson, 1992; Pan, 1990).

4.4 The SPEEDY Model and Documentation

Molteni (2003) and Kucharski et al. (2006), at the International Center for Theoretical Physics (ICTP) developed a "SPEEDY" model (from "Simplified Parameterizations, primitivE-Equation DYnamics") that is a remarkably fast and realistic general circulation model (see http://users.ictp.it/~kucharsk/speedy-net.html). This page contains links to excellent documentations of both the numerics and the subgrid scale physical parameterizations of two versions of the SPEEDY model, an older 5-level model, and a newer 8-level model. The climatological means of the 8 level model are compared to reanalyses, used as verification. The climatological performance of the SPEEDY is impressive, especially considering that it can be run for years in a few hours on a PC. As discussed in Chapter 5, the SPEEDY model has been adapted to create input/output of forecasts, observations and analyses (Miyoshi, 2005, https://github .com/takemasa-miyoshi/letkf), and thus has been used to test many new approaches for numerical weather prediction (Amezcua et al., 2011; Hotta et al., 2016) and data assimilation. SPEEDY has been also coupled with the dynamic vegetation and terrestrial carbon model of Zeng (2003); Zeng et al. (2005) (see Kucharski et al., 2012). More recently SPEEDY has been also coupled with the NEMO ocean model, with a one-way anomaly coupling is applied from the ocean to the atmosphere (Kröger and Kucharski, 2011), which corrects a cold bias in the East Pacific and allows for ENSO like variability to occur. The clear documentation of the SPEEDY model can serve as an introductory description of both the dynamical core and the physical parameterizations of a full spectral global general circulation model (GCM). Codes and instructions are available from F. Kucharski.

4.5 Cumulus Parameterizations and "Superparameterization"

In a paper entitled "The cumulus parameterization problem: Past, Present and Future," Arakawa (2004) reviewed virtues and problems of cumulus parameterizations since the early 1960s. As illustrated in schematic Figure 4.1.1, he pointed out that clouds and their associated physical processes strongly influence the climate system in several ways: by coupling dynamical and hydrological processes in the atmosphere through the heat of condensation and evaporation and through redistributions of sensible and latent heat and momentum; by influencing hydrological process in the ground through precipitation; and by influencing the couplings between the atmosphere and oceans (or ground) through modifications of radiation and planetary boundary layer (PBL) processes.

Cumulus convection thus plays a central role in most of the physical interactions, and the representation of subgrid scale cumulus convection, known as cumulus parameterization, has always been at the core of efforts to numerically model the atmosphere. In spite of the experience accumulated over the past decades, cumulus parameterization is still an evolving subject. During the research phase that took place in 1960–1990, many cumulus parameterization schemes were developed and used in different models. In spite of having quite different reasonings behind them, GCMs using these different schemes generally produce comparable climatologies when the geographical distribution of sea surface temperature (SST) is prescribed. Arakawa (2004) discusses several of the major parameterizations schemes, including the first parameterization of cumulus convection as a simple convective adjustment (Charney and Eliassen, 1964; Manabe et al., 1965), the most famous, Arakawa-Schubert (1974) and other widely used parameterizations.

For the future, Arakawa (2004) envisions two possible extremes in dealing with processes that are currently subgrid scale: (1) "parameterize everything," and (2) "resolve everything," where (1) requires the most thinking, and (2) requires the most computer power.

Grabowski and Smolarkiewicz (1999), and Grabowski (2001), proposed an alternative approach that replaces the cumulus parameterization within each grid vertical column with an embedded 2-dimensional Cloud System Resolving Model within each grid column (Figure 4.5.1a). The 2D CSRM have periodic boundary conditions (shown by grey bars), and are forced by the large-scale advective tendencies of the local GCM column at the center (big black dots). The GCM columns, in turn, are forced back through updated domain-wide average thermodynamic variables. Inspired by Grabowski's idea, Khairoutdinov and Randall (2001) applied this Cloud Resolving Convective Parameterization (CRCP), which they called "superparameterization" to a global climate model and obtained encouraging results with little tuning. Randall et al. (2003) summarized their joint results. The superparameterization can be considered as a compromise between the "parameterize everything" and "resolve everything" approaches.

We show here Figure 4.5.1, adapted from Arakawa (2004), where he proposed some important extensions to the original CRCP shown in Figure 4.5.1a, and which Arakawa refers to with the more general name Cloud System Resolving Model (CSRM).

The 2-dimensionality of the CSRM in Figure 4.5.1a has the advantage of allowing to explicitly resolve convection, but with the computational cost of convection being much smaller than a fully 3-dimensional cloud resolving model, although imposing

4.5 Cumulus Parameterizations and "Superparameterization"

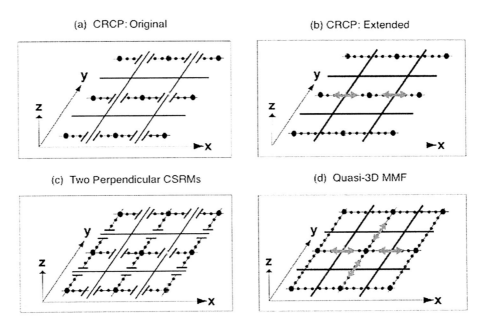

Figure 4.5.1 (a) An illustration of the framework used in Grabowski's Cloud Resolving Convective Parameterization (CRCP). (b) Same as in (a) but for a revised version in which a GCM and a 2D CRM are coupled at the large dots for scalar variables and at the arrows for velocity components. (c) Same as in (a) but for a revised version in which two perpendicular CRMs are embedded into a GCM grid box. (d) An illustration of a quasi-3D MMF. Adapted from Arakawa (1997, 2004). ©American Meteorological Society. Used with permission.

an "East-West" structure in the convection. Figure 4.5.1c shows a combination of 2D E-W and N-S CSRMs that ameliorates this directional bias but doubles the computational cost of the CSRM. Note that the periodic boundary conditions of the CSRM in Figure 4.5.1a do not allow contiguous CSRMs to interact except through the GCM. This restriction is relaxed as in Figure 4.5.1b by replacing the periodic boundary conditions with a coupling of the velocity components with the GCM at the arrow points of 4.5.1. Figure 4.5.1d combines the approaches of version b and c, in a "Quasi-3D multiscale modeling framework (MMF)." The Quasi-3D MMF can simulate interactions between the following three categories of scales: scales resolved by the GCM, scales of cloud organizations within a GCM grid box, and statistical samples of small-scale features. Most importantly, convergence to a global 3D CSRM as the GCM grid size becomes small is guaranteed with this framework. However, this promising multiscale modeling framework does not seem to have been adopted widely. The original proposal by Grabowski was tested by Khairoutdinov and Randall (2001) and Khairoutdinov et al. (2005).

In summary, superparameterization is a very promising approach and it uses much less computer resources than a Cloud System Resolving global model. But it increases the cost of the original GCM by orders of magnitude, and it has not yet displaced the use of other more conventional cumulus parameterizations.

5 Data Assimilation

5.1 Introduction

In previous chapters, we saw that NWP is an initial/boundary value problem: given an estimate of the present state of the atmosphere (*initial conditions*) and appropriate surface, top, and lateral boundary conditions, the model simulates (*forecasts*) the atmospheric evolution. Obviously, the more accurate the estimate of the initial conditions, the better the quality of the forecasts. Currently, operational NWP centers produce initial conditions through *a statistical combination of observations and short-range forecasts that accounts for the uncertainty associated with each source of information.* This approach has become known as "*data assimilation*," whose purpose is defined by Talagrand (1997) as "using all the available information to determine as accurately as possible the state of the atmospheric (or oceanic) flow." *Information* comes from observations and from the model forecasts based on the dynamics of the system and should include not only their values but their estimated error statistics.

Gandin (1963) described in detail, for the first time, how to "optimally interpolate" model forecasts and observations based on their *error statistics* to obtain an "optimal" analysis. This major advancement had a profound impact, and optimal interpolation (OI), a "sequential" approach, became widely used in operational centers in the 1980s, replacing previously used empirical methods such as the successive correction method (SCM). 3D-Var, also a statistical interpolation that is "variational" (minimizing a cost function), replaced OI in the early 1990s.

Jazwinski (1970) wrote a classic book on linear and nonlinear filtering (estimating the true state of a system from noisy measurements), based on the fundamental work of Kalman (1960) and Kalman and Bucy (1961). Ghil et al. (1981) first suggested that the Kalman filter could, in principle, be used for the optimal estimation of atmospheric states, since it evolves the forecast errors, rather than keeping them constant. However, for a large nonlinear system like an atmospheric model, the Kalman filter is computationally completely unfeasible, and 4D-Var was designed, e.g., (Lewis and Derber, 1985; Courtier and Talagrand, 1990), to approximate the Kalman filter solution more efficiently. Daley (1991) wrote the first modern book, *Atmospheric Data Analysis*, focused on how to optimally combine atmospheric models and observations. Since the first edition of Daley's book, others have been published on data assimilation methods (e.g., Lewis et al. (2006), focused on least squares methods, Evensen (2009) on the increasingly popular ensemble Kalman filter approach, and Law et al. (2015) with a strong mathematical orientation). Lahoz and Menard (2010) contains review articles on different important issues in data assimilation (available for download), and Park and Xu (2013) is a compilation of different geophysical applications. Carrassi et al. (2018) is an overview of data assimilation in geosciences.

In this chapter, we first review early approaches to objective analysis and empirical methods for data assimilation (Section 5.2). Using two very simple scalar models, Section 5.3 introduces the modern concepts and methods of data assimilation (variational and sequential). This section, written at an undergraduate level, allows intuitive understanding of the main concepts of data assimilation, so that the otherwise daunting matrix and vector equations for real (and huge) model and observation applications presented in Section 5.4 appear as simple and obvious extensions of the scalar toy model equations in Section 5.3 to vectors and matrices. We show that optimal interpolation and 3D-Var have the same mathematical solution, but since they are solved numerically in very different ways, their result can have different characteristics.

The rest of this chapter is mostly devoted to two advanced methods that take into account the fact that forecast errors evolve with time (they have "errors of the day"), whereas the classical 3D-Var and OI assume a constant (climatological) background error covariance. The two methods are 4D-Var (Section 5.5), an extension of 3D-Var that includes time, and ensemble Kalman filter (EnKF, Sections 5.6.1–5.6.4). Sections 5.6.5 and 5.6.6 discuss and compare problems common to both methods, new approaches, and the increasingly popular hybrid Var-EnKF methods. Section 5.6.7 introduces the powerful diagnostic tool ensemble forecast sensitivity to observations and its applications. In the last Section 5.6.8, we give a brief introduction to the promising non-Gaussian assimilation method called particle filter.

5.2 Empirical Analysis Schemes

In Subsection 5.2.1, we describe early methods of data assimilation used to produce "objective analyses" by simply interpolating the observations to the model grid. These simple schemes were later replaced by methods using a model forecast as a "background" to be modified with the observations, but without accounting for the uncertainty (error statistics) of the model and of the observations (e.g., the widely used successive correction method, Subsection 5.2.2). The nudging method, a simple way to force a model to be close to the observations still used in some applications, is discussed in 5.2.3.

5.2.1 Early Approaches to Objective Analysis

In the first two NWP experiments, Richardson (1922) and Charney et al. (1950) created initial conditions by performing hand interpolations of the available observations into a regular grid, and these fields of initial conditions were then manually digitized, a very time consuming procedure. The need for a computer-based "*objective analysis*" became quickly apparent (Charney, 1951), and interpolation methods fitting observations to a regular grid were soon developed. Panofsky (1949) developed the first objective analysis algorithm based on two-dimensional polynomial interpolation, a procedure that can be considered "global" since the same function is used to fit all the observations.

Gilchrist and Cressman (1954) developed a "local polynomial" interpolation scheme for the geopotential height. A quadratic polynomial in x and y was defined at each grid point:

$$z(x,y) = a_{00} + a_{10}x + a_{01}y + a_{20}x^2 + a_{11}xy + a_{02}y^2 \quad (5.2.1)$$

The six coefficients were determined by minimizing the mean square difference E between the polynomial and observations of geopotential height and wind, close to the grid point (within *a radius of influence* of the grid point, see Figure 5.2.1):

$$\min_{a_{ij}} E = \min_{a_{ij}} \sum_{k=1}^{K_z} p_k \left(z_k^o - z(x_k,y_k)\right)^2 + \sum_{k=1}^{K_v} q_k \{[u_k^o - u_g(x_k,y_k)]^2 + [v_k^o - v_g(x_k,y_k)]^2\} \quad (5.2.2)$$

Here E is the minimum value of the cost function that provides the solution $z(x,y)$, p_k, q_k are empirical weighting coefficients, and u_g, v_g are the geostrophic wind components computed from the gradient of the geopotential height $z(x,y)$ at the observation point k, and K is the total number of observations within the radius of influence. Note that although the field being analyzed is just the geopotential height, the wind observations are useful as well because they provide additional information about its gradient.

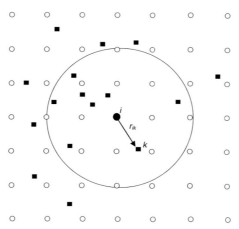

Figure 5.2.1 Schematic of grid points (circles), irregularly distributed observations (squares) and a radius of influence around a grid point marked with a black circle. In 4DDA, the grid point analysis is a combination of the forecast at the grid point (first guess) and the observational increments (observation minus first guess) computed at the observational points k. In certain analysis, schemes, like the SCM, some OIs, and the LETKF, only observations within a radius of influence, indicated by a circle, affect the analysis at the black grid point.

5.2 Empirical Analysis Schemes

These "analyses" interpolating from observations to model grids imitated the hand analyses of the observations made by expert analysts. However, for the initialization of operational models, it soon became clear that it was not enough to perform spatial interpolation of observations into regular grids, *because not enough data are available to initialize the models*. For example, the number of degrees of freedom in a modern NWP model is of the order of 10^9, whereas the total number of *conventional* observations of the variables used in the models (e.g., from rawinsondes, which were the core of the observing system until 1979, is of the order of 10^4). There are many new types of data currently available, including remotely sensed data such as satellite and radar observations, but they do not measure directly the variables used in the models (wind, temperature, moisture, and surface pressure). Moreover, the distribution of conventional observations in space and time is very nonuniform (Figure 1.4.1), with regions like North America and Eurasia that are relatively data-rich, and others that are much more poorly observed.

For these reasons, it became clear rather early in the history of NWP that, in addition to the new observations, creating a new analysis also required having a complete *first guess* estimation of the state of the atmosphere at all the grid points (Bergthórsson and Döös, 1955). The first guess (also known as *background field* or *prior information*) should be our best estimate of the state of the atmosphere prior to the use of the observations. Initially climatology, or a combination of climatology and a short forecast were used as a first guess, e.g., Gandin (1963); Bergthórsson and Döös (1955). As forecasts became better, the use of short-range forecasts as a first guess was universally adopted in operational systems in what is called an "analysis cycle" (Figure 1.1.4).

The analysis cycle is an intermittent data assimilation system that continues to be used in most global operational systems, typically with a 6-hr cycle performed four times a day. This time window of 6 hr was chosen because most rawinsondes, the heart of the observing system, are launched at 00 and 12UTC, and some, especially in Europe, are also launched at 06 and 18UTC. Here UTC stands for Universal Coordinated Time, and in meteorology 00UTC is usually abbreviated as 00Z.

The model forecast in the analysis cycle plays an extremely important role. Over data-rich regions, the analysis is strongly impacted by the information contained in the observations. In data-poor regions, the forecast benefits from the information coming from upstream. For example, 6-hr forecasts over the North Atlantic Ocean were quite good, even before the advent of satellite data, because of the information transported by the model from North America. Forecasts are thus able to transport information from data-rich to data-poor areas, and for this reason, data assimilation using a short-range forecast as a first guess has become known as four-dimensional data assimilation (4DDA) where the 4th dimension is time.

5.2.2 Successive Correction Method

The first analysis method using a first guess or background was based on an *empirical* approach known as the SCM, developed by Bergthórsson and Döös (1955) in Sweden, and by Cressman (1959) in the US Weather Service. In SCM, the first estimate of the gridded field is given by the background (or first guess) field:

$$f_i^0 = f_i^b \qquad (5.2.3)$$

where f_i^b is the background field evaluated at the ith grid point, and f_i^0 the corresponding zeroth iteration estimate of the gridded field (Figure 5.2.1).

After this first estimate, the following iterations are obtained by "successive corrections":

$$f_i^{n+1} = f_i^n + \frac{\sum_{k=1}^{K_i^n} w_{ik}^n \left(f_k^0 - f_k^n \right)}{\sum_{k=1}^{K_i^n} w_{ik}^n + \varepsilon^2} \qquad (5.2.4)$$

where f_i^n is the nth iteration estimation at the grid point i, f_k^0 is the kth observation surrounding the grid point i, f_k^n is the value of the nth field estimate evaluated at the observation point k (obtained by interpolation from the surrounding grid points), and ε^2 is an estimate of the ratio of the observation error variance to the background error variance. The weights w_{ik}^n can be defined in different ways. Cressman (1959) defined the weights in the SCM as

$$\left. \begin{array}{ll} w_{ik}^n = \dfrac{R_n^2 - r_{ik}^2}{R_n^2 + r_{ik}^2} & \text{for} \quad r_{ik}^2 \leq R_n^2 \\[3mm] w_{ik}^n = 0 & \text{for} \quad r_{ik}^2 > R_n^2 \end{array} \right\} \qquad (5.2.5)$$

where r_{ik}^2 is the square of the distance between an observation point $\mathbf{r}_k$ and a grid point at $\mathbf{r}_i$.

The radius of influence R_n is allowed to vary with the iteration, and K_i^n is the number of observations within a distance R_n of the grid point i. For example, in the 1980s the Swedish operational system used $R_1 = 1,500$ km, $R_2 = 900$ km for upper air analyses, and $R_1 = 1,500$ km, $R_2 = 1200$ km, $R_3 = 750$ km, $R_4 = 300$ km for the surface pressure analysis. The reduction of the radius of influence results in a field that after the first iteration reflects only the large scales, but which includes more information from the smaller scales after the additional iterations.

In the Cressman SCM, the coefficient ε^2 is assumed to be zero. This results in a "credulous" analysis that more faithfully reflects the observations, and for a very small radius of influence the analysis *converges to the observation values* if the observations are located at the grid points. If the data are noisy (e.g., if an observation has gross errors, or if it contains an unrepresentative sample of subgrid-scale variability), this can lead to "bull's eyes" (many isolines crowded around an unrealistic grid-point value) in the analysis. Including $\varepsilon^2 > 0$ assumes that the observations have errors, and gives some weight to the background field.

Barnes (1964, 1978) developed another empirical version of the SCM that is still used for analyses of radar data or other small-scale observations for which there is no available background field from a model. Since we have no information on the

5.2 Empirical Analysis Schemes

background field, the background error variance can be considered to be very large, so that $\varepsilon^2 = 0$. The weights are given by $w_{ik}^n = e^{-r_{ik}^2/2K_n^2}$. The radii of influence are changed by a constant factor at each iteration: $R_{n+1}^2 = \gamma R_n^2$. If $\gamma = 1$, only the large scales are captured. For $\gamma < 1$ more details in the observations are reproduced in the analysis as more iterations are performed.

Although the SCM method is empirical, it is simple and economical, and it provides reasonable analyses. Bratseth (1986) showed that if the weights are chosen in a statistically appropriate way instead of using the empirical formulas presented above, the SCM can be made to converge to a proper statistical interpolation (optimal interpolation or OI).

5.2.3 Nudging

Another empirical and fairly widely used method for data assimilation is Newtonian relaxation or nudging (Hoke and Anthes, 1976; Kistler, 1974). This consists of adding to the prognostic equations a term that nudges the solution toward the observations (interpolated to the model grid). For example, for a primitive equation model, the zonal velocity forecast equation is written as

$$\frac{\partial u}{\partial t} = -\mathbf{v} \cdot \nabla u + fv - \frac{\partial \phi}{\partial x} + \frac{u_{obs} - u}{\tau_u} \tag{5.2.6}$$

and similarly for the other equations.

The relaxation time scale, τ, is chosen based on empirical considerations and may depend on the variable. If τ is very small, the solution converges toward the observations too fast, and the dynamics do not have enough time to adjust. If τ is very large, the errors in the model can grow too much before the nudging becomes effective. Hoke and Anthes (1976) indicated that τ should be chosen so that the last term is similar in magnitude to the less dominant terms. They used a very short time scale (about 20 minutes) in their experiments. Stauffer and Seaman (1990) used about 1 hr in experiments assimilating synoptic observations, and reported a fair amount of success. Zou et al. (1992) made optimal parameter estimations of the nudging time scale using 4D-Var. Kaas et al. (1999) performed an interesting experiment, nudging a model toward a 15 year reanalysis from the ECMWF, and by averaging the mean forcing introduced by nudging, empirically determined corrections to reduce model deficiencies. van Leeuwen (2010) used nudging within a type of ensemble data assimilation (EDA) known as particle filters (Section 5.6.8) to force the ensemble members (particles) to be close to the observations.

Although this method is not generally used for large-scale assimilation, some groups use it for assimilating small-scale observations (e.g., radar observations) when there are no available statistics to perform a statistical interpolation. Interestingly, Schraff et al. (2016) only recently replaced the nudging system that had been used in the German weather service (DWD) operational data assimilation for the COSMO model, with the LETKF (Hunt et al., 2007). They still included latent heat nudging as an effective way to assimilate radar observed precipitation.

156   **Data Assimilation**

5.3   Introduction to Statistical Estimation Methods through the Use of Toy Models

We have described in Section 5.2 several empirical methods for data assimilation. In this section, we present methods that are based on statistical estimation theory. Citing again Talagrand (1997):

Assimilation of meteorological or oceanographical observations can be described as the process through which all the available information is used in order to estimate as accurately as possible the state of the atmospheric or oceanic flow. The available information essentially consists of the *observations* proper, and of the *physical laws* that govern the evolution of the flow. The latter are available in practice under the form of a *numerical model*. The existing assimilation algorithms can be described as either *sequential* or *variational*.

In statistical estimation, the best estimate of the real field is obtained by a *statistical interpolation* of the background field and the observations using the statistical information about their respective errors. In this section, we give toy examples of both *sequential* and *variational* approaches that provide a very simple introduction to the approaches used in data assimilation. The methodology and equations derived through these simple examples will then provide the reader with a clear understanding of the full equations for the multivariate data assimilation methods known as sequential, such as OI, Kalman filtering, ensemble Kalman filter (EnKF), and variational, such as 3D-Var and 4D-Var.

Historical note: The estimate that combines a forecast with the observations to give the initial conditions for the next forecast is known in meteorology as "objective analysis" or simply "analysis" because, historically, it replaces the hand-drawn analyses prepared by skillful analysts.

5.3.1   Sequential (or Least Squares) Method

Toy example 1: Obtain the best estimate (analysis) of the true temperature of an object given two observations, or, alternatively a forecast and an observation.

The best estimate of the state of the atmosphere (analysis) is obtained, as stated by Talagrand (1997), from combining *prior* information about the atmosphere (background or first guess) with observations, but in order to combine them optimally we also need statistical information about the respective errors in these "pieces of information." In this simple example, which serves as an introduction to statistical estimation, we will determine the best estimate of the true value of a scalar (e.g., the true temperature T_t) given two independent pieces of information, a forecast (prior) T_b and an observation T_o:

$$\left. \begin{array}{l} T_b = T_t + \varepsilon_b \\ T_o = T_t + \varepsilon_o \end{array} \right\} \tag{5.3.1}$$

The forecasts and observations have errors $\varepsilon_b, \varepsilon_o$ that we do not know. Let $E()$ represent the *expected value*, i.e., the average that one would obtain if making many similar measurements. We assume that T_b and T_o are unbiased: $E(T_b - T_t) = E(T_o - $

$T_t) = 0$, or equivalently,

$$E(\varepsilon_b) = E(\varepsilon_o) = 0 \tag{5.3.2}$$

and that we know their error variances:

$$E(\varepsilon_b^2) = \sigma_b^2 \quad E(\varepsilon_o^2) = \sigma_o^2 \tag{5.3.3}$$

In (5.3.3), we have made use of the assumption of unbiased errors (5.3.2), so that $Var\,(\varepsilon) = E\left(\varepsilon^2\right) - (E\,(\varepsilon))^2 = E\left(\varepsilon^2\right) = \sigma^2$ for both the forecast and the observation. We also assume that *the forecast and observation errors are uncorrelated*:

$$E(\varepsilon_b \varepsilon_o) = 0 \tag{5.3.4}$$

Equations (5.3.2), (5.3.3), and (5.3.4) represent the statistical information that we need about the actual observations. We try to optimally estimate T_t from a linear combination of the two observations since they represent all the information that we have about the true value of T:

$$T_a = a_b T_b + a_o T_o \tag{5.3.5}$$

The analysis T_a should be unbiased:

$$E(T_a) = E(T_t) \tag{5.3.6}$$

which implies

$$a_o + a_b = 1 \tag{5.3.7}$$

In addition, the coefficients of T_a should be chosen to minimize the mean square error of T_a, given by

$$\sigma_a^2 = E[(T_a - T_t)^2] = E[(a_b(T_b - T_t) + a_o(T_o - T_t))^2] \tag{5.3.8}$$

subject to the constraint (5.3.7). Substituting $a_o = 1 - a_b$, the minimization of σ_a^2 with respect to a_b gives

$$a_b = \frac{1/\sigma_b^2}{1/\sigma_b^2 + 1/\sigma_o^2} \quad a_o = \frac{1/\sigma_o^2}{1/\sigma_b^2 + 1/\sigma_o^2} \tag{5.3.9}$$

indicating that the weights of the forecast and the observation are proportional to their *precision* (defined as the inverse of their variances, and sometimes referred as accuracy or information). These weights can be also written as

$$a_b = \frac{\sigma_o^2}{\sigma_b^2 + \sigma_o^2} \quad a_o = \frac{\sigma_b^2}{\sigma_b^2 + \sigma_o^2} \tag{5.3.10}$$

indicating that the weights are inversely proportional to the error variances. Using these weights in (5.3.5), we obtain that the best estimate (analysis) of the true temperature given the forecast and the observation is given by

$$T_a = \frac{\sigma_o^2}{\sigma_b^2 + \sigma_o^2}T_b + \frac{\sigma_b^2}{\sigma_b^2 + \sigma_o^2}T_o \tag{5.3.11}$$

This relation between T_a, on the one hand, and T_b and T_o on the other, is more conveniently written as "the analysis is equal to the forecast plus an optimal weight w multiplied by the *innovation*," where the innovation $(T_o - T_b)$, is the *new information* brought in by the observation:

$$T_a = T_b + w(T_o - T_b) \tag{5.3.12}$$

where

$$w = \frac{\sigma_b^2}{\sigma_b^2 + \sigma_o^2} \tag{5.3.13}$$

Moreover, substituting the coefficients (5.3.10) into (5.3.8), we obtain a relationship between the analysis error variance and the forecast and observation error variances:

$$\frac{1}{\sigma_a^2} = \frac{1}{\sigma_b^2} + \frac{1}{\sigma_o^2} \tag{5.3.14}$$

This important equation indicates that, if the assumptions are correct, the coefficients are optimal and the statistics of the errors are exact, then the "precision" of the analysis (i.e., the inverse of the analysis variance) is the sum of the precision of the forecast and of the observation. Another relationship between the analysis, background and observation errors is given by $\sigma_a^2 = [\sigma_b^2 \times \sigma_o^2]/(\sigma_b^2 + \sigma_o^2)$. Both equations indicate that the analysis error variances are smaller than the error variances of either the forecast or the observation.

Toy example 2: A stone in space whose temperature is estimated by remotely measuring its emitted radiance.

In this example, we estimate the temperature of an object in space (e.g., a stone) having a forecast and a measurement of the radiance it emits. Note that we *do not* measure the temperature T (K), only the emitted radiance y (W/m^2). This example is much more representative of a "real life" situation in which model variables are different from the observed variables. In this case, we need to have an *observation operator* (also known as *forward model*) to transform variables from "*model space*" to "*observation space*" and vice versa.

For this toy example, we need to have an observation operator that transforms the temperature into the variable that we measure:

$$y_o = h(T_t) + \varepsilon_0 \tag{5.3.15}$$

where T_t is the true temperature, $h(T_t)$ is the exact observation operator that converts temperature into the corresponding radiance, and ϵ_0 as in equation (5.3.15) is the observation error. In the case of the stone, the observation operator would be the Stefan-Boltzmann's law: $y = h(T) = \sigma T^4$.

We also need to have a forecast model for the *temperature*:

$$T(t_{i+1}) = m[T(t_i)] \tag{5.3.16}$$

5.3 Introduction to Statistical Estimation Methods 159

For example, we could estimate the change in temperature using a simple forward time scheme: $T(t_{i+1}) = T(t_i) + \Delta t$ [SW heating + LW cooling]. The model forecast will give us the *prior* or *background* information on the temperature at time t_{i+1} starting from the analysis at the previous time step:

$$T_b(t_{i+1}) = m[T_a(t_i)] \tag{5.3.17}$$

For this simple example, we will derive (with a good understanding of their simple meaning!) the data assimilation equations for both OI/KF and Var. In Section 5.4.4 we will do the same for the general multivariate case that in operational models involves huge vectors and matrices. We will see that the multivariate equations have exactly the same form and interpretation as the simple scalar equations for the stone in space.

The new information (or innovation) introduced by the observation at time t_{i+1}, is the *observational increment*, namely the difference between the observation of the radiances and the forecast transformed into radiances with the observation operator. Since all the variables are defined at time t_{i+1} we omit the time index of t_{i+1}:

$$y_o - h(T_b) = h(T_t) + \varepsilon_o - h(T_t + \varepsilon_b) \approx \varepsilon_o - H\varepsilon_b \tag{5.3.18}$$

Here we used a Taylor expansion of the second observation operator on the RHS and linearized it assuming that the prior T_b is close to the truth T_t. $H = dh/dT$ is the gradient of the nonlinear observation operator at $T_t \approx T_b$ and therefore has units $W/(m^2 K)$.

The formula for the analysis temperature at time t_n is

$$T_a = T_b + w(y_o - h(T_b)) \tag{5.3.19}$$

and the corresponding analysis error is obtained subtracting the true temperature from both sides of (5.3.19) and using (5.3.18):

$$\varepsilon_a = \varepsilon_b + w(\varepsilon_o - H\varepsilon_b) \tag{5.3.20}$$

The equation for the analysis temperature (5.3.19) is very similar to that in toy model 1, c.f. (5.3.12), except that now the optimal weight w involves the gradient of the observation operator H that transforms ϵ_b, the forecast error in model space, into the forecast error in observation space, $H\epsilon_b$. The corresponding optimal weight is given by

$$w = \sigma_b^2 H (\sigma_o^2 + H\sigma_b^2 H)^{-1} \tag{5.3.21}$$

Exercise 5.3.1 Derive equation (5.3.21) by computing $\partial \varepsilon_a^2 / \partial w = 0$ and finding the value of w that minimizes the analysis error variance assuming that the forecasts and observations are not biased and their errors are not correlated.

The analysis error variance is obtained from squaring equation (5.3.20) and taking its expected value:

$$\sigma_a^2 = \overline{\varepsilon_a^2} = (1 - wH)\sigma_b^2 = \frac{\sigma_o^2}{\sigma_o^2 + \sigma_b^2 H^2} \sigma_b^2 \tag{5.3.22}$$

160 **Data Assimilation**

which can also be written as

$$\frac{1}{\sigma_a^2} = \left(\frac{1}{\sigma_b^2} + \frac{H^2}{\sigma_o^2} \right) \tag{5.3.23}$$

Exercise 5.3.2 Derive equations (5.3.22) and (5.3.23).

5.3.2   Variational (Maximum Likelihood) Approach

In a variational approach, we obtain the same best estimate of T_t by minimizing a cost function of the temperature T defined as the sum of the square of the distance (or misfit) of the estimate T to the forecast and to the observations, weighted by their corresponding precisions:

$$J(T) = \frac{1}{2} \left[\frac{(T - T_b)^2}{\sigma_b^2} + \frac{(h(T) - y_o)^2}{\sigma_o^2} \right] \tag{5.3.24}$$

Exercise 5.3.3 Show that the minimum of the *cost function* J defined in (5.3.24) is obtained for the same value $T = T_a$ as in (5.3.19), with the same optimal weight w as in (5.3.21).[1] Hint: $\partial J / \partial T = 0$ for $T = T_a$. Find the increment in temperature $\delta T = T - T_b$ that will make $\partial J / \partial \delta T = 0$, linearizing $h(T_b + \delta T)$ with respect to δT. Finding the solution for δT is called the *incremental* variational approach.

This exercise shows that, as proved by Lorenc (1986), the least squares error approach (minimizing the analysis error covariance) and the variational approach (maximizing the likelihood of the analysis, discussed later), *give the same optimal analysis* even though they are ostensibly very different methods.

We now show that (5.3.24) can be formulated using the *maximum likelihood* approach, where we ask the question: Given the forecast T_b (obtained from past information) and the observation y_o, which are both assumed to have Gaussian errors with standard deviations σ_b and σ_o, what is the most likely value, (i.e., the most probable value), of the true temperature T? We *define the analysis as the most likely value of T given the forecast and observation and their statistical errors*.

The probability distribution of a forecast given a true value T and an observational standard deviation σ_b, is given by the Gaussian distribution

$$p(T_b \,|T) = \frac{1}{\sqrt{2\pi}\sigma_b} e^{-\frac{(T_b - T)^2}{2\sigma_b^2}}$$

Conversely, the *likelihood* of a true value T given a forecast T_b with an error with a standard deviation σ_b is the same as this probability ((Edwards, 1984)):

$$L(T||T_b) = p(T_b|T) = \frac{1}{\sqrt{2\pi}\sigma_b} e^{-\frac{(T_b - T)^2}{2\sigma_b^2}}$$

---

[1] While the incremental variational approach without outer loops leads to the same solution as the (non-iterative) KF/OI equation, its solution with multiple outer loops will be different from KF/OI solution, which is due to the nonlinear observation operator

5.3 Introduction to Statistical Estimation Methods 161

Similarly, the likelihood of a true value T given an observation y_o with a standard deviation σ_o is

$$L(T \mid y_o) = p(y_o \mid T) = \frac{1}{\sqrt{2\pi}\sigma_o} e^{-\frac{(h(T)-y_o)^2}{2\sigma_o^2}}$$

Therefore, the most likely value of T given the two independent measurements T_b and y_o is the one that maximizes the joint probability, i.e., their product:

$$\max_T L(T \| T_b, y_o) = p(T_b \mid T) p(y_o \mid T) = \frac{1}{2\pi\sigma_b\sigma_o} e^{-\frac{(T_b-T)^2}{2\sigma_b^2} - \frac{(y_o-h(T))^2}{2\sigma_o^2}}$$

Since the logarithm is a monotonic function, we can take the logarithm of the likelihood and obtain the same maximum likelihood temperature:

$$\max_T \ln L(T \| T_b, y_o) = \max_T \left[const. - \frac{(T_b - T)^2}{2\sigma_b^2} - \frac{(y_o - h(T))^2}{2\sigma_o^2} \right] \quad (5.3.25)$$

Equation (5.3.25) shows that minimizing the cost function (5.3.24) maximizes the logarithm of the likelihood and therefore the likelihood of the analysis T given the forecast T_b and the observation y_o.

5.3.2.1 Bayes Theorem Applied to Data Assimilation

For a more general derivation of the maximum likelihood[2] analysis (Purser, 1984), we can use Bayes theorem that allows the "inversion" of conditional probabilities. Assume we have made the forecast T_b valid at the analysis time t_{i+1}, which gives us *prior information about the true temperature T at this time*, and then we obtain the new observation y_o also valid at the analysis time. Using standard probability rules, the probability of the true temperature T given the new observation y_o is given by

$$p(T|y_o) = p(T \cap y_o)/p(y_o) \quad (5.3.26)$$

whereas the probability of making a new observation y_o given a true temperature T is given by

$$p(y_o|T) = p(T \cap y_o)/p(T) \quad (5.3.27)$$

Note that the denominators in (5.3.26) and (5.3.27) are computed after we know the prior information T_b. Since the observation y_o is made independently of the forecast T_b, $p(y_o)$ is simply the total probability of obtaining a measurement y_o, i.e., $p(y_o)$ is a *normalization constant*. On the other hand, $p(T)$ is the probability of the true temperature being T after we know the prior information from the forecast T_b. Therefore, $p(T)$ is *not* a constant, but *the probability distribution of the true T given the prior information T_b*.

---

[2] In the context of data assimilation (as well as atmospheric remote sensing (Rodgers, 2000)), the maximum likelihood (ML) method usually refers to the method that maximizes the *posterior* probability density function. Its solution is also called *maximum a posteriori* (MAP) solution.

162 Data Assimilation

From (5.3.26) and (5.3.27), we derive Bayes theorem:

$$p(T|y_o) = p(T)p(y_o|T)/p(y_o) \tag{5.3.28}$$

which says

"The *posterior* probability of the true temperature T given the *prior information* T_b, and after receiving new information y_o, is given by the *prior probability* of T (based on the forecast T_b) multiplied by *the likelihood of T* given the observation y_o (which is the same as the probability of observing y_o given T, (Edwards, 1984)) normalized by the total probability of y_o."

This can be briefly read as *"posterior = prior * likelihood / normalization."* If we are looking for the temperature with maximum likelihood we only need to find the maximum of the product of the prior and the likelihood distributions, thus, it is not necessary to normalize it. It is important to note that although we have derived the variational analysis using maximum likelihood assuming a Gaussian distribution, Bayes' equation (5.3.28) is *much more general* and does not require the probability distributions of the errors to be normal. In Section 5.6.8 we will see how Bayes theorem has been applied in a new data assimilation method known as "particle filter."

Exercise 5.3.4 Derive the maximum likelihood cost function (5.3.24) using Bayes' rule, assuming that the error distributions of the forecast and the observation are Gaussian.

5.3.3 Analysis Cycle Equations for the "Stone in Space" Toy Model

The analysis cycle, as suggested in the previous section 5.3.1, has two stages: *a forecast stage*, started from the previous analysis valid at t_i that serves as initial conditions for the forecast valid at t_{i+1} and *an analysis stage* where the forecast (prior information) and the new observations are optimally combined to create the new analysis (*posterior*) valid at t_{i+1}. We will illustrate this process with the toy example of a stone in space.

The analysis can be obtained using sequential or variational approaches, since they are equivalent. The sequential OI is a simplification of the more advanced Kalman filter, and the variational 3D-Var a simplification of 4D-Var. In the advanced schemes, the background error variance σ_b^2 is forecasted, whereas in the simpler schemes *it is estimated once and for all*.

In the forecast phase of the analysis cycle, the background is first obtained through a forecast:

$$T_b(t_{i+1}) = M\,[T_a(t_i)] \tag{5.3.29}$$

which says "The forecast is obtained by running the model starting from the previous analysis." It is also necessary to estimate the error variance of the background. In OI, this is done by making some suitably simple assumption, such as that the model integration increases the initial error variance by a fixed amount, e.g., by a factor a not much greater than 1:

$$\sigma_b^2 = a\sigma_a^2 \qquad (5.3.30)$$

In Kalman filtering, (5.3.29) is the same as in OI, but instead of assuming a value for

$$\sigma_b^2\,(t_{i+1})$$

as in (5.3.30), we compute the forecast error covariance using the forecast model itself. Note that if we applied the model (5.3.19) to update the *true* temperature, there would be an error, since the model is not perfect: $T_t\,(t_{i+1}) = M\,[T_t\,(t_i)] - \varepsilon_M$. The model error is assumed to be unbiased (which unfortunately in general is not a good assumption) with an error variance $Q = E(\varepsilon_M^2)$. Then

$$\varepsilon_{b,i+1} = (T_b - T_t)_{i+1} = M(T_a)_i - M(T_t)_i + \varepsilon_M = \mathbf{M}\varepsilon_{a,i} + \varepsilon_M \qquad (5.3.31)$$

where $\mathbf{M} = \partial M/\partial T$ is the linearized or tangent linear model operator. From (5.3.31), we obtain the forecast for the background error covariance at the new time level:

$$\sigma_{b,i+1}^2 = E(\varepsilon_{b,i+1}^2) = \mathbf{M}\sigma_{a,i}^2\mathbf{M} + \mathbf{Q} \qquad (5.3.32)$$

Equation (5.3.32) says: "In Kalman Filter the forecast error variance is updated using the linear tangent model, and we may add the estimated model error variance $\mathbf{Q}$." Note that this "forecast of the error covariance" can be done on small models, but *it is computationally unfeasible for large models*. Only the introduction of ensemble Kalman filter (EnKF) made possible the implementation of Kalman filter on large models.

Exercise 5.3.5 Derive (5.3.31) and (5.3.32)

In the analysis phase (for both OI and Kalman filtering), we get the new observation $T_o(i + 1)$ and we derive the new analysis $T_a(i + 1)$ using (5.3.19), the estimates of w from either (5.3.30) for OI, or (5.3.32) for Kalman filtering, and the new analysis error variance using (5.3.18). After the analysis is done, the cycle for time t_{i+1} has been completed, and we can proceed to the next cycle.

5.4 Multivariate Statistical Data Assimilation Methods

We now generalize the least squares method from its application to a scalar observation and forecast, to the equations for the vectors of observations and vectors of the forecast fields (background) both of which have *huge* dimensions. We start with the OI equations, originally derived by Eliassen et al. (1954) (reproduced in Bengtsson et al. (1981)). However, this statistical interpolation approach was not adopted operationally until Gandin (1963) independently derived the multivariate OI equations and applied them to objective analysis in the former Soviet Union. Lev Gandin's work had a profound impact on the research and operational communities, and OI replaced the empirical Successive Corrections Scheme, becoming the operational analysis scheme of choice during the 1980s and early 1990s (e.g., Lorenc, 1981). In this discussion, we generally follow the notation proposed by Ide et al. (1997) for data assimilation methods, noting that this short paper, although ostensibly devoted to notation, is also

164   Data Assimilation

an excellent overview of data assimilation. We first write the OI equations and their interpretation in Section 5.4.1, so that they can be directly compared with the scalar equations of the previous Section 5.3.3, and then formally derive these equations in Section 5.4.2 (although the derivation in this section may now be skipped). In Section 5.4.3, we introduce 3D-Var and show that, as was the case for the scalar system, the 3D-Var solution is formally equivalent to the OI method. However, the method for solving 3D-Var (minimization of a cost function given by the negative of the log-likelihood of the analysis) is quite different from OI, and it has significant advantages for operational systems (Parrish and Derber, 1992), so that 3D-Var replaced the popular OI in the 1990s. Both 3D-Var and OI use a *constant background error covariance* **B**, whereas the more advanced (but still computationally feasible) methods, 4D-Var and ensemble Kalman filter (EnKF), discussed in Section 5.5, allow for evolving error covariances, i.e., they include "forecast errors of the day."

5.4.1   Multivariate Analysis Cycle: Equations and Their Interpretation

In Section 5.3, we studied the formulation of the optimal analysis of a scalar toy model and a single observation. We now consider the complete NWP operational problem of finding an optimal analysis of a field of model variables $\mathbf{x}_a$, given a background field $\mathbf{x}_b$ available at grid points in two or three space dimensions, and a set of p observations $\mathbf{y}^o$ available at irregularly spaced points $\mathbf{r}_i$ (see Figure 5.2.1). For an atmospheric model, for example, a forecast or analysis vector could be $\mathbf{x} = (p_s, T, q, u, v)$, a very long vector of the ordered atmospheric variables (surface pressure, temperature, moisture, zonal velocity, meridional velocity), each of which is in turn a vector ordered by grid point. So the model state vector $\mathbf{x}$ is then a single vector of length n, where n is the product of the number of horizontal and vertical points times the number of variables (with p_s being 2-dimensional). The (unknown) truth $\mathbf{x}_t$, discretized in the model space, is also a vector of length n. The vector $\mathbf{y}_o$ of length p contains all the observations we want to include in the analysis. The observed variables are, in general, different from the model variables by: (a) being located at different points and (b) possibly being *indirect* measures of the model variables, as in the second toy model, where we only measured the emitted radiations. Examples of such measurements are radar reflectivities and Doppler shifts, satellite radiances, and global positioning system (GPS) radio occultation of atmospheric refractivities.

We will now cast and interpret the multivariate OI equations by using the scalar toy model equations (Section 5.3.3) and simply replacing the scalars (forecasts, analyses, observations) with the corresponding vectors:

$$T_b \rightarrow \mathbf{x}_b; \quad T_a \rightarrow \mathbf{x}_a; \quad y_o \rightarrow \mathbf{y}_o \tag{5.4.1}$$

and their error variances with the corresponding error covariances (we will generally use the shorter symbols **B,A,R** for the covariances, as suggested by Ide et al. (1997)):

$$\sigma_b \rightarrow \mathbf{P}_b = \mathbf{B}; \quad \sigma_a \rightarrow \mathbf{P}_a = \mathbf{A}; \quad \sigma_o \rightarrow \mathbf{P}_o = \mathbf{R} \tag{5.4.2}$$

5.4 Multivariate Statistical Data Assimilation Methods

The scalar model m becomes a multidimensional model M, the linear tangent model dm/dT becomes the matrix $\mathbf{M}$, the multidimensional optimal weight w becomes the Kalman gain matrix $\mathbf{K}$, and the multidimensional observation operator or forward model h becomes H, with its linear tangent (Jacobian) model $\mathbf{H}$:

$$m \to M; \quad dm/dT \to \partial M/\partial \mathbf{x} = \mathbf{M}; \quad w \to \mathbf{K};$$
$$h \to H; \quad dh/dT \to \partial H/\partial \mathbf{x} = \mathbf{H} \tag{5.4.3}$$

So, by simply making these replacements into the scalar equations in Section 5.3.3, the multivariate equations for the analysis cycle become a forecast step and an analysis step:

Forecast step:

$$\mathbf{x}_b(t_{i+1}) = M\,[\mathbf{x}_a(t_i)] \tag{5.4.4}$$

Equation (5.4.4) says: "We use the nonlinear model to forecast $\mathbf{x}_b(t_{i+1})$ starting from the analysis at time t_i."

$$\mathbf{B}(t_{i+1}) = \mathbf{M}\,[\mathbf{A}(t_i)]\,\mathbf{M}^T \tag{5.4.5}$$

Equation (5.4.5) says: "We *could* use the tangent linear model $\mathbf{M}$ and its transpose (the adjoint model $\mathbf{M}^T$) to forecast the background error covariance at time t_{i+1} (where we have ignored the model error $\mathbf{Q}$)," as is done in the *extended Kalman filter* (EKF) approach. However, this step is so computationally expensive that it is **completely unfeasible**.

In optimal interpolation and 3D-Var, the background error covariance $\mathbf{B}(t_{i+1})$ is estimated by the "NMC method" of Parrish and Derber (1992) or similar approaches, and is kept constant in time (see Section 5.4.4). Ensemble Kalman filter (EnKF) instead estimates an evolving $\mathbf{B}(t_{i+1})$ from an ensemble of forecasts, assuming that the ensemble average is an estimate of the truth (Section 5.6). Once we obtain $\mathbf{B}$, we can derive two identical formulations of the Kalman gain matrix $\mathbf{K}$, which does not depend on the new observations:

$$[\mathbf{K}]_{i+1} = \left[\mathbf{B}\mathbf{H}^T\left(\mathbf{R} + \mathbf{H}\mathbf{B}\mathbf{H}^T\right)^{-1}\right]_{i+1}$$
$$or \tag{5.4.6}$$
$$[\mathbf{K}]_{i+1} = \left[\mathbf{A}\mathbf{H}^T\mathbf{R}^{-1}\right]_{i+1} = \left[\left(\mathbf{B}^{-1} + \mathbf{H}^T\mathbf{R}^{-1}\mathbf{H}\right)^{-1}\mathbf{H}^T\mathbf{R}^{-1}\right]_{i+1}$$

The first formula for $\mathbf{K}$ is derived with the sequential (Kalman/OI) approach. It says "The Kalman gain matrix $\mathbf{K}$ is given by the background error covariance scaled by the transpose of $\mathbf{H}$, divided by the total error covariance (sum of observation error covariance and the background error covariance in observation space)." The second formula for $\mathbf{K}$, derived within the variational approach finding the minimum of the cost function, is sometimes more useful. It says "$\mathbf{K}$ is given by the analysis covariance (i.e., the inverse of the sum of the precisions of the model and of the observations scaled to be in model space) divided by the observation error covariance."

Exercise 5.4.1 Show that the two formulations of $\mathbf{K}$ in (5.4.6) are equivalent. Hint: Left multiply both equations by $\left(\mathbf{B}^{-1} + \mathbf{H}^T\mathbf{R}^{-1}\mathbf{H}\right)$ and right multiply them by $\left(\mathbf{R} + \mathbf{HBH}^T\right)$.

Analysis step: Once the new observations at time t_{i+1} become available, we can compute the observational increments and proceed to the analysis (or posterior):

$$\mathbf{x}_a(t_{i+1}) = \mathbf{x}_b(t_{i+1}) + \mathbf{K}_{i+1}\left[\mathbf{y}_o - h(\mathbf{x}_b)\right]_{i+1} \tag{5.4.7}$$

which says "The analysis is equal to the forecast (background or prior) plus the Kalman gain matrix multiplying the innovation (observational increment)."

As indicated before, "the analysis precision is the sum of the forecast and observations precisions":

$$\mathbf{A}^{-1}(t_{i+1}) = \left(\mathbf{B}^{-1} + \mathbf{H}^T\mathbf{R}^{-1}\mathbf{H}\right)_{i+1} \tag{5.4.8}$$

The analysis error covariance can be written also as

$$\mathbf{A}(t_{i+1}) = (\mathbf{I} - \mathbf{KH})\,\mathbf{B}(t_{i+1}) \tag{5.4.9}$$

which says "The analysis error covariance is equal to the background error covariance reduced by the identity matrix minus the Kalman gain scaled so that it is in model space."

A comparison with the interpretation of the analysis cycle equations for the space stone toy model in 5.3.3 shows that they are essentially identical, and that they have exactly the same meaning.

5.4.2 Derivation of OI and 3D-Var Analysis Equations

In Section 5.4.1, we presented the equations of an analysis cycle using OI by simply using the corresponding equations derived for toy scalars in Section 5.3.3 and replacing scalars by vectors and variances by covariances, with the same interpretation as the equations in the stone in space toy scalar model. We will now formally derive these same equations, as used in the complete NWP operational problem of finding an optimum analysis $\mathbf{x}_a$ of a field of model variables, given a background field $\mathbf{x}_b$ available at grid points in two or three dimensions, and a set of p observations $\mathbf{y}_o$ available at irregularly spaced points $\mathbf{r}_i$ (Figure 5.2.1).

As we did in (5.3.19) for a scalar, the analysis is cast as the background plus the innovation weighted by optimal weights which we will determine from statistical interpolation,

$$\begin{aligned}\mathbf{x}_t - \mathbf{x}_b &= \mathbf{K}[\mathbf{y}_o - H(\mathbf{x}_b)] - \boldsymbol{\varepsilon}_a = \mathbf{Kd} - \boldsymbol{\varepsilon}_a \\ \boldsymbol{\varepsilon}_a &= \mathbf{x}_a - \mathbf{x}_t\end{aligned} \tag{5.4.10}$$

but now the truth, the analysis, and the background are vectors of length n (the total number of grid points times the number of model variables) and the optimal weight matrix has dimension $n \times p$. The forward observational operator H converts the background field into first guesses (forecasts) of the observations $\mathbf{y}_b$. H can be nonlinear

5.4 Multivariate Statistical Data Assimilation Methods

(e.g., the radiative transfer equations that go from temperature and moisture vertical profiles to the satellite observed radiances). The observation field $\mathbf{y}_o$ is a vector of length p, the number of observations. The vector $\mathbf{d}$, also of length p, is the innovation or observational increments vector:

$$\mathbf{d} = \mathbf{y}_o - H(\mathbf{x}_b) \qquad (5.4.11)$$

5.4.2.1 Some Mathematical Remarks

(a) The optimal $n \times p$ weight matrix $\mathbf{K}$ is also called the "Kalman gain matrix," and is the same matrix that appears in Kalman filtering.

(b) An error covariance matrix is obtained by multiplying a vector error

$$\boldsymbol{\varepsilon} = \begin{bmatrix} e_1 \\ \vdots \\ e_n \end{bmatrix}$$

by its transpose $\boldsymbol{\varepsilon}^T = [e_1, \ldots, e_n]$ and averaging over many cases, to obtain the expected value:

$$P = \overline{\boldsymbol{\varepsilon}\boldsymbol{\varepsilon}^T} = \begin{bmatrix} \overline{e_1 e_1} & \overline{e_1 e_2} & \cdots & \overline{e_1 e_n} \\ \overline{e_2 e_1} & \overline{e_2 e_2} & \cdots & \overline{e_2 e_n} \\ \vdots & \vdots & & \vdots \\ \overline{e_n e_1} & \overline{e_n e_2} & \cdots & \overline{e_n e_n} \end{bmatrix} \qquad (5.4.12)$$

where the overbar represents the expected value (i.e., is the same as $E(\cdot)$). A covariance matrix is symmetric and semi-positive definite. The diagonal elements are the variances of the vector error components $\overline{e_i e_i} = \sigma_i^2$.

(c) The transpose of matrix products is given by the product of the transposes, but in reverse order; a similar rule applies to the inverse of a product:

$$[\mathbf{AB}]^T = \mathbf{B}^T \mathbf{A}^T; \quad [\mathbf{AB}]^{-1} = \mathbf{B}^{-1} \mathbf{A}^{-1} \qquad (5.4.13)$$

(d) The general form of a quadratic function (a scalar) is

$$F(\mathbf{x}) = \frac{1}{2} \left[\mathbf{x}^T \mathbf{A} \mathbf{x} \right] + \mathbf{d}^T \mathbf{x} + c$$

where $\mathbf{A}$ is a symmetric matrix, $\mathbf{d}$ is a vector and c a scalar. To find the gradient of this scalar function $\nabla_{\mathbf{x}} F = \partial F / \partial \mathbf{x}$ (a column vector), we use the following properties of the gradient with respect to $\mathbf{x}$: $\nabla \left(\mathbf{d}^T \mathbf{x} \right) = \nabla \left(\mathbf{x}^T \mathbf{d} \right) = \mathbf{d}$ (since $\nabla_{\mathbf{x}} \mathbf{x}^T = \mathbf{I}$, the identity matrix), and $\nabla \left(\mathbf{x}^T \mathbf{A} \mathbf{x} \right) = 2 \mathbf{A} \mathbf{x}$. Therefore,

$$\nabla F(\mathbf{x}) = \mathbf{A}\mathbf{x} + \mathbf{d}, \quad \nabla^2 F(\mathbf{x}) = \mathbf{A} \quad \text{and} \quad \delta F = (\nabla F)^T \delta \mathbf{x} \qquad (5.4.14)$$

(e) Formulation of multiple regression or best linear unbiased estimation (BLUE). Assume we have two time series of vectors

$$\mathbf{x}(t) = \begin{bmatrix} x_1(t) \\ x_2(t) \\ \vdots \\ x_n(t) \end{bmatrix} \quad \mathbf{y}(t) = \begin{bmatrix} y_1(t) \\ y_2(t) \\ \vdots \\ y_p(t) \end{bmatrix} \tag{5.4.15}$$

centered about their mean value, $E(\mathbf{x}) = 0, E(\mathbf{y}) = 0$, i.e., vectors of *anomalies*.

We derive now the best linear unbiased estimation (BLUE) of $\mathbf{x}$ in terms of $\mathbf{y}$, i.e., the optimal value of the weight matrix $\mathbf{W}$ of dimension $n \times p$ in the multiple linear regression

$$\mathbf{x}_a(t) = \mathbf{W}\mathbf{y}(t) \tag{5.4.16}$$

which approximates the true relationship $\mathbf{x}(t) = \mathbf{W}\mathbf{y}(t) - \boldsymbol{\varepsilon}(t)$.

Here $\boldsymbol{\varepsilon}(t) = \mathbf{x}_a(t) - \mathbf{x}(t)$ is the linear regression (analysis) error, and $\mathbf{W}$ is an $n \times p$ matrix that minimizes the mean squared error $E\left(\boldsymbol{\varepsilon}^T\boldsymbol{\varepsilon}\right)$. In order to derive optimal $\mathbf{W}$ we write the regression equation matrix components explicitly:

$$x_i(t) = \sum_{k=1}^{p} w_{ik}\,y_k(t) - \varepsilon_i(t) \tag{5.4.17}$$

Then the error that we want to minimize is $\sum_{i=1}^{n}\varepsilon_i^2(t) = \sum_{i=1}^{n}\left[\sum_{k=1}^{p}w_{ik}y_k(t) - x_i(t)\right]^2$ and the derivative with respect to the weight matrix components is

$$\frac{\partial \sum_{i=1}^{n}\varepsilon_i^2(t)}{\partial w_{ij}} = 2\left[\sum_{k=1}^{p}w_{ik}y_k(t) - x_i(t)\right]\left[y_j(t)\right] = 2\left[\sum_{k=1}^{p}w_{ik}y_k(t)y_j(t) - x_i(t)y_j(t)\right]$$

In matrix form, this is $\frac{\partial \boldsymbol{\varepsilon}^T\boldsymbol{\varepsilon}}{\partial w_{ij}} = 2\{[\mathbf{W}\mathbf{y}(t)\mathbf{y}^T(t)]_{ij} - [\mathbf{x}(t)\mathbf{y}^T(t)]_{ij}\}$ so that if we take a long time mean and choose $\mathbf{W}$ to minimize the mean squared error, we get $\mathbf{W}E\left(\mathbf{y}\mathbf{y}^T\right) - E\left(\mathbf{x}\mathbf{y}^T\right) = 0$ or the normal equation

$$\mathbf{W} = E\left(\mathbf{x}\mathbf{y}^T\right)\left[E\left(\mathbf{y}\mathbf{y}^T\right)\right]^{-1} \tag{5.4.18}$$

which gives the best linear unbiased estimation (BLUE) for the linear regression $\mathbf{x}_a(t) = \mathbf{W}\mathbf{y}(t)$.

5.4.2.2 Statistical Assumptions and Derivation of OI and 3D-Var Formulas

We define the background error and the analysis error as vectors of length n:

$$\begin{aligned} \boldsymbol{\varepsilon}_b(x, y) &= \mathbf{x}_b(x, y) - \mathbf{x}_t(x, y) \\ \boldsymbol{\varepsilon}_a(x, y) &= \mathbf{x}_a(x, y) - \mathbf{x}_t(x, y) \end{aligned} \tag{5.4.19}$$

The p observations available at irregularly spaced points $\mathbf{y}_o(\mathbf{r}_i)$ have observational errors

$$\boldsymbol{\varepsilon}_{oi} = \mathbf{y}_o(\mathbf{r}_i) - \mathbf{y}_t(\mathbf{r}_i) = \mathbf{y}_0(\mathbf{r}_i) - H[\mathbf{x}_t(\mathbf{r}_i)] \tag{5.4.20}$$

We don't know the truth $\mathbf{x}_t$, thus we don't know the errors of the available background and observations, but we can make a number of assumptions about

5.4 Multivariate Statistical Data Assimilation Methods 169

their statistical properties. The background and observations are assumed to be unbiased:

$$\left.\begin{aligned} E\{\boldsymbol{\varepsilon}_b(x,y)\} &= E\{\mathbf{x}_b(x,y)\} - E\{\mathbf{x}_t(x,y)\} = \mathbf{0} \\ E\{\boldsymbol{\varepsilon}_c(\mathbf{r}_i)\} &= E\{\mathbf{y}_o(\mathbf{r}_i)\} - E\{\mathbf{y}_t(\mathbf{r}_i)\} = \mathbf{0} \end{aligned}\right\} \tag{5.4.21}$$

If the forecasts (background) and the observations are biased, in principle, we can and should correct the bias before proceeding. Dee and Da Silva (1998) showed how the model bias can be estimated as part of the analysis cycle.

The error covariance matrices for the analysis $\mathbf{A}$, background $\mathbf{B}$, and observations $\mathbf{R}$, are defined as

$$\left.\begin{aligned} \mathbf{P}_a &= \mathbf{A} = E\{\boldsymbol{\varepsilon}_a\boldsymbol{\varepsilon}_a^T\} \\ \mathbf{P}_b &= \mathbf{B} = E\{\boldsymbol{\varepsilon}_b\boldsymbol{\varepsilon}_b^T\} \\ \mathbf{P}_o &= \mathbf{R} = E\{\boldsymbol{\varepsilon}_o\boldsymbol{\varepsilon}_o^T\} \end{aligned}\right\} \tag{5.4.22}$$

The nonlinear observation operator H that transforms model variables into observed variables can be linearized with respect to a perturbation $\delta\mathbf{x}$ as

$$H(\mathbf{x} + \delta\mathbf{x}) = H(\mathbf{x}) + \mathbf{H}\delta\mathbf{x} \tag{5.4.23}$$

where $\mathbf{H}$ is a $p \times n$ matrix known as the linear observation operator or *Jacobian of the observation operator*, with elements $h_{i,j} = \partial H_i / \partial x_j$. We also assume that the background (usually a model forecast) is a good approximation of the truth, so that the analysis and the observations are equal to the background values plus small increments. Therefore, the innovation vector (5.4.22) can be written as

$$\begin{aligned} \mathbf{d} &= \mathbf{y}_o - H(\mathbf{x}_b) = \mathbf{y}_o - H(\mathbf{x}_t + (\mathbf{x}_b - \mathbf{x}_t)) \\ &= \mathbf{y}_o - H(\mathbf{x}_t) - H(\mathbf{x}_b - \mathbf{x}_t) = \boldsymbol{\varepsilon}_o - \mathbf{H}\boldsymbol{\varepsilon}_b \end{aligned} \tag{5.4.24}$$

The $\mathbf{H}$ matrix transforms vector perturbations in model space into their corresponding values in observation space. Its transpose or adjoint $\mathbf{H}^T$ transforms vector perturbations in observation space to vectors in model space. In OI, the background error covariance $\mathbf{B}$ (a matrix of dimension $n \times n$) and the observation error covariance R (a matrix of dimension $p \times p$) *are assumed to be known and constant*. In addition, we assume that the observation and background errors are uncorrelated:

$$E\{\boldsymbol{\varepsilon}_o\boldsymbol{\varepsilon}_b^T\} = 0 \tag{5.4.25}$$

We will now use the best linear unbiased estimation (BLUE) formula (5.4.18) to derive the optimal weight matrix $\mathbf{K}$, which approximates the true relationship $\mathbf{x}_t - \mathbf{x}_b = \mathbf{K}\mathbf{d} - \boldsymbol{\varepsilon}_a$. From (5.4.24), $\mathbf{d} = \boldsymbol{\varepsilon}_o - \mathbf{H}\boldsymbol{\varepsilon}_b$, and from (5.4.18), the optimal weight matrix $\mathbf{K}$ (or Kalman gain) that minimizes $\boldsymbol{\varepsilon}_a^T\boldsymbol{\varepsilon}_a$ is given by

$$\begin{aligned} \mathbf{K} &= E\{(\mathbf{x}_t - \mathbf{x}_b)[\mathbf{y}_o - H(\mathbf{x}_b)]^T\}(E\{[\mathbf{y}_o - H(\mathbf{x}_b)][\mathbf{y}_o - H(\mathbf{x}_b)^T]\})^{-1} \\ &= E[(-\boldsymbol{\varepsilon}_b)(\boldsymbol{\varepsilon}_o - \mathbf{H}\boldsymbol{\varepsilon}_b)^T]\{E[(\boldsymbol{\varepsilon}_o - \mathbf{H}\boldsymbol{\varepsilon}_b)(\boldsymbol{\varepsilon}_o - \mathbf{H}\boldsymbol{\varepsilon}_b)^T]\}^{-1} \end{aligned} \tag{5.4.26}$$

Recall that in (5.4.25), we assumed that the background errors are not correlated with the observational errors, i.e., that their covariance is equal to zero. Substituting

170 **Data Assimilation**

the definitions of background error covariance $\mathbf{B}$ and observational error covariance $\mathbf{R}$ from (5.4.22) into (5.4.26), we obtain the optimal weight or Kalman gain matrix

$$\mathbf{K} = \mathbf{B}\mathbf{H}^T \left(\mathbf{R} + \mathbf{H}\mathbf{B}\mathbf{H}^T\right)^{-1} \qquad (5.4.27)$$

Finally, we derive the analysis error covariance:

$$\begin{aligned}
\mathbf{A} &= E\{\boldsymbol{\varepsilon}_a \boldsymbol{\varepsilon}_a^T\} \\
&= E\{\boldsymbol{\varepsilon}_b \boldsymbol{\varepsilon}_b^T + \boldsymbol{\varepsilon}_b(\boldsymbol{\varepsilon}_o - H\boldsymbol{\varepsilon}_b)^T \mathbf{K}^T + \mathbf{K}(\boldsymbol{\varepsilon}_o - H\boldsymbol{\varepsilon}_b)\boldsymbol{\varepsilon}_b^T \\
&\quad + \mathbf{K}(\boldsymbol{\varepsilon}_o - H\boldsymbol{\varepsilon}_b)(\boldsymbol{\varepsilon}_o - H\boldsymbol{\varepsilon}_b)^T \mathbf{K}^T\} \\
&= \mathbf{B} - \mathbf{B}\mathbf{H}^T \mathbf{K}^T - \mathbf{K}\mathbf{H}\mathbf{B} + \mathbf{K}\mathbf{R}\mathbf{K}^T + \mathbf{K}\mathbf{H}\mathbf{B}\mathbf{H}^T \mathbf{K}^T \qquad (5.4.28)
\end{aligned}$$

and substituting (5.4.27) we obtain

$$\mathbf{A} = (\mathbf{I} - \mathbf{K}\mathbf{H})\,\mathbf{B} \qquad (5.4.29)$$

An alternative formulation for the analysis error covariances showing (as in the scalar case) the additive properties of the precisions (if all the statistical assumptions hold true) can be derived from the variational form of $\mathbf{K}$ in (5.4.6), using (5.4.10) and (5.4.25). We can show that

$$\begin{aligned}
\boldsymbol{\varepsilon}_a &= \boldsymbol{\varepsilon}_b + [\mathbf{B}^{-1} + \mathbf{H}^T \mathbf{R}^{-1} \mathbf{H}]^{-1} \mathbf{H}^T \mathbf{R}^{-1}(\boldsymbol{\varepsilon}_o - H\boldsymbol{\varepsilon}_b) \\
&= [\mathbf{B}^{-1} + \mathbf{H}^T \mathbf{R}^{-1} \mathbf{H}]^{-1} [\mathbf{B}^{-1}\boldsymbol{\varepsilon}_b + \mathbf{H}^T \mathbf{R}^{-1} \boldsymbol{\varepsilon}_o]
\end{aligned} \qquad (5.4.30)$$

If we compute $\mathbf{A} = E(\boldsymbol{\varepsilon}_a \boldsymbol{\varepsilon}_a^T)$, we obtain

$$\mathbf{A}^{-1} = \mathbf{B}^{-1} + \mathbf{H}^T \mathbf{R}^{-1} \mathbf{H} \qquad (5.4.31)$$

Note that these estimations of the analysis uncertainties are dependent on the assumption that the statistical estimates of the errors are accurate. If the observations and/or the background error covariances are poorly estimated, if there are observation or model biases, or if the observations and background errors are correlated, the analysis precision can be considerably worse than implied by these analysis equations.

5.4.3 Numerical Solutions of OI and 3D-Var

We have seen that 3D-Var and OI are equivalent so that analytically they solve the same problem (Exercise 5.4.1, see also Lorenc (1986)). However, since the numerical methods to solve these equations numerically are very different, the numerical solutions for the analyses can also have different characteristics. If, as occurs in reality, the statistics are only approximations of the true statistics, then the OI equations

$$\mathbf{x}_a - \mathbf{x}_b = \mathbf{K}\left[\mathbf{y}_o - H(\mathbf{x}_b)\right]$$

or

$$\delta\mathbf{x}_a = \mathbf{K}\mathbf{d}$$

and

$$\mathbf{K} = \mathbf{B}\mathbf{H}^T (\mathbf{H}\mathbf{B}\mathbf{H}^T + \mathbf{R})^{-1}$$

provide a *statistical interpolation*, not necessarily an *optimal interpolation*.

5.4.3.1 Remarks: Errors of Representativeness, Error Correlations, and Super Observations

(a) It is important to note that observation errors come from two different sources: one is the instrumental error variances proper, the second is the presence in the observations of subgrid-scale variability not represented in the grid-average values of the model and analysis. The second type of error is denoted *error of representativeness*. By performing a grid average similar to the Reynolds average discussed in Chapter 4, we obtain that the observational error variance $\mathbf{R}$ is the sum of the instrument error variance $\mathbf{E}$ and the representativeness error variance $\mathbf{F}$, assuming that these errors are not correlated. If in addition we allow for errors in the observation operator H with observation error covariance $\mathbf{R}_H$, these can also be included in the observation error covariance (Lorenc, 1986):

$$\mathbf{R} = \mathbf{E} + \mathbf{F} \quad (+\mathbf{R}_H)$$

(b) Perhaps the most important advantage of statistical interpolation schemes such as OI and 3D-Var over empirical schemes such as SCM is that the error correlation between observational increments is taken into account. Recall that in SCM, the weights of the observational increments depend only on their distance to the grid point. Therefore, in SCM, all observations will be given similar weight even if a number of them are "bunched up" in one quadrant (Figure 5.2.1), with just a single observation in a different quadrant. In OI (or 3D-Var), by contrast, the isolated observational increment will be given more weight in the analysis than observations that are close together and therefore less independent. The fact that isolated observations have more independent information than observations close together is a result of the fact that the forecast error correlations $b_{jk} / \sqrt{b_{jj} b_{kk}}$ at the observation points j, k are large if the observation points are close together.

(c) When several observations are too close together, then the solution of (5.4.29) becomes an ill-posed problem. In those cases, it is common to create a "superobservation" combining the close individual observations. This has the advantage of removing the ill-posedness, while at the same time reducing the error of the "superob" by averaging the random errors of the individual observations. The superob should be a weighted average that takes into account the relative observation errors of the original observations. An alternative to creating superobs is to "thin" the observations, simply deleting many of them, in order to avoid the use of many observations with correlated errors which are not accounted for in the observation error covariance $\mathbf{R}$, usually assumed to be diagonal (Miyoshi et al., 2013; Kotsuki et al., 2017b).

5.4.3.2 Optimal Interpolation

In order to make it computationally feasible, OI has been solved in limited regions in physical (model) space, either grid point by grid point (e.g., McPherson et al., 1979; DiMego et al., 1985) or over limited volumes (Lorenc, 1981). Here we will show how to solve OI in a simple grid point example (Figure 5.4.1). Recall that $\mathbf{H}$ is the linear perturbation (Jacobian) of the forward observational model H, and $\mathbf{H}^T$ is its transpose or adjoint. Multiplying by $\mathbf{H}$ on the left transforms grid-point increments into

Figure 5.4.1 Simple system with three grid points (black dots) e, f, g, and two observation points, 1 and 2.

observation increments (e.g., by linear interpolation), and right-multiplying observation increments by $\mathbf{H}^T$ transforms them back to grid point increments. There are n grid points, or if we are considering several variables, n is the product of the number of grid points and the variables. Consider a specific grid point with the subscript g. The subscripts j and k represent particular observations affecting the grid point g, and there are p such observations. $\mathbf{B}$ is the background error covariance, so that if the background error is $\varepsilon_b(x, y) = \mathbf{x}_b(x, y) - \mathbf{x}_t(x, y)$, then $b_{jk} = E[\varepsilon_b(x_j, y_j)\varepsilon_b^T(x_k, y_k)]$, estimated or averaged over many cases.

As an illustration, let us write these equations for the case of three grid points e, f, g, and two observations, 1 and 2 (Figure 5.4.1).

In this simple case, $\delta \mathbf{x}_a = (\delta x_e^a, \delta x_f^a, \delta x_g^a)^T$, the observation vector increment is $\mathbf{d} = (y_1^o - y_1^b, y_2^o - y_2^b)^T$, and the background values at the observation points are

$$\mathbf{H}\mathbf{x}^b = \begin{pmatrix} h_{1e} & h_{1f} & h_{1g} \\ h_{2e} & h_{2f} & h_{2g} \end{pmatrix} \begin{pmatrix} x_e^b \\ x_f^b \\ x_g^b \end{pmatrix} = \mathbf{y}^b$$

The coefficients of the observation operator $\mathbf{H}$ are obtained from linear or higher order interpolation of the grid location to the observation location. We are assuming that the observed and analyzed variables are the same, so that the coefficients of the matrix $\mathbf{H}$ are simply interpolation coefficients. For example, if we used linear interpolation, $\mathbf{H}$ would be

$$\mathbf{H} = \begin{pmatrix} \frac{x_f - x_1}{x_f - x_e} & \frac{x_1 - x_e}{x_f - x_e} & 0 \\ 0 & \frac{x_g - x_2}{x_g - x_f} & \frac{x_2 - x_f}{x_g - x_f} \end{pmatrix}$$

The background error covariance matrix elements are the error covariances between grid points:

$$\mathbf{B} = \begin{pmatrix} b_{ee} & b_{ef} & b_{eg} \\ b_{fe} & b_{ff} & b_{fg} \\ b_{ge} & b_{gf} & b_{gg} \end{pmatrix}$$

so that

$$\mathbf{B}\mathbf{H}^T = \begin{pmatrix} b_{e1} & b_{e2} \\ b_{f1} & b_{f2} \\ b_{g1} & b_{g2} \end{pmatrix}$$

is an approximation by interpolation of the background error covariances between grid to observation points, e.g., $b_{g2} = b_{ge}h_{2e} + b_{gf}h_{2f} + b_{gg}h_{2g}$. Then,

$$\mathbf{HBH}^T = \begin{pmatrix} b_{11} & b_{12} \\ b_{21} & b_{22} \end{pmatrix}$$

is an approximation by back interpolation of the background error covariance between observation points. The observation error covariance for this case is

$$\mathbf{R} = \begin{pmatrix} r_{11} & r_{12} \\ r_{21} & r_{22} \end{pmatrix}$$

It is usually a reasonable assumption that measurement errors made at different locations are uncorrelated, in which case R is a diagonal matrix. (Measurement errors could be correlated, but only within small groups of observations made by the same instrument, in which case R is a block diagonal matrix, where the blocks are still easily invertible.)

From this simple example, it is apparent that, in general, we can write the OI equation for each grid point g *influenced by p observations* as an equation for the optimal weights k_{gj}, the elements of the optimal weight matrix $\mathbf{K}$, solving the equation $\mathbf{K}(\mathbf{R} + \mathbf{HBH}) = \mathbf{BH}^T$ for the grid point:

$$\sum_{j=1}^{p} k_{gj}(b_{jk} + r_{jk}) = b_{gk} \qquad k = 1, \ldots, p \tag{5.4.32}$$

and then using these optimal weights to solve $\delta \mathbf{x}_a = \mathbf{Kd}$:

$$\delta x_g^a = \sum_{j=1}^{p} k_{gj} d_j \tag{5.4.33}$$

Note the fundamental role that the background error covariance plays in the determination of the optimal weights (5.4.32). The background error covariance determines the scale and the structure of the corrections to the background. Therefore, the practical implementation of (5.4.32) and (5.4.33) requires an estimation of the elements of the background error covariance $\mathbf{B}$, where a number of additional simplifications are usually made as discussed in Section 5.4.4.

5.4.3.3   3D-Var

We discuss in more detail the numerical solution of 3D-Var because it is more complex than that of OI. As indicated in the previous two sections, the formulation of 3D-Var is based on minimizing a scalar cost function proportional to the negative of the log-likelihood, so that the minimizing solution $\mathbf{x}$ is the most likely analysis given the prior forecast and the new observations.

$$J(\mathbf{x}) = \frac{1}{2} \left[(\mathbf{x} - \mathbf{x}_b)^T \mathbf{B}^{-1}(\mathbf{x} - \mathbf{x}_b) + (H(\mathbf{x}) - \mathbf{y}_o)^T \mathbf{R}^{-1}(H(\mathbf{x}) - \mathbf{y}_o) \right] \tag{5.4.34}$$

The cost function in the brackets has two terms. The first is J_b, measuring the distance or misfit between the solution we are seeking and the forecast $\mathbf{x}_b$, scaled by the background error covariance $\mathbf{B}$. The second is J_o, measuring the misfit between the model solution transformed into observation space $H(\mathbf{x})$ and the observations $\mathbf{y}_o$, scaled by the observation error covariance $\mathbf{R}$.

We can write this cost function in *incremental form* by linearizing with respect to $\delta\mathbf{x} = \mathbf{x} - \mathbf{x}_b$ (Courtier et al., 1994):

$$J(\delta\mathbf{x}) = \frac{1}{2}\left[\delta\mathbf{x}^T\mathbf{B}^{-1}\delta\mathbf{x} + (\mathbf{H}\delta\mathbf{x} - \mathbf{d})^T\mathbf{R}^{-1}(\mathbf{H}\delta\mathbf{x} - \mathbf{d})\right] \tag{5.4.35}$$

where $\quad \mathbf{d} = \mathbf{y}_o - H(\mathbf{x}_b), \quad H(\mathbf{x}) - \mathbf{y}_o \simeq \mathbf{H}\delta\mathbf{x} - \mathbf{d}.$

Exercise 5.4.2 Derive (5.4.35) from (5.4.34).

Exercise 5.4.3 The Laplacian of the cost function is known as the *Hessian*. Show that the Hessian of the 3D-Var cost function is equal to the inverse of analysis error covariance $\mathbf{A}$.

The cost function is now a quadratic function of the analysis increments $\delta\mathbf{x}$ and so we can use Remark 5.4.2.1(d): Given a quadratic function $F(\mathbf{x}) = \frac{1}{2}\mathbf{x}^T\mathbf{A}\mathbf{x} + \mathbf{d}^T\mathbf{x} + c$, where $\mathbf{A}$ is a symmetric matrix, $\mathbf{d}$ is a vector and c a scalar, its gradient is given by $\nabla F(\mathbf{x}) = \mathbf{A}\mathbf{x} + \mathbf{d}$. Therefore, the gradient of the cost function J with respect to $\mathbf{x}$ (or with respect to $\delta\mathbf{x}$) is

$$\nabla J(\delta\mathbf{x}) = \mathbf{B}^{-1}\delta\mathbf{x} + \mathbf{H}^T\mathbf{R}^{-1}\mathbf{H}\delta\mathbf{x} - \mathbf{H}^T\mathbf{R}^{-1}\mathbf{d} \tag{5.4.36}$$

To obtain the minimum of J at $\mathbf{x} = \mathbf{x}_a$, we set $\nabla J(\delta\mathbf{x}) = 0$ and solve for $\delta\mathbf{x} = \delta\mathbf{x}_a$:

$$\delta\mathbf{x}_a = (\mathbf{B}^{-1} + \mathbf{H}^T\mathbf{R}^{-1}\mathbf{H})^{-1}\mathbf{H}^T\mathbf{R}^{-1}\mathbf{d} = \mathbf{A}^{-1}\mathbf{H}^T\mathbf{R}^{-1}\mathbf{d} \tag{5.4.37}$$

Formally, this is the solution of the variational (3D-Var) analysis problem, but the presence of $\mathbf{B}^{-1}$ multiplying the control variable $\mathbf{x}$ would make it extremely difficult to solve. In practice, the solution is obtained through minimization algorithms for a *preconditioned* version of the cost function, using iterative methods for minimization such as the conjugate gradient or quasi-Newton methods. As we saw before, the OI analysis solution (which minimizes the analysis error covariance and finds the optimal weight matrix through a least squares approach), and the 3D-Var analysis (which minimizes a cost function measuring the distance to the observations weighted by the inverse of the error variance) are mathematically the same. However, since the two algorithms are solved using very different methods and different approximations, the numerical solutions to the analysis increments can be quite different.

Parrish and Derber (1992, PD92 hereafter) pioneered the first operational 3D-Var system at NCEP (at that time known as NMC) in 1991, followed by ECMWF in 1996, Meteo-France in 1998 and the UKMO in 1999 (the latter three centers implemented 4D-Var a few years after). PD92 implemented for the first time a number of ideas, including the preconditioning of the control variable for the minimization, a new approach to estimate the background error covariance $\mathbf{B}$ that became widely used, known as "the NMC method" (see next Subsection 5.4.4.3), and the implementation of the statistical interpolation on a spectral model. PD92 were inspired by Phillips (1986) in noting that if error covariances are *homogeneous* and *isotropic*, (i.e., they don't change if shifted or rotated horizontally), then the background error covariance

5.4 Multivariate Statistical Data Assimilation Methods

B in *spectral* form is *diagonal*. This simplified and accelerated the global solution of the minimization. PD92 called this 3D-Var approach Spectral Statistical Interpolation (SSI). Wu et al. (2002) extended this approach to a grid-point-based 3D-Var, allowing for a more realistic horizontal variability in the covariances. This approach, known as the Gridpoint Statistical Interpolation (GSI), was implemented at NCEP in 2007 (Kleist et al., 2009). The GSI allowed for spatially varying error correlations and estimated the background error correlation in space using a very efficient "recursive filter" (Purser et al., 2003a,b) discussed later.

As an example, we briefly describe the 3D-Var algorithm as implemented by Miyoshi (2005), following Barker et al. (2004) and PD92. Since Miyoshi applied 3D-Var to the SPEEDY intermediate global primitive equations model (Molteni, 2003), this implementation is relatively simple but contains all the essential elements of 3D-Var. The cost function is preconditioned by a transformation of the control variable:

$$\delta \mathbf{x} = \mathbf{U} \delta \mathbf{v}; \quad \delta \mathbf{x}^T = \delta \mathbf{v}^T \mathbf{U}^T \tag{5.4.38}$$

where

$$\mathbf{B} = \mathbf{U} \mathbf{U}^T \tag{5.4.39}$$

and $\delta \mathbf{v}$ is a non dimensional vector with the same number of degrees of freedom as $\delta \mathbf{x}$. The generation of the background error covariance **B** is the most fundamental problem in data assimilation, even when using a climatological, constant estimation, as in OI and 3D-Var. How to construct the estimated background error covariance is discussed in the next section.

Substituting (5.4.38) into the cost function (5.4.35), we obtain

$$J(\delta \mathbf{v}) = \frac{1}{2} \left[\delta \mathbf{v}^T \delta \mathbf{v} + (\mathbf{H} \mathbf{U} \delta \mathbf{v} - \mathbf{d})^T \mathbf{R}^{-1} (\mathbf{H} \mathbf{U} \delta \mathbf{v} - \mathbf{d}) \right] \tag{5.4.40}$$

The transformation of variables (5.4.38) *preconditions* the cost function J by making it depend like a "hyper sphere" on the control variable $\delta \mathbf{v}$, rather than the original, very elongated "hyper ellipsoid" dependence on $\delta \mathbf{x}$. This greatly accelerates the minimization since the gradient of the cost function points much closer to the minimum if the cost function is spherical than if it is ellipsoidal. The gradient of the cost function becomes

$$\nabla J(\delta \mathbf{v}) = \delta \mathbf{v} + \mathbf{U}^T \mathbf{H}^T \mathbf{R}^{-1} (\mathbf{H} \mathbf{U} \delta \mathbf{v} - \mathbf{d}) \tag{5.4.41}$$

The construction of the variable transformation **U** is the key in the 3D-Var algorithm. Following PD92, **U** can be separated into an error standard deviation, **A**, a spatial error correlation **C**, and an inter-variable correlation **V** that includes, for example, the geostrophic forecast error correlation between wind and height/temperature forecast errors:

$$\mathbf{U} = \mathbf{V} \mathbf{C} \mathbf{A} \tag{5.4.42}$$

5.4.3.4 Computation of A, C, and V from the "NMC Method"

Background Error Covariance B

The construction of the climatological background error covariance **B** with the "NMC method" is described in detail in the next subsection. Here we indicate how to construct the components of **U** as given by (5.4.42).

(a) Standard deviation

The first component of **U** is the error standard deviation **A** in (5.4.42). In the SSI of PD92, the analysis is done in spectral space, so that the standard deviation is given by the square root of the coefficients of **B** for each spectral component. Since the spectral **B** is assumed to be diagonal (no correlation of errors between different spectral modes) the error correlation is homogeneous and isotropic (i.e., the standard deviations are the same at different locations and the correlations depend only on distance), which is not realistic. For the SPEEDY model, since the 3D-Var analyzes not in spectral space but in the physical grid space, **B** is computed in model grid space, and the components of **A** are the square roots of the **B** values.

(b) Spatial correlation and the recursive filter

The spatial error correlation **C** is assumed to be separable into horizontal and vertical error correlations, and also assumed to be Gaussian. Their correlation length scale can be directly estimated from **B** and stored. Furthermore, these length scales are assumed to be only a function of vertical level. The vertical error correlation could also be included, but for the SPEEDY model with only seven vertical levels, no vertical error correlations were included. Computing the spatial Gaussian correlation functions for each control variable and each grid point would be computationally unfeasible, since the dimension n is of order of $10^7 - 10^9$. Instead, a clever application of recursive filters (Purser et al., 2003a,b) allows to quickly approximate the local two-dimensional Gaussian shape for each grid point. A recursive filter (RF) is a space (or time) filter where values from its previous outputs are used as inputs. Miyoshi (2005) uses the RF filter with four terms, and we follow his derivation, based on Purser et al. (2003a).

If the input is a delta function $A(x) = \delta(x) = \frac{1}{2\pi} \int \tilde{A}(k)e^{ikx}dk$, then the output $C(x)$ of the RF with infinite terms is a Gaussian function with length scale $(\sigma\Delta x)$, where Δx is the grid size:

$$C(x) = D_{(\infty)}^{-1}\delta(x) = \frac{1}{2\pi} \exp\left(-\frac{x^2}{2(\sigma\Delta x)^2}\right) \tag{5.4.43}$$

If instead of an infinite number of terms, we use only n, the output of a delta function will be an approximation of the same Gaussian function (equation 5.4.43). Wu et al. (2002) and Miyoshi (2005) used $n=4$ and showed that this is already a rather accurate representation of the Gaussian filter, transforming the input signal $A(x) = \delta(x)$ into a good approximation of a Gaussian. The filter (5.4.43) and its approximations with n terms can be separated into two terms, one going forward in space, with input A and

5.4 Multivariate Statistical Data Assimilation Methods

output B and another going backward in space, with input B and output C. For the 4th order RF these steps are written as

$$B_i = \beta A_i + \alpha_1 B_{i-1} + \alpha_2 B_{i-2} + \alpha_3 B_{i-3} + \alpha_4 B_{i-4} \tag{5.4.44}$$

$$C_i = \beta C_i + \alpha_1 C_{i+1} + \alpha_2 C_{i+2} + \alpha_3 C_{i+3} + \alpha_4 C_{i+4} \tag{5.4.45}$$

In equation (5.4.44, the forward process in ascending order of i), A is the input and the filtered value B is obtained. In equation (5.4.45, the backward process in descending order of i), B is the input and the filtered value C is obtained. Equations (5.4.44) and (5.4.45) are mutually adjoint, so (5.4.44) may be used in $\mathbf{U}$ whereas (5.4.45) may be used in $\mathbf{U}^T$. The coefficients β, α_i are obtained by solving a 4th degree equation and depend on the correlation length scale Purser et al. (2003a). In two dimensions, a similar transformation is carried out in the y direction, and multiplied by the transformation in the x direction, so that using the RF to get a two-dimensional quasi-Gaussian of the input at a given grid point only doubles the computations compared to one dimension.

(c) Inter-variable correlation errors

The forecast errors of winds and heights (and temperatures) are strongly correlated because of the quasi-geostrophic balance. It is important to cast the minimization problem for 3D-Var with control variables that are not correlated (PD92). Therefore, an important component of 3D-Var is the expression of the inter-variable correlation $\mathbf{V}$, which transforms prognostic variables into control variables that can be assumed to be independent of each other. Some prognostic variables are strongly dependent on each other mostly because of the geostrophic balance. In low resolution global models, only the geostrophic balance is considered since it is the strongest and the most important balance in the system. However, cyclostrophic balance, including the centrifugal force due to flow with strong curvature, like in hurricanes, should be included with high resolution models (Barker et al., 2004). Also, frictional effects near the surface can introduce error correlations and should be accounted for.

Miyoshi (2005) included only the geostrophic balance by replacing the wind variables with the unbalanced wind components:

$$u_u = u - r_1 u_g(p_s, T), \quad v_u = v - r_2 v_g(p_s, T) \tag{5.4.46}$$

where r_1 and r_2 are regression coefficients determined from statistics, and u_g and v_g are the zonal and meridional components of the geostrophic wind computed from p_s and T using the geostrophic balance equation on the sigma coordinate system, see equation (2.6.27) in Chapter 2:

$$f \mathbf{k} \times \mathbf{v}_g = -RT\nabla \ln p_s - \nabla \phi \tag{5.4.47}$$

where f, $\mathbf{k}$, R, and ϕ are the Coriolis parameter, vertical unit vector, gas constant, and geopotential height, respectively. The other control variables (T, q, and p_s) are assumed to be independent of the unbalanced wind control variables (u_u and v_u). Thus, only the geostrophic balance is considered for inter-variable correlation. Figure

3.1 in Miyoshi (2005) shows that the regression coefficients between the forecast error in the geostrophic and the unbalanced wind components are, as could be expected, very small in the tropics and increase to over 0.7 in mid latitudes and upper levels. These coefficients are obtained from the climatological "NMC method" background error covariance, described in next subsection.

Once these equations have been coded into the 3D-Var, it is possible to use (5.4.40) and/or (5.4.41) and an iterative minimization routine (e.g., conjugate gradient or quasi-Newton) and obtain the incremental solution $\delta \mathbf{v}$. Then the transform (5.4.38) is used to transform back the analysis increments into the model variables $\delta \mathbf{x}_a = \mathbf{U} \delta \mathbf{v}$.

A powerful tool to evaluate 3D-Var and other data assimilation systems is used to examine the analysis increment resulting from equation (5.4.40) using *a single observation* (e.g., a temperature observational increment chosen to be 1K, or a zonal (or meridional) wind increment of 1m/sec). This allows checking what are the effective forecast error correlations of the 3D-Var scheme, and whether they are physically meaningful. Figures 2 and 3 in PD98, for example, show that the SSI, given a single temperature observational increment, produces the expected analysis increments both for the temperature, and for the wind components, not only in mid-latitudes but at the Equator as well. The latter is an important property, since the geostrophic balance itself breaks down at the Equator, and the 3D-Var wind increments solution is far better than the zero wind increments that OI would provide at the Equator. This is due to the adoption of the "NMC method" to estimate the background error covariance $\mathbf{B}$ (next Subsection 5.4.4).

5.4.4 Estimation of the Background Error Covariance B

5.4.4.1 Introduction

In both, OI and 3D-Var the background error covariance; $\mathbf{B}$ is assumed to be constant in time: it is a climatology of the real, evolving background error covariance that more advanced methods (4D-Var and ensemble Kalman filter) account for. The importance of the background error covariance $\mathbf{B}$ in the resulting analysis can hardly be overemphasized since it has a fundamental impact in determining the characteristics of the analysis increments. Essentially, *the analysis increments can only occur within the subspace spanned by* $\boldsymbol{B}$. As indicated in the previous subsection forecast errors of winds and heights (or temperatures) are strongly correlated because of the quasi-geostrophic balance. This means that the analysis increments, which take place in the same subspace as the background error covariance, will be properly correlated only if the elements of $\mathbf{B}$ are properly correlated.

For example, if we consider *a single observation* of a model variable at a grid point g, as discussed in the previous section, using the OI formulation of $\mathbf{K}$,

$$\delta \mathbf{x}^a = \mathbf{K} \mathbf{d}$$

where $\mathbf{K} = \mathbf{B} \mathbf{H}^T (\mathbf{R} + \mathbf{H} \mathbf{B} \mathbf{H}^T)^{-1}$ and $\mathbf{d} = \mathbf{y}^o - H(\mathbf{x}^b)$. For a single observation, for which the matrix $\mathbf{H}$ of dimension $p \times n$ is $1 \times n$,

5.4 Multivariate Statistical Data Assimilation Methods 179

$$\mathbf{H} = (0, \ldots, 1, \ldots 0); \qquad \mathbf{B} = \begin{pmatrix} b_{11} & \ldots & b_{1g} & \ldots & b_{1n} \\ \ldots & & & & \ldots \\ b_{g1} & \ldots & b_{gg} & \ldots & b_{gn} \\ \ldots & & & & \ldots \\ b_{n1} & \ldots & b_{ng} & \ldots & b_{nn} \end{pmatrix} \qquad \mathbf{BH}^T = \begin{pmatrix} b_{1g} \\ \ldots \\ b_{ng} \end{pmatrix};$$

$$\left(\mathbf{R} + \mathbf{HBH}^T \right)^{-1} = \frac{1}{\sigma_o^2 + \sigma_b^2}.$$

Therefore, the analysis increment for a single observation of a model variable is

$$\delta \mathbf{x}^a = \mathbf{Kd} = \frac{y_g^o - x_g^b}{\sigma_o^2 + \sigma_b^2} \begin{pmatrix} b_{1g} \\ \ldots \\ b_{ng} \end{pmatrix} \qquad (5.4.48)$$

In other words, the analysis increment for a single observation at grid point g will be proportional to the gth column of $\mathbf{B}$ multiplied by the observational increment at grid point g and divided by the sum of the observation plus the model error variances. Note that if any of the values of b_{ig} is zero (i.e., the forecast error covariance between i and g is estimated to be negligible), then the analysis increment δx_{ig} is always going to be zero, no matter how big the observational increment $y_g^o - x_g^b$.

Exercise 5.4.4 Derive equation (5.4.48)

More generally, consider the case of a background error covariance $\mathbf{B} = \sum_{i=1}^{k} \mathbf{b}_i \mathbf{b}_i^T$, spanning a subspace of dimension $k < n$, smaller than the dimension of the model. The 3D-Var cost function

$$J(\mathbf{x}) = \frac{1}{2} \left[(\mathbf{x} - \mathbf{x}_b)^T \mathbf{B}^{-1} (\mathbf{x} - \mathbf{x}_b) + (H(\mathbf{x}) - \mathbf{y}_o)^T \mathbf{R}^{-1} (H(\mathbf{x}) - \mathbf{y}_o) \right]$$

shows that the analysis increment has to be within the k-dimensional subspace spanned by the vectors $\mathbf{b}_i$. This is because outside this subspace, the inverse of the covariance matrix is infinitely large, and therefore increments not within the k-dimensional subspace are "forbidden" because they would result in large increases of the value of the cost function.

5.4.4.2 Estimations of B Used in OI before the "NMC Method"

The $\mathbf{B}$ covariance used in OI for many years was assumed to be homogeneous and isotropic (i.e., the covariance does not change with a rigid translation or rotation). In that case, the background error correlation of the geopotential height at two points depends only on the distance between the two points. Gandin (1963), Schlatter (1975), Thiébaux and Pedder (1987) and others have used a Gaussian exponential function for the geopotential error correlation μ_{ij} between two grid points i, j:

$$\mu_{ij} = e^{-r_{ij}^2 / 2L_\phi^2} \qquad (5.4.49)$$

where $r_{ij}^2 = (x_i - x_j)^2 + (y_i - y_j)^2$ is the square of the distance between two points i and j, and L_ϕ, typically of the order of 500 km, defines the background error correlation scale. Gaussian functions have also been used for the vertical correlation functions.

Data Assimilation

These assumptions only qualitatively begin to reflect the true structure of the background error correlation. For example, in the real atmosphere the background error correlation length should depend on the Rossby radius of deformation, and therefore should be a function of latitude, with longer horizontal error correlations in the tropics than in the extratropics (Balgovind et al., 1983; Baker et al., 1987). It should also depend on the data density: At the boundaries between data-rich and data-poor regions, the correlations of the forecast errors should not be isotropic (Cohn and Parrish, 1990). We refer the reader to the discussions in Daley (1991) and Thiébaux and Pedder (1987) for further details and references.

Another important assumption usually made in the OI analysis of large-scale flow is that the background wind error correlations are geostrophically related to the geopotential height error correlations. This has two advantages: It avoids having to estimate independently the wind error correlation, and it imposes an approximate geostrophic balance of the wind and height analysis increments, and therefore improves the balance of the analysis. Once a functional assumption for the background error correlation of the height like (5.4.49) is made, then the multivariate correlation between heights and winds can be obtained from the height correlations. For example, consider the background error correlation between two horizontal wind components:

$$E(\delta u_i \delta v_j) = -\frac{g}{f_i} \frac{g}{f_j} E\left(\frac{\partial \delta z_i}{\partial y_i} \frac{\partial \delta z_j}{\partial x_j}\right) \tag{5.4.50}$$

Now, since the geopotential error at the point x_j is independent of y_i and vice versa, we can combine the derivatives, use (5.4.49) and write

$$\begin{aligned} E(\delta u_i \delta v_j) &= -\frac{g}{f_i} \frac{g}{f_j} \frac{\partial^2 E(\delta z_i \delta z_j)}{\partial y_i \partial x_j} = -\frac{g}{f_i} \frac{g}{f_j} \frac{\partial^2 b_{ij}}{\partial y_i \partial x_j} \\ &= -\frac{g^2}{f_i} \frac{\sigma_z^2}{f_j} \frac{\partial^2 \mu_{ij}}{\partial y_i \partial x_j} \end{aligned} \tag{5.4.51}$$

The standard deviation of the wind increments can also be derived from the geostrophic relationship, $E(\delta u_i^2)^{1/2} = (g\sigma_z/f_i)$, $E(\delta v_j^2)^{1/2} = (g\sigma_z/f_j)$, so that we obtain the correlation of the increments of the two wind components by dividing (5.4.51) by these standard deviations: $\rho_{u,v} = -\partial^2 \mu_{ij}/\partial y_i \partial x_j$. Similarly, we can obtain the correlations between the increments of any two of the three variables at two points i, j: $\rho_{h,h} = \mu_{ij}$, $\rho_{h,u} = -\frac{\partial \mu_{ij}}{\partial y_i}$, $\rho_{u,h} = -\frac{\partial \mu_{ij}}{\partial y_j}$, etc.

Figure 5.4.2 shows schematically the shape of typical wind/height correlation functions used in OI (e.g., Schlatter, 1975). Note that the u–h correlations and the h–u correlations have the opposite sign because the first and second variables correspond to the first and second points i and j respectively. For example, a height observation leading to a positive analysis increment of h will result in positive increments of u to its north. Conversely, a positive increment of u will lead to negative increments of h to its north.

Equations (5.4.51) are not valid at the equator, and additional approximations have to be made in the tropics to allow for a smooth decoupling of wind and height increments (Lorenc, 1981).

5.4 Multivariate Statistical Data Assimilation Methods

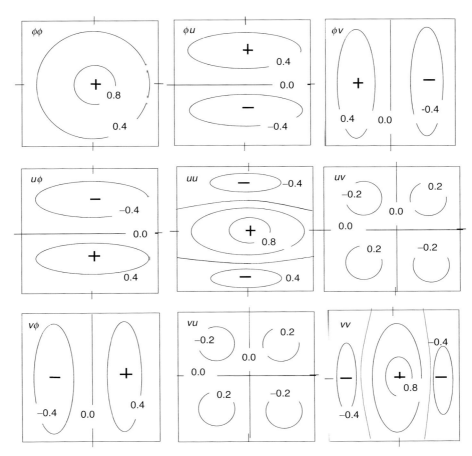

Figure 5.4.2 Schematic illustration of the correlation and cross-correlation functions for multivariate OI analysis derived using the geostrophic increment assumption (after Gustafsson, 1981). Both x- and y-axes go from $-\sqrt{2}L_\phi$ to $+\sqrt{2}L_\phi$. ©Springer Nature. Used with permission.

In addition, it is common *to select the observations to be included* in solving the linear system for the weight coefficients (5.4.32), depending on the computer resources available for the analysis, allowing for a maximum number of observations affecting each grid point. Rules for the selection of the subset of observations to be used typically depend on the distance to the grid point (within a maximum radius of influence), the types of observations (giving priority to the most accurate) and their distribution (trying to "cover" all quadrants, and choosing the closest stations).

Lorenc (1981) gave a comprehensive description of the OI implemented at the ECMWF. Although later improvements were implemented in the error covariances (Hollingsworth and Lönnberg, 1986) and other components of the analysis cycle, it remained the backbone of the ECMWF analysis until its replacement by 3D-Var in 1996.

5.4.4.3 The "NMC Method"

The methods that were in use in the 1980s to estimate $\mathbf{B}$ in OI applications were based on estimations of the horizontal and vertical correlations between forecast errors, estimated from the differences between the short-range forecasts and the rawinsonde observations (Thiébaux and Pedder, 1987; Hollingsworth and Lönnberg, 1986). In contrast, for 3D-Var (and even 4D-Var), the method that has been almost universally adopted does not depend on measurements at all, but is defined based on the structure of differences between short range forecasts valid at the same time. Based on the fact that, for example, a 24 hr and a 48 hr forecast valid at the same time; both represent the same verification global map; their differences are assumed to have the typical characteristics of forecast errors. Therefore the average forecast covariance as in (5.4.52) is a reasonable *proxy* for the structure of forecast errors. Although very simple, this approach has been shown to produce significantly better results than previous estimates computed from forecast minus observation estimates of $\mathbf{B}$ used in OI.

So, in the "NMC method" for estimating the forecast error covariance (PD92) based on the difference between (for example) 24 and 48 hr forecasts, the covariance is computed as:

$$\mathbf{B} \approx \alpha E\{[\mathbf{x}_f(48 \text{ h}) - \mathbf{x}_f(24 \text{ h})][\mathbf{x}_f(48 \text{ h}) - \mathbf{x}_f(24 \text{ h})]^T\} \tag{5.4.52}$$

As indicated in (5.4.52), in the NMC (now NCEP) method, the structure of the forecast or background error covariance is estimated as an average over many (a month or longer) differences between two short-range model forecasts verifying at the same time. The magnitude of the covariance is then tuned multiplying (5.4.52) by an appropriate scalar so that the variances are similar to those of the short range forecasts errors. As shown in Subsection 5.4.3.3, in the NCEP SSI, Parrish and Derber (1992) assumed homogeneity and isotropy, so that $\mathbf{B}$ is diagonal and has no covariance between different spectral modes, but in the GSI, this assumption is lifted through the use of recursive filter that can efficiently compute local Gaussian functions (Purser et al., 2003a). In the vertical, PD92 use an empirical orthogonal function expansion of (5.4.52).

Although very simple, this approach has been shown to produce significantly better results than previous estimates computed from forecast minus observation estimates of $\mathbf{B}$ used in OI.

There are several reasons for this improvement. One is that the rawinsonde network does not have enough density to allow a proper estimate of the global structures used in OI (PD92, Rabier et al. (1998)), whereas (5.4.52) provides a global representation of the forecast error balance structures, including the errors in the tropics. As a result, the analysis increments are much more balanced than they were in OI. Another reason is that, by construction, the forecast differences in (5.4.52) are dominated by the fastest growing perturbations, and are akin to leading Lyapunov vectors or bred vectors (Chapter 6). These are also the perturbations that dominate the forecast errors (Corazza et al., 2003). By averaging over many cases, the NMC estimate of the 3D-Var forecast error covariance is thus dominated by the climatology of the growing errors that also dominate the forecast errors.

5.4 Multivariate Statistical Data Assimilation Methods

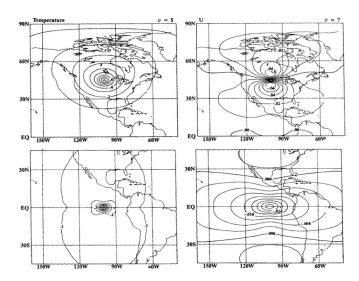

Figure 5.4.3 Top: Analysis increment for a single temperature observation at layer 5 and 45°N, 100°W. Bottom: Same but at the Equator, 100°W. Left is the temperature analysis increment at layer 5, and right is the U increment at layer 7 (Parrish and Derber, 1992). ©American Meteorological Society. Used with permission.

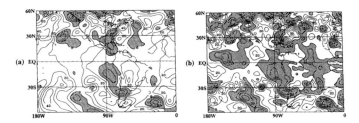

Figure 5.4.4 250 hPa height analysis increments for (a) SSI and (b) operational (optimal interpolation). Both were created using the operational background. Contour interval is 10m (Parrish and Derber, 1992). ©American Meteorological Society. Used with permission.

Figure 5.4.3 from PD92 compares the SSI analysis response to a single temperature observational increment on the temperature itself and on the u-wind for both a midlatitude and an equatorial observation, showing how the "NMC" background error covariance reflects the different types of balances at different latitudes. Figure 5.4.4 also from PD92 compares analysis increments obtained with the SSI, and the OI it replaced, using the same operational background and the same observations. It is very clear that the SSI is less noisy and has much smaller changes in height near the equator than the OI. As a result, the SSI initial conditions were in much better balance than the OI and did not require the balancing step essential for the OI, performed with nonlinear normal mode initialization. This better balance of the SSI analysis was frequently attributed to the fact that 3D-Var assimilates all observations simultaneously, with a single global minimization of the cost function, whereas OI solves the optimal

184   **Data Assimilation**

interpolation problem grid-point by grid-point (or volume by volume), introducing noise in the boundaries. However, the fact that the advanced method known as local ensemble transform Kalman filter (LETKF, Hunt et al. (2007), Section 5.6) solves the Kalman equations with a local approach similar to OI without introducing significant imbalances indicates that the main deficiency in OI is due to the imperfect structure of the OI error covariance, rather than the use of local observations.

5.4.5 Physical-Space Statistical Analysis Scheme, and Its Relationship to 3D-Var and OI

Cohn et al. (1998) introduced the Physical-space Statistical Analysis System (PSAS, pronounced "Pizzaz") related to 3D-Var and OI, in which the minimization is performed in the (physical) space of the observations, rather than in the model space as in the 3D-Var scheme (grid points or spectral variables in the case of the NCEP scheme, PD92), motivated by the fact that there are generally much fewer observations than model degrees of freedom.

They proposed to solve the OI/3D-Var analysis equation, written with $\mathbf{K} = (\mathbf{BH}^T)(\mathbf{R} + \mathbf{HBH}^T)^{-1}$ as in the OI formulation:

$$\delta\mathbf{x}_a = (\mathbf{BH}^T)(\mathbf{R} + \mathbf{HBH}^T)^{-1}\mathbf{d} \tag{5.4.53}$$

but separated into two steps. Here $\delta\mathbf{x}_a = \mathbf{x}_a - \mathbf{x}_b$; $\mathbf{d} = \mathbf{y}_o - H(\mathbf{x}_b)$.

PSAS solves first

$$\mathbf{w} = (\mathbf{R} + \mathbf{HBH}^T)^{-1}\mathbf{d} \tag{5.4.54}$$

and then

$$\delta\mathbf{x}_a = (\mathbf{BH}^T)\mathbf{w} \tag{5.4.55}$$

The first step is the most computer intensive, and is solved by minimization of a cost function F:

$$F(\mathbf{w}) = \frac{1}{2}\mathbf{w}^T(\mathbf{R} + \mathbf{HBH}^T)\mathbf{w} - \mathbf{w}^T\mathbf{d} \tag{5.4.56}$$

Recall from Exercise 5.4.1 that

$$\mathbf{BH}^T(\mathbf{R} + \mathbf{HBH}^T)^{-1} = (\mathbf{B}^{-1} + \mathbf{H}^T\mathbf{R}^{-1}\mathbf{H})^{-1}\mathbf{H}^T\mathbf{R}^{-1}$$

so that the solutions to 3D-Var and OI are formally the same. Therefore, the PSAS analysis solution of (5.4.53) should also be the same as 3D-Var. The main difference of PSAS is that the size of the control vector $\mathbf{w}$ is p, the number of observations, generally much smaller than the size of the control vector for 3D-Var $\delta\mathbf{x}_a$, which is n, the dimension of the model. Therefore, if $p << n$, PSAS has the important advantage that it solves the minimization problem in a space much smaller than 3D-Var.

In summary, OI, 3D-Var and PSAS are all formally solving the same equation

$$\delta\mathbf{x}_a = \mathbf{Kd}$$

5.4 Multivariate Statistical Data Assimilation Methods

185

All three methods assume a constant error covariance $\mathbf{B}$, so that the structure of the errors and their correlations are only a "climatology" of the "errors of the day" that in reality evolve with the flow. However, the methods used to solve the three algorithms are quite different.

OI (the oldest of the three methods) has been solved grid point by grid point (or small volume by small volume) selecting the closest and most important observations for each grid point analysis. This, combined with the use of a constant background error covariance $\mathbf{B}$ estimated from statistics of observations minus forecasts, led to some noisiness, making essential to have an "initialization" step (at that time a nonlinear normal mode initialization approach), to reduce the gravity waves induced by the analysis. As we noted, the local ensemble transform Kalman filter (LETKF, Section 5.6) also performs an analysis grid point by grid point, with a local selection of observations similar to OI, but it is not affected by significant noisiness, because the analysis is based on solving the ensemble Kalman filter equations, without approximating the background error covariance as done in OI.

3D-Var was implemented operationally by PD92, who also introduced the "NMC method" for estimating $\mathbf{B}$. They assumed that the structure of forecast errors would be similar to that obtained from the difference between a 24 hr and a 48 hr forecast verifying at the same time t_i. Note that these structures could be also obtained from the difference of 12 and 24 hr forecasts or any similar difference of forecasts verifying at the same time, except for the differences in diurnal errors. These global differences $\delta\mathbf{x}(t_i)$ were then multiplied by their transpose, averaged over a month or longer, and finally scaled by a factor α in order to have an amplitude similar to that expected in the 6 hr error covariance:

$$\mathbf{B} = \frac{\alpha}{N} \sum_{i=1}^{N} \delta\mathbf{x}(t_i)\delta\mathbf{x}^T(t_i)$$

The NMC method has been very successful and widely used, although some improvements have been made to PD92. As indicated before, the Hessian of the 3D-Var cost function is $J''(\delta\mathbf{x}) = \mathbf{B}^{-1} + \mathbf{H}^T\mathbf{R}^{-1}\mathbf{H} = \mathbf{A}^{-1}$, equal to the inverse of the analysis error covariance matrix. By contrast, the Hessian of the PSAS cost function is $F''(\mathbf{w}) = \mathbf{R} + \mathbf{HBH}^T$, equal to the total error covariance (sum of the observational and forecast error covariances) in observation space.

PSAS has been preconditioned by a change of variables $\mathbf{u} = \mathbf{R}^{1/2}\mathbf{w}$. A remarkable study by El Akkraoui et al. (2008) showed that the solutions of the 3D-Var (known as the primal approach) and of PSAS (known as the dual approach) for a real data case are indeed identical, but only *after they achieve minimization*. The solutions after minimization are such that $J(\mathbf{v}_a) = -F(\mathbf{u}_a)$. However, 3D-Var reduces the gradient of the cost function ∇J relatively monotonically, whereas PSAS takes many more iterations to converge because the gradient ∇F initially increases by an order of magnitude. El Akkraoui et al. (2008) further showed that with additional preconditioning based on the singular vectors of the two Hessians, both methods converge at the same rate. PSAS was implemented operationally at NASA GMAO (Cohn et al., 1998) and at the US Naval Research Laboratory (the system NAVDAS. Daley and Barker, 2001).

Data Assimilation

There are other applications where the faster minimization in observation space of PSAS, can be used (Penny, pers. comm., 2017)

In conclusion, the advantages of 3D-Var over OI that led to its popularity among operational and research centers in the 1990s are:

(a) 3D-Var (like PSAS) is minimized using global minimization algorithms, and as a result some of the simplifying approximations required by OI become unnecessary (Parrish and Derber, 1992; Derber et al., 1991; Courtier et al., 1998; Rabier et al., 1998; Andersson et al., 1998) . Since in 3D-Var there is no data selection, all available data are used simultaneously. This reduces the jumpiness in the boundaries between regions that have selected different observations.

(b) It is possible to add constraints to the cost function without increasing the cost of the minimization. For example, PD92 included a "penalty" term into the 3D-Var cost function forcing *simultaneously* the analysis increments to approximately satisfy the linear global balance equation. In OI, the imposition of the geostrophic constraint on the increments ensured only an approximate balance in the analysis. In practice it was found necessary to follow the OI analysis with a *nonlinear normal mode initialization*. With the global balance equation added as a weak constraint to the cost function, the NCEP global model spin up (indicated for example by the change of precipitation over the first 12 hr of integration) was reduced by more than an order of magnitude compared with the results obtained with OI. In other words, with the implementation of 3D-Var it became much less necessary to perform a separate initialization step in the analysis cycle.

(c) It is also possible to incorporate important nonlinear relationships between observed variables and model variables in the H operator in the minimization of the cost function (5.5.1) by performing "inner" iterations with the linearized $\mathbf{H}$ observation operator kept constant and "outer" iterations in which it is updated. The use of this *outer loop* has become an essential tool in the success of both 3D-Var and 4D-Var (see Section 5.5.3).

(d) The introduction of 3D-Var has allowed three-dimensional variational assimilation of radiances (Derber and Wu, 1998). In this approach, there is no attempt to perform retrievals and, instead, each satellite sensor radiance measurement is taken as an independent observation with uncorrelated errors. As a result, for each satellite observation, even if some channel measurements are rejected because of cloud contamination, others may still be used. In addition, because all the data are assimilated simultaneously, information from one channel at a certain location can influence the use of satellite data at a different geographical location. The quality control of the observations becomes easier and more reliable when it is made in the space of the observations than in the space of the retrievals.[3]

(e) It is also possible to include a quality control (i.e., variational quality control) of the observations within the 3D-Var analysis (Anderson and Järvinen, 1999).

---

[3] Joiner and Da Silva (1998) pointed out that the use of retrievals from remotely sensed observations is a viable option within the variational analysis approach, as long as the innovation vector is computed consistently with the use of retrievals from radiances. If Dy^o is the (linearized) retrieval algorithm applied to satellite radiances to obtain, e.g., temperature and moisture profiles, then the innovation vector should be computed consistently as $Dy^o - DFx^b$, where F is the forward (linearized radiative transfer) algorithm that converts model variables into model radiances. In other words, the forward observational operator in (5.4.53) is H = DF, and the observational error covariance for the retrievals becomes $E[(D\delta y^o)(D\delta y^o)^T] = DE[\delta y^o \delta y^{oT}]D^T = DRD^T$ instead of R, which is the observation error covariance when radiances are directly assimilated. If this method is applied to OI, the weight matrix becomes $K = BF^T B^T (DFBF^T D^T + DRD^T)^{-1}$.

5.5 Advanced Data Assimilation Methods with Evolving Covariance: 4D-Var

5.5.1 Introduction: "Errors of the Day"

In Section 5.4, we have discussed three types of statistical interpolation methods: OI, a sequential method, 3D-Var, a variational method, and the closely related PSAS variational method, and how they are numerically solved. We saw that these sequential and variational methods solve the same problem, so that formally their exact solution should be the same, but because the approximations and methods used to solve them numerically are very different, their resulting analysis increments are also significantly different.

A very important assumption made in these three methods is that the forecast error covariance matrix $\mathbf{B}$ can be estimated *once and for all*, as an average of the estimated daily error covariance, so that in these methods, the forecast errors are assumed to be statistically stationary (equal to their climatology). The question we should address is whether this assumption is reasonably accurate in reality. In order to assess this, consider Figure 5.5.1 showing the daily RMS 6-hr forecast errors in 1996 over Eastern USA from the NCEP/NCAR Reanalysis (R1, (Kalnay and Toth, 1996) and (Kistler et al., 2001)), estimated from the analysis increments, i.e., from the difference between the forecasts and the analysis. This is a good estimate of the forecast errors since this region is data-rich, and therefore the analyses are close to the truth.[4] Although the mean error (7 m) is dominant, the striking feature in Figure 5.5.1 is *the large day-to-day variability* in the forecast RMS error (with a period of about 2 to 5 days), which is almost as large as the mean error, and which did not change much through the year. This figure emphasizes the obvious importance of the "errors of the day" (Kalnay et al., 1997), which in this area are dominated by baroclinic instabilities of synoptic time scales, and are completely ignored when the forecast error covariance $\mathbf{B}$ is assumed to be constant as in OI and 3D-Var.

In order to understand why the error covariance should be flow dependent, rather than constant in time and isotropic, as in OI and 3D-Var, consider schematic Figure 5.5.2, showing the impact of a single observation of temperature at a station location (black square) through which a cold front is passing. As we saw before, the analysis increment is proportional to $\mathbf{B}$. If we assume $\mathbf{B}$ is constant in time, homogeneous and isotropic, as in the 3D-Var of Parrish and Derber (1992), the analysis increments for a single observational increment of, say, 1C, will be essentially circles near the observation (schematic a), with a maximum of about 1C at the observation location, and an amplitude decreasing with distance. This is similar to Figure 5.4.3a reproduced

---

[4] The R1 reanalysis uses a 3D-Var data assimilation system, based on Parrish and Derber (1992, PD92), using a forecast model that remains constant throughout the many decades of reanalysis, so that the average error changes are only due to changes in the observing system. Over four decades (from 1958 to 1996) the improvement in the mean 6 hr forecast error due the observing system over Eastern USA was about 30% (3 m), with the average analysis increment reduced from about 10 m in 1958 (not shown) to about 7 m in 1996. We note that the NCEP/NCAR reanalysis uses satellite temperature retrievals, not direct assimilation of radiances, which would have resulted in larger improvements.

Data Assimilation

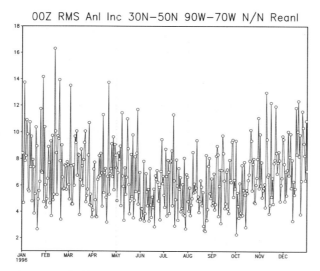

Figure 5.5.1 Daily variation of the RMS of the analysis increment (difference between the 6-hr forecast of the 500 hPa geopotential height and the corresponding analysis) at 00Z over Eastern USA, from the NCEP/NCAR reanalysis (Kistler et al., 2001) in 1996. This proxy of the forecast errors highlights the importance of the "errors of the day." (Courtesy of Wesley Ebisuzaki and Robert Kistler.)

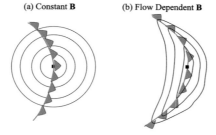

Figure 5.5.2 Schematic of the analysis increment of a single temperature observation on a cold front, proportional to the background error covariance. (a) 3D-Var, using the NMC method to estimate a constant **B**. (b) Evolving **B** with "errors of the day." (Courtesy of Takemasa Miyoshi.)

from PD92. If instead, **B** represents the real forecast "errors of the day" associated with the frontal passage, the error covariance of the temperature forecasts, and therefore the analysis increments for the temperature, will be stretched along the front and very narrow across the front (schematic b).

Exercise 5.5.1 Explain why are the error correlations stretched along the front?

We saw in Section 5.4 that, unlike 3D-Var and OI, the Kalman filter formally does account for the day-to-day variability of **B** by advancing in time the analysis error covariance with the linear tangent and the adjoint models, $\mathbf{B}_{i+1} = \mathbf{L}\mathbf{A}_i\mathbf{L}^T$. However, actually computing this evolution would be equivalent to integrating the linear tangent

5.5 Advanced Data Assimilation Methods with Evolving Covariance: 4D-Var

model for the length of the assimilation window about $n/2$ times, where n is the dimension of the model. Since in current operational models, $n \sim 10^9$ or more, evolving **B** is completely unfeasible from a computational point of view (except for toy models, where the number of degrees of freedom n is small). In order to address this problem, two advanced methods with evolving "errors of the day" have been developed to obtain analyses approximately similar to Kalman filter at a cost that is computationally feasible. In this section, we discuss 4D-Var, a variational approach, and in Section 5.6 ensemble Kalman filter.

5.5.2 4D-Var Extension of 3D-Var and Its Relationship to Kalman Filter

4D-Var is an important extension of the 3D-Var method that allows for the assimilation of observations that are distributed in time within an assimilation window (t_0, t_N), rather than assuming that they are all at the end of the window, as in 3D-Var.[5] As will be shown in this subsection, 4D-Var also has the important advantage that it *implicitly* evolves the background error covariance **B** within the assimilation window.

In 4D-Var (Figure 5.5.3), the cost function (5.5.1) includes a background term J_b measuring the distance of the minimizing solution to the background *valid at the beginning of the assimilation window* t_0, and scaled by the initial background error

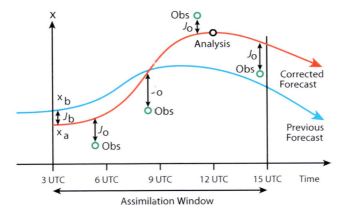

Figure 5.5.3 Schematic of 4D-Var. $\mathbf{x}_b$ is the previous cycle analysis from which a background (previous or prior) forecast is obtained. $\mathbf{x}_a$ is the 4D-Var new analysis at the beginning of the assimilation window. J_b is the cost function associated with the difference between the background analysis and the new 4D-Var analysis at the beginning of the assimilation window. J_o are the cost functions associated with the difference between each observation at time t_i and the forecast started from $\mathbf{x}_a$ and valid at t_i. Note that 4D-Var finds the initial conditions $\mathbf{x}_a$ that lead to a forecast that best fits the observations. The new analysis at the end of the assimilation window is given by the forecast started from $\mathbf{x}_a$. ©ECMWF. Used with permission.

---

[5] Some of the pioneering work done in 4D-Var techniques appeared in Lewis and Derber (1985); Hoffman (1986); Talagrand and Courtier (1987); Courtier and Talagrand (1990); Derber (1989); Daley (1991); Zupanski (1993); Bouttier and Rabier (1997); Bouttier et al. (1997); Rabier and Courtier (1992).

covariance $\mathbf{B}_0$. This term is similar to the J_b term in 3D-Var, but in 3D-Var J_b is valid at the analysis time. The observational J_o term in the 4D-Var cost function is also similar to the observation cost function J_o of 3D-Var, except that it includes *a summation over time* of the cost function for each observational increment, computed as the difference between the observation and *the model forecast valid at the time of the observation*, i.e., the model has to be integrated from t_0 to the time t_i of the i-th observation:

$$J[\mathbf{x}_0] = \frac{1}{2}\left[\mathbf{x}(t_0) - \mathbf{x}^b(t_0)\right]^T \mathbf{B}_0^{-1}\left[\mathbf{x}(t_0) - \mathbf{x}^b(t_0)\right]$$

$$+ \frac{1}{2}\sum_{i=0}^{N}\left[H(\mathbf{x}_i) - \mathbf{y}_i^o\right]^T \mathbf{R}_i^{-1}\sum_{i=0}^{N}\left[H(\mathbf{x}_i) - \mathbf{y}_i^o\right], \qquad (5.5.1)$$

where

$$\mathbf{x}_i = M_{0\to i}(\mathbf{x}_0)$$

The *control variable* $\mathbf{x}(t_0)$, i.e., the variable with respect to which the cost function is minimized, is the *initial* state of the model within the time window. The *analysis at the end of the assimilation window* $\mathbf{x}_N$, which is the analysis that we are actually seeking, is given by the *model integration* starting from the minimization solution $\mathbf{x}_0$ at the initial time:

$$\mathbf{x}_N = M_{0\to N}(\mathbf{x}_0) \qquad (5.5.2)$$

Thus, the model is used as *a strong constraint* (Sasaki, 1970) in 4D-Var, i.e., the analysis solution is assumed to exactly satisfy the model equations. In other words, *4D-Var seeks an initial condition such that the forecast best fits the observations within the assimilation interval*, with the observations assimilated at their right time. The analysis is this forecast at the end of the window.

An important advantage of 4D-Var is that *if we assume* that $\mathbf{B}_0$, the background error covariance used at the beginning of the assimilation window represents the error covariance of the initial background $\delta\mathbf{x}_0^b\delta\mathbf{x}_0^{bT}$ correctly, as would be the case in Kalman filter, then the solution that minimizes the cost function will coincide with the solution that would be obtained with the Kalman filter. In other words, in 4D-Var the background error covariance is implicitly evolved with time within the assimilation window.

Proof. We will now prove this important property, namely that for a linear model, *if* $\mathbf{B}_0$ at t_0 is equal to the error covariance given by the Kalman filter, 4D-Var generates the same analysis as Kalman filter. We write the cost function (5.5.1) in an incremental form, where the initial increment of the solution is $\delta\mathbf{x}_0 = \left[\mathbf{x}_0 - \mathbf{x}_0^b\right]$, the increment at t_i is $\delta\mathbf{x}_i = \mathbf{x}_i - \mathbf{x}_i^b = \mathbf{L}_{0\to i}\delta\mathbf{x}_0$, and the observation increment with respect to the minimizing solution is $\mathbf{y}_i^o - H(\mathbf{x}_i) = \mathbf{y}_i^o - H(\mathbf{x}_i^b) + H(\mathbf{x}_i^b) - H(\mathbf{x}_i) = \mathbf{d}_i - \mathbf{HL}_{0\to i}\delta\mathbf{x}$ with $\mathbf{d}_i = \mathbf{y}_i^o - H(\mathbf{x}_i^b)$ being the observational increment with respect to the background (and thus independent of $\mathbf{x}_i$).

5.5 Advanced Data Assimilation Methods with Evolving Covariance: 4D-Var

Then we can write the 4D-Var incremental cost function (5.5.1) as

$$J(\delta\mathbf{x}_0) = \frac{1}{2}\delta\mathbf{x}_0^T\mathbf{B}_0^{-1}\delta\mathbf{x}_0 + \frac{1}{2}\sum_{i=1}^{N}[\mathbf{HL}_{0\to i}\delta\mathbf{x} - \mathbf{d}_i]^T\mathbf{R}_i^{-1}[\mathbf{HL}_{0\to i}\delta\mathbf{x} - \mathbf{d}_i] \quad (5.5.3)$$

For simplicity, we consider only one observation $N = 1$, but the demonstration can be easily generalized to the N observations available at multiple times by replacing J_0 in (5.4.35) by the summation over all observations as in (5.5.3).

For a single observation, the 4D-Var incremental cost function (5.5.3) becomes

$$J(\delta\mathbf{x}_0) = \frac{1}{2}\delta\mathbf{x}_0^T\mathbf{B}_0^{-1}\delta\mathbf{x}_0 + \frac{1}{2}[\mathbf{HL}\delta\mathbf{x}_0 - \mathbf{d}_1]^T\mathbf{R}_1^{-1}[\mathbf{HL}\delta\mathbf{x}_0 - \mathbf{d}_1] \quad (5.5.4)$$

The solution to (5.5.4) is obtained by setting its gradient equal to zero:

$$\nabla J(\delta\mathbf{x}_0) = \mathbf{B}_0^{-1}\delta\mathbf{x}_0 + \left[\mathbf{L}^T\mathbf{H}^T\mathbf{R}_1^{-1}\mathbf{HL}\delta\mathbf{x}_0 - \mathbf{L}^T\mathbf{H}^T\mathbf{R}_1^{-1}\mathbf{d}_1\right] = 0$$

which gives the 4D-Var solution, $(\mathbf{B}_0^{-1} + \mathbf{L}^T\mathbf{H}^T\mathbf{R}_1^{-1}\mathbf{HL})\delta\mathbf{x}_0 = \mathbf{L}^T\mathbf{H}^T\mathbf{R}_1^{-1}\mathbf{d}_1$, or

$$\delta\mathbf{x}_0 = \left(\mathbf{B}_0^{-1} + \mathbf{L}^T\mathbf{H}^T\mathbf{R}_1^{-1}\mathbf{HL}\right)^{-1}\mathbf{L}^T\mathbf{H}^T\mathbf{R}_1^{-1}\mathbf{d}_1 \quad (5.5.5)$$

Now we write the Kalman filter solution, which gives $\delta\mathbf{x}_1$ at the analysis time t_1 at the end of the assimilation window. From (5.4.28), we have the Kalman filter analysis increment $\delta\mathbf{x}_1 = \mathbf{K}\mathbf{d}_1$, where the Kalman gain is given by $\mathbf{K} = \mathbf{B}_1\mathbf{H}^T(\mathbf{R} + \mathbf{HB}_1\mathbf{H}^T)^{-1}$ and $\mathbf{B}_1 = \mathbf{LB}_0\mathbf{L}^T$, since the background error covariance is advanced from the beginning to the end of the assimilation window. In addition, in order to compare with the 4D-Var solution $\delta\mathbf{x}_0$ we have to take into account that $\delta\mathbf{x}_1 = \mathbf{L}\delta\mathbf{x}_0$. So the Kalman filter solution is $\delta\mathbf{x}_1 = \mathbf{B}_1\mathbf{H}^T(\mathbf{R}_1 + \mathbf{HB}_1\mathbf{H}^T)\mathbf{d}_1$ or

$$\delta\mathbf{x}_0 = \mathbf{L}^{-1}\mathbf{LB}_0\mathbf{L}^T\mathbf{H}^T(\mathbf{R}_1 + \mathbf{HLB}_0\mathbf{L}^T\mathbf{H}^T)^{-1}\mathbf{d}_1, \quad (5.5.6)$$

which concludes the proof.

Exercise 5.5.2 Prove that the 4D-Var solution (5.5.5) and of the Kalman filter solution (5.5.6) are the same, *provided the initial $\mathbf{B}_0$ used in 4D-Var is the same as that used in Kalman filter*. Hint: show that the matrices that multiply $\mathbf{d}_1$ in (5.5.5) and (5.5.6) are the same by multiplying both matrices on the left by $\left(\mathbf{B}_0^{-1} + \mathbf{L}^T\mathbf{H}^T\mathbf{R}_1^{-1}\mathbf{HL}\right)$, and on the right by $(\mathbf{R}_1 + \mathbf{HLB}_0\mathbf{L}^T\mathbf{H}^T)$.

Unfortunately, in reality, this is not the case, since the $\mathbf{B}_0$ used in 4D-Var is a constant matrix, usually the same one used in 3D-Var (e.g., using the NMC method) but with a reduced amplitude. The amplitude of $\mathbf{B}_0$ is reduced because, when 4D-Var is cycled in time, the final analysis from the previous assimilation window (5.5.2) becomes the background $\mathbf{x}_0^b$ in the next window. Since 4D-Var is more accurate than 3D-Var, this background field is more accurate than the 3D-Var background, and the amplitude of the climatological $\mathbf{B}_0$ has to be tuned to a value smaller than the one used in 3D-Var (e.g., Kalnay et al., 2007a). In 4D-Varm it is advantageous to use longer windows, since, for long enough windows, the analysis "forgets" about the initial $\mathbf{B}_0$, and then 4D-Var asymptotes to the Kalman analysis. In that case, 4D-Var, like Kalman filter, is able to find the best linear unbiased estimation, but not its error covariance.

192   Data Assimilation

5.5.3   Numerical Solution of 4D-Var: Inner and Outer Loops

In this subsection, we describe how the minimization of the 4D-Var cost function is carried out efficiently.

When the control variable $\mathbf{x}(t_0)$ is changed by a small amount $\delta\mathbf{x}(t_0)$, the variation in the cost function $J[\mathbf{x}(t_0)]$ is given by

$$\delta J = J[\mathbf{x}(t_0) + \delta\mathbf{x}(t_0)] - J[\mathbf{x}(t_0)] \approx \left[\frac{\partial J}{\partial \mathbf{x}(t_0)}\right]^T \cdot \delta\mathbf{x}(t_0) \qquad (5.5.7)$$

where the gradient of the cost function $[\partial J/\partial\mathbf{x}(t_0)]_j = \partial J/\partial x_j(t_0)$ is a column vector. As suggested by (5.5.7), iterative minimization schemes require the estimation of the cost function gradient. In the simplest scheme, the steepest descent method, the change in the control variable after each iteration is chosen to be in the direction opposite to the gradient:

$$\delta\mathbf{x}(t_0) = -a\nabla_{\mathbf{x}(t_0)}J = -a\,\partial J/\partial\mathbf{x}(t_0)$$

and the amplitude a is obtained by a line minimization. More efficient methods, such as the conjugate gradient (Parrish and Derber, 1992) or the quasi-Newton methods (Navon and Legler, 1987), also require the use of the gradient, so that in order to solve this minimization problem efficiently, we need to be able to compute the gradient of J with respect to the elements of the control variable.

As we saw in Remark 5.4.2.1, given a symmetric matrix $\mathbf{A}$ and a functional $J = \frac{1}{2}\mathbf{x}^T\mathbf{A}\mathbf{x}$, the gradient is given by $\partial J/\partial\mathbf{x} = \mathbf{A}\mathbf{x}$. If $J = \mathbf{y}^T\mathbf{A}\mathbf{y}$, and $\mathbf{y} = \mathbf{y}(\mathbf{x})$, then the chain rule for the gradient is

$$\frac{\partial J}{\partial\mathbf{x}} = \left[\frac{\partial\mathbf{y}}{\partial\mathbf{x}}\right]^T\mathbf{A}\mathbf{y} \qquad (5.5.8)$$

where $[\partial\mathbf{y}/\partial\mathbf{x}]_{k,l} = \partial y_k/\partial x_l$ are the elements of the gradient matrix.

We write the two terms of the 4D-Var cost function (5.5.1) as $J = J_b + J_o$. From the rules discussed above, the gradient of the *background* component of the cost function $J_b = \frac{1}{2}[\mathbf{x}(t_0) - \mathbf{x}^b(t_0)]^T\mathbf{B}_0^{-1}[\mathbf{x}(t_0) - \mathbf{x}^b(t_0)]$ with respect to $\mathbf{x}(t_0)$ is given by

$$\frac{\partial J_b}{\partial\mathbf{x}(t_0)} = \mathbf{B}_0^{-1}[\mathbf{x}(t_0) - \mathbf{x}^b(t_0)] \qquad (5.5.9)$$

The gradient of the second term of (5.5.1), i.e., the *observational* cost function term, is

$$J_o = \frac{1}{2}\sum_{i=0}^{N}\left[H(\mathbf{x}_i) - \mathbf{y}_i^o\right]^T\mathbf{R}_i^{-1}\sum_{i=0}^{N}\left[H(\mathbf{x}_i) - \mathbf{y}_i^o\right]$$

is more complicated because it involves the model integration, from the analysis at the beginning of the assimilation window to the time of the observation, $\mathbf{x}_i = M_{0\to i}(\mathbf{x}_0)$. If we introduce a perturbation to the initial state $\delta\mathbf{x}_0$, then $\delta\mathbf{x}_i = \mathbf{L}(t_0, t_i)\delta\mathbf{x}_0$, so that

$$\frac{\partial(H(\mathbf{x}_i) - \mathbf{y}_i^o)}{\partial\mathbf{x}(t_0)} = \frac{\partial H}{\partial\mathbf{x}_i}\frac{\partial M}{\partial\mathbf{x}_o} = \mathbf{H}_i\mathbf{L}(t_0, t_i) \qquad (5.5.10)$$

5.5 Advanced Data Assimilation Methods with Evolving Covariance: 4D-Var

As indicated by (5.5.10), the matrices $\mathbf{H}_i, \mathbf{L}(t_0,t_i)$ are the Jacobians $\partial H/\partial \mathbf{x}_i$, $\partial M/\partial \mathbf{x}_o$. The *linear tangent model*, $\mathbf{L}(t_0,t_i)$ is a matrix that integrates a perturbation from t_0 to t_i, and it consists of the successive application of matrices that integrate forward in time for shorter steps:

$$\mathbf{L}(t_0,t_i) = \mathbf{L}(t_{i-1},t_i)\ldots\mathbf{L}(t_1,t_2)\mathbf{L}(t_0,t_1) \quad (5.5.11)$$

When the linear tangent model (5.5.11) is transposed, we obtain the *adjoint model* that integrates a perturbation *backwards* in time:

$$\mathbf{L}^T(t_i,t_0) = \mathbf{L}^T(t_1,t_0)\mathbf{L}^T(t_2,t_1)\ldots\mathbf{L}^T(t_i,t_{i-1}) \quad (5.5.12)$$

The rules for coding of the linear tangent model and the adjoint model are discussed in Sections 5.5.5, 6.3, and Appendix A.

From (5.5.10), the gradient of the observation cost function is given by

$$\left[\frac{\partial J_o}{\partial \mathbf{x}(t_0)}\right] = \sum_{i=0}^{N} \mathbf{L}^T(t_i,t_0)\mathbf{H}_i^T \mathbf{R}_i^{-1}[H(\mathbf{x}_i) - \mathbf{y}_i^o] \quad (5.5.13)$$

Equation (5.5.13) shows that every iteration of the 4D-Var minimization requires the computation of the gradient of J_o, i.e., computing the increments $[H(\mathbf{x}_i) - \mathbf{y}_i^o]$ at the observation times t_i during a forward integration, multiplying them by $\mathbf{H}_i^T \mathbf{R}_i^{-1}$ and integrating these weighted increments back to the initial time using the adjoint model. Since parts of the backward adjoint integration are common to several time intervals, the summation in (5.5.13) can be rearranged more efficiently. Assume, for example that the assimilation window is from 00 Z to 12 Z, and that there are observations every 3 hr (Figure 5.5.4). We compute during the forward integration the weighted negative observation increments $\bar{\mathbf{d}}_i = \mathbf{H}_i^T \mathbf{R}_i^{-1}[H(\mathbf{x}_i) - \mathbf{y}_i^o] = -\mathbf{H}_i^T \mathbf{R}_i^{-1}\mathbf{d}_i$. The adjoint model $\mathbf{L}^T(t_i,t_{i-1}) = \mathbf{L}_{i-1}^T$ applied on a vector "advances" it from t_i to t_{i-1}. Then we can write (5.5.13) in the example shown in Figure 5.5.4 as

$$\frac{\partial J_o}{\partial \mathbf{x}_o} = \bar{\mathbf{d}}_o + \mathbf{L}_0^T\{\bar{\mathbf{d}}_1 + \mathbf{L}_1^T[\bar{\mathbf{d}}_2 + \mathbf{L}_2^T(\bar{\mathbf{d}}_3 + \mathbf{L}_3^T\bar{\mathbf{d}}_4)]\} \quad (5.5.14)$$

In summary, from Equations (5.5.9) and (5.5.13) we obtain the gradient of the cost function, and using a minimization algorithm (e.g., quasi-Newton or conjugate gradient) we can find the optimal change that minimizes the cost function, and modifies appropriately the control variable $\mathbf{x}(t_0)$. This minimization is usually performed in the incremental form as in schematic Figure 5.5.4, using linear tangent and adjoint models that have much lower horizontal resolution than the operational model. Therefore,

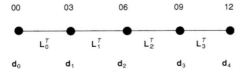

Figure 5.5.4 Schematic of the computation of the gradient of the observational cost function for a period of 12 hr, observations every 3 hr and the adjoint model that integrates backwards in time within each interval. Note that we start from the last weighted increment $\bar{\mathbf{d}}_4$ and go backwards in time.

194 Data Assimilation

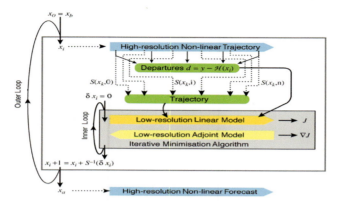

Figure 5.5.5 Schematic of the inner and outer loops from the ECMWF 4D-Var. ©ECMWF. Used with permission.

the preliminary analysis increments obtained by minimization may not be accurate enough, and it is common to use this initial analysis to update the initial state of the nonlinear model and recompute the nonlinear forecast at high resolution, and the observational increments $\mathbf{d}_i = \mathbf{y}_i^o - H(\mathbf{x}_i^b)$ within the assimilation window. The linear minimization with a low resolution model that may take 50–100 iterations, is usually referred to as the *inner loop*, and the nonlinear high resolution updated forecast and observational increments are known as *outer loop*. ECMWF typically performs three outer loops with increasing resolution (see their schematic Figure 5.5.5 for the outer and inner loops). The use of an outer loop that updates the nonlinear trajectory from which the observational increments are obtained has been found to be significantly beneficial. A similar outer loop is also used in 3D-Var (e.g., in the NCEP Global Statistical Interpolation, GSI) by adding to the background field the optimal increment $\delta \mathbf{x}(t_0)$ and recomputing the Jacobians needed to solve for a new, more accurate $\delta \mathbf{x}(t_0)$. It has been found that the outer loop improves substantially the accuracy of the observational increments and the analysis for 3D-Var as well as 4D-Var.

5.5.4 Further Remarks on 4D-Var

(1) The Hybrid ECMWF Ensemble of Data Assimilations (EDA)

As we saw, the two most important advantages of 4D-Var are (1) the observations are assimilated at their right time; and (2) if we assume that: (a) the model is perfect, and (b) the *a priori* error covariance at the initial time $\mathbf{B}_0$ is correct, then the 4D-Var analysis at the final time is identical to that of the extended Kalman filter (Lorenc, 1986; Daley, 1991), as we showed for the case of a single observation (Exercise 5.5.2). This means that *implicitly 4D-Var is able to evolve the forecast error covariance from* $\mathbf{B}_0$ *to the final time* (Thépaut et al., 1993). Unfortunately, this implicit covariance *is not* available at the end of the cycle, and neither is the new analysis error

5.5 Advanced Data Assimilation Methods with Evolving Covariance: 4D-Var

covariance. In other words, 4D-Var, like Kalman filter, is able to find the best linear unbiased estimation but not its error covariance.

A simplified reduced rank Kalman filter algorithm was proposed to estimate the evolution of the analysis errors in the subspace of the dynamically most unstable modes (Fisher and Courtier, 1995; Cohn and Todling, 1996) but was never implemented operationally.

Instead, ECMWF implemented in 2011 a method that allows for a flow dependent background error covariance $\mathbf{B}_0$ but still within the 4D-Var framework: the Ensemble of Data Assimilations (EDA),which can be considered a "hybrid 4D-Var" (see Section 5.6.5, where hybrid methods combining ensemble and variational methods are discussed). An ensemble of 10 4D-Var data assimilations were carried out (increased to 25 in 2014), where the ECMWF model physics, the sea surface temperature (SST) and the observations themselves were perturbed differently for each 4D-Var. The differences between the forecasts and analyses of the different ensemble members can be considered as representative of the statistical structure of the forecast and analyses "errors of the day" allowing for the construction of a flow-dependent $\mathbf{B}_0$. This may be considered a generalization of the approach used in the "NMC method" to estimate the statistical structure of a constant $\mathbf{B}_0$. The ensemble variance in the EDA is then calibrated online (Bonavita et al., 2012). At ECMWF, Hamrud et al. (2015) and Bonavita et al. (2015a) have also tested a hybrid of the 4D-Var with the LETKF with encouraging results (Section 5.6.5).

4D-Var has been very successfully implemented at ECMWF, MeteoFrance, JMA, Canada, and has resulted in significantly improved forecast scores compared to 3D-Var, attributed principally to the fact that observations are assimilated at the time they were made, rather than at the analysis time.

(2) Long Windows and Model Errors

Long windows are advantageous for 4D-Var because the longer the window, the more observations are used, and as a result the final analysis becomes less dependent on the estimated (constant) background error covariance, which can then be "forgotten" (Talagrand, 1997). On the other hand, the fact that 4D-Var with a strong constraint assumes a perfect model is also a disadvantage for long windows, since, for example, it will give the same credence to older observations at the beginning of the interval as to newer observations at the end of the interval (Ménard and Daley, 1996). Derber (1989) and Zupanski (1993) suggested methods to correct for model errors. Trémolet (2006) has developed a *weak constraint* version of 4D-Var at ECMWF, in which there is an additional *model error term* in the cost function allowing for the solution to be different from the model integration solution.

$$J[\mathbf{x}_0] = \frac{1}{2}\left[\mathbf{x}_0 - \mathbf{x}_0^b\right]^T \mathbf{B}_0^{-1}\left[\mathbf{x}(t_0) - \mathbf{x}^b(t_0)\right] + \frac{1}{2}\sum_{i=0}^{N}\left[H(\mathbf{x}_i) - \mathbf{y}_i^o\right]^T \mathbf{R}_i^{-1}\sum_{i=0}^{N}\left[H(\mathbf{x}_i) - \mathbf{y}_i^o\right]$$
$$+ \frac{1}{2}\sum_{i=1}^{N}[\mathbf{x}_i - M_i(\mathbf{x}_{i-1})]^T \mathbf{Q}_i^{-1}[\mathbf{x}_i - M_i(\mathbf{x}_{i-1})]$$

$$(5.5.15)$$

Data Assimilation

4D-Var can also be written in an incremental form with the cost function defined by

$$J(\delta\mathbf{x}_0) = \frac{1}{2}(\delta\mathbf{x}_0)^T \mathbf{B}_0^{-1}(\delta\mathbf{x}_0)$$

$$+ \frac{1}{2}\sum_{i=0}^{N}[H_i\mathbf{L}(t_0,t_i)\delta\mathbf{x}_0 - \mathbf{d}_i^o]^T\mathbf{R}_i^{-1}[\mathbf{H}_i\mathbf{L}(t_0,t_i)\delta\mathbf{x}_0 - \mathbf{d}_i^o] \qquad (5.5.16)$$

and the observational increment defined as in (5.5.3). Within the incremental formulation, it is possible to choose a *"simplification operator"* that solves the problem of minimization in a lower dimensional space $\mathbf{w}$ than that of the original model variables $\mathbf{x}$:

$$\delta\mathbf{w} = \mathbf{S}\delta\mathbf{x} \qquad (5.5.17)$$

$\mathbf{S}$ is meant to be rank deficient (as would be the case, for example, if a lower resolution spectral truncation was used for $\mathbf{w}$ than for $\mathbf{x}$), so that its inverse doesn't exist, and we have to use a generalized inverse $\mathbf{S}^{-I} = [\mathbf{S}\mathbf{S}^T]^{-1}\mathbf{S}^T$. Then the minimum of the problem is obtained for

$$J(\delta\mathbf{w}), \mathbf{x}_0^b = \mathbf{x}_0^g + \mathbf{S}^{-I}\delta\mathbf{w}_0 \qquad (5.5.18)$$

and a new *"outer iteration"* at the full model resolution can be carried out (Lorenc, 1997).

The iteration process can also be accelerated through the use of *"preconditioning,"* a change of control variables that makes the cost function more "spherical," and therefore each iteration can get closer to the center (minimum) of the cost function (e.g., Parrish and Derber, 1992; Lorenc, 1997).

5.5.5 Introduction to the Construction of the Tangent Linear and Adjoint Models

We have seen that 4D-Var assumes an initial constant background error covariance that represents the error of the previous analysis $\mathbf{B}_0$ and then minimizes the 4D-Var cost function (5.5.1) using (5.5.13) and the calculation (5.5.14) schematically described in Figure 5.5.4. This requires access to both the tangent linear model $\mathbf{L}$ and its adjoint (also linear) $\mathbf{L}^T$. Here we discuss how these models are constructed using a very simple and clear example from Prof. Shu-Chih Yang using the Lorenz (1963a) (pers. comm., 2006).

The properties of the famous Lorenz (1963a) model, widely used in the theory of chaos, are discussed in Section 6.2. The model has three variables, (x_1, x_2, x_3), and three equations for their time derivatives. Consider just the third equation of this nonlinear model:

$$\frac{dx_3}{dt} = x_2x_1 - bx_3. \qquad (5.5.19)$$

To integrate this *nonlinear* component of the model we need to choose a finite differences time scheme (Chapter 3, Table 3.2.1). We choose the Euler forward scheme,

5.5 Advanced Data Assimilation Methods with Evolving Covariance: 4D-Var

which is the simplest scheme to advance the model by one time step:

$$x_3(t + \Delta t) = x_3(t) + [x_1(t)x_2(t) - bx_3(t)]\Delta t. \tag{5.5.20}$$

Now we create the tangent linear model by adding to (x_1, x_2, x_3) small perturbations $(\delta x_1, \delta x_2, \delta x_3)$:

$$x_3(t + \Delta t) + \delta x_3(t + \Delta t) = x_3(t) + \delta x_3(t) + [(x_1(t) + \delta x_1(t))(x_2(t) + \delta x_2(t))$$
$$- b(x_3(t) + \delta x_3(t))]\Delta t. \tag{5.5.21}$$

Subtracting (5.5.20) from (5.5.21) and neglecting terms quadratic on the perturbations δx, we obtain the Linear Tangent model ($\mathbf{L}$):

$$\delta x_3(t + \Delta t) = \delta x_3(t) + [x_2(t)\delta x_1(t) + x_1(t)\delta x_2(t) - b\delta x_3(t)]\Delta t. \tag{5.5.22}$$

$\mathbf{L}$ gives the linear evolution in time of perturbations $(\delta x_1, \delta x_2, \delta x_3)$ small enough to neglect their quadratic or higher terms.

The Adjoint model $\mathbf{L}^T$ is the transpose of $\mathbf{L}$, so we have to write a matrix version of (5.5.22) including equations *for all the "active" variables*, namely all the perturbations that appear in this equation that we can transpose. There are four "active" variables in (5.5.22): $\delta x_3(t + \Delta t)$, $\delta x_1(t)$, $\delta x_2(t)$, $\delta x_3(t)$, so we need to write a matrix version of (5.5.22) that *explicitly* indicates that $\delta x_1(t)$, $\delta x_2(t)$, $\delta x_3(t)$ remain the same after applying (5.2.2):

$$\begin{bmatrix} \delta x_3(t + \Delta t) \\ \delta x_1(t) \\ \delta x_2(t) \\ \delta x_3(t) \end{bmatrix} = \begin{bmatrix} 0 & x_2(t)\Delta t & x_1(t)\Delta t & (1 - b\Delta t) \\ 0 & 1 & 0 & 0 \\ 0 & 0 & 1 & 0 \\ 0 & 0 & 0 & 1 \end{bmatrix} \begin{bmatrix} \delta x_3(t + \Delta t) \\ \delta x_1(t) \\ \delta x_2(t) \\ \delta x_3(t) \end{bmatrix} \tag{5.5.23}$$

This is the matrix version of the tangent linear model (5.5.22). We now write the adjoint matrix $\mathbf{L}^T$ of (5.5.23), transposing the matrix and using asterisks on the active variables to indicate that they are now adjoint variables:

$$\begin{bmatrix} \delta x_3^*(t + \Delta t) \\ \delta x_1^*(t) \\ \delta x_2^*(t) \\ \delta x_3^*(t) \end{bmatrix} = \begin{bmatrix} 0 & 0 & 0 & 0 \\ x_2\Delta t & 1 & 0 & 0 \\ x_1(t)\Delta t & 0 & 1 & 0 \\ (1 - b\Delta t) & 0 & 0 & 1 \end{bmatrix} \begin{bmatrix} \delta x_3^*(t - \Delta t) \\ \delta x_1^*(t) \\ \delta x_2^*(t) \\ \delta x_3^*(t) \end{bmatrix} \tag{5.5.24}$$

This is the adjoint model in matrix form, which integrates backwards in time. So the individual equations from the adjoint matrix should be executed in reverse order. Therefore, the tangent linear ($\mathbf{L}$) model single line,

$$\delta x_3(t + \Delta t) = \delta x_3(t) + [x_2(t)\delta x_1(t) + x_1(t)\delta x_2(t) - b\delta x_3(t)]\Delta t,$$

becomes the following several lines of code in the adjoint model ($\mathbf{L}^T$):

$$\delta x_3^*(t) = \delta x_3^*(t) + (1 - b\Delta t)\delta x_3^*(t + \Delta t)$$
$$\delta x_2^*(t) = \delta x_2^*(t) + (x_1(t)\Delta t)\delta x_3^*(t - \Delta t)$$
$$\delta x_1^*(t) = \delta x_1^*(t) + (x_2(t)\Delta t)\delta x_3^*(t - \Delta t)$$
$$\delta x_3^*(t - \Delta t) = 0 \tag{5.5.25}$$

The full adjoint code is executed in reversed order with respect to the tangent linear model, as shown here with this one line linear tangent model example (5.5.22).

Since this conversion from $\mathbf{L}$ to $\mathbf{L}^T$ is an arduous and long job, compilers have been created to generate the $\mathbf{L}$ and $\mathbf{L}^T$ models, such as the Tangent and Adjoint Models Compiler (TAMC, Giering and Kaminski, 1998) and the Tapenade (Hascoët and Pascual, 2013, online version available at www-tapenade.inria.fr:8080/tapenade/index.jsp). Despite this, creating and updating the exact tangent and adjoint models of an operational model with full nonlinear physics is very difficult, if not impossible. For example, if the parameterization of precipitation in a model code contains a statement such as "if the parcel becomes supersaturated, then condensate the excess vapor and rain it out," the linear tangent and the adjoint of this statement cannot be constructed because the "if statement" is inherently nonlinear. This problem with the creation of an adjoint code does not appear when the tangent linear model is approximated as the difference between two nonlinear solutions with small perturbations which saturate in time, as it happens with ensemble perturbations, or bred vector perturbations (Section 6.5).

Despite the difficulties in creating exact $\mathbf{L}$ and $\mathbf{L}^T$ models, 4D-Var has been extremely successful in operational centers. Figure 5.5.6 (Yang et al., 2006) comparing

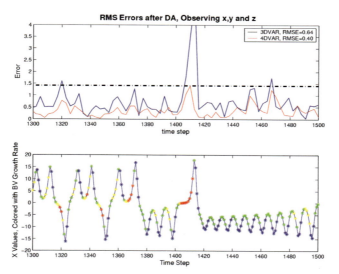

Figure 5.5.6 Top: Comparison of the errors made by 3D-Var (in blue) and by 4D-Var (in red) applied to the Lorenz (1963a) model for the same time period, assimilating observations every eight steps in (x, y, z) with Gaussian errors with $\sigma = \sqrt{2}$. Bottom: Perfect Lorenz (1963a) model showing the x variable with dots colored according to the bred vector growth rate, with blue indicating decay and red indicating strong growth, a predictor of regime change (Evans et al., 2004). Note the similarity between the periods of large bred vector growth (instability) at the top figure, and the large error growth in the 3D-Var analysis (in blue), whereas 4D-Var (in red) is able to substantially reduce this dynamical error growth. (Figure courtesy of Shu-Chih Yang.)

data assimilation schemes with synchronization) shows an example of the improvements brought by 4D-Var compared with 3D-Var. The largest error growths take place at the time the model state is undergoing growing instabilities, as shown in the bottom panel by large bred vector growth (red stars), which are a precursor of regime change (Evans et al., 2004).

5.6 Advanced Data Assimilation Methods with Evolving Covariance: Ensemble Kalman Filter

5.6.1 Introduction

In this section, we introduce different types of ensemble Kalman filter (EnKF), describe the local ensemble transform Kalman filter (LETKF) as a representative prototype of these methods, and provide several examples of how advanced properties and applications that had been explored and developed for 4D-Var can be adapted to the LETKF without requiring an adjoint model. Although ensemble Kalman filter is less mature than 4D-Var, its simplicity (it does not need an adjoint model) and competitive performance with respect to 4D-Var has made it quickly popular. In addition, the "hybrid" approach of combining an EnKF with a variational approach (either 3D-Var or 4D-Var) has become increasingly common.

We saw in Section 5.4 that variational and sequential methods (3D-Var and optimal interpolation) basically solve the same problem using the same climatological (constant) background error covariance **B** (equation (5.4.6), Lorenc, 1986). Similarly, in Section 5.5 we showed that 4D-Var with long windows is an approximation of the sequential Kalman filter method (Fisher et al., 2005), with an implicitly evolving **B**.

In this section, we will see how ensemble Kalman filter is used to approximate the evolving forecast error covariance **B** of the Kalman filter method. In the same way that 3D-Var and OI are approximations to solve the same problem using a constant **B**, 4D-Var and EnKF are approximations to solving the same problem, Kalman filter, with an *evolving* **B**. The methods used in their numerical solution for large atmospheric and oceanic problems are very different, so that it is particularly interesting to compare both methods and their properties whenever possible.

In Subsection 5.6.2, we present an insightful derivation of the Kalman filter due to Hunt et al. (2007) and introduce the extended Kalman filter (EKF). In 5.6.3, we introduce the ensemble Kalman filter (EnKF) and its two basic approaches, stochastic and square root filters. In Subsection 5.6.4, we show, as an example of a square root filter, the algorithm used in the local ensemble transform Kalman filter (LETKF, Hunt et al., 2007). Subsection 5 6.5 contains a discussion of methods known as "hybrids" that combine the advantages of EnKF with those of variational methods, including the methods known as "Covariance Hybrid" and "Gain Hybrid," and the increasingly popular "4DEnVar."

Because 4D-Var has a much longer history (e.g., Talagrand and Courtier, 1987; Courtier and Talagrand, 1990; Thépaut and Courtier, 1991), and has been implemented in several operational centers (e.g., Rabier et al., 2000), there are many innovative

200   Data Assimilation

methods that were developed and explored in the context of 4D-Var. We present in Subsections 5.6.6 and 5.6.7 examples of how advanced approaches created in the context of 4D-Var can be adapted and optimized for the EnKF (and vice versa).

5.6.2   Introduction to the Kalman Filter and Extended Kalman Filter Equations

Here we summarize key points of an alternative derivation of the Kalman filter equations for a linear perfect model (Kalman, 1960) due to Hunt et al. (2007) that provides much insight on the role that the background term plays in the variational cost function.

We start by assuming that the analysis $\bar{\mathbf{x}}_{n-1}^a$ valid at time t_{n-1} has Gaussian errors with covariance $\mathbf{P}_{n-1}^a$ so that the likelihood $L(\mathbf{x} - \bar{\mathbf{x}}_{n-1}^a)$ of the true state $\mathbf{x}$ is proportional to $\exp[-\frac{1}{2}(\mathbf{x} - \bar{\mathbf{x}}_{n-1}^a)^T (\mathbf{P}_{n-1}^a)^{-1}(\mathbf{x} - \bar{\mathbf{x}}_{n-1}^a)]$. Here the overbar represents the expected value. Previous observations $\mathbf{y}_j^o$ from time t_1 to t_{n-1} are also assumed to have a Gaussian distribution with error covariances $\mathbf{R}_j$, so that the likelihood of a trajectory of states $\mathbf{x}(t_j)$ given the past observations is proportional to $\exp\{ \prod_{j=1}^{n-1}[-\frac{1}{2}(\mathbf{y}_j^o - \mathbf{H}_j\mathbf{M}_{t_{n-1},t_j}\mathbf{x})^T \mathbf{R}_j^{-1}(\mathbf{y}_j^o - \mathbf{H}_j\mathbf{M}_{t_{n-1},t_j}\mathbf{x})]\}$, where $\mathbf{H}_j$ is the linear observation operator that transforms a model prediction into the corresponding observation, and $\mathbf{M}_{t_{n-1},t_j}$ is the linear forecast model that evolves a state from $\mathbf{x}(t_{n-1})$ to $\mathbf{x}(t_j)$.

The analysis $\bar{\mathbf{x}}_{n-1}^a$ and its covariance $\mathbf{P}_{n-1}^a$ are the mean and covariance of a Gaussian probability distribution representing the relative likelihood of a state $\mathbf{x}$ given all previous observations, so that taking logarithms of the likelihoods, for some constant c,

$$\sum_{j=1}^{n-1} [\mathbf{y}_j^o - \mathbf{H}_j\mathbf{M}_{t_{n-1},t_j}\mathbf{x}]^T \mathbf{R}_j^{-1} [\mathbf{y}_j^o - \mathbf{H}_j\mathbf{M}_{t_{n-1},t_j}\mathbf{x}] = [\mathbf{x} - \bar{\mathbf{x}}_{n-1}^a]^T(\mathbf{P}_{n-1}^a)^{-1} [\mathbf{x} - \bar{\mathbf{x}}_{n-1}^a] + c$$

$$(5.6.1)$$

The Kalman filter determines $\bar{\mathbf{x}}_n^a$ and $\mathbf{P}_n^a$ such that an equation analogous to (5.6.1) holds at time t_n. In the **forecast step** of the Kalman filter, the analysis $\bar{\mathbf{x}}_{n-1}^a$ and its covariance are propagated to time t_n with the linear forecast model $\mathbf{M}_{t_{n-1},t_n}$ and its adjoint $\mathbf{M}_{t_{n-1},t_n}^T$ creating the background state and its covariance:

$$\bar{\mathbf{x}}_n^b = \mathbf{M}_{t_{n-1},t_n} \bar{\mathbf{x}}_{n-1}^a,$$
$$\mathbf{P}_n^b = \mathbf{M}_{t_{n-1},t_n} \mathbf{P}_{n-1}^a \mathbf{M}_{t_{n-1},t_n}^T$$

$$(5.6.2)$$

(Recall that in (5.6.2), the second equation to forecast the background error covariance in the Kalman filter is actually *computationally unfeasible*, except for toy-sized models.)

Propagating equation (5.6.1), and using equation (5.6.2), we get a relationship valid for states at time t_n (see Hunt et al., 2007 for further details), showing that the back-

5.6 Advanced Data Assimilation Methods with Evolving Covariance 201

ground term represents the Gaussian probability distribution of a state, given the past observations up to t_{n-1}:

$$\sum_{j=1}^{n-1} [\mathbf{y}_j^o - \mathbf{H}_j\mathbf{M}_{t_n,t_j}\mathbf{x}]^T \mathbf{R}_j^{-1} [\mathbf{y}_j^o - \mathbf{H}_j\mathbf{M}_{t_n,t_j}\mathbf{x}] = [\mathbf{x} - \bar{\mathbf{x}}_n^b]^T (\mathbf{P}_n^b)^{-1} [\mathbf{x} - \bar{\mathbf{x}}_n^b] + c$$

(5.6.3)

When the new observations at time t_n are available, we use equation (5.6.3) to obtain an expression equivalent to equation (5.6.1) valid at time t_n for another constant c':

$$[\mathbf{x} - \bar{\mathbf{x}}_n^b]^T (\mathbf{P}_n^b)^{-1} [\mathbf{x} - \bar{\mathbf{x}}_n^b] + [\mathbf{y}_n^o - \mathbf{H}_n\mathbf{x}]^T (\mathbf{R}_n^{-1}) [\mathbf{y}_n^o - \mathbf{H}_n\mathbf{x}]$$
$$= [\mathbf{x} - \bar{\mathbf{x}}_n^a]^T (\mathbf{P}_n^a)^{-1} [\mathbf{x} - \bar{\mathbf{x}}_n^a] + c'$$

(5.6.4)

The analysis state that minimizes the variational cost function

$$J(\mathbf{x}) = [\mathbf{x} - \bar{\mathbf{x}}_n^b]^T (\mathbf{P}_n^b)^{-1} [\mathbf{x} - \bar{\mathbf{x}}_n^b] + [\mathbf{y}_n^o - \mathbf{H}_n\mathbf{x}]^T (\mathbf{R}_n^{-1}) [\mathbf{y}_n^o - \mathbf{H}_n\mathbf{x}]$$

is the state with maximum likelihood given all the observations including the latest observation $\mathbf{y}_n^o$. Equation (5.6.3) shows that in this cost function the background term represents the Gaussian distribution of a state with the maximum likelihood trajectory (history), i.e., $\bar{\mathbf{x}}_n^b$ is the analysis/forecast trajectory that best fits the past data available until t_{n-1}.

We can now obtain the Kalman filter equations for the **analysis step** by equating the terms in equation (5.6.4) that are quadratic or linear in $\mathbf{x}$. First equating the coefficients of the quadratic terms we obtain the Kalman filter analysis error covariance:

$$(\mathbf{P}_n^a)^{-1} = (\mathbf{P}_n^b)^{-1} + \mathbf{H}_n^T \mathbf{R}_n^{-1} \mathbf{H}_n$$

(5.6.5)

As we saw in toy models, equation (5.6.5) says "the precision of the analysis (inverse of the analysis error covariance) is the sum of the precisions of the forecast (background) and of the observations written in model space."

Next, we equate all the first order terms and obtain

$$\bar{\mathbf{x}}_n^a = \mathbf{P}_n^a \left[(\mathbf{P}_n^b)^{-1} \bar{\mathbf{x}}_n^b + \mathbf{H}_n^T \mathbf{R}_n^{-1} \mathbf{y}_n^o \right].$$

(5.6.6)

Replacing $(\mathbf{P}_n^b)$ from (5.6.5) into (5.6.6), we obtain

$$\bar{\mathbf{x}}_n^a = \bar{\mathbf{x}}_n^b + \mathbf{P}_n^a \mathbf{H}_n^T \mathbf{R}_n^{-1} (\mathbf{y}_n^o - \mathbf{H}_n\bar{\mathbf{x}}_n^b),$$

(5.6.7)

which says "the analysis is equal to the forecast, plus the Kalman gain matrix $\mathbf{K} = \mathbf{P}_n^a \mathbf{H}_n^T \mathbf{R}_n^{-1}$ multiplied by the observational increment."

Exercise 5.6.1 Derive equations (5.6.5), (5.6.6) and (5.6.7) from equation (5.6.4)

The Kalman gain matrix that multiplies the observational increment $\delta\mathbf{y}_n^o = \mathbf{y}_n^o - \mathbf{H}_n\bar{\mathbf{x}}_n^b$ in equation (5.6.7) can be written also as

$$\mathbf{K}_n = \mathbf{P}_n^a \mathbf{H}_n^T \mathbf{R}_n^{-1} = \mathbf{P}_n^b \mathbf{H}_n^T (\mathbf{H}_n\mathbf{P}_n^b\mathbf{H}_n^T + \mathbf{R}_n)^{-1}$$

as shown in (5.4.6).

Data Assimilation

For nonlinear models M_{t_{n-1},t_n}, the *extended Kalman filter* (EKF) approximation uses a nonlinear model in the forecast step to advance the background state, but the covariance is advanced using the model linearized around the trajectory $\bar{\mathbf{x}}_n^b$, and its adjoint (e.g., Ghil and Malanotte-Rizzoli, 1991; Nichols, 2010):

$$\bar{\mathbf{x}}_n^b = M_{t_{n-1},t_n}\left(\bar{x}_{n-1}^a\right),$$
$$\mathbf{P}_n^b = M_{t_{n-1},t_n}\mathbf{P}_{n-1}^a \mathbf{M}_{t_{n-1},t_n}^T \tag{5.6.8}$$

The cost of advancing the background error covariance with the linear tangent and adjoint models in equation (5.6.8) is what makes the EKF computationally *unfeasible* for any atmospheric model of realistic size without major simplifications.

We note here that the Remark 1 of Ide et al. (1997) "[In sequential methods] observations are processed whenever available and then discarded" follows from the fact that the background term is the most likely solution given all the past data. *However, this assumes that the Kalman filter has assimilated enough observations that it already has spun-up from the initial conditions.* When the KF starts assimilating observations, there is a "spin-up period" until enough observations are used to ensure that the background state has become the most likely given the past data. Therefore, *during the spin-up period, the observations may be used more than once in order to accelerate and shorten the spin-up* (Section 5.6.6).

5.6.3 Introduction to Ensemble Kalman Filtering Methods: Stochastic and Square Root Filters

Evensen (1994) suggested that the extended Kalman filter equation (5.6.8) could be efficiently computed with an *ensemble Kalman filter* (EnKF) for nonlinear models. The ensemble is created running K forecasts, where the size of the forecast ensemble, K, is much smaller than n, the number of degrees of freedom of the model, $K \ll n$. Then equation (5.6.8) can be replaced by

$$\mathbf{x}_{n,k}^b = M_{t_{n-1},t_n}\left(\mathbf{x}_{n-1,k}^a\right), \qquad \bar{\mathbf{x}}_n^b = \frac{1}{K}\sum_{k=1}^{K}\mathbf{x}_{n,k}^b$$

$$\mathbf{P}_n^b \approx \frac{1}{K-1}\sum_{k=1}^{K-1}\left(\mathbf{x}_{n,k}^b - \bar{\mathbf{x}}_n^b\right)\left(\mathbf{x}_{n,k}^b - \bar{\mathbf{x}}_n^b\right)^T \tag{5.6.9}$$

where the overbar now represents the *ensemble average*.

At long enough distances, the ensemble forecast errors should not be correlated. However, because the background error covariance is estimated from a relatively small ensemble, sampling errors create spurious error correlations at long distances. For this reason, Houtekamer and Mitchell (2001), Bishop et al. (2001), and Hamill et al. (2001) introduced the idea of *localizing* $\mathbf{P}_n^b$, i.e., multiplying each term of the covariance by a Gaussian function $\exp(-r_{ij}^2/2L^2)$, where r_{ij} is the distance between two grid points, and L is the localization scale, and the effect of localization is that long distance correlations are damped to zero. Gaspari and Cohn (1999) introduced an efficient 5th order polynomial approximation of this Gaussian function that has been universally

5.6 Advanced Data Assimilation Methods with Evolving Covariance

adopted. Mitchell et al. (2002) pointed out that this localization introduces imbalances in the analysis. Hunt and Miyoshi in 2005 used an alternative localization, multiplying instead the inverse of the observation error covariance $\mathbf{R}^{-1}$ by the Gaspari–Cohn Gaussian function, thus assuming that long distance observations have larger errors and reducing their impact on the grid point analyses. Because, unlike $\mathbf{P}_n^b$, $\mathbf{R}$ is typically either diagonal or block diagonal, this "observation localization" may be less prone to generate imbalances (Greybush et al., 2011). Yoshida (2019) and Chang and Kalnay (2022) introduced a more general localization method based on the "error correlation cutoff" method (Yoshida and Kalnay, 2018) that can localize observations even between two coupled systems, such as the atmosphere and ocean. It allows the assimilation of observations from different systems, guiding the selection of observations in *strongly coupled data assimilation*.

There are two basic approaches to performing EnKF, stochastic and square-root filters. The *stochastic* EnKF (e.g., Burgers et al., 1998; Houtekamer and Mitchell, 1998; Keppenne, 2000; van Leeuwen, 2020, and others) uses ensembles of data assimilation systems with randomly perturbed observations.[6] Perturbing the observations assimilated in different ensemble members with random errors whose error covariance is equal to $\mathbf{R}$ is required in this approach in order to avoid an underestimation of the size of the analysis error covariance (Burgers et al., 1998), but it may introduce an additional source of sampling errors (Whitaker and Hamill, 2002). Evensen (2009) is an excellent book focused on stochastic EnKF.

An alternative to the *stochastic* (or *perturbed observations*) EnKF approach are the *ensemble square-root filters* that generate a single analysis ensemble mean and covariance, satisfying the Kalman filter equations for linear models (Tippett et al., 2003; Bishop et al., 2001; Anderson, 2001, 2003; Whitaker and Hamill, 2002; Ott et al., 2002, 2004; Hunt, 2005; Hunt et al., 2007). We will focus in the rest of the chapter on *square-root* (or *deterministic*) filters. Houtekamer and Mitchell (2001) pointed out that observations with uncorrelated errors could be assimilated serially (one at a time), with the background for a new observation being the analysis obtained when assimilating the previous observation. Tippett et al. (2003) discuss the differences between several square-root filters that derive computational efficiency by assimilating observations serially. Another Monte Carlo method that avoids using perturbed observations is due to Pham (2001). Nerger et al. (2012) showed that this scheme is equivalent to the ETKF of Bishop et al. (2001).

Although all the square-root filters compute the same Kalman filter background error covariance, different square-root filters are still possible because *different analysis ensemble perturbations can have the same analysis error covariance*. Of the three schemes discussed in Tippett et al. (2003), the ensemble adjustment Kalman filter (EAKF) of Anderson (2001, 2003), in which the observations are assimilated serially, has been implemented into the flexible Data Assimilation Research Testbed (DART)

---

[6] van Leeuwen (2020) shows that a more consistent approach given non-Gaussian observation likelihood is to perturb the model-simulated observations, which ensures the correct skewness of the posterior distributions.

infrastructure, and has been applied to many geophysical problems (www.image.ucar.edu/DAReS/Publications/).

The ensemble square root filter (EnSRF) of Whitaker and Hamill (2002) results in simple scalar assimilation equations when observations are assimilated serially, and have been adopted for a number of systems, such as the assimilation of only surface observations in the 20th Century Reanalysis (Whitaker et al., 2004; Compo et al., 2011), the regional EnKF of Torn and Hakim (Torn et al., 2006), the EnKF systems used by the group led by Fuqing Zhang at Penn State (Zhang et al., 2006), and the hybrid systems implemented at NCEP (Kleist and Ide, 2015a).

The third square-root filter discussed in Tippett et al. (2003) is the ensemble transform Kalman filter (Bishop et al., 2001), which introduced the computation of the analysis covariance by a transform method adapted by Hunt et al. (2007), fully discussed in the Section 5.6.4. Zupanski (2005) proposed the maximum likelihood ensemble filter where a 4D-Var cost function with possibly nonlinear observation operators, is minimized within the subspace of the ensemble forecasts. In this system, the control forecast is allowed to have higher resolution than the rest of the ensemble. A review of stochastic EnKF methods is presented in Evensen (2003), and a comparison of EnKF with 4D-Var results in Kalnay et al. (2007a).

Ott et al. (2002, 2004) and Hunt et al. (2007) developed an alternative type of square-root EnKF (without perturbed observations) by performing the analyses *locally in space*, as did Keppenne (2000) and Kalnay and Toth (1994). This is computationally very efficient because the analyses at different grid points are independent and thus they can be computed in parallel. Since observations are assimilated simultaneously, not serially as in EnSRF and EAKF, it is simple to account for observation error correlations. However, the square-root EnKF methods that assimilate observations serially can also gain computational efficiency by separating observations into sub-regions, so that observations from two different regions that are far enough from each other that their error correlations can be assumed to be negligible can also be assimilated in parallel.

5.6.4 Example of a Square-Root EnKF: Local Ensemble Transform Kalman Filter

As indicated in Section 5.6.3, for many years, Kalman filter, which (unlike 3-DVar and optimal interpolation) estimates the evolving "errors of the day," could not be applied to large forecast models because the estimation of the forecast error covariance is computationally unfeasible for large models with many degrees of freedom. Evensen (1994) introduced the idea of performing Kalman filter DA by running an *ensemble of forecasts*, rather than a single one, and estimating the forecast error covariance using the ensemble members and their differences with the ensemble mean, which represents a best estimate of the truth, so that the dimension of the estimated error covariance is only $K - 1$, where K is the number of ensemble members. This idea led to an explosion of different ensemble Kalman filters (EnKFs). In the Evensen (1994) filter, all the ensemble members used the same observations and this made it underdispersive. Burgers et al. (1998), and Houtekamer and Mitchell (1998) introduced independently

5.6 Advanced Data Assimilation Methods with Evolving Covariance

the "stochastic EnKF," which solves this problem by adding random perturbations to the observations, representing the random differences in the estimations of the true observations by each ensemble member. Several methods were proposed to avoid the need of extra sampling errors introduced by the stochastic approach, starting with Whitaker and Hamill (2002) who introduced EnSRF. Alternative approaches to square root filters were also introduced by Anderson (2001, ensemble adjustment Kalman filter (EAKF)), and by Bishop et al. (2001, Ensemble Transform Kalman filter (ETKF)). Both EnSRF and EAKF adopted a "serial" approach for the assimilation of observations, assimilating one observation at a time, which reduced substantially the computational costs.

There are many different algorithms for ensemble Kalman filtering, and a recent paper by Vetra-Carvalho et al. (2018), provides an excellent overview of not only characteristics of different types of EnKFs, but also of the newer approaches to particle filters (Section 5.6.8), which, unlike EnKFs, do not assume that the prior forecast and the likelihood of the new observations given the forecast, are both Gaussian, ensuring that the posterior (new analysis) is also Gaussian. Houtekamer and Zhang (2016) wrote an excellent review of EnKFs for atmospheric Data Assimilation, their advances and problems.

Different EnKFs have a rather similar performance (e.g., Bonavita et al., 2015a). Here we describe the formulation of a widely used square-root EnKF, the local ensemble transform Kalman filter (LETKF), introduced by Hunt et al. (2007). We summarize the properties of the LETKF following Yang et al. (2009), who proposed a method to reduce computational costs with little degradation by computing the analysis in a coarser grid, and interpolating the LETKF analysis weights rather than the analysis increments.

Let $\mathbf{x}^f$ denote a matrix whose columns are the background ensemble in a local region evolved from a set of perturbed initial conditions. The ensemble states can be represented by (5.6.10):

$$\mathbf{x}^f = \bar{\mathbf{x}}^f + \mathbf{X}^f, \tag{5.6.10}$$

where $\bar{\mathbf{x}}^f$ is a column vector containing the mean of the ensemble and $\mathbf{X}^f$ is a matrix whose columns are the background ensemble perturbations from the ensemble mean.

With K denoting the ensemble size, the background error covariance matrix is defined by $(K-1)^{-1}\mathbf{X}^f(\mathbf{X}^f)^T$. Similar definitions are given for the analysis ensemble mean and the perturbations: $\bar{\mathbf{x}}^a$ and $\mathbf{X}^a$, respectively.

The LETKF determines a *transform matrix* that converts the *local background ensemble perturbations* into the *analysis ensemble perturbations*. The local analysis error covariance can be written as (5.6.11), where $\mathbf{P}^a$ is the analysis error covariance matrix and $\tilde{\mathbf{P}}^a$ is the analysis error covariance matrix *in ensemble space*.

$$\mathbf{P}^a = \frac{1}{K-1}\mathbf{X}^a(\mathbf{X}^a)^T = \mathbf{X}^f\tilde{\mathbf{P}}^a(\mathbf{X}^f)^T. \tag{5.6.11}$$

The *transform* matrix $\tilde{\mathbf{P}}^a$ (Hunt et al., 2007) is efficiently computed as

$$\tilde{\mathbf{P}}^a = [(K - 1)I + (\mathbf{Y}^f)^T \mathbf{R}^{-1} \mathbf{Y}^f]^{-1}, \tag{5.6.12}$$

where $\mathbf{Y}^f$ are the background perturbations in observation space, and $\mathbf{R}$ is the observation error covariance, assumed to be diagonal. Note that (5.6.12) can also be interpreted as: "In ensemble space, the precision of the analysis is equal to the precision of the forecast plus the precision of the observations," since, in ensemble space, the ensemble perturbation matrix is the unit matrix. This simple *transform matrix*, (5.6.12), makes the LETKF computation of the analysis in (5.6.11) very efficient.

With the observation operator H, the background ensemble is converted from the model space to observation space. The background perturbations in observation space (5.6.10), are then approximated by computing the mean and the deviations in observation space, denoted as $\bar{\mathbf{y}}^f$ and $\mathbf{Y}^f$ in

$$H(\mathbf{x}^f) = H(\bar{\mathbf{x}}^f + \mathbf{X}^f) \approx \bar{\mathbf{y}}^f + \mathbf{Y}^f. \tag{5.6.13}$$

Given that the ensemble size is much smaller than the model dimension, the $K \times K$ matrix $\tilde{\mathbf{P}}^a$ is efficiently computed within the ensemble space simply representing the corresponding precision relationship. After the transform matrix $\tilde{\mathbf{P}}^a$ is obtained, the mean analysis at the central grid point of the local region is computed from the background ensemble mean according to

$$\begin{aligned}
\bar{\mathbf{x}}^a &= \bar{\mathbf{x}}^f + \mathbf{X}^f \tilde{\mathbf{P}}^a (\mathbf{Y}^f)^T \mathbf{R}^{-1} (\mathbf{y}^o - \bar{\mathbf{y}}^f) \\
&= \bar{\mathbf{x}}^f + \mathbf{X}^f \mathbf{w}^a.
\end{aligned} \tag{5.6.14}$$

In (5.6.14), the $K \times 1$ vector of weights $\mathbf{w}^a$ is derived from the information about observational increments, $\mathbf{y}^o - \mathbf{y}^f$. In the final step, the analysis ensemble perturbations at the central grid point are derived by multiplying the background ensemble perturbations by the symmetric square root of $(K - 1)\tilde{\mathbf{P}}^a$:

$$\mathbf{X}^a = \mathbf{X}^f [(K - 1)\tilde{\mathbf{P}}^a]^{1/2} = \mathbf{X}^f \mathbf{W}^a. \tag{5.6.15}$$

In (5.6.15), the weight matrix $\mathbf{W}^a$ is a multiple of the symmetric square root of the local analysis error covariance matrix in ensemble space. It is computed by singular vector decomposition (SVD):

$$\mathbf{W}^a = \mathbf{U}\mathbf{S}^{1/2}\mathbf{U}^T, \tag{5.6.16}$$

where $\mathbf{U}$ is the matrix whose columns are the left singular vectors of $(K - 1)\tilde{\mathbf{P}}^a$ and $\mathbf{S}$ is a diagonal matrix whose diagonal elements are the singular values.

The use of a symmetric square-root matrix ensures that the sum of the analysis ensemble perturbations is zero, and depends continuously on $\tilde{\mathbf{P}}^a$ (Hunt et al., 2007). Adjacent analysis points, whose corresponding local background ensembles have small differences, will have slightly different $\tilde{\mathbf{P}}^a$. The derived symmetric square-root matrix can carry such characteristics and thus yields similar analysis ensemble perturbations at adjacent points, necessary to ensure the smoothness of the analysis. This property of the symmetric square-root matrix also ensures that the analysis ensemble

5.6 Advanced Data Assimilation Methods with Evolving Covariance

perturbations are consistent with the background ensemble perturbation since the symmetric square root matrix makes $\mathbf{W}^a$ the matrix closest to the identity matrix, given the constraint of the analysis error covariance matrix (Ott et al., 2004). Harlim (2006) demonstrated that the symmetric solution has better performance than those obtained with a non-symmetric square root, given the same ensemble size.

Equations (5.6.14) and (5.6.15) show that the analysis ensemble at each grid point is simply derived through a linear combination of the background ensemble, with weighting coefficients given by $\bar{\mathbf{w}}^a$ (a $K \times 1$ vector) for the mean analysis, and $\mathbf{W}^a$ (a $K \times K$ matrix) for the analysis perturbations. Thus, the kth analysis ensemble member is given by

$$\mathbf{x}_k^a = \bar{\mathbf{x}}^f + \mathbf{X}_k^f [\bar{\mathbf{w}}^a + \mathbf{W}_k^a], \tag{5.6.17}$$

where $\mathbf{W}_k^a$ is the kth column of $\mathbf{W}^a$.

Let $\mathbf{u}$ denote a vector of K ones, $\mathbf{u} = (1, 1, \ldots, 1)^T$; $\mathbf{u}$ is an eigenvector of $\mathbf{X}^f$ because the sum of the perturbations has zero mean ($\mathbf{X}^f \mathbf{u} = \mathbf{0}$). Since the sum of the columns of $\mathbf{Y}^f$ is equal to zero, $\mathbf{u}$ is also an eigenvector of $(\tilde{\mathbf{P}}^a)^{-1}$ and, thus, of $\mathbf{W}^a$ as well (Hunt et al., 2007):

$$\mathbf{W}^a \mathbf{W}^a \mathbf{u} = (K - 1)\tilde{\mathbf{P}}^a \mathbf{u} = \mathbf{u}. \tag{5.6.18}$$

With $\mathbf{X}^f \mathbf{u} = \mathbf{0}$, the analysis perturbations also have a zero mean:

$$\mathbf{X}^a \mathbf{u} = \mathbf{X}^f \mathbf{W}^a \mathbf{u} = \mathbf{0}. \tag{5.6.19}$$

5.6.4.1 Analysis Weights Interpolation

Yang et al. (2009) developed a new approach to increase the computational efficiency of the LETKF. Instead of computing the LETKF analysis at every model grid point, they compute the analysis on the significantly coarser analysis weight grid, and then interpolate back the *analysis weights* (5.6.14) and (5.6.15) onto the high resolution grid. Because the weights vary on horizontal scales larger than those that appear in the analysis increments, there is little degradation in the quality of the weight interpolated analysis compared to the analyses derived from the original high-resolution grid, even when the density of the reduced grid point analysis is as small as 11% of the original grid points. Furthermore, the weight interpolation approach also improves the analysis accuracy in data void regions, where the standard LETKF with the high resolution grid gives no analysis corrections. This method has been widely adopted in both EnKF and particle filter systems (e.g., Potthast et al., 2019).

The other useful procedure in this paper follows Corazza et al. (2007) in adding to the analysis ensemble perturbations, random perturbation with a very small amplitude (2% of the observational error). It had been found that this procedure greatly accelerated the spinup of bred vector perturbations by encouraging the ensemble to capture more sub-growing directions and to avoid the ensemble tendency to collapse into similar directions. This procedure was named "keeping the bred vectors young" by Andrew Lorenc (pers. commun., 2005). It was found to be also very useful in the LETKF analysis ensemble, where it is called "rejuvenating the ensemble" (e.g., Potthast et al., 2019).

5.6.5 Hybrids of Ensemble and Variational Data Assimilation

A major advantage of EnKF is that its background error covariance $\mathbf{P}^b_{ens} = \mathbf{X}^b_{ens}\mathbf{X}^{bT}_{ens}$ is *flow dependent* reflecting "errors of the day" (e.g., see Figures 5.5.1 and 5.5.2). However, by construction, the perturbations $\mathbf{X}^b_{ens}$ used to create $\mathbf{P}^b_{ens}$ are the differences between each ensemble member and the ensemble mean, so that the ensemble perturbations add up to zero. Therefore, if there are K ensemble members, $\mathbf{P}^b_{ens}$ has a rank of only $K - 1$, which is extremely small compared with n, the number of degrees of freedom of the model, so that the ensemble error covariance is severely *rank-deficient*. This problem is ameliorated by the use of background covariance *localization* in EnKF. Localization not only reduces sampling errors by zeroing out error correlations at long distances, but it also increases the effective rank of the EnKF and the number of observations that can be assimilated (without localization only $K - 1$ observations could be assimilated). Since the horizontal localization is typically of order O(500 km), the number of "independent" localizations that fit into a global atmospheric model is of order O(100). Thus the use of localization can increase the effective rank of $\mathbf{P}^b_{ens}$ by only a few orders of magnitude, and its rank is still much smaller than that of a full atmospheric model.

By contrast, the 3D-Var background error covariance $\mathbf{B}_{3DVar}$ can have the full rank of the model because it is estimated from differences between forecasts (Parrish and Derber, 1992). Although the simplification made by Parrish and Derber (1992) that $\mathbf{B}_{3DVar}$ is zonally symmetric reduced its rank, it was still much higher rank than the EnKF, even after localization. Since the analysis increment has to be in the subspace spanned by $\mathbf{B}$, 3D-Var is more robust than the ensemble Kalman filters, which need to use inflation and localization to avoid filter divergence (Houtekamer and Zhang, 2016). However, the disadvantage of $\mathbf{B}_{3DVar}$ is that it is computed once and for all, so that unlike $\mathbf{P}^b_{ens}$, $\mathbf{B}_{3DVar}$ *cannot reflect the "errors of the day."* In 4D-Var, the background error covariance $\mathbf{B}_0$ represents the error covariance of the previous analysis and is also assumed to be constant, even though the 4D-Var solution then evolves $\mathbf{B}$ implicitly within the assimilation window. In 4D-Var, the initial background error covariance is typically chosen as $\mathbf{B}_0 \sim \gamma \mathbf{B}_{3DVar}$ where $\gamma < 1$ is tuned to reflect the fact that the previous 4D-Var analysis is more accurate than the 3D-Var analysis, and thus the background error is smaller (e.g., Kalnay et al., 2007a). By using a full rank background error covariance both 3D-Var and 4D-Var are more robust than EnKF, which suffers from filter divergence if either the number of ensemble members or the number of observations are too small (e.g., Penny, 2014).

5.6.5.1 Covariance Hybrid

Hamill and Snyder (2000) proposed to create a "hybrid" between 3D-Var and EnKF that would take advantage of both the accuracy of the EnKF and the robustness of 3D-Var. They minimize a *hybrid cost function*:

$$J(\mathbf{x}^a) = (\mathbf{x}^a - \mathbf{x}^b)^T [\alpha \mathbf{B}_{3DVar} + (1 - \alpha)\mathbf{P}^b_{ens}]^{-1}(\mathbf{x}^a - \mathbf{x}^b)$$
$$+ (\mathbf{y}^o - h(\mathbf{x}^a))^T \mathbf{R}^{-1}(\mathbf{y}^o - h(\mathbf{x}^a)) \tag{5.6.20}$$

This type of hybrid is called "**Covariance Hybrid**," since it is based on combining the covariances of the variational 3D-Var covariance $\mathbf{B}_{3DVar}$, and an ensemble covariance $\mathbf{P}^b_{ens}$. For $\alpha = 1$ in (5.6.20), the analysis becomes the same as 3D-Var, and for $\alpha = 0$ it becomes the same as EnKF. Hamill and Snyder (2000) tested this hybrid on a quasi-geostrophic model and found that the hybrid analysis was more accurate than both 3D-Var and EnKF, and that the best results were obtained with $\alpha \leq 0.5$.

A Covariance Hybrid scheme can also be constructed by combining an EnKF with 4D-Var rather than 3D-Var, by minimizing a 4D-Var cost function with a hybrid background error covariance:

$$J[\mathbf{x}^a(t_0)] = (\mathbf{x}^a(t_0) - \mathbf{x}^b(t_0))^T [\alpha \mathbf{B}_{4DVar} + (1-\alpha)\mathbf{P}^b_{ens}]^{-1} (\mathbf{x}^a(t_0) - \mathbf{x}^b(t_0))$$

$$+ \sum_{i=0}^{N} (\mathbf{y}^o_i - H(\mathbf{x}^a(t_i)))^T \mathbf{R}^{-1} (\mathbf{y}^o_i - H(\mathbf{x}^a(t_i))) \qquad (5.6.21)$$

Here the analysis $\mathbf{x}^a(t)$ is the model trajectory that starts at $\mathbf{x}^a(t_0)$ and that minimizes the 4D-Var-EnKF hybrid cost function (5.6.21).

In 2003, Lorenc published two fundamental papers. Lorenc (2003a) is devoted to 4D-Var and the properties of its error covariance. In the second one, Lorenc (2003b) focused on EnKF, its differences with respect to 4D-Var, and on the advantages and disadvantages of the two methods, although at that time less was known about potential new methodologies based on EnKF. Lorenc (2003b) also introduced a new approach to the generation of the EnKF Covariance Hybrid, by augmenting the state vector with an additional set of control variables preconditioned with the square root of the ensemble covariance. Wang et al. (2007) proved that the two approaches to Covariance Hybrid, Hamill and Snyder (2000) and Lorenc (2003b), are actually mathematically equivalent, but the Lorenc (2003b) approach is simpler to incorporate ensemble information into operational 3D-Var schemes. Indeed, this version of the hybrid method was adopted by a number of operational centers (UK Met Office, Clayton et al., 2013 NCEP, Kleist and Ide, 2015a,b, who replaced the GSI (a 3D-Var) with a GSI-EnSRF hybrid, and obtained an impressive reduction of errors). At Environment Canada, where both a 4D-Var and an EnKF operational data assimilations were operational, Buehner et al. (2010a,b) showed that for the Canadian system, 4D-Var and EnKF had similar performance. Canada became the first country to replace 4D-Var with a new hybrid 4D EnVar (see Subsection 5.6.5.3, Buehner et al. (2013)).

5.6.5.2 Gain Hybrid

Penny (2014) proposed a different type of hybrid. Instead of augmenting the covariance of a variational system with the dynamical error covariance from an ensemble, as done in the Covariance Hybrid approach, Penny proposed augmenting an EnKF analysis with information from a variational system. Rather than combining the two types of covariances, Penny combined the gain matrices from the EnKF and a variational system (in his example he used 3D-Var), so this method is labeled "Gain Hybrid." Penny (2014) compared both the Covariance Hybrid and the Gain Hybrid methods to the 3D-Var, and, when the observations are scarce, they both reduce the analysis errors even compared to the more accurate EnKF. Moreover, both hybrids provide more robustness to the EnKF by allowing the analysis solution to be found outside the subspace spanned by the ensemble members, so that the hybrid systems remain stable with fewer ensemble members than the pure EnKF approach. An advantage of the Gain Hybrid is that, if both a variational system and an EnKF are available, the Gain Hybrid implementation requires minimal modifications: The EnKF analysis $\bar{\mathbf{x}}_{EnKF}^a$ is computed first and used as first guess (background) in the 3D-Var analysis:

$$J(\mathbf{x}_{3DVar}^a) = (\mathbf{x}_{3DVar}^a - \bar{\mathbf{x}}_{EnKF}^a)^T \mathbf{B}^{-1} (\mathbf{x}_{3DVar}^a - \bar{\mathbf{x}}_{EnKF}^a)$$
$$+ (\mathbf{y}^o - H\mathbf{x}_{3DVar}^a)^T \mathbf{R}^{-1} (\mathbf{y}^o - H\mathbf{x}_{3DVar}^a). \quad (5.6.22)$$

Then the Gain Hybrid analysis is defined as

$$\bar{\mathbf{x}}_{HG}^a = \alpha \bar{\mathbf{x}}_{EnKF}^a + (1 - \alpha) \mathbf{x}_{3DVar}^a, \quad (5.6.23)$$

and the ensemble is recentered around the hybrid mean $\bar{\mathbf{x}}_{HG}^a$. Penny et al. (2015) implemented the Gain Hybrid ocean assimilation system hybrid-GODAS, which showed superior performance than the pure EnKF and 3D-Var systems (Figure 5.6.1).

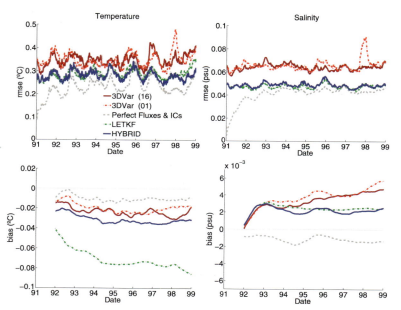

Figure 5.6.1 (Top) Global analysis-minus-truth RMS errors (RMSEs) and (bottom) 12-month moving average of biases in (left) temperature and (right) salinity for the top 700 m in an OSSE with the hybrid-GODAS system. Results are shown for the hybrid (blue), LETKF (green), 3D-Var with surface forcing perturbations from ensemble member 01 (red) and 16 (dark red), and the reference perfect-forcing case (gray). The ensemble methods produce lower RMSEs than 3D-Var and approach the RMSE levels of the reference perfect-forcing case. A small cold bias is produced by LETKF and salty bias produced by 3D-Var, both gradually increasing throughout the experiment period. The hybrid reduces these biases while maintaining a similar RMSE accuracy with LETKF (Penny et al., 2015). ©American Meteorological Society. Used with permission.

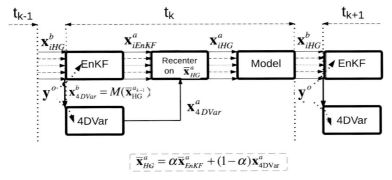

Figure 5.6.2 Schematic of the 4D-Var version of the Gain Hybrid (Penny, 2014; Hamrud et al., 2014, 2015; Bonavita et al., 2015a) experimentally tested at ECMWF.

At ECMWF, Hamrud et al. (2014) implemented this type of Gain Hybrid by combining their well-developed and tested 4D-Var and the LETKF, which were shown to give forecasts of similar skill. A schematic of the simple Gain Hybrid algorithm extended to 4DVar-LETKF Gain Hybrid is shown on Figure 5.6.2.

As indicated in Figure 5.6.2, the background initial trajectory used by the 4D-Var is given by the model evolution of the ensemble mean from the previous analysis (alternatively, it could have been chosen as the ensemble mean of the forecasts). In the standalone 4D-Var this background would be the forecast from the 4D-Var analysis also obtained from the previous analysis window, t_{k-1}. Hamrud et al. (2014) also chose to use the estimation of the standard deviations of the background errors produced by the EnKF short forecasts rather than from the EDA system as in the operational system.

Even when combining an ensemble Kalman filter with 3D-Var, it is a common practice to assign constant and even equal weights ($\alpha = 0.5$) to the variational and the ensemble components of the hybrids. de Azevedo et al. (2018) proposed instead to define a dynamic weight for the variational method using the ensemble spread, considered to be a measure of uncertainty in the EnKF. They define alpha, which multiplies the 3D-Var analysis, as the normalized spread for each variable at each level. Then, $(1 - \alpha)$, which decreases with increasing spread, becomes the factor that multiplies the ensemble analysis. The results of an Observing Systems Simulation Experiment (OSSE) with a perfect model are very encouraging: the dynamically weighted Hybrid Gain (HG) analyses are more balanced than the original HG analyses, and the forecasts are consistently improved throughout the 5-day forecasts.

At the resolution of T399 for EnKF, the Gain Hybrid EnKF-4DVar, using 100 ensemble members for EnKF, and T399 outer loops and two T95/T159 inner loops, the Gain Hybrid EnKF-4DVar forecasts were found to be substantially better than those of either 4D-Var or EnKF. The results of the new Gain Hybrid were also comparable to those of the operational hybrid system at ECMWF (4DVar-EDA).

5.6.5.3 4D-Var and 4DEnVar

Another promising hybrid approach is 4DEnVar (Liu et al., 2008, 2009; Buehner et al., 2013, 2015; Gustafsson and Bojarova, 2014; Lorenc et al., 2015; Kleist and

Ide, 2015b). It is based on the 4D-Var formulation, but it is solved taking advantage of an available ensemble of forecasts, and it is significantly more efficient than 4D-Var and essentially similar or better in accuracy, but less accurate that hybrid 4D-Var (Lorenc et al., 2015).

4DEnVar is developed starting from the incremental 4D-Var cost function for an assimilation window $(t_0, \ldots, t_K)$ given in Section 5.5.2, with $\mathbf{M}$ being the linear tangent model:

$$J(\delta\mathbf{x}) = J_b + J_o = \frac{1}{2}(\delta\mathbf{x})^T \mathbf{B}^{-1}\delta\mathbf{x} + \frac{1}{2}\sum_{t_k=t_0}^{t_K}(\mathbf{H}_k\mathbf{M}_k\delta\mathbf{x} - \mathbf{d}_k)^T \mathbf{R}_k^{-1}(\mathbf{H}_k\mathbf{M}_k\delta\mathbf{x} - \mathbf{d}_k),$$

where $\delta\mathbf{x}$ is the assimilation increment, $\mathbf{B}$ the background error covariance, $\mathbf{d}_k = \mathbf{y} - H_k(M_k(\mathbf{x}_b))$ are the innovations at time t_k, and $\mathbf{x}_b$ is the model background at t_0. Since the matrix $\mathbf{B}$ is huge, variational methods introduce a preconditioning matrix $\mathbf{U}$ such that $\mathbf{B} = \mathbf{U}\mathbf{U}^T$. The transformation matrix $\mathbf{U}$ is usually given as a series of simpler transform operators (e.g., Parrish and Derber, 1992; Derber and Bouttier, 1999; Barker et al., 2004).

Then, if we define $\delta\mathbf{x} = \mathbf{U}\mathbf{X}$, where $\mathbf{X}$ is the preconditioned assimilation increment, the 4D-Var cost function becomes

$$J(\mathbf{X}) = J_b + J_o = \frac{1}{2}\mathbf{X}^T\mathbf{X} + \frac{1}{2}\sum_{t_k=t_0}^{t_K}(\mathbf{H}_k\mathbf{M}_k\mathbf{U}\mathbf{X} - \mathbf{d}_k)^T \mathbf{R}_k^{-1}(\mathbf{H}_k\mathbf{M}_k\mathbf{U}\mathbf{X} - \mathbf{d}_k),$$

and its gradient is given by

$$\nabla_\mathbf{X} J = \mathbf{X} + \sum_{t_k=t_0}^{t_K}\mathbf{U}^T\mathbf{M}_k^T\mathbf{H}_k^T\mathbf{R}_k^{-1}(\mathbf{H}_k\mathbf{M}_k\mathbf{U}\mathbf{X} - \mathbf{d}_k). \tag{5.6.24}$$

Note that the gradient calculation (5.6.24) computed at each iteration of the minimization of the 4D-Var cost function requires the (forward) integration of the linear tangent model $\mathbf{M}_k$, and the backward integration of the adjoint model $\mathbf{M}_k^T$, and this is not only computationally very expensive, but the requirement of developing and maintaining the linear tangent and adjoint models are also onerous.

Liu et al. (2008) introduced the idea of minimizing the same 4D-Var cost function, but instead of using linear tangent and adjoint models, to compute the minimization based on an ensemble Kalman filter. The ensemble estimates the background error covariance:

$$\mathbf{B} \simeq \mathbf{X}_b'\mathbf{X}_b'^T, \tag{5.6.25}$$

where, as before, $\mathbf{X}_b'$ is the scaled matrix whose columns are the ensemble perturbations with respect to the ensemble mean: $\mathbf{X}_b' = \frac{1}{\sqrt{N-1}}(\ldots, \mathbf{x}_{bn} - \overline{\mathbf{x}}_b, \ldots)$. Equation (5.6.25) indicates that we can apply $\mathbf{U} = \mathbf{X}_b'$ for preconditioning, so that the cost function gradient (5.6.24) becomes:

$$\nabla_\mathbf{X} J = \mathbf{X} + \sum_{t_k=t_0}^{t_K}(\mathbf{H}_k\mathbf{M}_k\mathbf{X}_b')^T \mathbf{R}_k^{-1}\left((\mathbf{H}_k\mathbf{M}_k\mathbf{X}_b')\mathbf{X} - \mathbf{d}_k\right). \tag{5.6.26}$$

5.6 Advanced Data Assimilation Methods with Evolving Covariance

As pointed out by Liu et al. (2008), in equation (5.6.26) the background error in observation space ($\mathbf{H}_k \mathbf{M}_k \mathbf{X}_b'$) and its transpose, only need to be computed once, using ensemble forecasts outside the minimization iteration loop, *so that the coding and computational costs of 4DEnVar are greatly reduced compared to 4D-Var* (5.6.24). In addition, 4DEnVar does not require to create and maintain a linear tangent model and its adjoint.

Environment Canada, which had one of the most successful 4D-Var systems, replaced it completely with a 4DEnVar, taking advantage the existence of the operational ensemble Kalman filter system used to initialize ensemble forecasts (Buehner et al., 2013, 2015; Houtekamer and Mitchell, 1998). This was the first time that a fully operational 4D-Var has ceased to be operational.

The disadvantage of 4DEnVar is that (as in EnKF), the number of ensemble members is orders of magnitude smaller than the dimension of the model. This may be addressed by the use of covariance localization, and by combining 4DEnVar with 3D-Var into a hybrid system, as in Gustafsson and Bojarova (2014). Another disadvantage of 4DEnVar is that, although it is more accurate and computationally more efficient than the (non-hybrid) 4D-Var, *it is less accurate than the Hybrid 4D-Var* (Lorenc et al., 2015) . After a careful series of experiments, Lorenc et al. (2015) concluded that the fact that 4DEnVar does not represent the evolution of the errors from the climatological covariance, whereas 4D-Var does represent them, was the main reason for this deficiency.

5.6.6 Running in Place: A No-Cost Smoother

4D-Var and EnKF are essentially solving the same problem since they minimize the same cost function using different computational methods. These differences lead to some advantages and disadvantages for each of the two methods (see, for example, Lorenc, 2003a; Table 7 of Kalnay et al., 2007a). A major difference between 4D-Var and the EnKF is the dimension of the subspace of the analysis increments (analysis minus background). 4D-Var corrects the background forecast in a subspace that has the dimension of the adjoint model used in the minimization algorithm, which typically has half the resolution of the nonlinear model. The 4D-Var increment subspace is thus much larger than the local subspace of corrections in the EnKF, which has dimension of only $K - 1$, determined by the ensemble size K. It would be impractical to try to overcome this EnKF disadvantage by using a very large ensemble size. Fortunately, the localization of the error covariances carried out in the EnKF in order to reduce long distance covariance sampling errors, substantially addresses this problem by greatly increasing the number of degrees of freedom available to fit the data. As a result, experience has been that the quality of the EnKF analyses with localization increases with the number of ensemble members, but that their improvement is much slower when the size of the ensemble is increased much beyond 100. The observation that $O(100)$ ensemble members are, to some extent, sufficient for the EnKF seems to hold for atmospheric problems ranging from the storm- and meso-scales to the global-scales (Fuqing Zhang, pers comm., 2007). However, Miyoshi et al. (2014) performed a perfect model experiment of data assimilation with the SPEEDY model (resolution T30/L7) and found that, using 10,240 ensemble members, localization of the covariance was not longer needed. With this huge number of ensemble members, there were statistically significant long-distance error correlations, and as a result the analysis was improved by the large ensemble size (Kondo and Miyoshi, 2016). They found that increasing the size of the ensemble by an order of magnitude (from 40 to 320), also gave much improved correlations but was still noisier than with 10,240 members.

A number of very useful properties and extensions of 4D-Var have been explored over the years. They include the ability to assimilate observations at their right time (Talagrand and Courtier, 1987); the fact that within the data assimilation window 4D-Var acts as a smoother (Thépaut and Courtier, 1991); the availability of an adjoint model allowing the estimation of the impact of observations on the analysis (Cardinali et al., 2004) and on the forecasts (Langland and Baker, 2004); the ability to use long assimilation windows (Pires et al., 1996); *outer loops* that correct the background state when computing non-linear observation *operators*; and the possibility of accounting for model errors by using the model as a weak, not a strong, constraint (Trémolet, 2006). In the rest of this section we discuss how methods developed and implemented for 4D-Var can also be adapted and used in the LETKF, a prototype of EnKF.

5.6.6.1 4D-LETKF and No-Cost Smoother

Hunt et al. (2004) developed a 4D extension of the local ensemble Kalman filter (LEKF, Ott et al., 2004), taking advantage of the fact that the observational increments are expressed as linear combinations of the forecast ensemble perturbations at the time of the observation. Since the forecasts ensemble perturbations are available throughout the analysis window, using the same linear combination the observational increments can be "transported" forward or backward in time to the time of the analysis. This also allows to account for observation errors correlated in time, as Järvinen et al. (1999) did within 4D-Var. Hunt et al. (2007) showed that the 4D extension is particularly simple within the LETKF framework. Note that 4D-LETKF *determines the linear combination of ensemble forecasts valid at the end of the assimilation window that best fits the data throughout the assimilation window.*

This property allows creating a "no-cost" smoother for the LETKF with analogous smoothing properties as 4D-Var (Figure 5.6.3): the same weighted combination of the forecasts with weights given by the vector $\bar{\mathbf{w}}^a$ *is valid at any time of the assimilation interval.* It provides a smoothed analysis mean that (as in 4D-Var) is more accurate than the original analysis because it uses the future data available within the assimilation window (Kalnay et al., 2007b; Yang et al., 2012). As in 4D-Var, the smoothed analysis at the beginning of the assimilation window is an improvement over the filtered analysis computed using only past data. At the end of the assimilation interval only past data is used so that (as in 4D-Var) the smoother coincides with the analysis obtained with the filter.

It should be noted that in the same way we can use the weights $\bar{\mathbf{w}}^a$ to provide a mean smoother solution as a function of time, we can use the matrix $\mathbf{W}^a$ and apply it to the forecast perturbations $\mathbf{X}^b \mathbf{W}^a$ to provide an associated uncertainty evolving with

5.6 Advanced Data Assimilation Methods with Evolving Covariance

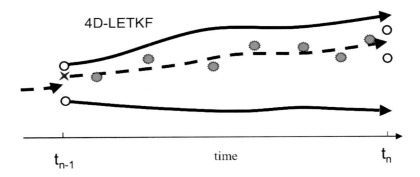

Figure 5.6.3 Schematic showing that the 4D-LETKF finds the linear combination of the ensemble forecasts at t_n that best fits the observations *throughout* the assimilation window $t_{n-1} - t_n$. The white circles represent the ensemble of analyses (whose mean is the analysis $\bar{\mathbf{x}}^a$), the full lines represent the ensemble forecasts, the dashed line represents the linear combination of the forecasts whose final state is the analysis, and the grey stars represent the asynchronous observations. The cross at the initial time of the assimilation window t_{n-1} is a *no-cost Kalman smoother*, i.e., an analysis at t_{n-1} improved using the information of "future" observations within the assimilation window by weighting the ensembles at t_{n-1} with the weights obtained at t_n. The smoothed analysis ensemble at t_{n-1} (not shown in the schematic) can also be obtained at no cost using the same linear combination of the ensemble forecasts valid at t_n given by the LETKF weight matrix $\mathbf{W}^a$ (Adapted from Kalnay et al., 2007a).

time (*Ross Hoffman, pers. comm.*, 2008). The updating of the uncertainty is critical for the "Running in Place" method described next, but the uncertainty is not updated in the "Quasi-Outer Loop" approach.

5.6.6.2 Use of the No-Cost Smoother to Accelerate the Spin-Up (Running in Place and Quasi Outer Loop)

4D-Var has been observed to spin up faster than EnKF (Caya et al., 2005), presumably because of its smoothing properties that allow finding the initial conditions at the beginning of the assimilation window that will best fit all the observations. The no-cost smoother allowed the development of a new algorithm, *Running in Place* (RIP), Kalnay and Yang (2010), that is useful to accelerate spinning up in rapidly evolving situations, such as severe storms or tornado predictions. At the start of severe storm convection, the dynamics of the system changes substantially, and the statistics of the processes become non-stationary. In this case, as in the spin-up case in which there are no previous observations available, the RIP algorithm ignores the rule "use the data and then discard it" and re-assimilates the new observations to accelerate the spin-up. Figure 5.6.4 shows how the standard LETKF (black line), like other EnKFs, has a very slow spin-up, but after applying Running in Place (RIP), the LETKF (red line) spins up even faster than 4D-Var (blue line). A computationally more efficient version of the RIP algorithm is the "Quasi-Outer Loop" (QOL) in which only the ensemble mean is updated. Both the RIP and the QOL algorithms accelerate the spin-up substantially so that the LETKF spins-up at least as fast as 4D-Var (Yang

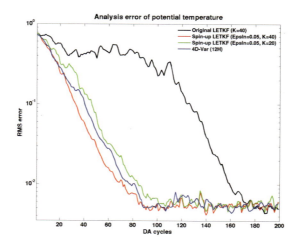

Figure 5.6.4 Comparison of the spin-up of a quasi-geostrophic model simulated data assimilation when starting from random initial conditions. Observations (simulated radiosondes) are available every 12 hr, and the analysis root-mean-square (RMS) errors are computed by comparing with a nature run (NR). Black line: original LETKF with 40 ensemble members, and no prior statistical information, blue line: optimized 4D-Var, red line: LETKF "running in place" with $\varepsilon = 5\%$ and 40 ensemble members, green line: as the red line but with 20 ensemble members.

et al., 2013; Penny et al., 2013). A procedure similar to RIP was developed, by Wang et al. (2013a) to use with the EnSRF, which, unlike the LETKF, does not provide the weights of the ensemble perturbations at the analysis time to successfully accelerate the spin up of tornado predictions.

5.6.7 Ensemble Forecast Sensitivity to Observations and Proactive Quality Control

5.6.7.1 FSO, EFSO, and HFSO

Langland and Baker (2004, hereafter LB04) developed a powerful adjoint-based method, which they named forecast sensitivity to observations (FSO), that estimates the impact of any observation on short-range forecasts, without the need to carry out data-denial experiments. This procedure can evaluate the impact of any one observation, or any group of them, that was assimilated in the data assimilation and forecast system, using a measure, such as dry or moist total energy of the short-range forecast error. It can also be used as a diagnostic tool to monitor the quality of observations, showing which observations make the forecast worse, and can also give an estimate of the relative importance of observations from different sources (Chen and Kalnay, 2019, hereafter CK19; and Chen and Kalnay, 2020, hereafter CK20). Following a similar procedure, Gelaro and Zhu (2009, hereafter GZ09) showed that FSO provides accurate assessments of the forecast sensitivity with respect to most of the observations assimilated. However, this powerful and efficient method to estimate observation

5.6 Advanced Data Assimilation Methods with Evolving Covariance

impact requires the adjoint of the forecast model, which is complex to develop and not always available, as well as the adjoint of the data assimilation algorithm. A first ensemble equivalent of the adjoint FSO was developed by Ancell and Hakim (2007) and Torn and Hakim (2008). The ensemble formulation was simplified in Kalnay et al. (2012). The simplified Ensemble FSO (EFSO) maps, using the ensemble forecasts readily available from the DA system, the forecast error changes between two consecutive forecasts by DA corresponding to each individual observation, with low computational cost (see next subsection for a brief derivation of EFSO). Recently, a hybrid approach, HFSO, was developed by Buehner et al. (2018), that projects the forecast sensitivity to analysis using the ensemble forecasts as in EFSO, but computes the observational impact through minimization of a cost function as in the adjoint FSO, consistent with the EnVar DA systems. Several operational centers and research groups have implemented one of the approaches of FSO to compare the impact of different observing systems on modern DA systems (e.g., Gelaro and Zhu, 2009; Cardinali, 2009; Gelaro et al., 2010; Lorenc and Marriott, 2014; Ota et al., 2013; Sommer and Weissmann, 2016), and other studies have explored the applications of FSO impacts. Lien et al. (2018) showed that the long-term averaged EFSO impact provides the detailed information needed to optimize data selection and the design of quality control procedures. Kotsuki et al. (2017a) found that using EFSO impact as an ordering method in a serial ensemble square root filter for the Lorenz (1996) model significantly improved the analysis accuracy. Chen (2018) showed that EFSO can be used to monitor the impact of every single observation, even at the level of hyperspectral channels.

5.6.7.2 Brief Derivation of EFSO, and a "Bridging" Example with a Low-Resolution GFS Model and PrepBUFR Observations

We briefly derive the EFSO formulation of Kalnay et al. (2012). Following LB04 we define a cost function to measure the impact of the observations assimilated at time $t = 0$ on the forecast at time t as the difference between the squares of the forecast errors with and without assimilating the observations $\mathbf{y}_0$:

$$ J = \Delta e^2 = (e_{t|0}^2 - e_{t|-6}^2) = [\mathbf{e}_{t|0}^T \mathbf{C} \mathbf{e}_{t|0} - \mathbf{e}_{t|-5}^T \mathbf{C} \mathbf{e}_{t|-6}], \qquad (5.6.27) $$

where the matrix of weights $\mathbf{C}$ defines the squared norm to be used (Dry Total energy in LB04 and GZ09, Moist Total Energy in CK20). We assume that the forecast length (6 hr) is short enough to allow the use of the linear tangent model $(\overline{\mathbf{x}}_{t|0}^f - \overline{\mathbf{x}}_{t|-6}^f) \approx \mathbf{M}(\overline{\mathbf{x}}_0^a - \overline{\mathbf{x}}_{0|-6}^b)$:

$$ \begin{aligned} \Delta e^2 &= \mathbf{e}_{t|0}^T \mathbf{C} \mathbf{e}_{t|0} - \mathbf{e}_{t|-6}^T \mathbf{C} \mathbf{e}_{t|-6} = (\mathbf{e}_{t|0} - \mathbf{e}_{t|-6})^T \mathbf{C} (\mathbf{e}_{t|0} + \mathbf{e}_{t|-6}) \\ &= (\overline{\mathbf{x}}_{t|0}^f - \overline{\mathbf{x}}_{t|-6}^f)^T \mathbf{C} (\mathbf{e}_{t|0} + \mathbf{e}_{t|-6}) \\ &\approx \left[\mathbf{M}(\overline{\mathbf{x}}_0^a - \overline{\mathbf{x}}_{0|-6}^b) \right]^T \mathbf{C} (\mathbf{e}_{t|0} + \mathbf{e}_{t|-6}) \\ &= [\mathbf{M} \mathbf{K} \delta \mathbf{y}_0]^T \mathbf{C} (\mathbf{e}_{t|0} + \mathbf{e}_{t|-6}) \end{aligned} \qquad (5.6.28) $$

In equation (5.6.28) we use the definition of the Kalman gain matrix $\mathbf{K} = \frac{1}{K-1} \mathbf{X}_0^a (\mathbf{X}_0^a)^T \mathbf{H}^T \mathbf{R}^{-1}$, and the matrix $\mathbf{Y}_0^a = \mathbf{H} \mathbf{X}_0^a$, composed of the analysis ensemble

218 Data Assimilation

perturbations in observation space. Here $\mathbf{P}_0^a = \frac{1}{K-1}\mathbf{X}_0^a(\mathbf{X}_0^a)^T$ is the analysis ensemble error covariance. We can then write $\mathbf{MK} = \frac{1}{K-1}\mathbf{MX}_0^a(\mathbf{X}_0^a)^T\mathbf{H}^T\mathbf{R}^{-1} \approx \frac{1}{K-1}\mathbf{X}_{t|0}^f(\mathbf{Y}_0^a)^T\mathbf{R}^{-1}$. Therefore the cost function (5.6.28) representing the square of the difference of the forecast error squared assimilating the observation at time $t = 0$ minus the forecast without assimilating this observation becomes

$$\Delta e^2 \approx \frac{1}{K-1}\delta\mathbf{y}_0^T\mathbf{R}^{-1}\mathbf{Y}_0^a(\mathbf{X}_{t|0}^f)^T\mathbf{C}(\mathbf{e}_{t|0} + \mathbf{e}_{t|-6}). \tag{5.6.29}$$

Due to the limited size of the ensemble, a localization for error covariance is required to suppress the spurious long distance correlations. Applying such localization to equation (5.6.29) gives the EFSO impact formulation

$$\Delta e^2 \approx \frac{1}{K-1}\delta\mathbf{y}_0^T\mathbf{R}^{-1}\left[\rho \circ (\mathbf{Y}_0^a(\mathbf{X}_{t|0}^f)^T)\right]\mathbf{C}(\mathbf{e}_{t|0} + \mathbf{e}_{t|-6}), \tag{5.6.30}$$

where ρ is a localization matrix and $\circ$ represents an element-wise multiplication (Schur product).

We can then obtain the impact of each observation by decomposing the sum of the inner product of the innovation vector $\delta\mathbf{y}_0$ and the error sensitivity vector $\frac{\partial\Delta(e^2)}{\partial\delta\mathbf{y}_0} = \frac{1}{K-1}\mathbf{R}^{-1}\left[\rho \circ (\mathbf{Y}_0^a(\mathbf{X}_{t|0}^f)^T)\right]\mathbf{C}(\mathbf{e}_{t|0} + \mathbf{e}_{t|-6})$ into elements that correspond to each observation, so that $\frac{1}{K-1}\delta\mathbf{y}_{0,l}^T\left[\mathbf{R}^{-1}\left[\rho \circ (\mathbf{Y}_0^a(\mathbf{X}_{t|0}^f)^T)\right]\mathbf{C}(\mathbf{e}_{t|0} + \mathbf{e}_{t|-6})\right]_l$ is the impact estimated impact of the l-th observation. This estimated impact is represented by the short-term forecast error changes due to the assimilation of the corresponding observation. We define the observations with positive EFSO impact, which increase the forecast error, as *detrimental* observations. Conversely, the observations with negative EFSO impact, which decrease the forecast error, are defined as *beneficial* observations.

5.6.7.3 Results of EFSO/PQC with a Low-Resolution GFS Model and PrepBUFR Observations

Proactive quality control (PQC; Ota et al., 2013; Hotta et al., 2017) was proposed to allow the rejection of detrimental observations based on their EFSO impact, in order to improve the quality of both the analyses and the forecasts. Using the spectral Global Forecast System (GFS) from the NCEP, Ota et al. (2013) demonstrated with many independent cases, using 24h and 6h respectively as the EFSO verifying leadtime, that denying the detrimental observations identified by their regional EFSO impact significantly reduced the resulting forecast error. Hotta et al. (2017) showed that even using just 6h as EFSO verifying time similarly reduced the forecast error, allowing to apply PQC at every analysis cycle. Chen et al. (2017), using the same cases as in Hotta et al. (2017), found that rejecting detrimental observations based on *global EFSO impact*, rather than *regional impact*, provided significantly more improvement. CK19 (Chen and Kalnay, 2019) further explored cycling PQC in an idealized environment using the Lorenz-96 model. The idealized experiments showed that among all the deficiencies tested in the DA system, the cycling PQC improvement responded only to those present in the observations, indicating that PQC effectively removes the impact from the detrimental observations.

CK20 (Chen and Kalnay, 2020) used a low-resolution spectral GFS model coupled with the local ensemble transform Kalman filter (LETKF; Hunt et al., 2007) and assimilated real observations from the NCEP PrepBUFR dataset (i.e., all observations except for radiances), as in Lien et al. (2016b, 2018). This study was a "bridge" between the idealized Lorenz-96 model used by CK19 and the full implementation of EFSO/PQC that would be required in a complex operational system. This system with intermediate complexity allows an efficient exploration of the properties of PQC in a realistic model assimilating real observations, as well as testing whether EFSO/PQC could be used in operations and improve the forecasts.

PQC is a fully-flow dependent QC based on EFSO. Past studies showed in several independent cases that GFS forecasts can be improved by rejecting observations identified as detrimental by EFSO. However, the impact of cycling and accumulating the impact of PQC in sequential data assimilation has, so far, only been examined using the simple Lorenz'96 model. The results demonstrate the major benefit of cycling PQC in a sequential data assimilation framework through the accumulation of improvements from previous PQC updates. Such accumulated PQC improvement is much larger than the "current" PQC improvement that would be obtained at each analysis cycle using "future" observations implicitly. As a result, it is unnecessary to use "future" information, and hence this allows the operational implementation of cycling PQC. The results show that the analysis and forecast are improved the most by rejecting all of the observations estimated to be detrimental by EFSO, but that major improvements also come from rejecting just the most detrimental 10% observations. The forecast improvements brought by PQC are observed throughout the 10 days of integration and provides the equivalent of more than 12–18 hr forecast leadtime gain (Figure 9 of CK20). An important finding is that PQC not only reduces substantially the root-mean-squared forecast errors but also the forecast biases. During the month of the experiment (January 2008) there is an episode of "skill dropout" in the control 5-day forecast. At the analysis time the differences between the TRUTH (CFSR), the CTL and the PQC-30 are very similar. By 3 days an unstable low pressure system starts developing in the Truth and deepens by day 5, but its features are not apparent in the CTL 5-day forecast. By contrast, the accumulated PQC-update in the analysis recovers the intensification of both the surface and the upper-level trough and signficantly improves the forecast skill (Figures 13 and 14 of CK20).

5.6.8   Particle Filter

We conclude this section on EDA with a brief introduction to perhaps the most novel and potentially promising approach: particle filter (PF). This approach has become widely explored in the last decade, and it continues to evolve rapidly.

Recall that EnKF assumes that the analysis ensemble perturbations are Gaussian, and that during the assimilation window the perturbations of the model forecasts evolve *linearly*, so that the background ensemble perturbations remain Gaussian.

The observation errors are also assumed to be Gaussian, so that the next analysis perturbations are Gaussian as well. The incremental 4D-Var also assumes linearity, and errors are usually assumed to be Gaussian. In reality, models can be highly nonlinear, and several iterative approaches for EnKF have been developed to improve the accuracy when the forecast perturbations grow nonlinearly, making them non-Gaussian (e.g., Kalnay and Yang, 2010; Bocquet and Sakov, 2012; Yang et al., 2012; Sakov et al., 2012). In addition, observational errors can also be non-Gaussian, especially for variables such as precipitation or relative humidity that are themselves far from Gaussian (e.g., Lien et al., 2013, 2016a,b).

The PF take a very different approach (see review by van Leeuwen, 2009 and van Leeuwen et al., 2019, and references therein). The ensemble sample is not assumed to have a Gaussian distribution since both the model and the analysis can be nonlinear. Instead, the ensemble forecasts (referred to as "particles") are compared with the observations, and weights are given to the particles depending on how well they fit the observations. Unfortunately this results in very few particles getting a large weight, and lots of particles getting very small or zero weights, which leads to a degeneration of the ensemble. Metropolis and Ulam (1949) already found this problem, and proposed a resampling technique, reintroduced by Gordon et al. (1993). The idea of resampling is simply that particles with very low weights are abandoned, while multiple copies of particles with high weights are kept in the posterior pdf.

Bayes' theorem (Section 5.3.2.1) also holds for pdf's if they are continuously differentiable (the derivative is continuous). The *posterior* pdf at time t_{i+1}, (i.e., the pdf of the new analysis), given the prior (forecast pdf, which depends on previous observations up to $\mathbf{y}_i^o$), and given the new observation $\mathbf{y}_{i+1}^o$, is equal to the pdf of the prior multiplied by the likelihood of the observation given the state, and divided by a constant:

$$p(\mathbf{x}_{i+1}|\mathbf{y}_{i+1}^o) \sim p(\mathbf{x}_{i+1}^b)p(\mathbf{y}_{i+1}^o|\mathbf{x}_{i+1})/\text{constant}.$$

Computing the full pdf of the prior is essentially impossible since it involves convolution of the forecast and the previous analysis over all possible values of the model. The basic idea of particle filters is to replace this step by sampling the model pdf with a sufficiently large number of model states $\mathbf{x}_1, \ldots, \mathbf{x}_N$ ("particles"), with corresponding weights $w_1, \ldots, w_N$ that add up to 1. This ensemble represent the model pdf as the sum of weighted delta functions centered at the particle locations, initially with equal weights $1/N$ adding up to 1:

$$p(\mathbf{x}) = \sum_{n=1}^{N} w_n \delta(\mathbf{x} - \mathbf{x}_n)$$

The PF step is conceptually simple (see schematic Figure 5.6.5, adapted from van Leeuwen (2010)):

At the beginning of the analysis cycle (stage 1), the pdf of the analysis is represented by the dashed curve. We choose N model states (with $N = 7$ in the figure) that sample the pdf with equal weights ($w_n = 1/N$). We then compute the forecasts (represented

5.6 Advanced Data Assimilation Methods with Evolving Covariance

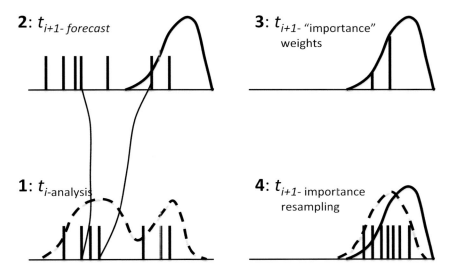

Figure 5.6.5 Schematic of the standard particle filter, also known as SIR (dashed): In stage 1 the analysis pdf at time t_i is indicated with a dashed line and is sampled by $N = 7$ equally weighted ensemble members or particles, represented by the vertical bars. In stage 2 (at the end of the assimilation window, t_{i+1}) the thin lines represent the forecasts for each particle, and the full line the pdf of the new observation. In this case most of the forecasts "miss" the observation pdf, and therefore will have negligible weight in the estimation of their "importance" (stage 3). In stage 4 the "important" particles are resampled to replace the lost particles so that we have a pdf of the analysis (dashed) and 7 equal weight particles. (Adapted from van Leeuwen, 2010, Figure 1. ©John Wiley and Sons. Used with permission.)

by the thin lines) for each of the particles, using the full nonlinear model plus possibly some random perturbations $\mathbf{R}_{i+1}$,

$$\mathbf{x}_{i+1}^b = M(\mathbf{x}_i^a) + \mathbf{R}_{i+1}.$$

At this time t_{i+1} we obtain the new observations $\mathbf{y}_{i+1}^o$ with a pdf shown by the full line (stage 2). As suggested in the figure, some of the forecasts fall outside the observation pdf so that they are then considered essentially irrelevant and dropped. This is done in the stage 3, where the "importance" of the particles is estimated based on Bayes' theorem:

$$p(\mathbf{x}|\mathbf{y}^o) = \sum_{n=1}^{N} w_n \delta(\mathbf{x} - \mathbf{x}_n),$$

with the weights now proportional to their "importance," i.e., to the likelihood of the observation given the forecast, $w_n \sim p(\mathbf{y}^o|\mathbf{x}_n)$ under the constraint that they should add up to 1. The likelihood (conditional probability of the observations given the particle $\mathbf{x}_n$) may be computed assuming observational errors are Gaussian, $w_n \sim p(\mathbf{y}^o|\mathbf{x}_n) \sim \exp[-\frac{1}{2}(\mathbf{y}^o - H(\mathbf{x}_n))^T \mathbf{R}^{-1}(\mathbf{y}^o - H(\mathbf{x}_n))]$ (shown in stage 3 as a full line) so that the importance weights can be calculated. Since many particles are lost, in stage 4 the particles are resampled with equal weights, taking several copies of the most important (most probable) particles and ignoring those with negligible importance. These particles are now sampling the posterior (or analysis) pdf (dashed line). Note that the analysis pdf is not known or calculated, but inferred from the particles sampling. At this point the analysis cycle can be continued.

It is clear that in this process, many particles may be lost, and only a few particles will retain all the weight, indicating "filter degeneracy." Especially for high dimensional systems, there is a tendency for the maximum weight to become close to 1, with other weights becoming very small. This is because $p(\mathbf{x})$ is a function in a space of dimension dim($\mathbf{x}$) so that the space of the problem grows exponentially with the dimension of the model. For example, if dim($\mathbf{x}$) = 100 and there are 10 grid points for each of the model variables $(x_1, x_2, \ldots, x_{100})$, the number of particles required for $p(\mathbf{x})$ is 10^{100} (example from Snyder et al., 2008). The complexity of the problem thus grows exponentially with the space dimension, something known as "the curse of dimensionality."

van Leeuwen (2010) proposed an approach to deal with the curse of dimensionality by "nudging" the forecasts toward future observations, adding to the forecast equations a nudging term (see Section 5.2.3),

$$\mathbf{x}_{i+1}^b = M(\mathbf{x}_i^a) + \mathbf{R}_{i+1} + \nu(\mathbf{y}_{i+1}^o - H(M(\mathbf{x}_i^a))).$$

The nudging should ensure that the forecasts are all "pulled" toward the observations so that large differences in their importance weights do not occur. Van Leeuwen shows promising results with this approach with the Lorenz-63 model using a particle ensemble of 3 members, and with a Lorenz-95 model with 40 variables or 1000 variables and only 20 ensemble members, suggesting that nudging could address the "curse of dimensionality" problem.

In the last decade there has been many papers with methods that address the "curse of dimensionality" of particle filters that for large dimensional systems require very large number of particles. For example, Snyder et al. (2008), showed that a PF with a 200 dimensional system would require 10^{11} ensemble members. One of the several promising methods recently reviewed by van Leeuwen et al. (2019) is based on the localization widely used in ensemble Kalman filtering, where only observations within a limited distance of a model grid point can be assimilated (Houtekamer and Zhang, 2016).

Poterjoy (2016) introduced a localized PF where observations are assimilated sequentially, as in the Anderson (2001), and in the Whitaker and Hamill (2002) ensemble Kalman filters. This makes the Poterjoy algorithm adapt naturally to those algorithms. Poterjoy and Anderson (2016) performed the first very successful application of a localized PF for data assimilation in a high dimensional geophysical model. The local PF was found to provide stable filtering results throughout a full year, using only 25 particles, and outperformed the ensemble Kalman filters, which are Gaussian, when observation networks included non-Gaussian errors, or related nonlinearly to the model state.

Penny and Miyoshi (2016) developed a local particle filter (LPF) in a form that maintained the ensemble transform matrix of the local ensemble transform Kalman filter (LETKF, Hunt et al., 2007). They tested the LPF on the Lorenz-96 model, and compared the LPF with the LETKF. The results showed that (1) the accuracy of the LPF surpasses LETKF as the forecast length increases (thus increasing the degree of nonlinearity; (2) the cost of LPF is significantly lower than LETKF; (3) LPF prevents filter divergence experienced by LETKF in cases of non-Gaussian observation error distributions.

Because of their similarity, Miyoshi and his Data Assimilation Team at RIKEN, Japan, adapted the original LETKF code developed by Miyoshi (2005) based on an intermediate AGCM known as the SPEEDY model. They developed the code design of the transform matrix based particle filter LAPF (Potthast et al., 2019) and its Gaussian mixture extension LMCPF, with only minor modifications to the existing LETKF code (Miyoshi et al., 2019). With 40 ensemble members, preliminary experiments showed that the LMCPF outperformed the LETKF also with 40 members (T. Miyoshi, personal communication, 2019).

6 Atmospheric Predictability and Ensemble Forecasting

6.1 Introduction to Atmospheric Predictability

In his 1951 paper on NWP, Charney indicated that he expected that even as models improved there would still be a limited range to skillful atmospheric predictions, but he attributed this to inevitable model deficiencies and finite errors in the initial conditions.* Lorenz (1963a,b) discovered the fact that the atmosphere, like any dynamical system with instabilities, has a *finite limit of predictability* (which he estimated to be about two weeks) *even if the model is perfect, and even if the initial conditions are known almost perfectly*. He did so by performing what is now denoted an "identical twin" experiment: he compared two runs made with the same model but with initial conditions that differed only very slightly. Just from round-off errors, he found that after a few weeks the two solutions were as different from each other as two random trajectories of the model.

Lorenz (1993) described how this fundamental discovery took place: His original goal had been to show that statistical prediction could not match the accuracy attainable with a nonlinear dynamical model, and therefore that NWP had a potential for predictive skill beyond that attainable purely through statistical methods. He had acquired a Royal-McBee LGP-30 computer, with a memory of 4K words and a speed of 60 multiplications per second, which for the late 1950s was very powerful. He developed and programmed in machine language a "low-order" atmospheric model (i.e., a model whose evolution was described by only 12 variables) driven by external heating and damped by dissipation. During 1959, he changed parameters in the model for several months trying to find a nonperiodic solution (since a periodic solution would be perfectly predictable from past statistics, and that would have defeated his purpose). He submitted a preliminary title, "The statistical prediction of solutions of dynamical equations," to the NWP conference that was going to take place during 1960 in Tokyo, gambling that he would indeed be able to find, for the first time in history, a nonperiodic numerical solution. After making the external heating a function of both latitude and longitude, he finally found the nonperiodic behavior that he was seeking. He rounded off and printed the evolution of the variables with three significant digits, which seemed sufficient to define the state of the model with plenty of accuracy. After running the model for several simulated years and satisfying himself that the solution had no periodicities, he decided to repeat part of an integration in more detail. When he came back from a coffee break, Lorenz found that the new

6.1 Introduction to Atmospheric Predictability

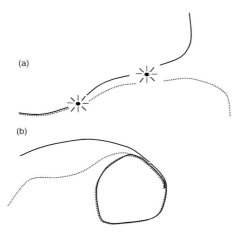

Figure 6.1.1 Schematic illustrating trajectories of: (a) a dynamical system with instabilities, which no matter how close they initially are, inevitably drift apart, and (b) a stable system with stationary or periodic orbits: after a possible transient stage, the trajectories stay close to each other, i.e., they become infinitely predictable.

solution was completely different from the original run. Before calling service for the computer, he checked the results and found that at the beginning, the new run did coincide with the original printed numbers, but than after a few days, the last digit became different, and then the next one, and after about two months, any resemblance with the original integration disappeared (Lorenz, 1993):

Lorenz (1963a,b) thus discovered the fundamental theorem of predictability: Unstable systems have a finite limit of predictability, and conversely, stable systems are infinitely predictable (since they are either stationary or periodic), as suggested by the schematic Figure 6.1.1.

Since the early days of Lorenz's momentous discovery, which gave impetus to the new science of chaos,[1] additional progress has been made, but his findings have not been changed in any fundamental way. In NWP, substantial progress has been made through the realization that the chaotic behavior of the atmosphere requires the replacement of single "deterministic" forecasts by "ensembles" of forecasts with differences in the initial conditions and in the model characteristics that realistically reflect the uncertainties in our knowledge of the atmosphere. This realization led to the introduction of operational ensemble forecasting at both NCEP and ECMWF in December 1992. It also led to work on extending the usefulness of NWP forecasts through a systematic exploitation of the chaotic nature of the atmosphere.

---

[1] It should be noted that Poincaré (1897, see Alligood et al., 1996) had already discovered that the planetary system is chaotic, i.e., that the orbits of the planets cannot be predicted well beyond a certain number of (millions of) years. He showed this for the simplest three-body problem of two stars with circular orbits moving on a plane around their center of mass, and a third "asteroid" with negligible mass in comparison with the first two, moving in the same plane. He found that the motion of the third body was *sensitively dependent on the initial conditions*, the hallmark of chaos (Alligood et al., 1996).

6.2 Brief Review of Fundamental Concepts about Chaotic Systems

Lorenz (1963a) introduced a three-variable model that is a prototypical example of chaos theory. These equations were derived as a simplification of Saltzman (1962) nonperiodic model for convection. Like Lorenz's (1960) original 12-variable model, the three-variable model is a dissipative system. This is in contrast to **Hamiltonian** systems, which **conserve total energy** or some other similar property of the flow. The system is **nonlinear** (it contains products of the dependent variables) but **autonomous** (the coefficients are time-independent). Sparrow (2012) wrote a whole book on the Lorenz three-variable model that provides a nice introduction to the subject of chaos, bifurcations, and strange attractors. Lorenz (1993) is a superbly clear introduction to chaos with a very useful glossary of the nomenclature used in today's literature. Alligood et al. (1996) is also a very clear introduction to dynamical systems and chaos. In this section, we use bold type to introduce some of the words used in the dynamical system vocabulary.

The Lorenz (1963a) equations are

$$\left.\begin{aligned} \frac{dx}{dt} &= \sigma(y - x) \\ \frac{dy}{dt} &= rx - y - xz \\ \frac{dz}{dt} &= xy - bz \end{aligned}\right\} \qquad (6.2.1)$$

The solution obtained by integrating the differential equations in time is called a **flow**. The **parameters** σ, b, r are kept constant within an integration, but they can be changed to create a **family of solutions** of the **dynamical system** defined by the differential equations. The particular parameter values chosen by Lorenz (1963a), $\sigma = 10$, $b = 8/3$, $r = 28$, result in **chaotic** solutions (sensitively dependent on the initial conditions), and since this publication they have been widely used in many papers. The solution of a time integration from a given **initial condition** defines a **trajectory** or **orbit** in **phase space**. The coordinates of a **point in phase space** are defined by the simultaneous values of the independent variables of the model, $x(t), y(t), z(t)$. The **dimension of the phase space** is equal to the number of independent variables (in this case three). The dimension of the subspace actually visited by the solution after an initial transient period (i.e., the **dimension of the attractor**) can be much smaller than the dimension of the phase space. A **volume** in phase space can be defined by a set of points in phase space, such as a hypercube $V = \delta x \delta y \delta z$, a hypersphere $V = \{\delta \mathbf{r}; \ |\delta \mathbf{r}| \leq \varepsilon\}$, etc.

The fact that the Lorenz system (6.2.1) is dissipative can be seen from the **divergence** of the **flow**:

$$\frac{\partial \dot{x}}{\partial x} + \frac{\partial \dot{y}}{\partial y} + \frac{\partial \dot{z}}{\partial z} = -(\sigma + b + 1) \qquad (6.2.2)$$

which shows that an original volume V contracts with time to $Ve^{-(\sigma+b+1)t}$. This proves the existence of a **bounded globally attracting set of zero volume** (i.e., an attractor

6.2 Brief Review of Fundamental Concepts about Chaotic Systems   227

of **dimension** smaller than n, the dimension of the phase space). A solution may start from a point away from the attracting set but it will eventually settle on the **attractor**. This initial portion of the trajectory is known as a **transient**. The **attracting set** (the set of points approached again and again by the trajectories after the transients are over) is called the **attractor** of the system. The attractor can have several components: **stationary** points (equilibrium or steady state solutions of the dynamical equations), **periodic orbits**, and more complicated structures known as **strange attractors** (which can also include periodic orbits). The different components of the attractor have corresponding **basins of attraction** in the phase space, which are all the initial conditions that will evolve to the same attractor. The fact that any initial volume in phase space contracts to zero with time is a general property of dissipative bounded systems, including atmospheric models with friction. Hamiltonian systems, on the other hand, are **volume-conserving**.

If we change the **parameters** of a dynamical system (in this example σ, b, r) and obtain families of solutions, we find that there is a point at which the behavior of the flow changes abruptly. The point at which this sudden change in the characteristics of the flow occurs is called a **bifurcation point**. For example, in Lorenz's equations the origin is a stable, stationary point for $r < 1$, as can be seen by investigating the local stability at the origin. The **local stability** of a point can be studied by linearizing the flow about the point and computing the eigenvalues of the linear flow. For $r < 1$, the stationary point is stable: All three eigenvalues are negative. This means that all orbits near the origin tend to get closer to it. At $r = 1$, there is a bifurcation, and for $r > 1$, two new additional stationary points $C_{\pm}$ are born, with coordinates $(x, y, z)_{\pm} = (\pm \sqrt{b(r-1)}, \pm \sqrt{b(r-1)}, r-1)$. For $r > 1$, the origin becomes **nonstable**: one of the three eigenvalues becomes positive (while the other two remain negative), indicating that the flow diverges locally from the origin in one direction. For $1 < r < 24.74\ldots$, C_+ and C_- are stable, and at $r = 24.74\ldots$ there is another bifurcation so that above that critical value C_+ and C_- also become unstable. As discussed by Lorenz (1993), a ubiquitous phenomenon is the occurrence of bifurcations of periodic motion leading to **period doubling**, and **sequences of period doubling bifurcations** leading to chaotic behavior (see Sauer, 1997)

A solution of a dynamical system can be defined to be **stable** if it is bounded, and if any other solution, once sufficiently close to it, remains close to it for all times. This indicates that a bounced stable solution must be **periodic** (repeat itself exactly) or at least **almost periodic**, since once the trajectory approaches a point in its past history, the trajectories will remain close forever (Figure 6.1.1(b)). A solution that is not periodic or almost periodic is therefore **unstable**: Two trajectories that start very close will eventually diverge completely (Figure 6.1.1(a)).

The long-term stability of a dynamical system of n-variables is characterized by the **Lyapunov exponents**. Consider a point in a trajectory, and introduce a (hyper)sphere of small perturbations about that point. If we apply the model to evolve each of those perturbations, we find that after a short time the sphere will be deformed into a (hyper)ellipsoid. In an unstable system, at least one of the axes of the ellipsoid will become larger with time, and once nonlinear effects start to be significant

228   Atmospheric Predictability and Ensemble Forecasting

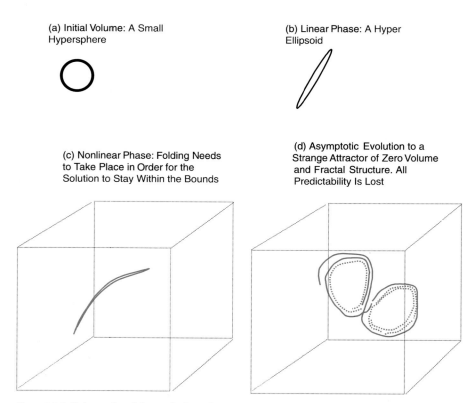

Figure 6.2.1 Schematic of the evolution of a small spherical volume in phase space in a bounded dissipative system. Initially (during the linear phase) the volume is stretched into an ellipsoid while the volume decreases. The solution space is bounded, and a bound is schematically indicated in the figure by the hypercube. The ellipsoid continues to be stretched in the unstable directions, until (because the solution phase space is bounded) it has to fold through nonlinear effects. This stretching and folding continues again and again, evolving into an infinitely foliated (fractal) structure. This structure, of zero volume and fractal dimension, is called a "strange attractor." The attractor is the set of states whose vicinity the system will visit again and again (the "climate" of the system). Note that in phases (a), (b), and (c), there is predictive knowledge: we know where the original perturbations generally are. In (d), when the original sphere has evolved into the attractor, all predictability is lost: We only know that each original perturbation is within the climatology of possible solutions, but we don't know where, or even in which region of the attractor it may be.

the ellipsoid will be deformed into a "banana" (Figure 6.2.1). Consider the linear phase, during which the sphere evolves into an ellipsoid. We can maintain the linear phase for an infinitely long period by taking an infinitely small initial sphere, or, alternatively, by periodically scaling down the ellipsoid dimensions dividing all its dimensions by the same scalar. Each axis j of the ellipsoid grows or decays over the long term by amounts given by $e^{\lambda_j t}$, where the λ_js are the Lyapunov exponents ordered by size $\lambda_1 \geq \lambda_2 \geq \ldots \geq \lambda_n$. The total volume of the ellipsoid will evolve like $V_0 e^{-(\lambda_1 + \lambda_2 + \cdots \lambda_n)t}$. Therefore, a Hamiltonian (volume-conserving) system is characterized by a sum of Lyapunov exponents equal to zero, whereas for a dissipative system the sum is negative.

Because the attractor of a dissipative system is bounded (the trajectories are enclosed within some hyperbox), if the first Lyapunov exponent is greater than zero, at least one of the axes of the ellipsoid keeps getting longer with time. The ellipsoid will eventually be distorted into a banana shape: it has to be folded in order to continue fitting into the box. The banana will be further stretched along the unstable axis and then necessarily folded again and again onto itself in order to continue fitting into the box. Since the volume of the ellipsoid eventually goes to zero for a dissipative system, the repeated stretching and folding of the ellipsoid of a chaotic system eventually converges to a zero-volume attractor with an infinitely folated structure (a process similar to the stretching and folding used to make "phyllo" dough!). This structure is known as "**strange attractor**" (Ruelle, 1989). It has a **fractal** structure: A dimension d which in general is not an integer and is smaller than the original space dimension n, estimated by Kaplan and Yorke (1979) to be

$$d = k + (\lambda_1 + \cdots + \lambda_k)/|\lambda_{k+1}| \tag{6.2.3}$$

where the sum of the first k Lyapunov exponents is positive, and the sum of the first $k + 1$ exponents is negative. If the system is Hamiltonian, its invariant manifold has the same dimension as the phase space.

In summary, a **stable** system has all Lyapunov exponents less than or equal to zero. A **chaotic** system has *at least one Lyapunov exponent greater than zero:* if at least $\lambda_1 > 0$, chaotic behavior will take place because at least one axis of the ellipsoid will be continuously stretched, leading to the separation of orbits originally started closely along that axis. Note that a **chaotic bounded flow** must also have *a Lyapunov exponent equal to zero*, with the corresponding local Lyapunov vector parallel to an orbit. This can be understood by considering two initial conditions such that the second is equal to the first after applying the model for one time step. The solutions corresponding to these initial conditions will remain close together, since the second orbit will always be the same as the first orbit shifted by one time step, and on the average, the distance between the solutions will remain constant. If we add a tiny perturbation, though, the second solution will diverge from the first one because there is a positive Lyapunov exponent.

6.3 Tangent Linear Model, Adjoint Model, Singular Vectors, and Lyapunov Vectors

In 1965, Lorenz published another paper based on a low-order model that behaved like the atmosphere. It was a quasi-geostrophic two-level model in a periodic channel, with a "Lorenz" vertical grid (velocity and temperature variables defined at the same two levels, see Section 3.3.5), and a spectral (Fourier) discretization in longitude and latitude. By keeping only two Fourier components in latitude and three in longitude, and choosing appropriate values for the model parameters, he was able to find

230   **Atmospheric Predictability and Ensemble Forecasting**

a model able to reproduce baroclinic instability and nonlinear wave interactions with just 28 variables. In this fundamental paper, Lorenz introduced for the first time (without using their current names) the concepts of the tangent linear model, adjoint model, singular vectors, and Lyapunov vectors for the low-order atmospheric model, and their consequences for ensemble forecasting. He also pointed out that the predictability of the model is not constant with time: It depends on the stability of the evolving atmospheric flow (the basic trajectory or reference state). In the following introduction to these subjects, we follow Lorenz (1965), Szunyogh et al. (1997), and Pu et al. (1997b).

6.3.1   Tangent Linear Model and Adjoint Model

Consider a nonlinear model. Once it has been *discretized in space* using, for example, finite differences or a spectral expansion leading to n independent variables (or degrees of freedom), the model can be written as a set of n nonlinear coupled ordinary differential equations:

$$\frac{d\mathbf{x}}{dt} = \mathbf{F}(\mathbf{x}) \qquad \mathbf{x} = \begin{bmatrix} x_1 \\ \vdots \\ x_n \end{bmatrix} \qquad \mathbf{F} = \begin{bmatrix} F_1 \\ \vdots \\ F_n \end{bmatrix} \tag{6.3.1}$$

This is the model in differential form. Once we choose a time-difference scheme (e.g., Crank–Nicholson, see Table 3.2.1), it becomes a set of nonlinear-coupled *difference* equations. Typically, an atmospheric model consists of one such system of difference equations which, for example, using a two-time level Crank–Nicholson scheme would be of the form

$$\mathbf{x}^{n+1} = \mathbf{x}^n + \Delta t \mathbf{F} \left(\frac{\mathbf{x}^n + \mathbf{x}^{n+1}}{2} \right) \tag{6.3.2}$$

A numerical solution of (6.3.1) starting from an initial time t_0 can be readily obtained by integrating the model numerically using (6.3.2) between t_0 and a final time t (i.e., "running the model"). This gives us a *nonlinear model solution* that depends only on the initial conditions:

$$\mathbf{x}(t) = M[\mathbf{x}(t_0)] \tag{6.3.3}$$

where M is the time integration of the numerical scheme from the initial condition to time t. A small perturbation $\mathbf{y}(t)$ can be added to the basic model integration $\mathbf{x}(t)$:

$$M[\mathbf{x}(t_0) + \mathbf{y}(t_0)] = M[\mathbf{x}(t_0)] + \frac{\partial M}{\partial \mathbf{x}} \mathbf{y}(t_0) + O[\mathbf{y}(t_0)^2]$$
$$= \mathbf{x}(t) + \mathbf{y}(t) + O[\mathbf{y}(t_0)^2] \tag{6.3.4}$$

At any given time, the linear evolution of the small perturbation $\mathbf{y}(t)$ will be given by

$$\frac{d\mathbf{y}}{dt} = \mathbf{J}\mathbf{y} \tag{6.3.5}$$

where $\mathbf{J} = \partial \mathbf{F}/\partial \mathbf{x}$ is the Jacobian of $\mathbf{F}$.

6.3 Tangent Linear and Adjoint Model, Singular and Lyapunov Vectors 231

This system of linear ordinary differential equations is the *tangent linear model* in differential form. Its solution between t_0 and t can be obtained by integrating (6.3.5) in time using the same time difference scheme used in the nonlinear model (6.3.3)

$$\mathbf{y}(t) = \mathbf{L}(t_0,t)\mathbf{y}(t_0) \tag{6.3.6}$$

Here $\mathbf{L}(t_0,t) = \partial M/\partial \mathbf{x}$ is an $(n \times n)$ matrix known as the *resolvent* or *propagator* of the tangent linear model: it propagates an initial perturbation at time t_0 into the final perturbation at time t. Because it is linearized over the flow from t_0 to t, $\mathbf{L}$ depends on the *basic trajectory* $\mathbf{x}(t)$ (the solution of the nonlinear model), but it does not depend on the perturbation $\mathbf{y}$. (The original nonlinear model is autonomous since $\mathbf{F}(\mathbf{x})$ depends on $x(t)$ but not explicitly on time, but the linear tangent model is nonautonomous.) Lorenz (1965) introduced the concept of the tangent linear model of an atmospheric model, but he actually obtained it directly from (6.3.4) neglecting terms quadratic or higher order in the perturbation $\mathbf{y}$:

$$M[\mathbf{x}(t_0)] + \mathbf{L}(t_0,t)\mathbf{y}(t_0) = \mathbf{x}(t) + \mathbf{y}(t) \approx M[\mathbf{x}(t_0) + \mathbf{y}(t_0)] \tag{6.3.7}$$

He did so by creating as initial perturbations a "sphere" of small perturbations of size ε along the n unit basis vectors $\mathbf{y}_i(t_0) = \varepsilon \mathbf{e}_i$ and applying (6.3.7) to each of these perturbations. With this choice of initial perturbations, subtracting (6.3.3) he obtained the matrix that defines the tangent linear model:

$$\mathbf{L}(t_0,t)[\varepsilon \mathbf{e}_1,\ldots,\varepsilon \mathbf{e}_n] = \varepsilon \mathbf{L}(t_0,t) = [\mathbf{y}_1(t),\ldots,\mathbf{y}_n(t)] \tag{6.3.8}$$

The Euclidean *norm* of a vector is the inner product of the vector with itself:

$$\|\mathbf{y}\|^2 = \mathbf{y}^T \mathbf{y} = \langle \mathbf{y},\mathbf{y} \rangle \tag{6.3.9}$$

The Euclidean norm of $\mathbf{y}(t)$ is therefore related to the initial perturbation by

$$\|\mathbf{y}(t)\|^2 = (\mathbf{L}y(t_0))^T \mathbf{L}y(t_0) = \langle \mathbf{L}y(t_0),\mathbf{L}y(t_0) \rangle = \langle \mathbf{L}^T \mathbf{L}y(t_0),y(t_0) \rangle \tag{6.3.10}$$

The *adjoint* of an operator $\mathbf{K}$ is defined by the property $\langle \mathbf{x},\mathbf{Ky} \rangle \equiv \langle \mathbf{K}^T\mathbf{x},\mathbf{y} \rangle$. In this case of a model with real variables, the *adjoint* of the tangent linear model $\mathbf{L}(t_0,t)$ is simply the *transpose* of the tangent linear model.

Now assume that we separate the interval (t_0, t) into two successive time intervals. For example, if $t_0 < t_1 < t$,

$$\mathbf{L}(t_0,t) = \mathbf{L}(t_1,t)\mathbf{L}(t_0,t_1) \tag{6.3.11}$$

Since the **adjoint** of the tangent linear model is the transpose of the TLM, the property of the transpose of a product is also valid:

$$\mathbf{L}^T(t_0,t) = \mathbf{L}^T(t_0,t_1)\mathbf{L}^T(t_1,t) \tag{6.3.12}$$

Equation (6.3.11) shows that the tangent linear model can be cast as a product of the tangent linear model matrices corresponding to short integrations, or even single time steps. Equation (6.3.12) shows that the adjoint of the model can also be separated into single time steps, but they are executed backwards in time, starting from the last time step at t, and ending with the first time step at t_0. For low-order models the

Atmospheric Predictability and Ensemble Forecasting

tangent linear model and its adjoint can be constructed by repeated integrations of the nonlinear model for small perturbations, as done by Lorenz (1965), equation (6.3.7), and by Molteni and Palmer (1993) with a global quasi-geostrophic model.

For large NWP models, this approach is too time consuming, and instead it is customary to develop the linear tangent and adjoint codes from the nonlinear model code following some rules discussed in Appendix A. An example of a FORTRAN code for a nonlinear model, and the corresponding tangent linear model and adjoint models are also given in Appendix A. A simple example is also given by Shu-Chih Yang, for the adjoint of the third equation of the Lorenz-63 model in Chapter 5.

6.3.2 Singular Vectors

Recall that for a given basic trajectory and an interval (t_0, t_1) the tangent linear model is a matrix that when applied to a small initial perturbation $\mathbf{y}(t_0)$ produces the final perturbation $\mathbf{y}(t_1)$:

$$\mathbf{y}(t_1) = \mathbf{L}(t_0, t_1)\mathbf{y}(t_0) \tag{6.3.13}$$

Singular value decomposition theory (e.g., Golub and Van Loan, 1996) indicates that for any matrix $\mathbf{L}$ there exist two orthogonal matrices $\mathbf{U}$, $\mathbf{V}$ such that

$$\mathbf{U}^T\mathbf{L}\mathbf{V} = \mathbf{S} \tag{6.3.14}$$

where

$$\mathbf{S} = \begin{bmatrix} \sigma_1 & 0 & \cdots & 0 \\ 0 & \sigma_2 & \cdots & 0 \\ \vdots & \vdots & & \vdots \\ 0 & 0 & \cdots & \sigma_n \end{bmatrix}$$

and

$$\mathbf{U}\mathbf{U}^T = \mathbf{I} \qquad \mathbf{V}\mathbf{V}^T = \mathbf{I} \tag{6.3.15}$$

$\mathbf{S}$ is a diagonal matrix whose elements are the *singular values* of $\mathbf{L}$.

If we left multiply (6.3.14) by $\mathbf{U}$, we obtain

$$\mathbf{L}\mathbf{V} = \mathbf{U}\mathbf{S} \quad \text{i.e.,} \quad \mathbf{L}(\mathbf{v}_1, \ldots, \mathbf{v}_n) = (\sigma_1\mathbf{u}_1, \ldots, \sigma_n\mathbf{u}_n) \tag{6.3.16}$$

where $\mathbf{v}_i$ are the columns of $\mathbf{V}$ and $\mathbf{u}_i$ the columns of $\mathbf{U}$. This implies that

$$\mathbf{L}\mathbf{v}_i = \sigma_i\mathbf{u}_i \tag{6.3.17}$$

Equation (6.3.17) defines the $\mathbf{v}_i$s as the *right singular vectors of* $\mathbf{L}$, hereafter referred to as *initial singular vectors*, since they are indeed valid at the beginning of the optimization interval over which $\mathbf{L}$ is defined.

We now right multiply (6.3.14) by $\mathbf{V}^T$ and obtain:

$$\mathbf{U}^T\mathbf{L} = \mathbf{S}\mathbf{V}^T \tag{6.3.18}$$

6.3 Tangent Linear and Adjoint Model, Singular and Lyapunov Vectors 233

Transposing (6.3.18), we obtain

$$\mathbf{L}^T\mathbf{U} = \mathbf{V}\mathbf{S} \quad \text{i.e.,} \quad \mathbf{L}^T(\mathbf{u}_1,\dots,\mathbf{u}_n) = (\sigma_1\mathbf{v}_1,\dots,\sigma_n\mathbf{v}_n) \tag{6.3.19}$$

so that

$$\mathbf{L}^T\mathbf{u}_i = \sigma_i\mathbf{v}_i \tag{6.3.20}$$

The $\mathbf{u}_i$s are the *left singular vectors* of $\mathbf{L}$ and will be referred to as *final* (or *evolved*) *singular vectors*, since they correspond to the end of the interval of optimization.

From (6.3.17) and (6.3.20) we obtain

$$\mathbf{L}^T\mathbf{L}v_i = \sigma_i\mathbf{L}^T\mathbf{u}_i = \sigma_i^2\mathbf{v}_i \tag{6.3.21}$$

Therefore, the initial singular vectors can be obtained as the eigenvectors of $\mathbf{L}^T\mathbf{L}$, a normal matrix whose eigenvalues are the squares of the singular values. Since $\mathbf{U}$, $\mathbf{V}$ are orthogonal matrices, the vectors $\mathbf{v}_i$ and $\mathbf{u}_i$ that form them constitute orthonormal bases, and any vector can be written in the following form:

$$\mathbf{y}(t_0) = \sum_{i=1}^{n}\langle\mathbf{y}_0,\mathbf{v}_i\rangle\mathbf{v}_i \tag{6.3.22a}$$

$$\mathbf{y}(t_1) = \sum_{i=1}^{n}\langle\mathbf{y}_1,\mathbf{u}_i\rangle\mathbf{u}_i \tag{6.3.22b}$$

where $\langle\mathbf{x},\mathbf{y}\rangle$ is the inner product of two vectors $\mathbf{x}$, $\mathbf{y}$. Therefore, using (6.3.22a) and (6.3.17)

$$\mathbf{y}(t_1) = \mathbf{L}(t_0,t_1)\mathbf{y}(t_0) = \sum_{i=1}^{n}\langle\mathbf{y}_0,\mathbf{v}_i\rangle\sigma_i\mathbf{u}_i \tag{6.3.23}$$

If we now take the inner product of (6.3.23) with $\mathbf{u}_i$, we obtain

$$\langle\mathbf{y}(t_1),\mathbf{u}_i\rangle = \sigma_i\langle\mathbf{y}(t_0),\mathbf{v}_i\rangle \tag{6.3.24}$$

This indicates that by applying the tangent linear model $\mathbf{L}$ each initial vector $\mathbf{v}_i$ component will be stretched by an amount equal to the singular value σ_i (or contracted if $\sigma_i < 1$), and the direction will be rotated to that of the evolved vector $\mathbf{u}_i$. Similarly applying the adjoint of the tangent linear model, $\mathbf{L}^T$, each final vector $\mathbf{u}_i$ will be stretched by an amount equal to the singular value σ_i and rotated to the initial vector $\mathbf{v}_i$.

Exercise 6.3.1 Use (6.3.20) and (6.3.22b) to show that $\langle\mathbf{y}(t_0),\mathbf{v}_i\rangle = \sigma_i\langle\mathbf{y}(t_1),\mathbf{u}_i\rangle$.

If we consider all the perturbations $\mathbf{y}(t_0)$ of size 1, from (6.3.24) we obtain that for each of them

$$\sum_{i=1}^{n}\left(\frac{\langle\mathbf{y}(t),\mathbf{u}_i\rangle}{\sigma_i}\right)^2 = \sum_{i=1}^{n}\langle\mathbf{y}(t_0),\mathbf{v}_i\rangle^2 = \|\mathbf{y}(t_0)\|^2 = 1 \tag{6.3.25}$$

so that an initial sphere of radius 1 becomes a hyperellipsoid of semiaxes σ_i. The first initial singular vector $\mathbf{v}_1$ is also called an "optimal vector" since it gives the direction

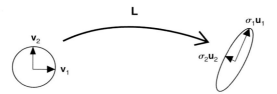

Figure 6.3.1 Schematic of the application of the tangent linear model to a sphere of perturbations of size 1 for a given interval (t_0, t_1).

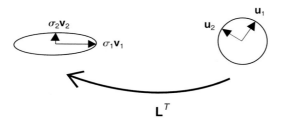

Figure 6.3.2 Schematic of the application of the adjoint of the tangent linear model to a sphere of perturbations of size 1 at the final time.

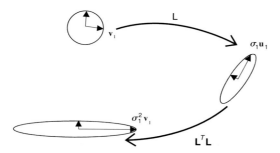

Figure 6.3.3 Schematic of the application of the adjoint of the tangent linear model to a sphere of perturbations of size 1 at the final time.

in phase space (i.e., the shape in physical space) of the perturbation that will attain maximum growth σ_1 in the interval (t_0, t_1) (Figure 6.3.1).

Note that applying $\mathbf{L}$ is the same as running the tangent linear model forward in time, from t_0 to t_1. Applying $\mathbf{L}^T$ is like running the adjoint model backwards, from t_1 to t_0. From (6.3.21) we see that if we apply the adjoint model to a sphere of final perturbations of size 1 (expanded on the basis formed by the evolved or left singular vectors), they also become stretched and rotated into a hyperellipsoid of semiaxes in the directions of the $\mathbf{v}_i$ with length σ_i (Figure 6.3.2).

Therefore, if we apply $\mathbf{L}^T\mathbf{L}$ (i.e., run the tangent linear model forward in time, and then the adjoint backwards in time, the first initial singular vector will grow by a factor σ_1^2 (see Figure 6.3.3), and the other initial singular vectors will grow or decay by their corresponding singular value squared σ_i^2. In other words, the (initial) singular vectors $\mathbf{v}_i$ are the eigenvectors of $\mathbf{L}^T\mathbf{L}$ with singular values σ_i^2. Conversely, if we apply the adjoint model first (integrate the adjoint model backwards from the final

6.3 Tangent Linear and Adjoint Model, Singular and Lyapunov Vectors

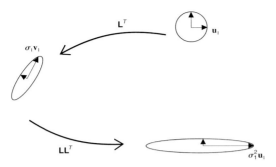

Figure 6.3.4 Schematic of the application of the tangent linear model forward in time followed by the adjoint of the tangent linear model to a sphere of perturbations of size 1 at the initial time.

to the initial time), followed by the tangent linear model (integrate forward to the final time), the final singular vectors $\mathbf{u}_i$ will grow both backward and forward, by a total factor also equal to σ_i^2 (Figure 6.3.4). In other words, the final singular vectors are the eigenvectors of $\mathbf{LL}^T$, and again they have eigenvalues equal to the square of the singular values of $\mathbf{L}$. Alternatively, once the initial singular vectors are obtained using, for example, the Lanczos algorithm, the final singular vectors can be derived by integrating the tangent linear model (6.3.17).

If we apply $\mathbf{L}^T\mathbf{L}$ repeatedly over the same interval (t_0, t), we obtain the leading initial singular vector, or first optimal vector. Additional leading singular vectors can be obtained by a generalization of the power method (Lanczos algorithm, Golub and Van Loan, 1996), which requires running the tangent linear model and its adjoint about three times the number of singular vectors required. For example, to get the leading 30 singular vectors optimized for $t_1 - t_0 = 36$ hr, the ECMWF performed 100 iterations, equivalent to running the tangent linear model for about 300 days (Molteni et al., 1996).

It is important to note that the adjoint model and the singular vectors are defined with respect to a given norm. So far we have used an Euclidean norm in which the weight matrix that defines the inner product is the identity matrix:

$$\|\mathbf{y}\|^2 = \mathbf{y}^T\mathbf{y} = \langle \mathbf{y}, \mathbf{y} \rangle \tag{6.3.26}$$

The leading (initial) singular vectors are the vectors of equal size (initial norm equal to one $\|\mathbf{y}(t_0)\|^2 = \mathbf{y}(t_0)^T\mathbf{y}(t_0) = \langle \mathbf{y}(t_0), \mathbf{y}(t_0) \rangle = 1$), that grow fastest during the optimization period (t_0, t_1), i.e., the initial vectors that maximize the norm at the final time:

$$J(\mathbf{y}(t_0)) \equiv \|\mathbf{y}(t_1)\|^2 = [\mathbf{L}\mathbf{y}(t_0)]^T \mathbf{L}\mathbf{y}(t_0) = \langle \mathbf{L}^T\mathbf{L}\mathbf{y}(t_0), \mathbf{y}(t_0) \rangle \tag{6.3.27}$$

If we define a norm using any other weight matrix $\mathbf{W}$ applied to $\mathbf{y}$, then the requirement that the initial perturbations be of equal size implies:

$$\|\mathbf{y}(t_0)\|^2 = (\mathbf{W}\mathbf{y}(t_0))^T \mathbf{W}\mathbf{y}(t_0) = \mathbf{y}(t_0)^T \mathbf{W}^T \mathbf{W}\mathbf{y}(t_0) = 1 \tag{6.3.28}$$

We can use a different norm to define the size of the perturbation to be maximized at the final time than the norm $\mathbf{W}$ used for the initial time (6.3.28). For example, the final norm could be a projection operator $\mathbf{P}$ at the end of the interval. Then the function that we want to maximize is, instead of (6.3.27):

$$J(\mathbf{y}(t_0)) = [\mathbf{PLy}(t_0)]^T \mathbf{PLy}(t_0) = \mathbf{y}(t_0)^T \mathbf{L}^T \mathbf{P}^T \mathbf{PLy}(t_0) \tag{6.3.29}$$

subject to the *strong constraint* (6.3.28).

From the calculus of variations, the maximum of (6.3.29) subject to the strong constraint (6.3.28) can be obtained by the unconstrained maximum of another function:

$$
\begin{aligned}
K(\mathbf{y}(t_0)) &= J(\mathbf{y}(t_0)) + \lambda[1 - \mathbf{y}(t_0)^T \mathbf{W}^T \mathbf{W}\mathbf{y}(t_0)] \\
&= \mathbf{y}(t_0)^T \mathbf{L}^T \mathbf{P}^T \mathbf{PLy}(t_0) + \lambda[1 - \mathbf{y}(t_0)^T \mathbf{W}^T \mathbf{W}\mathbf{y}(t_0)]
\end{aligned}
\tag{6.3.30}
$$

where the λ are the Lagrange multipliers multiplying the square brackets (equal to zero due to the constraint (6.3.28)).

The unconstrained minimization of K is obtained by computing its gradient with respect to the control variable $\mathbf{y}(t_o)$ and making it equal to zero. From Remark 5.4.2.1(d), we can compute this gradient as:

$$\nabla_{y(t_0)} K = \mathbf{L}^T \mathbf{P}^T \mathbf{PLy}(t_0) - \lambda \mathbf{W}^T \mathbf{W}\mathbf{y}(t_0) = 0 \tag{6.3.31}$$

It is convenient, given the constraint (6.3.28), to change variables:

$$\mathbf{W}\mathbf{y}(t_0) = \hat{\mathbf{y}}(t_0) \qquad \text{or} \qquad \mathbf{y}(t_0) = \mathbf{W}^{-1}\hat{\mathbf{y}}(t_0) \tag{6.3.32}$$

Then, (6.3.31) becomes

$$(\mathbf{W}^{-1})^T \mathbf{L}^T \mathbf{P}^T \mathbf{PLW}^{-1}\hat{\mathbf{y}}(t_0) = \lambda \hat{\mathbf{y}}(t_0) \tag{6.3.33}$$

subject to the constraint

$$\hat{\mathbf{y}}^T(t_0)\hat{\mathbf{y}}(t_0) = 1 \tag{6.3.34}$$

Therefore, the transformed vectors $\hat{\mathbf{y}}(t_0)$ are the eigenvectors of the matrix $(\mathbf{W}^{-1})^T \mathbf{L}^T \mathbf{P}^T \mathbf{PLW}^{-1}$ in (6.3.33), with eigenvalues equal to the Lagrange multipliers λ_i. After the leading eigenvectors $\hat{\mathbf{y}}(t_0)$ are obtained (using, for example, the Lanczos algorithm), the variables are transformed back to $\mathbf{y}(t_0)$ using (6.3.32). The eigenvalues of this problem are the square of the singular values of the tangent linear model: $\lambda_i = \sigma_i^2$.

This allows great generality (as well as arbitrariness[2]) in the choice of initial norm and final projection operator. Errico and Vukicevic (1992), showed that the singular

---

[2] Jon Ahlquist (2000, pers. communication) showed, given a linear operator $\mathbf{L}$, a set of arbitrary vectors $\mathbf{x}_i$, and a set of arbitrary nonnegative numbers σ_i arranged in decreasing order, how to construct an inner product and a norm such that the σ_i and the $\mathbf{x}_i$ are ith singular values and singular vectors of $\mathbf{L}$. He pointed out that "Because anything not in the null space can be a singular vector, even the leading singular vector, one cannot assign a physical meaning to a singular vector simply because it is a singular vector. Any physical meaning must come from an additional aspect of the problem. Said in another way, nature evolves from initial conditions without knowing which inner products and norms the user wants to use."

6.3 Tangent Linear and Adjoint Model, Singular and Lyapunov Vectors

vectors are very sensitive to both the choice of norm and the length of the optimization interval (the interval from t_0 to t_1). In another experiment, Palmer et al. (1998) tested different weight matrices $\mathbf{W}$ defining the initial norm. They used "streamfunction," "enstrophy," "kinetic energy," and "total energy" norms, which measured, as the "initial size" the square of the perturbation streamfunction, vorticity, wind speed and weighted temperature, wind and surface pressure, respectively. They found that the use of different initial norms resulted in extremely different initial singular vectors, and concluded that the total energy was the norm of choice for ensemble forecasting. In 1995, ECMWF included in their ensemble system a projection operator $\mathbf{P}$ that measures only the growth of perturbations north of $30°$ N, i.e., a matrix that multiplies variables that correspond to latitudes greater than or equal to $30°$ N by the number 1, and by 0 otherwise) (Buizza and Palmer, 1995). One could use any other pair of initial $\mathbf{W}$ and final $\mathbf{P}$ weights (norms) to answer the related question of *forecast sensitivity*. An example of a forecast sensitivity problem is: "What is the optimal (minimum size) initial perturbation (measured by the square of the change in surface pressure over the states of Oklahoma and Texas) that produces the maximum final change after a 1-day forecast (measured by the change in vorticity between surface and 500 hPa over the eastern USA)?" ECMWF has been routinely carrying out experiments to find out "What is the change in the initial conditions from 3 days ago that would lead to the best verification of today's analysis?" (see Errico, 1997; Rabier et al., 1996; Pu et al., 1997a,b, for more details)

6.3.3 Lyapunov Vectors

As we saw in Section 6.2, if we start a set of perturbations on a sphere of very small size, it will evolve into an ellipsoid. The growth of the axis of the hyperellipsoid after a finite interval s is given by the singular values $\sigma_i(t_0 + s)$. The (global) Lyapunov exponents describe the linear *long-term* growth of the hyperellipsoid:

$$\lambda_i = \lim_{s \to \infty} \frac{1}{s} \ln[\sigma_i(t_0 + s)] \tag{6.3.35}$$

In other words, the Lyapunov exponents describe the long-term average exponential rate of stretching or contraction in the attractor. (We call the Lyapunov exponents "global" to distinguish them from the finite time or "local" Lyapunov exponents which are useful in predictability applications.) There are as many Lyapunov exponents as the dimension of the model (number of independent variables or degrees of freedom). If the model has at least one λ_i greater than zero, then the system can be called chaotic, i.e., there is exponential separation of trajectories. In other words, there is at least one direction of the ellipsoid that continues to be stretched, and therefore two trajectories will diverge in time and eventually become completely different. Conversely, a system with all negative Lyapunov exponents is stable, and will remain predictable at all times. The first Lyapunov exponent can be estimated by running the tangent linear model for a long time starting from any randomly chosen initial perturbation $\mathbf{y}(t_0)$.

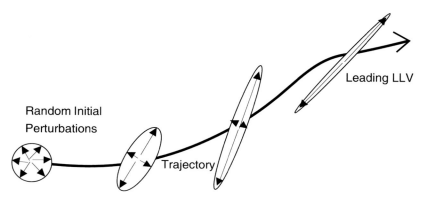

Figure 6.3.5 Schematic of how all perturbations will converge toward the leading LLV.

During a long integration, the growth rate of any random perturbation will converge to the first Lyapunov exponent:

$$\lambda_1 = \lim_{s \to \infty} \frac{1}{s} \ln \left[\frac{\|\mathbf{y}(t_0 + s)\|}{\|\mathbf{y}(t_0)\|} \right] \tag{6.3.36}$$

which is independent of the norm. In practice, the first Lyapunov exponent is obtained by running the tangent linear model for a long period from random initial conditions, and renormalizing the perturbation vector periodically in order to avoid computational overflow.

When we are dealing with atmospheric predictions, we are not really interested in the *global* growth properties, which correspond to the atmosphere's attractor (climatology), i.e., relevant average properties over many decades. Instead, in predictability problems we are interested in the growth rate of perturbations at a given time and space: we need to know the *local* stability properties in space and time, which are related to our ability to make skillful forecasts. We can define the leading local Lyapunov vector (LLV) at a certain time t, as the vector toward which *all* random perturbations $\mathbf{y}(t - s)$ started a long time s before t will converge (Figure 6.3.5).

$$\mathbf{l}_1(t) = \lim_{s \to \infty} \mathbf{L}(t - s, t)\mathbf{y}(t - s) \tag{6.3.37}$$

Once a perturbation has converged to the leading LLV $\mathbf{l}_1(t)$, the leading local Lyapunov exponent can be computed from the rate of change of its norm. In practice, the local leading Lyapunov exponent, also known as finite time Lyapunov exponent, can be estimated over a finite period τ:

$$l_1 \approx \frac{1}{\tau} \ln \left[\frac{\|\mathbf{l}_1(t + \tau)\|}{\|\mathbf{l}_1(t)\|} \right] \tag{6.3.38}$$

The argument of the logarithm is defined as the amplification rate $A(t, \tau)$.

The *first LLV is independent of the definition of norm*, and represents the direction in which maximum sustainable growth (or minimum decay) can occur in a system without external forcing. In fact, after a finite transition period T takes place,

6.3 Tangent Linear and Adjoint Model, Singular and Lyapunov Vectors 239

every initial perturbation will turn in the direction of the LLV at every point of the trajectory. This also includes the final singular vectors $\mathbf{u}_i$ for a sufficiently long optimization interval.

Trevisan and Legnani (1995) introduced the notion of the leading LLV. Additional LLVs can be obtained by Gramm–Schmidt orthogonalization, and this would seem to indicate that they are norm-dependent. However, Trevisan and Pancotti (1998) showed that it is also possible, at least in theory, to define additional LLVs (denoted *characteristic* vectors by Legras and Vautard, 1996) *without the use of norms*. The LLVs are therefore a fundamental characteristic of dynamical systems. It should be noted that regrettably, at this time, there is not a universally accepted nomenclature for LLVs. Legras and Vautard (1996) call the LLVs "backward Lyapunov vectors," since they were started an infinitely long time in the past. Unfortunately, this name is extremely confusing, since they represent forward evolution rather than backward evolution as this name would imply. The LLVs are also the final singular vectors optimized for an infinitely long time, i.e., the eigenvectors (valid at time t) of $\mathbf{L}(t-T,t)\mathbf{L}^T(t-T,t)$ for $T \to \infty$. Similarly, Legras and Vautard define as "forward Lyapunov vectors" the *initial* singular vectors obtained from a very long *backward* integration with the adjoint of the model, i.e., they are the eigenvectors (valid at time t) of $\mathbf{L}^T(t,t+T)\mathbf{L}(t,t+T)$ for very large T.

Legras and Vautard (1996) showed (as did Trevisan and Pancotti, 1998) that a complete set of LLVs (which they denote *characteristic* Lyapunov vectors) can be defined from the intersection of the subspaces spanned by the "forward" and "backward" Lyapunov vectors. The (characteristic) LLVs are therefore independent of the norm, and grow in time with a rate given by the local Lyapunov exponents. As such, they are a fundamental characteristic of dynamical systems.

Several authors have shown that the leading (first few) LLVs of low-dimensional dynamical systems span the attractor, i.e., they are parallel to the hypersurface in phase space that the dynamical system visits again and again ("realistic solutions"). Leading singular vectors, on the other hand, have very different properties. They can grow much faster than the leading LLVs, but are initially *off* the attractor: they point to areas in the phase space where solutions do not naturally occur (e.g., Legras and Vautard, 1996; Trevisan and Legnani, 1995; Trevisan and Pancotti, 1998; Pires et al., 1996), see also next section.

For ensemble forecasting, Ehrendorfer and Tribbia (1997) showed that if $\mathbf{V}$ is the initial analysis error covariance (which unfortunately we don't know and can only estimate), then the initial singular vectors defined with the norm $\mathbf{W} = \mathbf{V}^{-1/2}$ evolve into the eigenvectors of the evolved error covariance matrix. This implies that the leading singular vectors, defined using the initial error covariance, are optimal in describing the forecast errors at the end of the optimization period. The initial error covariance norm yields singular vectors quite different from those derived using the energy norm. Barkmeijer et al. (1998) used the ECMWF estimated 3D-Var error covariance as the initial norm (instead of the total energy norm) and obtained initial perturbations with structures closer to the bred vectors (i.e., leading LLVs) used at NCEP (see Section 6.5.1).

6.3.4 Simple Examples of Singular Vectors and Eigenvectors

In order to get a more intuitive feeling of the relationship between singular vectors and Lyapunov vectors, we consider a simple linear model in two dimensions:

$$\begin{bmatrix} x_1(t+T) \\ x_2(t+T) \end{bmatrix} = \mathbf{M}_T[x(t)] = \begin{bmatrix} 2x_1(t) + 3x_2(t) + 7 \\ 0.5x_2(t) - 4 \end{bmatrix} \tag{6.3.39}$$

We compute the two-dimensional tangent linear model, constant in time:

$$\mathbf{L} = \begin{bmatrix} \dfrac{\partial M_1}{\partial x_1} & \dfrac{\partial M_1}{\partial x_2} \\ \dfrac{\partial M_2}{\partial x_1} & \dfrac{\partial M_2}{\partial x_2} \end{bmatrix} = \begin{bmatrix} 2 & 3 \\ 0 & 0.5 \end{bmatrix} \tag{6.3.40}$$

The propagation or evolution of any perturbation (difference between two solutions) over a time interval $(t, t + T)$ is given by

$$\delta\mathbf{x}(t+T) = \mathbf{L}\delta\mathbf{x}(t) \tag{6.3.41}$$

Note that the translation terms in (6.3.39) do not affect the perturbations. The eigenvectors of L (which for this simple constant tangent linear model are also the Lyapunov vectors) are proportional to

$$\mathbf{l}_1 = \begin{pmatrix} 1 \\ 0 \end{pmatrix} \qquad \mathbf{l}_2 = \begin{pmatrix} -2 \\ 1 \end{pmatrix}$$

corresponding to the eigenvalues $\lambda_1 = 2$, $\lambda_2 = 0.5$, respectively, which in this case are the two Lyapunov numbers (their logarithms are the Lyapunov exponents). If we normalize them, so that they have unit length, the Lyapunov vectors are

$$\mathbf{l}_1 = \begin{pmatrix} 1 \\ 0 \end{pmatrix} \qquad \mathbf{l}_2 = \begin{pmatrix} -0.89 \\ 0.45 \end{pmatrix} \tag{6.3.42}$$

The Lyapunov vectors are not orthogonal, they are separated by an angle of $153.4°$ (Figure 6.3.6(a)). We will see that because they are not orthogonal, it is possible to find linear combinations of the Lyapunov vectors that grow faster than the leading Lyapunov vector. We will also see that the leading Lyapunov vector is the attractor of the system, since repeated applications of $\mathbf{L}$ to any perturbation make it evolve toward $\mathbf{l}_1$.

Applying first $\mathbf{L}$ and then its transpose $\mathbf{L}^T$, we obtain the symmetric matrix

$$\mathbf{L}^T\mathbf{L} = \begin{bmatrix} 4 & 6 \\ 6 & 9.25 \end{bmatrix} \tag{6.3.43}$$

whose eigenvectors are the *initial singular vectors*, and whose eigenvalues are the squares of the singular values. The initial singular vectors (eigenvectors of $\mathbf{L}^T\mathbf{L}$) are

$$\mathbf{v}_1 = \begin{pmatrix} 0.55 \\ 0.84 \end{pmatrix} \mathbf{v}_2 = \begin{pmatrix} 0.84 \\ -0.55 \end{pmatrix} \tag{6.3.44}$$

with eigenvalues $\sigma_1^2 = 13.17$, $\sigma_2^2 = 0.076$. As indicated before, the *singular values* of $\mathbf{L}$ are the square roots of the eigenvalues of $\mathbf{L}^T\mathbf{L}$, i.e., $\sigma_1 = 3.63$, $\sigma_2 = 0.275$.

6.3 Tangent Linear and Adjoint Model, Singular and Lyapunov Vectors

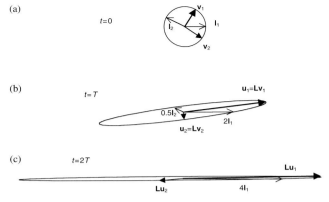

Figure 6.3.6 Schematic of the evolution of the two nonorthogonal Lyapunov vectors (thin arrows $\mathbf{l}_1$ and $\mathbf{l}_2$), and the corresponding two initial singular vectors (thick arrows $\mathbf{v}_1(0)$ and $\mathbf{v}_2(0)$), optimized for the interval $(0, T)$, for the tangent linear model $\mathbf{L} = \begin{bmatrix} 2 & 3 \\ 0 & 0.5 \end{bmatrix}$ with eigenvalues 2 and 0.5. (a) Time $t = 0$, showing the initial singular vectors $\mathbf{v}_1(0)$ and $\mathbf{v}_2(0)$, as well as the Lyapunov vectors $\mathbf{l}_1$ and $\mathbf{l}_2$. (b) Time $t = T$, evolved singular vectors, $\mathbf{u}_1(T) = \mathbf{L}\mathbf{v}_1(0)$, $\mathbf{u}_2(T) = \mathbf{L}\mathbf{v}_2(0)$ at the end of the optimization period; the Lyapunov vectors have grown by factors of 2 and 0.5, respectively, whereas the leading singular vector has grown by 3.63. The second evolved singular vector has grown by 0.275 and is still orthogonal, to the first singular vector. (c) Time $t = 2T$. Beyond the optimization period T, the evolved singular vectors $\mathbf{u}_1(2T) = \mathbf{L}\mathbf{u}_1(T)$, $\mathbf{u}_2(2T) = \mathbf{L}\mathbf{u}_2(T)$ are not orthogonal, and they approach the leading Lyapunov vector with similar growth rates.

Note that this implies that during the optimization period $(0, T)$ the leading singular vector grows almost twice as fast as the leading Lyapunov vector (3.63 vs. 2). The angle that the leading *initial singular vector* has with respect to the leading Lyapunov vector is 56.82°, whereas the second initial singular vector is perpendicular to the first one (Figure 6.3.6(a)).

The *final* or *evolved SVs at the end of the optimization period* $(0, T)$ are the eigenvectors of

$$\mathbf{L}\mathbf{L}^T = \begin{bmatrix} 13 & 1.5 \\ 1.5 & 0.25 \end{bmatrix} \quad (6.3.45)$$

and after normalization, they are

$$\mathbf{u}_1 = \begin{pmatrix} 0.99 \\ 0.12 \end{pmatrix} \quad \mathbf{u}_2 = \begin{pmatrix} 0.12 \\ -0.99 \end{pmatrix} \quad (6.3.46)$$

Note again that the operators $\mathbf{L}^T\mathbf{L}$ and $\mathbf{L}\mathbf{L}^T$ are quite different, and the final singular vectors are different from the initial singular vectors, but they have the same singular values $\sigma_1^2 = 13.17$, $\sigma_2^2 = 0.076$.

Alternatively, the evolved singular vectors at the end of the optimization period can also be obtained by applying $\mathbf{L}$ to the initial singular vectors, which is computationally inexpensive. In this case,

$$\mathbf{u}_1(T) = \mathbf{L}v_1(0) = \begin{bmatrix} 3.6 \\ 0.42 \end{bmatrix} \qquad \mathbf{u}_2(T) = \mathbf{L}v_2(0) = \begin{bmatrix} 0.03 \\ -0.27 \end{bmatrix}$$

which is the same as (6.3.46) but without normalization.

The final leading singular vector has strongly rotated toward the leading Lyapunov vector: At the end of the optimization period, the angle between the leading singular vector and the leading Lyapunov vector is only 6.6° (Figure 6.3.6(b)), and because the singular vectors have been optimized for this period, the final singular vectors are still orthogonal.

To obtain the evolution of the singular vectors *beyond the optimization period* $(0, T)$ we apply $\mathbf{L}$ again to the evolved singular vector valid at $t = T$ and obtain

$$\mathbf{u}_1(2T) = \mathbf{L}\mathbf{u}_1(T) = \begin{bmatrix} 8.47 \\ 0.21 \end{bmatrix} \qquad \mathbf{u}_2(2T) = \mathbf{L}\mathbf{u}_2(T) = \begin{bmatrix} -0.76 \\ -0.14 \end{bmatrix}$$

During the interval $(T, 2T)$ the leading singular vector grows by a factor of just 2.33, which is not very different from the growth rate of the leading Lyapunov vector. At the end of this second period (Figure 6.3.6(c)) the angle with the leading Lyapunov vector is only 1.41°. The angle of the second evolved singular vector at time T, after applying the linear tangent model $\mathbf{L}$ and the leading Lyapunov vector is also quite small (10.24°), and because it was further away from the attractor, the second singular vector (whose original, transient, singular value was 0.5), grows by a factor of 2.79. This example shows how quickly *all perturbations, including all singular vectors, evolve toward the leading Lyapunov vector, which is the attractor of the system*. It is particularly noteworthy that during the optimization period $(0, T)$, the first singular vector grows very fast as it rotates toward the attractor, but once it gets close to the leading Lyapunov vector, its growth returns to the normal leading Lyapunov vector's growth.

Let us now choose as the tangent linear model another matrix

$$\mathbf{L} = \begin{bmatrix} 2 & 30 \\ 0 & 0.5 \end{bmatrix}$$

with the same eigenvalues 2 and 0.5, i.e., with eigenvectors (Lyapunov vectors) that still grow at a rate of $2/T$ and $0.5/T$ respectively. However, now the angle between the first and the second Lyapunov vector is 177°, i.e., the Lyapunov vectors are almost antiparallel. In this case, the first singular vector grows by a factor of over 30 during the optimization period, but beyond the optimization period it essentially continues evolving like the leading Lyapunov vector.

These results do not depend on the fact that one Lyapunov vector grows and the other decays. As a third example, we choose

$$\mathbf{L} = \begin{bmatrix} 2 & 3 \\ 0 & 1.5 \end{bmatrix}$$

with two Lyapunov vectors growing with rates $2/T$ and $1.5/T$. The Lyapunov vectors are almost parallel, with an angle of 170°, and the leading singular vector grows during the optimization period by a factor of 3.83. Applying the tangent linear model again to the evolved singular vectors we obtain that at time $2T$ the leading singular vector has grown by a factor of 2.9 and its angle with respect to the leading Lyapunov vector is $1°$. Because it is not decaying, the second Lyapunov vector is also part of the attractor, but only those perturbations that are exactly parallel to it will remain parallel, all others will move toward the first Lyapunov vector.

These examples illustrate the fact that the fast growth of the singular vectors during the optimization period depends on the lack of orthogonality between Lyapunov vectors. *A very fast "supergrowth" of singular vectors is associated with the presence of almost parallel Lyapunov vectors*, and it takes place when the initial singular vector, which is not in the attractor, rotates back toward the attractor. At the end of the optimization period, the leading singular vector tends to be much closer to the attractor, more parallel to the leading Lyapunov vector. The second (trailing) singular vector is also moving toward the leading Lyapunov vector.

Finally, we point out that this introductory discussion is appropriate for relatively low-dimensional systems. For extremely high-dimensional systems like the atmosphere, there may be multiple sets of Lyapunov exponents corresponding to different types of instabilities. For example, as pointed out by Toth and Kalnay (1993), convective instabilities have very fast growth but small amplitudes, whereas baroclinic instabilities have slower growth but much larger amplitudes, and each of these can lead to different types of Lyapunov vectors. If we are interested in the predictability characteristics associated with baroclinic instabilities, then the analysis of growth rates of infinitesimally small Lyapunov vectors over infinitely long times may not be appropriate for the problem (Lorenz, 1996). In that case, it may be better to consider the finite amplitude, finite time extension of Lyapunov vectors introduced by Toth and Kalnay (1993, 1997) as bred vectors. Bred vectors are discussed in Section 6.5.1, and their relationship to Lyapunov vectors in (Kalnay et al., 2002).

6.4   Ensemble Forecasting: Early Studies

We saw in previous sections that Lorenz (1963a,b, 1965) showed that the forecast skill of atmospheric models depends not only on the accuracy of the initial conditions and on the realism of the model (as it was generally believed at the time), but also on the instabilities of the flow itself (See Figure 6.3.1). He demonstrated that any nonlinear dynamical system with instabilities, like the atmosphere, has a finite limit of predictability. The growth of errors due to instabilities implies that the smallest imperfection in the forecast model or the tiniest error in the initial conditions, will *inevitably* lead to a total loss of skill in the weather forecasts after a *finite* forecast length. Lorenz estimated this *limit of weather predictability* as about two weeks. With his simple model he also pointed out that predictability is strongly dependent on the evolution of the atmosphere itself: some days the forecasts can remain accurate for a week or longer, and on other days the forecast skill may break down after only 3 days. This discovery made inevitable the realization that NWP needs to account for the stochastic nature of the evolution of the atmosphere (Figure 6.4.1). As previously discussed, Lorenz (1965) studied the error growth of a complete "ensemble" of

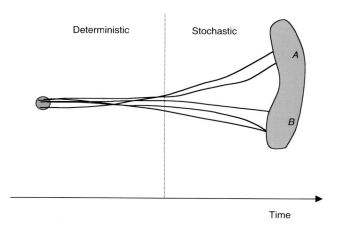

Figure 6.4.1 Schematic of ensemble prediction, with individual trajectories drawn for forecasts starting from a representative set of perturbed initial conditions within a circle representing the uncertainty of the initial conditions (ideally the analysis error covariance) and ending within the range of possible solutions. For the shorter range, the forecasts are close to each other, and they may be considered deterministic, but beyond a certain time, the equally probable forecasts are so different that they must be considered stochastic. The transition time is of the order of 2–3 days for the prediction of large-scale flow, but can be as short as a few hours for mesoscale phenomena like the prediction of individual storms. The transition time is shorter for strongly nonlinear parameters: even for large-scale flow, precipitation forecasts show significant divergence faster than the 500-hPa fields. The forecasts may be clustered into subsets A and B (Tracton and Kalnay, 1993). Published (1993) by the American Meteorological Society.

perturbed forecasts, with the ensemble size equal to the dimension of the phase space (one perturbation for each of the 28 model variables). In this paper, he introduced for the first time concepts related to singular vectors and LLVs discussed in the previous section. This was followed by several early approaches to the problem of accounting for the variable predictability of the atmosphere reviewed in this section.

6.4.1 Stochastic-Dynamic Forecasting

Historically, the first forecasting method to explicitly acknowledge the uncertainty of atmospheric model predictions was developed by Epstein (1969), who introduced the idea of **stochastic-dynamic forecasting**. He derived a continuity equation for the probability density $\varphi(X; t)$ of a model solution X of a dynamical model $\dot{X} = G(X(t))$, where the model has dimension D:

$$\frac{\partial \varphi}{\partial t} + \mathbf{\nabla}_D \cdot (\dot{X}\varphi) = 0 \qquad (6.4.1)$$

This equation indicates that in an ensemble of forecast solutions, "no member of the ensemble may be created or destroyed." An ensemble starting from an infinite number of perturbed integrations spanning the analysis uncertainty gives the "true" probability distribution (with all its moments), but even for a simple low-order model, the integration of (6.4.1) is far too expensive. Therefore, Epstein introduced an approximation

6.4 Ensemble Forecasting: Early Studies   245

to predict only the first and second moments of the probability distribution (expected means and covariances) rather than the full probability distribution. Epstein assumed that the model equations are of the form

$$\dot{x}_i = \sum_{j,k} a_{ijk} x_j x_k - \sum_j b_{ij} x_j + c_i \tag{6.4.2}$$

The forecast equations for the expected first and second moments are

$$\left.\begin{aligned} \dot{\mu}_i &= E(\dot{x}_i) \\ \dot{\rho}_{ij} &= E(x_i \dot{x}_j + \dot{x}_i x_j) \end{aligned}\right\} \tag{6.4.3}$$

The covariances ρ_{ij} are related to the second order moments by $\sigma_{ij} = E[(x_i - \mu_i)(x_j - \mu_j)]$. Substituting (6.4.2) into (6.4.3) gives rise to forecast equations for μ and σ that contain triple moments $(x_i x_j x_k)$. As done in turbulence models with a second-order closure for the triple products (Chapter 4), Epstein introduced a closure assumption for the third order moments around the mean $\tau_{ijk} = E[(x_i - \mu_i)(x_j - \mu_j)(x_k - \mu_k)]$. He assumed that $\sum_{kl} \left(a_{jkl} \tau_{ikl} + a_{ikl} \tau_{jkl} \right) = 0$, which then gives a closed set of equations for the means and covariances:

$$\left.\begin{aligned} \dot{\mu}_i &= \sum_{jk} a_{ijk}(\sigma_{jk} + \mu_j \mu_k) - \sum_j b_{ij} \mu_j + c_i \\ \dot{\sigma}_{ij} &= \sum_{kl} \left[a_{jkl}(\mu_k \sigma_{il} + \mu_l \sigma_{ik}) + a_{ikl}(\mu_k \sigma_{jl} + \mu_l \sigma_{jk}) \right] \\ &\quad - \sum_k (b_{ik}\sigma_{jk} + b_{jk}\sigma_{ik}) \end{aligned}\right\} \tag{6.4.4}$$

Epstein tested these "approximate" stochastic equations for a Lorenz three-variable model. The "true" probability distribution was computed from a Monte Carlo ensemble of 500 members, and the comparison indicated good agreement, at least for several simulated days. Note that in his case, the number of ensemble members was much larger than the number of degrees of freedom of the model, a situation that would be impossible to replicate with current models with millions of degrees of freedom. In his paper, Epstein also introduced the idea of using stochastic-dynamic forecasting in the *analysis cycle*, with the background forecast and error covariance provided by stochastic-dynamic forecasts combined with observations that also contain errors (cf. Sections 5.3–5.6).

Unfortunately, although the stochastic-dynamic forecasting method was introduced as a shortcut to an "infinite" Monte Carlo ensemble, in a model with N degrees of freedom, it requires $N(N + 1)/2 + N$ forecast equations, equivalent to making about $(N + 3)/2$ model forecasts. Although this was practical with a three-variable model, it is completely unfeasible for a modern model, with millions of degrees of freedom.

6.4.2   Monte Carlo Forecasting

In 1974b, Leith proposed the idea of performing ensemble forecasting with a limited number m of ensemble members instead of the conventional single (deterministic) control forecast. He also proposed performing an "optimal estimation" of the verification using linear regression on the dynamical forecasts, with optimal weights determined from forecast error covariances (cf. Sections 5.3–5.6). Since forecasts lose their skill at longer lead times, and individual forecasts eventually are further away from the verification than the climatology (cf. Equations (6.4.5) and ((6.4.6)), optimal estimation of the verification is equivalent to *tempering* (i.e., *hedging* the forecast toward climatology).

He cast his analysis using, instead of model variables, their deviation $\mathbf{u}$ with respect to climatology (also known as forecast *anomalies*). The true state of the atmosphere is denoted $\mathbf{u}_0$, and $\hat{\mathbf{u}}$ denotes an unbiased estimate of $\mathbf{u}_0$, whose expected value (average over many forecasts, represented by the angle brackets) is equal to zero: $\langle \hat{\mathbf{u}} \rangle = 0$.

We can compute the expected error covariance of a climatological forecast (i.e., a forecast of zero anomaly):

$$\langle (0 - \mathbf{u}_0)(0 - \mathbf{u}_0)^T \rangle = \langle \mathbf{u}_0 \mathbf{u}_0^T \rangle = \mathbf{U} \tag{6.4.5}$$

A single (deterministic) forecast $\hat{\mathbf{u}}$, on the other hand, has, on average, an error covariance given by

$$\langle (\hat{\mathbf{u}} - \mathbf{u}_0)(\hat{\mathbf{u}} - \mathbf{u}_0)^T \rangle = \langle \hat{\mathbf{u}}\hat{\mathbf{u}}^T + \mathbf{u}_0 \mathbf{u}_0^T - \hat{\mathbf{u}} \mathbf{u}_0^T - \mathbf{u}_0 \hat{\mathbf{u}}^T \rangle \underset{t \to \infty}{\longrightarrow} 2\mathbf{U} \tag{6.4.6}$$

This limit occurs because the last two terms in the second angle brackets go to zero as the forecasts become decorrelated with the true atmosphere at long lead times, and we assume that the model covariance is also unbiased. This equation indicates that for long lead times an individual deterministic forecast has twice the error covariance of a climatological forecast. Therefore, a "regressed" forecast, tempered toward climatology, must be better than a single deterministic forecast (in a least square error sense), with an error covariance that asymptotes to $\mathbf{U}$, and not to $2\mathbf{U}$.

A regressed forecast $\hat{\mathbf{u}}_0 = \hat{\mathbf{u}}\mathbf{A}$ is obtained by linear regression, minimizing the square of the regressed error $\varepsilon^T \varepsilon = \langle (\mathbf{u}_0 - \hat{\mathbf{u}}\mathbf{A})^T (\mathbf{u}_0 - \hat{\mathbf{u}}\mathbf{A}) \rangle$ with respect to the elements of the matrix of constant regression coefficients $\mathbf{A}$. As we did in the derivation of the optimal weight matrix for the observational increments in Section 5.4, we make use of the linear regression formulas: if the linear prediction equation is $\hat{\mathbf{y}} = \mathbf{x}\mathbf{A}$, then the error is given by $\varepsilon = \mathbf{y} - \mathbf{x}\mathbf{A}$. The matrix of the derivatives of the (scalar) squared error $\varepsilon^T \varepsilon$ with respect to each element of $\mathbf{A}$ is given by $\partial \varepsilon^T \varepsilon / \partial \mathbf{A} = -2\mathbf{x}^T (\mathbf{y} - \mathbf{x}\mathbf{A}) = 0$, which gives the normal equation $\mathbf{x}^T \mathbf{y} = \mathbf{x}^T \mathbf{x}\mathbf{A}$, or $\mathbf{A} = (\mathbf{x}^T \mathbf{x})^{-1}(\mathbf{x}^T \mathbf{y})$. Applying this to the regressed forecast, we obtain $\langle \hat{\mathbf{u}}^T (\mathbf{u}_0 - \hat{\mathbf{u}}\mathbf{A}) \rangle = 0$, or

$$\mathbf{A} = \langle \hat{\mathbf{u}}^T \hat{\mathbf{u}} \rangle^{-1} \langle \hat{\mathbf{u}}^T \mathbf{u}_0 \rangle \tag{6.4.7}$$

Estimating the required forecast statistics in (6.4.7) involves considerable work. The size of the regression matrix is usually large compared to the size of the sample available to estimate it, and in order to reduce the number of parameters to be estimated additional approximations are needed (e.g., by parameterizing error growth, Hoffman and Kalnay, 1983).

Now, instead of regression let's consider *an ensemble of m forecasts* computed from perturbations $\mathbf{r}_i$ to the initial best estimate (analysis) $\hat{\mathbf{u}}$. Ideally, the perturbations

6.4 Ensemble Forecasting: Early Studies
247

should be chosen so that their outer product is a good estimate of the initial error covariance (i.e., the analysis error covariance $\langle \mathbf{rr}^T \rangle = \mathbf{P}_a$, as suggested in Figure 6.4.1). In practice, however, the analysis error covariance can only be approximately estimated (e.g., Barkmeijer et al., 1998).

If $\bar{\mathbf{u}} = (1/m)\Sigma_{i=1}^{m}\mathbf{u}_i$ is the average of an ensemble of m forecasts, then its error covariance evolves like

$$\langle (\bar{\mathbf{u}} - \mathbf{u}_0)(\bar{\mathbf{u}} - \mathbf{u}_0)^T \rangle = \langle \bar{\mathbf{u}}\bar{\mathbf{u}}^T + \mathbf{u}_0\mathbf{u}_0^T - \bar{\mathbf{u}}\mathbf{u}_0^T - \mathbf{u}_0\bar{\mathbf{u}}^T \rangle \underset{t\to\infty}{\longrightarrow} \left(1 + \frac{1}{m}\right)\mathbf{U} \qquad (6.4.8)$$

since the last two terms in the second angle brackets go to zero at long time leads, and the first one evolves like

$$\langle \bar{\mathbf{u}}\bar{\mathbf{u}}^T \rangle = \left\langle \frac{1}{m}\sum_{i=1}^{m}\mathbf{u}_i\frac{1}{m}\sum_{j=1}^{m}\mathbf{u}_j^T \right\rangle \underset{t\to\infty}{\longrightarrow} \frac{m}{m^2}\mathbf{U} \qquad (6.4.9)$$

Equation (6.4.8) shows that averaging a Monte Carlo ensemble of forecasts approximates the tempering of the forecasts toward climatology, *without the need to perform regression*. It suggests that such tempering may be substantially achieved with a relatively small number of ensemble members (compare (6.4.8) with (6.4.5) and (6.4.6)). Leith (1974b) used an analytical turbulence model to test this hypothesis, and concluded that a Monte Carlo forecasting procedure represents a practical, computable approximation to the stochastic-dynamic forecasts proposed by (Epstein, 1969). He suggested that adequate accuracy would be obtained for the best estimate of the forecast (i.e., *the ensemble mean*) with sample sizes as small as eight, but that the estimation of *forecast errors* may require a larger number of ensemble members. Monte Carlo forecasting is thus a feasible approach for ensemble forecasting, requiring only a definition of the initial perturbations and m forecasts.

6.4.3 Lagged Average Forecasting

In 1983, Hoffman and Kalnay proposed *lagged average forecasting* (LAF) as an alternative to Monte Carlo forecasting, in which the forecasts initialized at the current initial time, $t = 0$, as well as at previous times, $t = -\tau, -2\tau, \ldots, -(N-1)\tau$ are combined to form an ensemble (see Figure 6.4.2). In an operational set up, τ is typically 6, 12, or 24 hr, so that the forecasts are already available, and the perturbations are generated automatically from the forecast errors. Since the ensemble comprises forecasts of different "age," Hoffman and Kalnay (1983) weighted them according to their expected error, which they estimated by parameterizing the observed error covariance growth. They compared the lagged average forecasting and Monte Carlo forecasting methods within a simulation system, using a primitive equations model as "nature," and a quasi-geostrophic model to perform the "forecasts." In this way, they allowed for model errors, unlike the previous "identical twin" experiments that assumed a perfect model. They "observed" the required variables and introduced random "observation errors" every 6 hr and performed many ensemble forecast experiments separated by 50 days of integration. They compared the results of single forecasts (ordinary dynamical

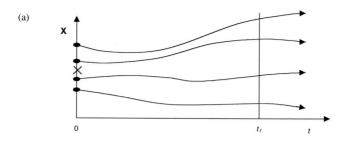

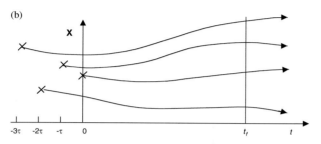

Figure 6.4.2 Schematic time evolutions of Monte Carlo forecasts (a) and lagged average forecasts (b). The abscissa is forecast time t, and the ordinate is the value of a forecast variable **X**. The crosses represent analyses obtained at time intervals τ, and the dots, randomly perturbed initial conditions; t_f is a particular forecast time. The initial "perturbation" for the lagged average forecast is the previous forecasts' error at the initial time (Adapted from Hoffman and Kalnay, 1983).

forecasts), Monte Carlo forecasting, lagged average forecasting and tempered ordinary dynamical forecasts, as well as persistence-climatology forecast (the most skillful baseline forecast).

Hoffman and Kalnay looked at the error growth of individual forecasts (Figure 6.4.3). Note in this figure that the individual forecast errors grow slowly and then at a certain time there is a rapid error growth until nonlinear saturation takes place (only the period of rapid growth is plotted). Note also that the forecast errors saturate around $\sqrt{2}$ of the climatological variability, as indicated by equation (6.4.6).

In these simulated forecasts, like in real weather forecasts, the forecast skill exhibits a lot of day-to-day variability. The rapid growth takes place at a time that varies from a minimum of 5 days to a maximum of 20 days. Hoffman and Kalnay tested the ability of the ensemble to predict the time at which the forecast error crossed 50% of the climatological standard deviation. They used as the predictor the spread of the ensembles (standard deviation with respect to their mean). They found that the lagged average forecasting ensemble average forecast was only slightly better than the Monte Carlo forecasting, but the advantage of lagged average forecasting in predicting forecast skill was much more apparent, with the correlation between predicted and observed time of crossing the 50% level being 0.68 for Monte Carlo forecasting and 0.79 for lagged average forecasting.

6.4 Ensemble Forecasting: Early Studies

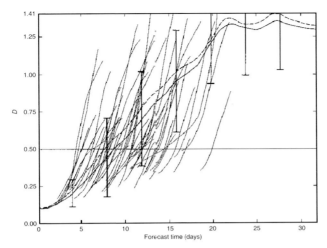

Figure 6.4.3 Time evolution of D, the individual forecast errors scaled by the climatological forecast error, plotted only during the period the forecast error crossed $D = 0.5$. Also plotted are two measures of average forecast error (Adapted from Hoffman and Kalnay, 1983).

The advantages of lagged average forecasting over Monte Carlo forecasting are probably due to the fact that lagged average forecasting perturbations in the initial conditions were not *randomly chosen* errors like in Monte Carlo forecasting but included dynamical influences and therefore contained "*errors of the day*" (Kalnay et al., 1997). This is because the perturbations are generated from actual forecast errors and therefore they are influenced by the evolution of the underlying background large-scale flow.

Lagged average forecasting has been frequently used for experimental ensemble forecasting, both for medium-range and climate prediction. However, the statistics required to estimate the weights of the members of the lagged average forecasting ensemble according to their "age" are very hard to obtain, so that except for the study by (Dalcher et al., 1988), all the lagged average forecasting members have been generally given equal weight. The advantages of lagged average forecasting are: (a) some of the forecasts are already available in operational centers; (b) it is very simple to perform and does not require special generation of perturbations; and (c) the perturbations contain "errors of the day" (Lyapunov vectors). Lagged average forecasting has also major disadvantages: (a) a large LAF ensemble would have to include excessively "old" forecasts; (b) without the use of optimal weights, the lagged average forecasting ensemble average may be tainted by the older forecasts.

Ebisuzaki and Kalnay (1991) introduced a variant of lagged average forecasting denoted *scaled lagged average forecasting* (SLAF) that reduces these two disadvantages. The perturbations are obtained by computing the forecast error of forecasts started at $t = -j\tau$, $j = 1, \ldots, N-1$, and multiplying these errors by $\pm 1/j$. This assumes that the errors grow approximately linearly with time during the first 2–3 days, and that the perturbations can be subtracted from and not just added to the analysis. The advantages of scaled lagged average forecasting are: (a) the initial perturbations of the ensemble members are all of approximately the same size (this can be enforced using a more sophisticated rescaling than linear growth), and (b) their number is doubled with respect to lagged average forecasting, so that only shorter-range forecasts are needed to create scaled lagged average forecasting. In practice, it has been observed that pairs of initial perturbations with opposite sign, as used in scaled lagged average forecasting, yield better ensemble forecasts, presumably because the Lyapunov vectors within the analysis errors can have either sign, whereas lagged average forecasting tends to maintain a single sign in the error. Experiments with the NCEP global model showed that scaled lagged average forecasting ensembles were better than lagged average forecasting ensembles (Ebisuzaki and Kalnay, 1991). This method is also easier to implement in regional ensemble forecasts, since it generates boundary condition perturbations consistent with the interior perturbations (Hou et al., 2001).

6.5   Operational Ensemble Forecasting Methods

Figure 6.5.1(a) shows the elements of a typical ensemble forecasting system: (1) the *control* forecast (labeled C) starts from the analysis (denoted by a cross), i.e., from the best estimate of the initial state of the atmosphere; (2) two *perturbed* ensemble forecasts (labeled P^+ and P^-) with initial perturbations added and subtracted from the control; (3) the *ensemble average* labeled A, and (4) the *true evolution* of the atmosphere (not known in real time), labeled T. This is an example of a "good" ensemble since the true evolution appears to be a plausible member of the ensemble. Figure 6.5.1(b) shows an example of a "bad" ensemble, in which the forecast errors are dominated by problems in the forecasting system (such as model deficiencies) rather than the chaotic growth of initial errors. In this case, the true evolution is quite different from the members of the ensemble, but the ensemble is still useful in identifying the presence of a deficiency in the forecasting system, which, with a single forecast, could not be distinguished from the growth of errors in the initial conditions.

Ensemble forecasting has essentially three basic goals. The first is *to improve the forecast by ensemble averaging*. The improvement is a result of the tendency of the ensemble average to filter out the components of the forecast that are uncertain (where the members of the ensemble differ from each other) and to retain those components that show agreement among the members of the ensemble. The filtering can take place only during the nonlinear evolution of the perturbations: if the perturbations are added to and subtracted from the analysis, the ensemble average forecast is equal to the control while the perturbations remain linear. The improvement of the ensemble average with respect to the control, shown schematically in Figure 6.5.1(a), is noticeable in Figure 1.6.2 after a few days of forecasts with the NCEP global ensemble. The second goal is to *provide an indication of the reliability of the forecast*: if the ensemble forecasts are quite different from each other, it is clear that at least some of them are wrong, whereas if there is good agreement among the forecasts, there is more reason to be confident about the forecast (cf. e.g., Figure 1.6.3(a) and (b)). The quantitative

6.5 Operational Ensemble Forecasting Methods

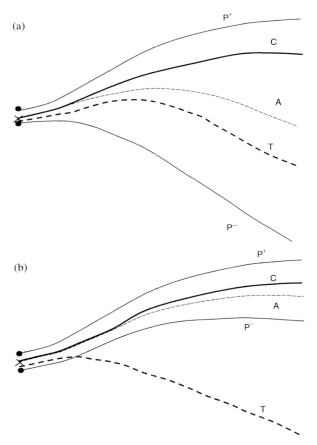

Figure 6.5.1 (a) Schematic of the components of a typical ensemble: (1) the control forecast (labeled C) which starts from the analysis (denoted by a cross), which is the best estimate of the true initial state of the atmosphere; (2) two perturbed ensemble forecasts (labeled P^+ and P^-) with initial perturbations added and subtracted from the control; (3) the ensemble average denoted A; and (4) the "true" evolution of the atmosphere labeled T. This is a "good" ensemble since the "truth" appears as a plausible member of the ensemble. Note that because of nonlinear saturation, the error of the ensemble member initially further away from the truth (in this case P^+) tends to grow more slowly than the error of the member initially closer to the truth. This results in a nonlinear filtering of the errors: the average of the ensemble members tends to be closer to the truth than the control forecast (Toth and Kalnay, 1997, also compare with Figure 1.6.2). (b) Schematic of a "bad" ensemble in which the forecast errors are dominated by system errors (such as model deficiencies). In this case, the ensemble is not useful for forecasting, but it helps to identify the fact that forecast errors are probably due to the presence of systematic errors, rather than to the chaotic growth of errors in the initial conditions.

relationship between the ensemble spread and the forecast error (or conversely, between the forecast agreement and the forecast skill) has yet to be firmly established, but is now routinely taken into consideration by human forecasters. The third goal of ensemble forecasting is to *provide a quantitative basis for probabilistic forecasting*.

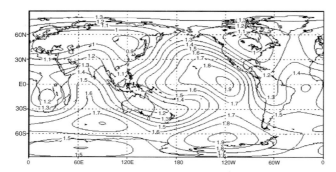

Figure 6.5.2 Estimation of the 500-hPa geopotential height analysis uncertainty obtained from running two independent analysis cycles, computing their rms difference, and using a filter to retain the planetary scales. The units are arbitrary. Note the minima over and downstream of rawinsonde-rich land regions and the maxima over the oceans (Courtesy I. Szunyogh, University of Maryland.)

In the example in Figure 6.4.1, one could claim that the ensemble indicates a 40% probability of cluster *A* and 60% for cluster *B*.

An ensemble forecasting system requires the definition of the initial amplitude and the horizontal and vertical structure of the perturbations. Typically, the initial amplitude is chosen to be close to the estimated analysis error. The amplitude of the analysis uncertainty depends on the distribution of the observations. Its statistical distribution can be estimated from the analysis error covariance (Chapter 5), which depends on the accuracy of the statistical assumptions, or empirically, from the rms differences between independent analysis cycles (Figure 6.5.2).

Ensemble forecasting methods differ mostly in the way the initial perturbations are generated and can be classified into essentially two classes: those that have random initial perturbations, and those where the perturbations depend on the dynamics of the underlying flow. In the first class, which we can denote Monte Carlo forecasting, the initial perturbations are chosen to be "realistic," i.e., they have horizontal and vertical structures *statistically* similar to forecast errors, and amplitudes compatible with the estimated analysis uncertainty. In the Monte Carlo ensembles, the amplitudes are realistic but the perturbations themselves are chosen randomly, without regard of the "dynamics of the day." For example, Errico and Baumhefner (1987) and Mullen and Baumhefner (1994) developed a Monte Carlo method that results in realistic perturbations compatible with the average estimated analysis error. However, by construction, this type of Monte Carlo forecast does not include finite-size "growing errors of the day" which are almost certainly present in the analysis. The experiments of Hollingsworth (1980), Hoffman and Kalnay (1983), Kalnay and Toth (1996) suggest that random initial perturbations do not grow as fast as the real analysis errors, even if they are in quasi-geostrophic balance. A second class of methods which includes errors of the day has been developed, tested, and implemented at several operational centers. The first two methods of this class implemented operationally are known as "breeding" and "singular vectors" (or optimal perturbations) methods.

In contrast to Monte Carlo forecasting, they are characterized by including in the initial perturbations growing errors that depend on the evolving underlying atmospheric flow. Two other methods in this class that are also very promising are based on ensembles of data assimilations, and ensembles based on operational systems from different centers, combining different models and data assimilations.

6.5.1 Breeding

Ensemble experiments performed at NCEP during 1991 showed that initial ensemble perturbations based on lagged average forecasting, scaled lagged average forecasting and on the forecast differences between forecasts verifying at the same initial time, grew much faster than Monte Carlo perturbations with the same overall size and statistical distribution (Kalnay and Toth, 1996). It was apparent that the differences in the growth rate were due to the fact that the first group included perturbations that, by construction, "knew" about the evolving underlying dynamics. Toth and Kalnay (1993, 1996, 1997) created a special operational cycle designed to "breed" fast growing "errors of the day" (Figure 6.5.3(a)). Given an evolving atmospheric flow (either a series of atmospheric analyses, or a long model run), a breeding cycle is started by introducing a random initial perturbation ("random seed") with a given initial size (measured with any norm, such as the rms of the geopotential height or the kinetic energy). It should be noted that the random seed is introduced only once. The same nonlinear model is integrated from the control and from the perturbed initial conditions. From then on, at fixed time intervals (e.g., every 6 hr or every 24 hr), the control forecast is subtracted from the perturbed forecast. The difference is scaled down so that it has the same amplitude (defined using the same arbitrary norm) as the initial perturbation, and then added to the corresponding new analysis or model state. It was found that beyond an initial transient period of 3–4 days after random perturbations were introduced, the perturbations generated in the breeding cycle (denoted bred vectors), acquired a large growth rate, faster than the growth rate for Monte Carlo forecasting or even scaled lagged average forecasting and forecast differences.

Toth and Kalnay (1993, 1997) also found that (after the transient period of 3–4 days) the shape or structure of the perturbation bred vectors did not depend on either the norm used for the rescaling or the length of the scaling period. The bred vectors did depend on the initial random seed in the sense that regional bred vector perturbations would have the same shape but different signs, and that in many areas two or more "competing bred vectors" appeared in cycles originated from different random seeds. The breeding method is a *nonlinear generalization of the method used to construct Lyapunov vectors* (performing two nonlinear integrations and obtaining the approximately linear perturbation from their difference). Since the bred vectors are related to Lyapunov vectors localized in both space and time, it is not surprising that they share their lack of dependence on the norm or on the scaling period. Toth and Kalnay (1993) have argued that breeding is similar to the analysis cycle. In the analysis cycle (represented schematically in Figure 6.5.3(b)) errors are evolved in time through the forecast used as background, and they are only partially corrected through the use of noisy data.

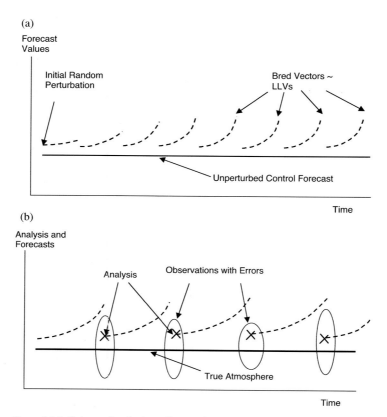

Figure 6.5.3 Schematic of a breeding cycle run on an unperturbed (control) model integration. The initial growth after introducing a random initial perturbation is usually very small, but with time, the perturbation is more dominated by growing errors. (a) The initial transient with slow growth lasts about 3–5 days. The difference of the complete perturbed (dashed line) and control (full line) forecasts is scaled back periodically (e.g., every 6 or every 24 hr) to the initial amplitude. The rescaling is done by dividing all the forecast differences by the same observed growth (typically about 1.5/day for mid-latitudes). In operational NWP, the unperturbed model integration is substituted by short-range control forecasts started from consecutive analysis fields. The breeding cycle is a nonlinear, finite-time, finite-amplitude generalization of the method used to obtain the leading Lyapunov vector. (Adapted from Kalnay and Toth, 1996.) (b) Schematic of the 6-hr analysis cycle. Indicated on the vertical axis are differences between the true state of the atmosphere (or its observational measurements, burdened with observational errors). The difference between the forecast and the true atmosphere (or the observations) increases with time in the 6-hr forecast because of the presence of growing errors in the analysis (Adapted from Kalnay and Toth, 1996.) ©American Meteorological Society. Used with permission.

Therefore, Toth and Kalnay argued that the analysis errors should project strongly on bred vectors. Corazza et al. (2003) compared bred vectors and background errors for a quasi-geostrophic model data assimilation system developed by Morss et al. (2001) Plate 1 shows a typical comparison, depicting that in fact there is a strong resemblance between the structure of the errors of the forecast used as a first guess and the

6.5 Operational Ensemble Forecasting Methods

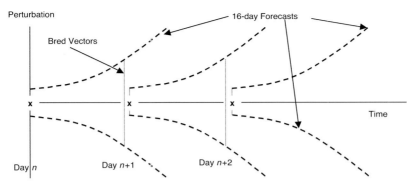

Figure 6.5.4 Schematic of a self-breeding pair of ensemble forecasts used at NCEP. Every day, the 1-day forecast from the negative perturbation is subtracted from the 1-day forecast from the positive perturbation. This difference is divided by 2, and then scaled down (by dividing all variables by the 1-day growth), so that difference is of the same size as the initial perturbation. The scaled difference is then added and subtracted from the new analysis, generating the initial conditions for the new pair of forecasts. This self-breeding is part of the extended ensemble forecast system, and does not require computer resources to generate initial perturbations beyond running the ensemble forecasts (Adapted from Toth and Kalnay, 1997). ©American Meteorological Society. Used with permission.

bred vectors valid at the same time (Corazza et al., 2003). Since the analysis errors are dominated by the background errors, especially when they are large, this resemblance indicates that the forecast and analysis errors do indeed project strongly on the bred vectors.

Figure 6.5.4 shows a schematic of how breeding cycles are self-propagated from the ensemble forecasts. The bred perturbations are defined every day from the difference between the one-day positive and the negative perturbation forecasts divided by 2, they are scaled down by their growth during that day, and added and subtracted to the new analysis valid at the time. This provides the initial positive and negative perturbations for the ensemble forecasts at no additional cost beyond that of computing the ensemble forecasts. Separate breeding cycles differ only in the choice of random initial perturbations (performed only once). It has been found empirically that for the atmosphere the finite amplitude bred vectors do not converge to a single "leading bred vector" (Kalnay et al., 2002).

Figures 6.5.5(a) and (b) show two out of five operational bred perturbations corresponding to March 5, 2000, at 00UTC. Figure 6.5.5(c) presents an estimate of the effective local dimension of the subspace of the five perturbations using the bred vector dimension defined in Patil et al. (2001).[3] Only the areas where the local dimension

---

[3] The local bred vector dimension is obtained as $\psi(\sigma_1, \ldots, \sigma_k) = \left(\sum_{i=1}^{k} \sigma_i\right)^2 / \sum_{i=1}^{k} \sigma_i^2$, where σ_i are the singular values corresponding to the k bred vectors within a region of about 10^6 m by 10^6 m, and it defines the effective local dimensionality. For example, if four out of five bred vectors lie along one direction, and one lies along a second direction, the bred vector dimension would be $\psi(\sqrt{4}, 1, 0, 0, 0) = 1.8$, less than 2 because one direction is more dominant than the other in representing the original data (Patil et al., 2001).

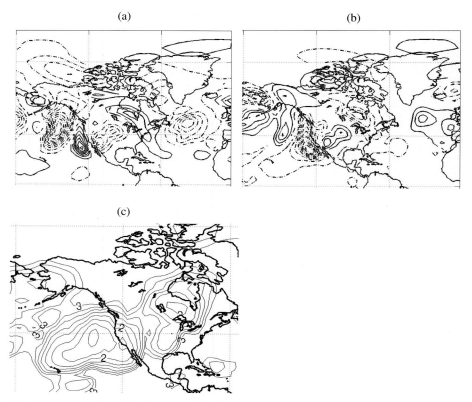

Figure 6.5.5 Examples of bred vectors (500-hPa geopotential height field differences, without plotting the zero contour) from the NCEP operational ensemble system valid at March 5, 2000: (a) bred vector 1; (b) bred vector 5. Note that over large parts of the eastern Pacific Ocean and western North America, the two perturbations have shapes that are very similar but of opposite signs and/or different amplitudes. In other areas the shape of the perturbations is quite different. (c) The bred-vector-local dimension of the five perturbations subspace (Patil et al., 2001). Only dimensions less than or equal to 3 are contoured with a contour interval 0.25. In these areas, the five independent bred vectors have aligned themselves into a locally low-dimensional subspace with an effective dimension less than or equal to 3. (Courtesy of D. J. Patil.)

has collapsed from the original five independent directions (shapes or structure of the bred vectors) to three or less are contoured. Note that these are the areas where the independently bred vectors aligned themselves into a smaller subspace. The collapse of the perturbations into fewer dimensions is what one could expect if there are locally growing dominant Lyapunov vectors expressing the regional dominant instability of the underlying atmospheric flow. These low-dimensional areas are organized into horizontal and vertical structures and have a lifetime of 4–7 days, similar to those of baroclinic developments (Patil et al., 2001).

The breeding ensemble forecasting system was introduced operationally in December 1992 at NCEP, with two pairs of bred vectors (Tracton and Kalnay, 1993). In 1994, seven pairs of self-breeding cycles replaced the original four perturbed forecasts.

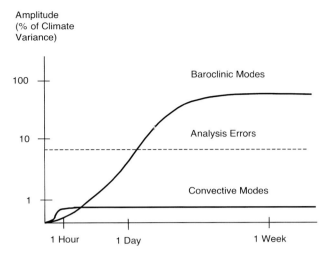

Figure 6.5.6 Schematic of the time evolution of the rms amplitude of high-energy baroclinic modes and low-energy convective modes. Note that although initially growing much faster than the baroclinic modes, convective modes saturate at a substantially lower level. These modes are therefore insignificant in the analysis/ensemble perturbation problem, since the errors in the analysis (dashed line) are much larger than the convective saturation level (Adapted from Toth and Kalnay, 1993). ©American Meteorological Society. Used with permission.

In addition, a regional rescaling was introduced that allowed larger perturbation amplitudes over ocean than over land proportionally to the estimate of the analysis uncertainty, Figure 6.5.2 (Toth and Kalnay, 1997).

Toth and Kalnay (1993) found that when the initial amplitude was chosen to be within the range of estimated analysis errors (i.e., between 1 m and 15 m for the 500-hPa geopotential height) the bred vectors developed faster in strong baroclinic areas. Their horizontal scale was that of short baroclinic waves, and their hemispherical average growth rate was about 1.5/day (similar to the estimated growth of analysis errors). However, if the initial amplitude was chosen to be much smaller than the estimated analysis errors (10 cm or less), then a different type of bred vector appeared, associated with convective instabilities, which grew much faster than baroclinic instabilities (at a rate of more than 5/day). The faster instabilities saturated at amplitudes much smaller than the analysis error range (Figure 6.5.6). Toth and Kalnay (1993) suggested that the use of nonlinear perturbations in breeding has the advantage of filtering Lyapunov vectors associated with fast growing but energetically irrelevant instabilities, like convection. This was confirmed by Lorenz (1996), who performed experiments with a low-order model containing large amplitude but slowly growing modes coupled with fast growing modes with very small amplitudes. Lorenz found that the use of breeding using finite amplitudes yielded the Lyapunov vectors of the large amplitude, slowly growing vectors, as desired, whereas for very small amplitudes the Lyapunov vectors of the fast system were recovered. In a complex system like the atmosphere, with multiple-scale instabilities, using breeding may thus be more appropriate than using Lyapunov vectors, which, in a model including *all* atmospheric instabilities, would yield vectors associated with Brownian motion, which are the fastest, though clearly irrelevant, instabilities present in the system (Kalnay et al., 2002). The nonlinear saturation of irrelevant fast growing modes is an advantage that suggests the use of breeding for other problems. For example, for seasonal and interannual forecasting using a coupled ocean–atmosphere system, the slower growing (but very large energy amplitude) coupled ENSO instabilities could perhaps be captured, while eliminating through nonlinear saturation the irrelevant details of weather perturbations (Cai et al., 2003).

Figure 1.6.3 shows two examples of one of the ways information on ensemble forecasts are presented to the users, the "spaghetti plots," or plots showing one contour line for each forecast. In one case, a 5-day forecast verifying on November 15, 1995, the agreement in intensity and location of the contours indicated to the forecasters that this was a very predictable snowstorm (see also the book cover). In the second case, a 2.5-day forecast verifying on October 21, 1995, the ensemble members show unusually strong divergence in the location of a winter storm, warning the human forecasters that this situation is intrinsically unpredictable. Note that although the ensemble forecasts show a wide divergence in the location of the storm, this is also a case in which, in perturbation space, there is very low dimensionality, since the perturbations align themselves along the same basic shape (the perturbations for the winter storm are a one-parameter family so that the local dimension is about 1).

The second example shows the potential value of ensembles in a new area of research: *targeted observations*. In cases like this in which the ensemble indicates a region of large uncertainty in the short-term forecasts, it should be possible to find the area that originated this region of uncertainty in time to launch new observations for the next analysis cycle, and thus decrease significantly the forecast error. Finding the area where the observations should be launched can be done through several approaches. They are the adjoint sensitivity approach, the use of singular vectors (Rabier et al., 1996; Langland et al., 1999; Pu et al., 1998, and others), the use of the quasiinverse of the tangent linear model (Pu et al., 1998), and ensemble-based singular value decomposition (Bishop and Toth, 1999). These methods were tested during FASTEX (January–February 1997 in the Atlantic) and NORPEX (winter of 1997–1998 in the North Atlantic (Langland et al., 1999; Pu and Kalnay, 1999)). The experience in the North Pacific has been so successful that targeted observations are now performed operationally over the Gulf of Alaska every winter (Szunyogh et al., 2000).

Plate 2 shows another example of how the massive amount of information contained in the ensemble forecast can be conveyed to the forecasters. It shows a probabilistic presentation of a 1-day and a 7-day forecast of precipitation above a threshold of 5 mm in 24 hr. The probabilities are simply computed as a percentage of the ensemble members with accumulated precipitation at least as large as the indicated threshold. They both verify on April 6, 2001. Note that the short-range forecast has many areas with probabilities equal to zero or above 95%, indicating that *all* the ensemble members agree that there will be no precipitation or at least 5 mm accumulated precipitation respectively. In the 7-day forecast, the areas with maximum probability of precipitation are generally in agreement with the short-range forecast, indicating the presence

6.5 Operational Ensemble Forecasting Methods 259

of skill. However, by this time there are few areas of consensus on either rain or no rain among the forecasts, since their solutions have dispersed significantly over a week.

6.5.2 Singular Vectors

ECMWF developed and implemented operationally in December 1992 an ensemble forecasting system based on initial perturbations that are linear combinations of the singular vectors of the 36-hr tangent linear model (Molteni et al., 1996; Molteni and Palmer, 1993; Buizza, 1997).

As discussed in Section 6.3, the singular vectors $\mathbf{y}_i(t_0)$ used to create the initial perturbations at the time t_0 are obtained as the leading eigenvectors $\hat{\mathbf{y}}_i(t_0)$ of

$$(\mathbf{W}^{-1})^T \mathbf{L}^T \mathbf{P}^T \mathbf{P} \mathbf{L} \mathbf{W}^{-1} \hat{\mathbf{y}}(t_0) = \sigma^2 \hat{\mathbf{y}}(t_0) \tag{6.5.1}$$

subject to

$$\hat{\mathbf{y}}^T(t_c)\hat{\mathbf{y}}(t_o) = 1 \quad \text{and} \quad \mathbf{y}(t_o) = \mathbf{W}^{-1}\hat{\mathbf{y}}(t_o) \tag{6.5.2}$$

ECMWF used as the projection operator $\mathbf{P}$, a symmetric projector operator that includes only forecast perturbations north of 30° N, and as the initial norm the total energy norm $\mathbf{W}^{-1}$. The linear tangent model $\mathbf{L}$ and its adjoint $\mathbf{L}^T$ are computed for the 36-hr forecasts (and more recently for the 48-hr forecast), which determine the length of the "optimization interval" (Section 6.3). Barkmeijer et al. (1998) tested the use of the analysis error covariance as the initial norm instead of the energy norm with good results. The singular vectors obtained with this norm were closer to bred vectors than those obtained with the total energy norm. They also found that the use of evolved vectors (also closer to Lyapunov or bred vectors) resulted in improved results. More recent experiments with a simplified Kalman filter also resulted in promising results (Fischer et al., 1998).

From (6.5.1) and (6.5.2), the initial singular vectors $\mathbf{y}_i(t_0)$ are the perturbations with maximum energy growth north of 30° N, for the time interval 0–36 hr (Buizza, 1994), or more recently, 0–48 hr. The method used to obtain the singular vectors is the Lanczos algorithm (Golub and Van Loan, 1996), which requires integrating forward with $\mathbf{L}$ for a period t (36 or 48 hr), and backward with $\mathbf{L}^T$. This forward–backward integration has to be performed about three times the number of singular vectors desired. Figure 6.5.7 shows an example of the horizontal structure corresponding to the initial and final singular vectors #1, 3, and 6. Figure 6.5.8 shows the corresponding initial and evolved vertical energy structure (Buizza, 1997). Singular vectors defined with the total energy norm tend to have a maximum initial energy at low levels (about 700 hPa), and their final (evolved) energy at the tropopause level. In 1996, when they were using 36-hr forecasts for the linear tangent model, ECMWF used 16 singular vectors selected from 38 leading singular vectors. This required a daily integration equivalent to about $3 \times 36 \times 2 \times 38$ hr of model integration with either $\mathbf{L}$ or $\mathbf{L}^T$ to create the perturbations. For this reason, the computation was done with a lower resolution

260 **Atmospheric Predictability and Ensemble Forecasting**

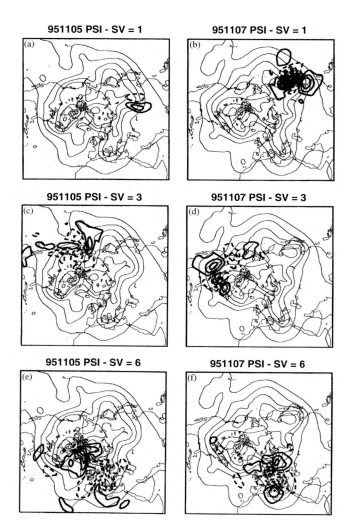

Figure 6.5.7 Singular vectors numbers 1 (top panels), 3 (middle panels), and 6 (bottom panels) at initial (left panels) and optimization time (right panels). Each panel shows the singular vector streamfunction at model level 11 (approximately 500 hPa), superimposed to the trajectory 500-hPa geopotential height field. Streamfunction contour interval is 0.5×10^{-8} m^2 s^{-1} for left panels and 20 times larger for the right panels; geopotential height contour interval is 80 m (Buizza, 1997. ©American Meteorological Society. Used with permission).

(T42/19 level) than the operational model. A second set of perturbations was added for the Southern Hemisphere, which originally had no perturbations, requiring additional computations.

The selection of 16 singular vectors is such that the first four are always selected, and from the fifth on, each subsequent singular vector is selected if 50% of its energy is located outside the regions where the singular vectors already selected are localized. Once the 16 singular vectors are selected, an orthogonal rotation in phase space and

6.5 Operational Ensemble Forecasting Methods 261

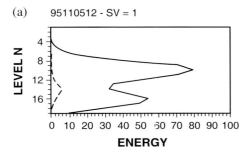

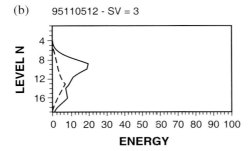

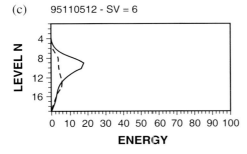

Figure 6.5.8 Total energy (m^2 s^{-2}) vertical profile of the (a) first, (b) third, and (c) sixth singular vector of November 5, 1995, at the initial (dashed line, values multiplied by 100) and optimization (solid line) times. Note that singular vectors are normalized to have unit initial total energy norm (Buizza, 1997. ©American Meteorological Society. Used with permission).

a final rescaling are performed to construct the ensemble perturbations. The purpose of the phase-space rotation is to generate perturbations that have the same globally averaged energy as the singular vectors but smaller local maxima and more uniform spatial distribution. The rotated singular vectors are characterized by having similar growth rates (at least to the period of optimization). The rotation is defined to minimize the local ratio between the perturbation amplitude and the amplitude of the analysis error estimate of the ECMWF OI analysis. The rescaling allows local amplitudes up to $\sqrt{1.5}$ larger than the OI error.

The 16 rotated perturbations are three-dimensional fields of temperature, vorticity, divergence, and surface pressure (no moisture, since the propagator **L** is "dry," although there has been more recent work to include physical processes in the tangent linear model and adjoint). They are added and subtracted to the control initial conditions to create 33 initial conditions (32 + control), from which the ensemble forecast is run with the nonlinear model at T63 resolution.

In 1997, ECMWF changed the system to an ensemble of 50 members (plus control) run at a resolution of T156 (with a linear Gaussian grid, since their use of a semi-Lagrangian scheme allows the use of a more efficient linear rather than quadratic grid). This increase in resolution had a major positive effect on the quality of the ECMWF ensemble forecasting system. In March 1998, ECMWF added to the initial perturbations the *evolved* (or final) singular vectors from 48 hr before the analysis time, which also resulted in improved results. The 2-day evolved singular vectors are much closer to the Lyapunov vectors (or bred vectors) (Barkmeijer et al., 1998).

Initially both NCEP and ECMWF considered in their ensembles only the errors generated by uncertainties in the initial conditions, and neglected the additional errors due to the models themselves. This is a reasonable (but not perfect) assumption only for the extratropics (Reynolds et al., 1994). In 1998, ECMWF tested an innovative way to account for the fact that the model has deficiencies (Buizza et al., 1999). The time derivatives of the physical parameterizations are multiplied by Gaussian random numbers with a mean of 1.0 and a standard deviation of 0.2, which have a time lag correlation of several hours and horizontal correlation of a few hundred kilometers. This introduction of randomness in the "physics" had a very good impact on the ensemble. It increased the ensemble spread to levels similar to those of the control forecast error, which is a necessary condition if "nature" (the verifying analysis) is to be a plausible member of the ensemble (Toth and Kalnay, 1993).

6.5.3 Ensembles Based on Multiple Data Assimilation

Houtekamer et al. (1996) and Houtekamer and Mitchell (1998) have developed a very promising ensemble forecasting system based on running an ensemble of data assimilation systems to create the initial conditions. In their different data assimilation systems they add random errors to the observations (in addition to the original observational errors) and include different parameters in the physical parameterizations of the model in different ensembles. This is a promising approach, related to but more general than breeding. One novel approach introduced by Houtekamer et al. (1996) is the use of perturbations in the physical parameterizations in the models used in different analysis cycles. Through a careful combination of changes in major parameterizations, it is possible to use the ensemble forecasts to isolate the impact of particular parameterizations. As indicated by the results of Miller et al. (1994), the introduction of uncertainty in the model should improve the efficiency of the ensemble.

Hamill et al. (2000) have shown in a quasi-geostrophic system that the multiple data assimilation ensemble system performs better than the singular vector or breeding approaches. The computational cost of creating the initial perturbations is comparable to that of the singular vector approach, whereas in the breeding method the perturbations are obtained as a by-product of the ensemble forecasts themselves and are therefore cost-free.

6.5.4 Multisystem Ensemble Approach

The ensemble forecasting approach should replicate in the initial perturbations the statistical uncertainty in the initial conditions: ideally, the initial perturbations should be the leading eigenvectors of the analysis error covariance (Ehrendorfer and Tribbia, 1997). Moreover, it should also reflect model imperfections and our uncertainty about model deficiencies. In the standard approaches discussed so far the uncertainty in the initial conditions is introduced through *perturbations added to the control analysis*, which is the best estimate of the initial conditions. As a result, the perturbed ensemble forecasts are, on the average, somewhat less skillful than the control forecast (see, e.g., Figure 1.6.2). Similarly, when perturbations are introduced upon the control model parameterizations (Buizza et al., 1999; Houtekamer et al., 1996), the model is made slightly worse, since the control model has been tuned to best replicate the evolution of the atmosphere.

A different approach that has become more popular recently is that of a multi-system ensemble. It has long been known that an ensemble average of operational global forecasts from different operational centers is far more skilful than the best individual forecast (e.g., Kalnay and Ham, 1989; Fritsch et al., 2000, and references therein). More recently, it has been shown that this is true also for shorter-range ensembles of regional models (Hou et al., 2001), and that the use of multisystems can therefore extend the utility of ensemble forecasting to the short-range. Krishna-murti et al. (2000) have shown that if the multisystem ensemble includes correction of the systematic errors by regression, the quality of the ensemble system is further significantly improved. Krishnamurti et al. (2000) call this multiple system approach a "superensemble."

The advantages of a multisystem ensemble are not surprising. Instead of adding per-turbations to the initial analysis, and introducing perturbations into the control model parameterizations, the multisystem approach takes the best (control) initial conditions and the best (control) model estimated at different operational centers that run compet-itive state-of-the-art operational analyses and model forecasts. Thus the multisystem probably samples the true uncertainty in both the initial conditions and the models better than any perturbation introduced *a posteriori* into a single operational system. The statistical correction of systematic errors introduced by Krishnamurti et al. (2000) is an added benefit of the method which can be considered an ideal "poor person's" approach to ensemble forecasting (Wobus and Kalnay, 1995).

6.6 Growth Rate of Errors and the Limit of Predictability in Mid-latitudes and in the Tropics

Lorenz (1963a, 1982) suggested that the limit of deterministic predictability was about 2 weeks. He obtained this empirical estimate from the doubling time of small errors derived from identical-twin model experiments, and from the rate of separation in time of atmospheric analogs (atmospheric states initially very similar to each other). Two weeks continues to be a good estimate of the limit of predictability despite the fact that different models provided different estimates (Charney et al., 1966). The analog method cannot give a precise answer because it would take an exceedingly long time to find atmospheric analogs close enough to estimate the growth of small errors (Van den Dool, 1994).

As indicated in the introduction to this chapter, the doubling time for small errors in the mid-latitude synoptic (weather) scales, for which the dominant instability is baroclinic, was estimated in the 1960s to be 3 days or longer. Modern models have much more resolution and are less sluggish than the early primitive equation models. Identical-twin experiments with these models and measurements of actual numerical forecast error growth have lowered the estimate of the doubling time of small errors from 3 days to about 2 days (Lorenz, 1982; Dalcher and Kalnay, 1987; Simmons et al., 1995; Toth and Kalnay, 1993).

Lorenz (1982) suggested a simple way to parameterize the evolution of small errors in a *perfect* model, in which the only source of errors is the unstable growth of small errors in the initial conditions, using the logistic equation:

$$\frac{d\varepsilon}{dt} = a\varepsilon(1 - \varepsilon) \tag{6.6.1}$$

Here, ε represents the rms average forecast error scaled so that at long forecast leads $\varepsilon \to 1$, i.e., it is the rms forecast error divided by the square root of twice the variance of the atmosphere (cf. Section 6.4). Equation (6.6.1) indicates that very small errors grow exponentially with a growth rate a. When they reach finite amplitude, the error growth rate is lowered by the last factor on the right-hand side, which slows it down until it saturates at $\varepsilon \approx 1$. The solution of the logistic equation (6.6.1) is

$$\varepsilon(t) = \frac{\varepsilon_0 e^{at}}{1 + \varepsilon_0(e^{at} - 1)} \tag{6.6.2}$$

where ε_0 is the initial error.

Exercise 6.6.1 Derive Equation (6.6.2) using separation of variables.

Figure 6.6.1 shows the solution for two values of the initial error, 10% and 1%, and an error growth rate $a = 0.35$/day, corresponding to a doubling time of about 2 days. The analysis error in the 500-hPa geopotential heights in current operational systems is of the order of 5–15 m, and the natural variability about 100 m, so that the current level of error in the initial conditions is $\sim 10\%$ or less. The upper limit for the best initial error achievable from data assimilation can be reasonably estimated to be no less than 1%. This is because, as pointed out by Lorenz, even if the observing system was essentially perfect at synoptic scales, errors in much smaller, unresolved scales would grow very fast and through nonlinear interactions quickly introduce finite errors in the initial synoptic scales of the model. The solution of the logistic equation for initial errors of 10% and 1% (Figure 6.6.1) suggests that 2 weeks is indeed a reasonable estimate of the time at which the forecast errors become so large that the ability to predict weather in mid-latitudes is lost. The range between the two curves can be taken as a simple upper estimate of how much forecasts could be improved by improving the initial conditions.

6.6 Growth Rate of Errors and the Limit of Predictability

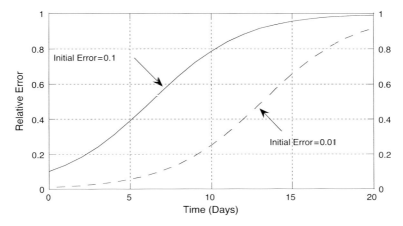

Figure 6.6.1 Time evolution of the rms forecast error divided by the square root of twice the climatological variance. It assumes that the forecast error growth satisfies the logistic equation (6.6.1), and that the growth rate of small errors is about 0.35/day, corresponding to a doubling of small errors in 2 days. Analysis errors in the initial conditions are estimated to be about 10% or less, but not smaller than 1%.

However, this is only an estimate of the *average* predictability in a perfect model. The actual predictability is quite variable and depends on the "atmospheric instabilities of the day." The 2-week "limit," which seemed huge during the 1960s when 2-day forecasts had little skill, is no longer large compared with what can be occasionally attained with current models. For example, during a very predictable period in December 1995, several numerical weather forecasts remained skillful for 15 days, with a pattern correlation between the forecast and observed anomalies (known as *anomaly correlation*) of more than 70% for the Northern Hemisphere extratropics (Toth and Kalnay, 1996). There are also periods in which the atmospheric predictability is much lower than average, as indicated by the fact that different operational global forecasts tend to show large dips in the medium-range forecast skill on the same days. Because of the day-to-day variability of atmospheric predictability, it is important to use forecast ensembles. They provide a tool for estimating the day-to-day variations in predictability and allow human forecasters to extend the range of the forecasts provided to the public during periods of high predictability. Ensembles based on multiple models have proven to be especially useful to the forecasters.

From (6.2.2) it is clear that the limit of predictability depends on the rate of error growth, which is the inverse of the *e*-folding or exponential time scale. Generally the time scales of instabilities are related to their spatial scales, so that small-scale instabilities grow much faster than those with larger scales. For this reason short synoptic waves are typically less predictable than longer waves (e.g., Dalcher and Kalnay, 1987), and mesoscale phenomena, such as fronts, squall lines, mesoscale convective systems and tornadoes, are intrinsically predictable for shorter time scales, of the order of a day or less (e.g., Droegemeier, 1997). Convection is a typical example of a short time scale phenomenon: cumulus clouds grow with an exponential time scale of the order of 10 min or so. It is therefore impossible to predict the precipitation associated with an individual thunderstorm for more than about an hour. Nevertheless, *if convective activity is organized or forced by the larger scales*, then convective precipitation can remain predictable much longer than individual thunderstorms. For example, summer convective precipitation is notoriously difficult to predict. However, when summer mesoscale convection is *forced* by a synoptic scale system, convection can be predicted to occur when and where forced by the larger scales, and therefore becomes predictable well beyond its own short predictability time scale. Similarly, mesoscale phenomena forced by the interaction of synoptic scales with surface topography have a much longer predictability than when they are not subjected to this organizing influence from the larger scales.

Two types of surprisingly regular progression of smaller scale phenomena have been discovered. As indicated above, mesoscale summer convection in the USA, when "unforced" by upward motion associated with synoptic-scale waves, is extremely difficult to predict. However, an examination of the Doppler radar reflectivities has shown that the area of maximum convection has a tendency to propagate eastward with considerable regularity, and with its intensity modulated by the diurnal cycle. The individual maxima of this wave-like propagation can be traced on radar reflectivities for 1–3 days (Carbone et al., 2000). This surprising discovery implies that such unforced convective activity, in principle, should be predictable for a day or two. Another example of regular propagation of convection is the Madden–Julian Oscillation (MJO, Madden and Julian, 1971, 1972). The MJO has a zonal wavenumber 1 with maximum amplitude in the deep tropics, and it moves eastward around the Equator with a period of 30–60 days. The MJO is not always present, but there are periods of several months in which it is very prominent (Weickmann et al., 1985). Although current models are not yet able to reproduce well the intensity and speed of propagation of the MJO, its regularity indicates that, in principle, there should be predictability in the convective precipitation associated with the MJO that could be exploited by dynamical or statistical methods for time scales of a month or longer.

Both of these are examples of small-scale convection organized into a regular, longer lasting propagation. It may take many years before dynamical models are able to fully reproduce the observed quasi-regular motion. In the mean time, a combination of statistical and dynamical methods may be the best way to exploit this latent longer time scale predictability.

So far we have mostly discussed the predictability of weather in mid-latitudes. The dynamics of mid-latitudes is dominated by synoptic-scale baroclinic instabilities (Holton, 1992), and the limit of deterministic weather predictability is a reflection of their baroclinic instability rates of growth.

In the tropics, the situation is quite different. Baroclinic instability is generally negligible in the tropics, and barotropic and convective instabilities, and their interactions are more dominant. Phenomena like easterly waves are a reflection of barotropic instability, and are less intense than baroclinic instabilities. Easterly waves are strongly modulated by convective precipitation, whereas in mid-latitudes, large-scale precipitation has a smaller effect on the evolution of the synoptic waves. Moreover, global atmospheric models are less accurate in the tropics, because their ability to

6.6 Growth Rate of Errors and the Limit of Predictability

parameterize realistically the subgrid scale processes such as convection, which are dominant in the tropics, is not as good as the numerical representation of the resolved baroclinic dynamics, which is dominant in the extratropics. In the tropics, the assumption of a perfect model is therefore much less justified than in the extratropics.

Equation (6.6.1) describes reasonably well error growth in a perfect model (an identical-twin experiment) The growth rates of random errors in an *imperfect* model have been parameterized by Dalcher and Kalnay (1987) and Reynolds et al. (1994) fitting operational forecast errors with an extension of the logistic equation (6.6.1) which includes growth of errors due not only to the presence of errors in the initial conditions but also to model deficiencies:

$$\frac{dv}{dt} = (bv + s)(1 - v) \qquad (6.6.3)$$

Here v is the variance of the random error (systematic or time averaged errors having been separated beforehand), b is the growth rate for small error variances due to instabilities ("internal" source) and s is an "external" source of random error variance due to model deficiencies. The solution of (6.6.3) is given by

$$v(t) = 1 - \frac{1+s}{1+\mu} \qquad (6.6.4)$$

where

$$\mu = \frac{v(0) + s/b}{1 - v(0)} e^{(b+s)t}$$

Reynolds et al. (1994) found that (6.6.4) fits observed errors fairly well, and estimated that the internal error growth rate is given by about 0.4/day in mid-latitudes and 0.1/day in the tropics. The external growth rate due to model deficiencies was found to be small, about 0.05/day in mid-latitudes, and considerably larger, about 0.1–0.2/day,

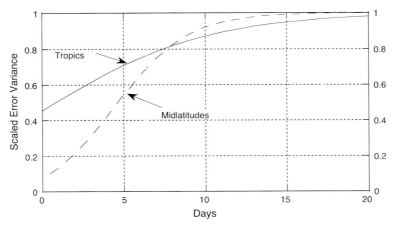

Figure 6.6.2 Parameterization of scaled forecast error variance in the presence of model deficiencies with values of the growth rate due to model deficiencies s and to instabilities b appropriate for mid-latitudes (($b = 0.4$, $s = 0.05$) and the tropics ($b = 0.1$, $s = 0.1$)). (From Reynolds et al., 1994.)

in the tropics (Figure 6.6.2). Although these estimates, which were obtained from fitting observed error growth, are only valid to the extent that the error variance growth for an imperfect model follows (6.6.3), they do reflect qualitatively the notion that in the tropics the synoptic scales are less unstable than in the extratropics, because of the absence of baroclinic instabilities. At the same time, the tropics are much harder to model, because of the difficulties associated with parameterizations of cumulus convection, which is much more influential in the tropics than in the extratropics. These results suggest that if convection did not play such a dominant role in the tropics, tropical weather forecasts would be skillful for longer periods than mid-latitudes predictions. In reality, the dominant role of tropical convection and the difficulties of its parameterization lead to the fact that currently tropical forecasts maintain useful skill only for about 3–5 days, whereas in the extratropics forecasts remain skillful on the average for 7 days or so.

6.7 The Role of the Oceans and Land in Monthly, Seasonal, and Interannual Predictability

It is well known that there is considerable atmospheric variability not only in the day-to-day weather but also in longer time scales, such as weekly, monthly, seasonal, interannual, and even decadal averages. For example, the atmospheric circulation averaged for January 1987 (an El Niño year) was substantially different than that of January 1989 (a La Niña year). Similarly, July 1987, during El Niño, was quite different from July 1988, during La Niña. In this case, it is clear that the differences were strongly influenced by the ENSO, a much longer lasting phenomenon than individual weather events. However, seasonal and interannual atmospheric variability can also take place due to unpredictable weather "noise," not just to longer-lasting surface forcings such as ocean SST anomalies, soil moisture or snow cover anomalies. The monthly or seasonally averaged atmospheric anomalies due to weather noise (e.g., a monthly averaged cold and wet January in the eastern USA because of the passage of two or three very strong cyclones) are unpredictable beyond the first two weeks because weather itself is not predictable. On the other hand, the variability due to long-lasting surface anomalies, of which SST anomalies are the most important, is predictable, if we can predict the SST anomaly. *Potential predictability* beyond the limit of deterministic weather predictability can be defined as the difference between the total variance of the anomalies averaged over a month or a season, minus the variance that can be attributed to weather noise (Madden, 1989). The predictability associated with information within the initial conditions is sometimes referred to as "predictability of the first kind," whereas that associated with information contained in the slowly evolving boundary conditions is referred to as "predictability of the second kind."

In a paper based on global model simulations of the growth of perturbations in the presence of SST anomalies, Charney and Shukla (1981) pointed out that the tropics have a shorter limit of weather predictability than the extratropics due to the factors discussed in the previous section. At the same time, they found that the tropics are much more responsive to the long-lasting ocean SST anomalies than the mid-latitudes. As a result, the potential predictability for the tropics at long time scales due to long-lasting ocean anomalies is much larger than that of the extratropics. The conclusions of this fundamental paper have been confirmed by many simulation experiments and actual dynamical and statistical forecasts.

This led to the search for methods to exploit the potential predictability associated with longer lasting lower boundary conditions, especially those associated with El Niño events. A major breakthrough occurred with the first successful prediction of El Niño by Cane et al. (1986), with a simplified coupled ocean–atmosphere model (Zebiak and Cane, 1987). ENSO is a complex interannual tropical oscillation due to the unstable coupling of the ocean and the atmosphere that has a profound effect in the global circulation even away from the tropics. As reviewed by Philander (1990), there are several unstable coupled modes in the tropical Pacific, which can explain the different ways El Niño (warm central and eastern Pacific events) and La Niña (cold central and eastern Pacific events) occur (Ruiz-Barradas et al., 2017; Bach et al., 2019).

Figure 6.7.1(a) shows a schematic of the interaction of the tropical Pacific ocean and atmosphere during "normal" years. On the average, the trade winds (easterlies) in the tropics produce westward advection of warm temperatures, so that the SSTs are much warmer in the tropical western Pacific than in the eastern Pacific. For the

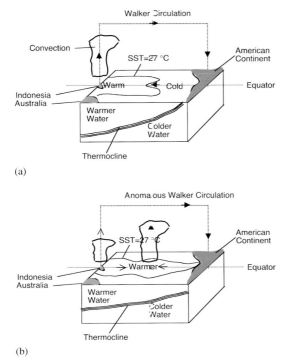

Figure 6.7.1 Schematic of the coupling of the ocean and atmosphere in the tropical Pacific: (a) normal conditions; (b) El Niño conditions.

same reason, the thermocline (an ocean layer of strong vertical gradients of temperature that separates the warm upper ocean from the colder lower layers) is also much deeper in the western Pacific. In the eastern Pacific, the colder deeper layer below the thermocline may even surface close to the American continent. The surface easterlies also produce a cold equatorial tongue due to "Ekman pumping" at the Equator, where the Coriolis force acting on westward currents creates a poleward acceleration in both hemispheres, producing horizontal divergence and strong upwelling of cold water. The strongest atmospheric convection takes place in the "warm pool" of water close to the Indonesian region. This convection drives the east–west atmospheric circulation known as the Walker circulation, after Sir Gilbert Walker, who discovered a strong negative correlation between the sea level pressure in Darwin, Australia, and in Tahiti, in the mid-Pacific. He named this relationship the "Southern Oscillation" (Walker, 1928). Bjerknes (1969) discovered that El Niño and the Southern Oscillation were part of a single, coupled ocean–atmosphere phenomenon.

During an El Niño (or warm event, Figure 6.7.1(b)), a weakening of the easterlies (westerly anomaly) will result in an eastward advection of warm SST by the eastward oceanic currents driven by the westerly atmospheric anomalies. The warm waters propagate eastward and in turn produce atmospheric low-level convergence, strengthening the warm anomaly depth and its eastward propagation. On other occasions the opposite effect occurs: La Niña (or cold event) takes place when the surface easterlies are stronger than normal, and because of this, the central equatorial Pacific is colder than normal (not shown). The equatorial region's SST and surface winds, as measured with TOGA (Tropical Ocean, Global Atmosphere) buoys deployed in the equatorial Pacific are shown in Plate 3 for normal, El Niño and La Niña conditions. Both the actual fields and their anomalies are shown for each case, demonstrating the complexity of the coupling.

An explanation of why the ENSO episodes alternate between warm and cold events was offered by Schopf and Suarez (1988), Suarez and Schopf (1988), and by Battisti and Hirst (1988), who independently suggested the "delayed oscillator" mechanism for simple coupled models. In this mechanism, a westerly/warm anomaly in the equatorial central Pacific deepens because of the unstable coupling, and in the process of adjustment the anomaly generates Rossby waves moving westward (Gill, 1980). The Rossby waves elevate the thermocline in the western region. When they reach the western boundary of the equatorial Pacific, the Rossby waves are reflected as eastward moving Kelvin waves, which also elevate the thermocline. When the Kelvin waves reach the central Pacific, they counteract the effect of the thermocline deepening by the unstable coupling. When this delayed negative feedback becomes sufficiently strong, it reverses the sign of the anomaly, and a cold (La Niña) episode starts. The process then starts again with the opposite sign.[4]

---

[4] The delayed oscillator mechanism has been illustrated with the simple equation $\dot{T} = T - T^3 - rT(t-d)$. Here $T(t)$ would represent the SST anomaly in the central equatorial Pacific, and the left-hand side is its rate of change. The first term on the right-hand side represents the unstable coupling with the atmosphere. The second term represents damping effects due to dissipation. The last term on the

6.7 The Role of the Oceans and Land   271

However, the observed ENSO episodes are much more complex, and are not well represented by a simple model. There are different "flavors" of El Niño, with some propagating the SST warm anomaly eastward, and others propagating it westward (Philander, 1990). Much research is taking place to understand and model these differences in the evolution of the coupled system better. Despite these difficulties, the fact that there are coupled oscillations that have time scales of 3–7 years provides us with the hope that the interannual variability of the tropical climate, dominated by the interactions of the tropical ocean with the tropical atmosphere, could be predictable for seasons through years. Moreover, numerical experiments and analysis of past observations have indicated that the tropical anomalies, especially the anomalous location of major centers of precipitation, have a profound influence on the extratropical circulation (e.g., Horel and Wallace, 1981). A scientific program, TOGA was created to study this problem. Wallace et al. (1998) reviewed the results of the first decade of the TOGA research.

The hope that long-time scale coupling could form the basis for seasonal to interannual prediction has begun to become a reality. A number of "hindcast" experiments using *observed* (not *predicted*) SST forcing of the global atmosphere have been made. They are usually known as AMIP experiments, since this was the setup used in the Atmospheric Model Intercomparison Project (Gates et al., 1999). In principle, they should give an upper limit to the predictability associated with SST anomalies, since the latter are "perfect." However, it is possible that the fact that uncoupled atmospheric runs using perfect SST do not include feedbacks from the atmosphere to the ocean may actually reduce the optimality of such an approach, since the atmosphere clearly has a profound effect on the oceans, especially in mid-latitudes (Peña et al., 2003).

The AMIP experiments show that, indeed, in most models the tropical SST anomalies produce a reasonably realistic atmospheric response, especially during El Niño or La Niña years. In the extratropics, the situation is more complicated. There are regions, such as Europe, where there is generally little predictability due to ENSO, and others, like the winter northern extratropical Pacific and North America, where the response is stronger, indicating significant potential predictability (Shukla et al., 2000). It should be noted that tropical ocean models, driven with observed surface wind stress, also give fairly realistic oceanic El Niño responses. The fact that both the tropical atmosphere and the tropical ocean respond realistically to one-way observed forcings (SST and wind stress, respectively), further corroborates that ENSO oscillations are

---

right-hand side represents the negative feedback of the thermocline elevation, delayed by the time it takes the Rossby waves generated by the anomaly T to reach the western boundary and return as Kelvin waves. This mechanism clearly dominates the Zebiak and Cane (1987) model used for the first successful ENSO forecasts (Cane et al., 1986). Cai et al. (2003) performed breeding experiments and showed that the perturbations of the forecasts grow fastest during the transitions between cold and warm episodes, and grow slowly or decay during the maxima of the warm and cold episodes, as would be expected from the linearized version of the delayed oscillator equation $\delta \dot{T} = (1 - 3T^2)\delta T - r\delta T(t - d)$. This suggests that the transitions between the ENSO episodes are the least predictable, at least for the Zebiak–Cane model.

the result of coupled ocean–atmosphere modes, either unstable, or marginally stable, forced by atmospheric stochastic noise due to the atmospheric weather or perhaps the MJO.

Two coordinated experiments with ensembles of global general circulation models were carried out in the late 1990s: PROVOST in Europe and Dynamical Seasonal Prediction (DSP) in the USA, both using "perfect" SSTs. Their results have been presented in a number of papers (Shukla et al., 2000; Fischer and Navarra, 2000; Branković and Palmer, 2000). Kobayashi et al. (2000) present the results obtained with similar experiments in Japan. Volume 126, No 567 of *The Quarterly Journal of the Royal Meteorological Society* was dedicated to this topic and also contains papers on statistical predictions. These papers, and the special issue of the *Journal of Geophysical Research* (Volume 107, Issue C7, 1998) dedicated to the TOGA program contain a wealth of descriptions of the (late 1990s) state-of-the-art understanding and ability to predict atmospheric anomalies beyond 2 weeks.

Several operational and research centers have started issuing seasonal to interannual forecasts based on these ideas (Barnston et al., 1994; Ji et al., 1996; Latif et al., 1994, 1998). At NCEP operational seasonal to interannual forecasts are based on coupled model integrations to predict SST anomalies, followed by ensembles of atmospheric forecasts forced with the predicted SST anomalies in the tropical Pacific, and with statistical predictions of SST in other oceans. NCEP also uses statistical prediction schemes (e.g., Van den Dool, 1994). The final "official forecast' is subjectively determined from both the dynamical and the statistical predictions. The ECMWF predictions are computed with coupled global ocean–atmosphere models, and run every week for 6 months. Several of these forecasts are available on the web (www.cpc.ncep .noaa.gov/, www.ecmwf.int/en/forecasts/documentation-and-support/long-range, and others).

In addition to the coupling of the atmosphere with the ocean, it is possible to have extended regional predictability from the coupling of the land and the atmosphere. The positive feedback within this coupling can be quite large (e.g., low precipitation results in low soil moisture, and this anomaly, in turn, reduces evaporation and precipitation during the spring and summer months). In subtropical regions associated with strong gradients in precipitation this mechanism can lead to long-lasting anomalies as large or larger than those due to SST anomalies (Koster et al., 2000). Because of this, it should be in principle possible to predict the long-lasting nature of these anomalies for several months (e.g., Atlas et al., 1993; Hong and Kalnay, 2000).

6.8   Decadal Variability and Climate Change

We conclude this chapter by pointing out that in addition to interannual variability, there is also considerable climate variability in the decadal and longer time scales. Climate variability may be due to either natural causes or to long-term changes that can be attributed to anthropogenic sources of pollution or changes in the land surface. The impact of mankind on our environment (usually referred to as "global warming")

6.8 Decadal Variability and Climate Change

is quite complex. Among the clearest examples of human impacts are the observed decreases of total ozone in Antarctic regions, and more recently in the Arctic regions, and the increase of CO_2 and other greenhouse gases to levels much higher than those reached in the past. The changes and expected impacts on climate change in the next decades have been reviewed by the IPCC, a body of experts from many countries, which has in 2001 issued its Third Assessment Report (Intergovernmental Panel on Climate Change, 2001).

Plate 4 shows the variations of the Earth's surface temperature over the last 140 years and over the last millennium. It is clear that there is climate variability at many time scales, and that the variability observed before the Industrial Revolution that took place last century is of natural and not of anthropogenic origin. Among the natural oscillations that would take place even in the absence of human forcings on long time scales are the North Atlantic Oscillation (NAO), associated with the Arctic Oscillation (AO), the Pacific Decadal Oscillation (PDO), the Antarctic Oscillation, and the Atlantic Subtropical Dipole.

We have seen that the variability associated with El Niño (ENSO) has a limit of predictability of the order of a few years, because the oscillations are the result of the chaotic dynamics of the coupled ocean–atmosphere with time scales of several years. Because of the long oceanic time scales associated with slow transports, it is possible that long-term coupling with the oceans may dominate decadal variability in, for example, the NAO. If this is true, then the NAO may be somewhat predictable, and its prediction will depend on the specification of the initial conditions of the ocean, and the ability of the models to reproduce the relevant physical mechanisms that dominate its coupled evolution. Rapid climate changes associated with transitions into ice ages may be completely unpredictable since they may be the result of unpredictable small changes, such as volcanic activity, resulting in major climate shifts.

Climate change of human origin is also predictable, but in a different sense than ENSO variability, which depends on the initial conditions of the coupled ocean–atmosphere. Anthropogenic climate change depends on an "external" forcing, such as the increase in greenhouse gases, rather than on the internal chaotic dynamics and the initial conditions. When the external forcing is known, the forced response of the climate change can be "predicted" fairly well with present-day models. Figure 6.8.1 shows the response of the global surface temperature to natural forcings (solar variations and volcanic activity), anthropogenic forcings (greenhouse gases and sulfate aerosols) and to both natural and anthropogenic forcings. It shows that some climate models are able to reproduce quite well the large-scale response to volcanic eruptions, and that the addition of greenhouse gas forcings results in fairly good agreement with observations. The fact that different climate models reproduce the global scale of the impact of increased greenhouse gases in a similar way suggests that their effects are to some extent predictable if we can predict the human forcing and its feedbacks on the environmental system. The impacts of climate change on regional scales are much harder to reproduce, because of the local influence of chaotic weather and climate dynamics. We discuss these frontiers of Earth system modeling and prediction in the next three sections.

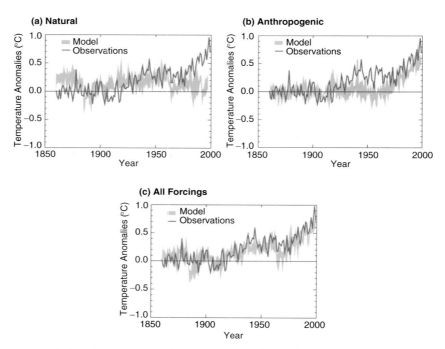

Figure 6.8.1 Simulation of the evolution of the annual global mean surface temperature and comparison with observations for different external forcings. The gray represents the range of four different runs with the same model. (a) Only natural forcings (solar variations and volcanic activity), (b) only anthropogenic forcings (greenhouse gases and an estimate of sulfate aerosols), (c) both. (From Intergovernmental Panel on Climate Change, 2001.)

6.9 Historical Development of Earth System Models: Progressive Coupling of New Components

The development of Earth system modeling throughout history demonstrates the necessity of bidirectional coupling between the subsystems within the model. In the 1960s, atmospheric scientists initiated the creation of mathematical models to explain the dynamics of the planetary climate system. Initially, these models were limited to the atmospheric system coupled with simple surface models. Later on, additional subsystems such as the ocean, land, sea-ice, clouds, vegetation, carbon, aerosols, and other chemical constituents were integrated in stages to make Earth system models (ESMs) more comprehensive (Figure 6.9.1). To ensure that the subsystems were accurately coupled, bidirectional feedbacks were progressively incorporated between them, as proposed by Manabe et al. in 1965.

The importance of implementing bidirectional coupling is demonstrated by the modeling of El Niño–Southern Oscillation (ENSO), which is a result of the coupled dynamics of the tropical ocean and atmosphere subsystems in the actual climate system, as discussed below. As an example, prior attempts to forecast El Niño, which is created by the interplay and feedbacks between the tropical atmosphere and ocean, had not been successful until Cane et al. (1986) and Zebiak and Cane (1987) developed

6.10 Domination of the Climate System by the Human System

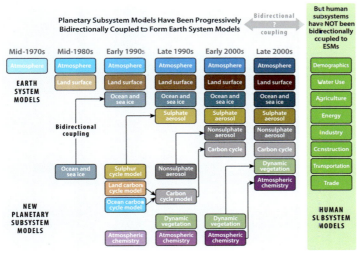

Figure 6.9.1 Progressive bidirectional coupling of new components in Earth system models from 1970s until present. However, the Bidirectional coupling with human subsystems is still missing (Mote et al., 2020). (The Earth system side of this figure was adapted from the Intergovernmental Panel on Climate Change (IPCC).)

bidirectionally coupled models of the tropical atmosphere and ocean. They were able to successfully predict the 1986 El Niño. Before that, atmospheric models and ocean models were only coupled through one-way interactions. The atmospheric models were influenced by sea surface temperatures (SSTs) but could not reciprocally change them. Similarly, the ocean models were impacted by atmospheric wind stress and surface heat fluxes, but could not alter them in turn. These "one-way" couplings were not capable of representing the positive, negative, and delayed feedbacks present in the real climate system that generate the El Niño phenomenon. The first prototype of the bidirectionally coupled ocean–atmosphere model was created by Zebiak and Cane (1987), which enabled the prediction of El Niño episodes several seasons in advance (Cane et al., 1986). Similarly, in order to advance the modeling and prediction of droughts, bidirectional coupling of the land and atmosphere subsystems was necessary (Koster et al., 2009). Currently, most global climate models implement bidirectionally coupled atmosphere–ocean–land–ice submodels, which is critical to accurately model climate change. The progression of coupled models reveals that it is crucial to include bidirectional feedbacks between the various components of the actual systems that the models represent, as important outcomes will be missed otherwise.

6.10 Domination of the Climate System by the Human System

Similarly, the Earth system cannot be realistically modeled without bidirectionally coupling it with the human system, which has become the main driver of climate

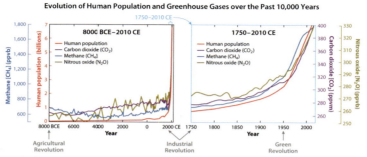

The abrupt and simultaneous upward trajectories of human population and greenhouse gases after the start of the Industrial Revolution (~1750), and the distinct acceleration after the start of the Green Revolution (~1950), show that the Human System has become the primary driver of these gases and the changes in the Earth System. Adapted from Fu & Li (2016), CC-BY, https://doi.org/10.1093/nsr/nww094.

Figure 6.10.1 We can observe from the evolution of world population and the atmospheric concentration of greenhouse gases over the past 10,000 years that human population has become the driving variable (Mote et al., 2020). Adapted from Fu and Li (2016).

change (Motesharrei et al., 2016). Figure 6.10.1 depicts how the growth of main greenhouse gases is driven by the human system's expansion. It is noteworthy that population growth experienced a sharp increase during the Industrial Revolution (~1750), which marked the large-scale usage of fossil fuels, and an even stronger acceleration with the "Green Revolution" (~1950) when major fossil fuel usage was introduced in agriculture, as elaborated below.

Although human activity had begun affecting the climate in the pre-Industrial Era via processes like forest clearing, irrigation agriculture, and soil depletion by agricultural societies (Ruddiman, 2003, 2005), the exploitation of fossil fuels at a massive scale during the Industrial Era has had a considerable impact on climate change. The human system's impact on the planet is not limited to resource consumption. Waste generated by the human system must also return to the Earth system. The human system's outputs (pollution) into the Earth system, that is, vast greenhouse gas emissions [mainly carbon dioxide (CO_2), methane (CH_4), and nitrous oxide (N_2O)] from fossil fuel consumption, are evident in Figure 6.10.1. The enormity of these emissions for the Earth system can be understood by the fact that it took nature hundreds of millions of years to store energy from the Sun by converting atmospheric stocks of carbon into underground fossil fuel stocks. However, humanity is now consuming these stocks and thus releasing the stored carbon back into the atmosphere within just a few hundred years. Therefore, in our current Fossil Fuel Era, humanity is returning those carbon stocks stored by nature back into the atmosphere at a rate approximately one million times faster than it took nature to store them away. This rapid depletion of nature's fossil fuel reserves is causing a significant change in atmospheric and oceanic carbon, modifying the composition of carbon in the atmosphere and oceans at a rate previously unseen in nature before industrial civilization.

This rapid consumption of fossil fuels and emission of greenhouse gasses has not only driven climate change but also led to the rapid growth of the human system, resulting in impacts on various planetary subsystems beyond climate change. Indeed,

6.10 Domination of the Climate System by the Human System

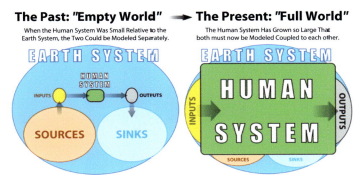

Figure 6.10.2 Growth of the human system has changed its relationship with the Earth system, and thus both must be modeled interactively to account for their feedback on each other (Mote et al., 2020).

human system forcing of the Earth system has accelerated continuously since the 1950s without much slowing (Steffen et al., 2006, 2015). The size of the human system has shifted from a minor influence on these subsystems to being the primary driver, posing a threat to the critical functions and ecosystem services of the Earth system. Motesharrei et al. (2016) provides numerous examples of the human system's dominance in various subsystems of the Earth system.

To summarize, the Earth system and the human system are bidirectionally coupled, with feedbacks. The real relationship between the human and the Earth systems can be schematically described as follows (Figure 6.10.2, Figure 5 from Mote et al., 2020): The human system is **within** the Earth system. The Earth system sources provide the Inputs needed by the human system. Then, the human system *Ouputs* (pollution, garbage) must be *absorbed* by the Earth system *Sinks*, such as the Atmosphere, Land and Ocean. However, in Climate Modeling, we do not have a representation of the human system, bidirectionally coupled with the Earth system, we just have the representation of the Earth system. So, we need to develop bidirectionally coupled Earth–human system models. This cannot be done with current Earth system models. This is especially important because the human system has become the main driver of change in the Earth system, but their bidirectional coupling is still required to model climate and climate change (Motesharrei et al., 2016). We argue this is also necessary for modeling the human system and carrying capacity (Mote et al., 2020). The Intergovernmental Panel on Climate Change (IPCC) is focused on changes of the Earth system driven by the human system. But it is not possible to make accurate predictions about the Earth system without bidirectional couplings (feedbacks) with the human system.

Current climate models and Earth system models, including the widely used Integrated Assessment Models (IAMs) and IPCC models do not incorporate these critical bidirectional feedbacks.

In particular, key human system variables, such as demographics, economic growth, disease transmission, and migration, are not dynamically driven by changes in the

Earth system models, but instead are exogenously projected into the future, often with statistical models such as UN population projections (estimated tables). Therefore currently existing models cannot capture the critical dynamic interactions between the human system and the Earth system.

6.11 Developing Data Assimilation Methods for Improving Human System Modeling

To address such shortcomings, we developed a coupled human and nature dynamics model, abbreviated as HANDY (Motesharrei et al., 2014). HANDY is a minimal model that for the first time modeled inequality dynamically. As a result, it became the first model to have irreversible collapses as a solution, as observed in history after the agricultural revolution about ten thousand years ago until the industrial revolution around 1750. HANDY has four state variables: population, divided into separate variables elites and commoners, nature, which represents regenerating natural resources, and Wealth. Commoners deplete nature to produce Wealth, which is consumed disproportionately by elites and commoners. Rapid growth of population and consumption leads to rapid depletion of nature. Resource scarcity eventually results in decline of Wealth but there is a time lag. Add one or a few sentences describing the HANDY minimal model.

The HANDY model shows that both economic inequality and resource overdepletion can lead to collapse. This result agrees with the historical evidence. The HANDY paper included numerical experiments for different types of societies. They showed that if total consumption does not surpass the carrying capacity by too much, reaching a sustainable state is feasible. But if the overshoot is too large, it will be hard to prevent collapse. Modern society has been able to grow far beyond Earth's carrying capacity because of abundant use of nonrenewable resources such as fossil fuels and groundwater, which is fossil water. However, the model indicates that an unsustainable scenario can be made sustainable by reducing per capita depletion and consumption rates, reducing inequality to decrease excessive consumption by the wealthiest, and reducing birth rates to slow population growth and stabilize population. The key question is whether these changes can be made in time. Designing the best policies to achieve this sustainable path requires modeling the Earth system and the human system bidirectionally coupled to each other.

HANDY provides a minimal example of a coupled Earth–human system model. There are many challenges to building these coupled models, as we discuss in Motesharrei et al. (2016). The first ingredient required for prediction is building reliable dynamic models that reflect the real world. Building good models requires experience with translating natural processes into mathematical equations, with attention to key variables and especially feedbacks among them.

Even if the human system is bidirectionally coupled to the Earth model in the coupled Earth–human system model, estimating the model states of the human system and the key parameters controlling the human processes and mechanisms will still be challenging, due to the less developed modeling of the human system processes and the less reliable nature of the observations of its variables. We can take advantage of data assimilation, especially those that can better handle non-Gaussian processes (e.g., particle filter), to infer the model states and parameters of the human system, while Earth system observations (impacted by the human system) could be used to improve the estimation of these human system model states and parameters. In the framework of Ensemble Strongly–Coupled Data Assimilation (EnSCDA, Kalnay et al., 2023), we can assimilate those more reliable observations related to different earth components (e.g., land), which are affected by the human system and thus show error correlations to the human system, to estimate the model states and parameters related to the human system more accurately. Besides, the ensemble data assimilation also provides us with uncertainties about those estimations.

Here we illustrate some examples about how to use DA methods for human system modeling.

To identify the key variables for the human system, we propose using sensitivity analysis methods from data assimilation, such as adjoint methods and ensemble sensitivity methods. To identify the observations that are able to constrain human system processes, we propose using Observing System Simulation Experiments (OSSEs, Atlas et al., 1985; Ross and Atlas, 2016) and ensemble forecast sensitivity to observations (EFSO, Kalnay et al., 2012, described in Section 5.6.7). Another challenge is uncertainty in parameters, so we propose using methods of Data Assimilation to overcome many of such challenges in parameter estimation (e.g., Liu et al., 2014a,b). For certain parameters, especially when they are related to processes that are poorly characterized, we propose using machine learning and deep learning. Finally, there are large uncertainties in these complex, coupled Earth–human systems. We propose using ensembles of models, and ensembles of runs for each model to quantify these uncertainties.

To summarize, we propose four avenues of Data Assimilation to improve modeling and prediction of the Earth system by coupling human system models:

- **Parameter estimation**. Besides challenges with model structure and identifying key dynamical variables, there are large uncertainties in determining values of parameters, especially those related to the human systems. Some of these parameters can be estimated from available empirical or historical data, but for many parameters a direct measurement is not available. Therefore, such parameters must be estimated indirectly, from comparing observations of dynamical or auxiliary variables to model output. To address this problem, we propose borrowing methods of Data Assimilation science. These methods are well developed within the atmospheric and oceanic sciences. Moreover, we can apply the latest methods from machine learning and artificial intelligence to estimate these hard-to-capture model parameters.
- **Identifying key human system variables**. A major challenge for coupled Earth–human systems is determining variables that will be represented in a model.

Including too many variables might make a model more accurate but will make it more complicated and including some variables often have a small marginal benefit. In order to have the right balance of simplicity and accuracy, we can take advantage of methods of sensitivity analysis. such as ensemble sensitivity or adjoint methods.

- **Identifying effective human system observations**. Once we identify key human system variables, it is critical to determine which observations are most effective to constrain those variables. Some of those key human system variables might not be directly observable, but could instead be estimated through indirect observations that are functions of multiple human system variables or functions of multiple human and Earth system variables. We propose using OSSE and EFSO to provide guidance for building better observing systems.
- **Quantifying uncertainties**. A main goal of modeling systems is to make predictions or forecasts for their future dynamics. Given the less developed modeling of the dynamics, limited capacity to represent the known dynamics due to computing constraints, as well as the more limited availability of observations, we often expect predictions to entail large uncertainties. Therefore, we propose two methods to represent and quantify uncertainties. First, running ensembles of different models allows for capturing the variability of representation of system structure and dynamics. Second, producing an ensemble of runs with different initial or boundary conditions allows estimating uncertainties arising due to nonlinearities and chaotic nature of the systems. Together, these two types of ensemble runs can help to constrain the uncertainty of predictions, and play a major role in decision making.

6.12   Controlling Chaos in Control Simulation Experiments

From the discussion of previous sections in Chapter 6, we know that the atmosphere is a chaotic system with limited predictability for phenomena with varying temporal and spatial scales. To make effective weather or climate predictions, we need to conduct data assimilation continually (Chapter 5) so that we can acquire initial conditions close to the reality as much as possible (e.g., see Penny (2017) about how to incorporate data assimilation to calculate Lyapunov exponents for dynamical systems). After decades of advances on data assimilation, observations, and numerical weather prediction models (Chapter 1), we can now make relatively accurate short-range predictions, at least for some atmospheric variables (e.g., Figure 1.4.5).

Given our ability to make relatively accurate short-range forecasts, Miyoshi and Sun (2022) proposed to modify the weather for disaster mitigation by taking advantage of chaotic behavior of the atmosphere, because the intrinsic chaotic nature of the atmosphere not only reveals limited predictability, but also implies the possibility of modifying the atmospheric evolution by imposing small perturbations on it. Previous attempts on weather modification have been made on rain enhancement through cloud seeding, and large-scale geoengineering. For example, Li et al. (2018) demonstrated

6.12 Controlling Chaos in Control Simulation Experiments

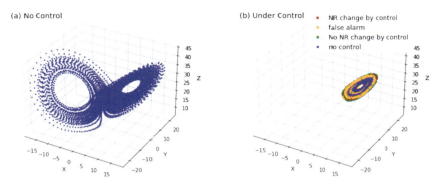

Figure 6.12.1 Phase space of the three-variable Lorenz model. (a) Lorenz's butterfly attractor from the NR without control; (b) the NR under control. Each dot shows every time step for 8000 steps (Miyoshi and Sun, 2022).

with a coupled general circulation model that deploying large-scale wind and solar farms in the Sahara can significantly increase the precipitation and vegetation there. Miyoshi and Sun's approach focuses on changing the short-term weather within its natural variability by utilizing the properties of chaotic systems.

Specifically, Miyoshi and Sun (2022) proposed a Control Simulation Experiment (CSE) for weather modification. This CSE can be viewed as an extension from the Observing Systems Simulation Experiments (OSSEs, Atlas et al., 1985; Ross and Atlas, 2016). In an OSSE, an independent model run serves as the nature run (NR), or the truth. Synthetic observations are then created by adding the prescribed observation errors to the simulated observational quantities using the NR. We then conduct assimilation experiments with initial conditions different from the NR. The accuracy of the analysis is then evaluated by comparing it to the NR. With OSSEs, the impact of new data assimilation methods or observing systems can be assessed by evaluating the variation of analysis accuracy introduced by those new components. In OSSEs, the NR is never altered by any components of the experiments. CSE goes one step further than OSSE by determining when and what perturbations to add to the NR so that the NR can evolve toward our expectations, and then adding those perturbations to the NR.

Miyoshi and Sun (2022) demonstrated the effectiveness of CSEs with the Lorenz 63 model by limiting the model trajectory to stay only in one wing of the "butterfly" attractor (Figure 6.12.1) Following the OSSEs conducted by Kalnay et al. (2007a) and Yang et al. (2012), Miyoshi and Sun (2022) adopted these additional steps to prevent regime shift:

1. Get the analysis ensemble by conducting DA at time step t.
2. Run ensemble forecasts initialized from the analysis ensemble for T steps (i.e., from time step t to $t + T$).
3. If the T-step forecast from at least one member shows the regime shift, then activate the control (step 4); otherwise, return to step 1 for the next DA at time $t + T_a$.

4. Deploy control by adding perturbations with Euclidean norm D to the NR for time steps $t + i$ ($i = 1, \ldots, T_a - 1$), where $D = \sqrt{dx^2 + dy^2 + dz^2}$.
5. At time $t + T_a$, the new NR is used to simulate the observations; go to step 1 for the next DA at time $t + T_a$.

The key step here is Step 4, because it requires adding perturbations that can prevent the NR from shifting its regime. Not surprisingly, Miyoshi and Sun (2022) found that random perturbations cannot achieve this goal. Instead, they adopted the following strategy: identify the member S that changes regime in its T-step forecast, and member N that stays the same regime. If all members shift regimes in this DA cycle, they will choose the member N from the previous DA cycle whose extended forecast shows no regime shift during the period from t to $t + T$. After acquiring members S and N, they rescaled the difference S-N for every time step from $t + 1$ to $t + T_a - 1$, and added these rescaled perturbations to the NR at the corresponding time steps. This strategy of control successfully prevented the regime shift (Figure 6.12.1(b)). Additional sensitivity experiments on parameters T and D show that control is more effective with a T longer than the mean transition time for the regime shift, and D that is neither too small nor too large ($0.02 < D < 0.5$). Interestingly, although only assimilating z is not sufficient to constrain the model states for DA, it is possible to control the NR by only perturbing z alone, with a perturbation magnitude of merely 3% of the observation error standard deviation.

As the frequency and intensity of extreme weather events increases (Knutson et al., 2020; Fischer et al., 2021), it becomes imperative to seek solutions to curb the impact of such events. The CSEs suggested by Miyoshi and Sun (2022) is a novel approach that has the potential to address this important problem. More research is needed to apply CSEs to various extreme weather scenarios and assess CSEs potential impact under each scenario. Once the scientific community acquires a clearer understanding of these experiments, we will be able to deploy these solutions in reality and hopefully make significant reduction to the harms of extreme weather events.

Appendix A Coding and Checking the Tangent Linear and the Adjoint Models

We have seen in Chapters 5 and 6 that given a nonlinear model $\mathbf{x}(t) = \mathcal{M}[\mathbf{x}(t_0)]$ integrated between t_0 and t, if we introduce a perturbation in the initial conditions $\delta\mathbf{x}(t_0)$, neglecting terms of order $O[\delta\mathbf{x}(t_0)]^2$, then

$$\mathbf{x}(t) + \delta\mathbf{x}(t) = \mathcal{M}[\mathbf{x}(t_0) + \delta\mathbf{x}(t_0)] \approx \mathcal{M}(\mathbf{x}(t_0)) + \frac{\partial\mathcal{M}}{\partial\mathbf{x}}\delta\mathbf{x}(t_0) \qquad (A.1.1)$$

so that the initial perturbation evolves like

$$\delta\mathbf{x}(t) \approx \mathbf{L}\delta\mathbf{x}(t_0) \qquad (A.1.2)$$

The Jacobian $\mathbf{L} = \partial\mathcal{M}/\partial\mathbf{x}$ is the tangent linear model that propagates the perturbation from t_0 to t. It is a matrix that for nonlinear models depends on the basic solution $\mathbf{x}(t)$. If there are n time steps between t_0 and $t_n = t$, this matrix is equal to the product of matrices corresponding to each time step:

$$\mathbf{L}(t_0, t_n) = \mathbf{L}(t_{n-1}, t_n)\ldots\mathbf{L}(t_1, t_2)\mathbf{L}(t_0, t_1) \qquad (A.1.3)$$

The adjoint model $\mathbf{L}^T$ is frequently introduced in the context of 4D-Var (Chapter 5), with a cost function measuring the misfit of the model solution to observations:

$$J(\mathbf{x}(t_0)) = \frac{1}{2}\sum_{i=1}^{N}\{\mathcal{H}[\mathbf{x}(t_i)] - \mathbf{y}_i^o\}^T \mathbf{R}_i^{-1}\{\mathcal{H}[\mathbf{x}(t_i)] - \mathbf{y}_i^o\} \qquad (A.1.4)$$

Here the observation error covariance $\mathbf{R}_i$ at a given time t_i is assumed to be symmetric. The *control* variables (which we vary in order to find the minimum of the cost function) are the initial conditions $\mathbf{x}(t_0) = \mathbf{x}_0$. If we take an increment in the initial condition $\delta\mathbf{x}(t_0)$, then

$$\delta J = (\nabla_{\mathbf{x}}J(\mathbf{x_0}), \delta\mathbf{x}(t_0)) = \sum_{i=0}^{N}(\mathbf{R}_i^{-1}\{\mathcal{H}[\mathbf{x}(t_i)] - \mathbf{y}_i^o\})^T \mathbf{H}[\delta\mathbf{x}(t_i)] \qquad (A.1.5)$$

Now, at a time t_i

$$\delta\mathbf{x}_i = \mathbf{L}(t_0, t_1)\delta\mathbf{x}(t_0) \qquad (A.1.6)$$

Therefore,

$$\delta J = [\nabla_{\mathbf{x}}J(\mathbf{x_0})]^T \delta\mathbf{x}(t_0) = \sum_{i=0}^{N}(\mathbf{L}^T\mathbf{H}^T\mathbf{R}_i^{-1}\{\mathcal{H}[\mathbf{x}(t_i)] - \mathbf{y}_i^o\})^T \delta\mathbf{x}(t_0) \qquad (A.1.7)$$

and the gradient of the cost function with respect to the initial conditions is

$$\nabla_{\mathbf{x}} J(\mathbf{x_0}) = \sum_{i=0}^{N} \mathbf{L}^T(t_i,t_0)(\mathbf{H}^T \mathbf{R}_i^{-1}\{\mathcal{H}[\mathbf{x}(t_i)] - \mathbf{y}_i^o\}) \qquad (A.1.8)$$

As we saw in Chapters 5 and 6, the transpose $\mathbf{L}^T(t_i,t_0)$ of the matrix of the tangent linear model is the adjoint model. If the tangent linear model matrix is complex, then the adjoint is the complex conjugate of the transpose of the tangent linear model. The observational increments are "forcings" of the adjoint model in computing the gradient of the cost function for 4D-Var.

In addition to 4D-Var (Chapter 5), the adjoint model has several other important applications such as computing the vectors that grow fastest in a period of optimization. These are the leading singular vectors, i.e., the leading eigenvectors of $\mathbf{L}^T\mathbf{L}$ (Chapter 6). The adjoint of a model can also be used to find optimal parameters in a model, e.g., the diffusion coefficient that produces forecasts closest to a verification field (Cacuci, 1981; Zou et al., 1992). All these applications require the definition of a norm, with respect to which the gradient is computed, and the choice of the appropriate control variables. In the case of singular vectors and 4D-Var, the control variables are the vectors of initial conditions; in the problem of finding an optimal parameter, the control variable is the parameter itself.

We now discuss briefly the rules for generating adjoint codes. More detailed discussions are available in Talagrand (1991), Talagrand and Courtier (1987), Navon et al. (1992), Yang and Navon (1996), and Giering and Kaminski (1998). There are presently two compilers available for the automatic generation of tangent linear and adjoint codes, given the FORTRAN code of the forward nonlinear model. (Odyssee, Rostaing et al., 1993, and the Tangent and Adjoint Model Compiler (TAMC), Giering, 1995, Giering and Kaminski, 1998, Tapenade, Hascoët and Pascual, 2013). Tapenade is available on the web (http://tapenade.inria.fr:8080/tapenade/index.jsp).

Taking the transpose of (A.1.3), we obtain that the adjoint model is

$$\mathbf{L}^T(t_n,t_0) = \mathbf{L}^T(t_1,t_0) \ldots \mathbf{L}^T(t_{n-1},t_{n-2})\mathbf{L}^T(t_n,t_{n-1}) \qquad (A.1.9)$$

Within a time step, the TLM code is also composed by a number of subcodes or steps applied in succession:

$$\mathbf{L}^T(t_{i-1},t_i) = \mathbf{S}_1 \ldots \mathbf{S}_{m-1}\mathbf{S}_m \qquad (A.1.10)$$

For example S_1 could be initialization of the new time step, S_2-S_7 could be the computation of the time tendencies coming from horizontal advection, vertical advection, convection, large-scale precipitation, S_8 radiation, etc., and S_9 the update of the model variables for the next time step. Each of these steps may contain several DO loops.

In constructing the adjoint, because of the transposition, these steps (and the DO loops within the step) are also reversed:

$$\mathbf{L}^T(t_i,t_{i-1}) = \mathbf{S}_m \ldots \mathbf{S}_2\mathbf{S}_1 \qquad (A.1.11)$$

We give now a simple example to illustrate the rules for constructing the adjoint from a nonlinear forward model. In the construction of the adjoint it is important to determine

Appendix A Coding and Checking the Tangent Linear and the Adjoint Models   285

which are the "active" variables updated in the model. Consider the simple forward model of the diffusion equation (in this case a linear model):

$$\frac{\partial u}{\partial t} = \sigma \frac{\partial^2 u}{\partial x^2} \tag{A.1.12}$$

It would be possible to derive first the analytic adjoint of this equation and then discretize it, but it is preferable to first discretize the original equation, code it, and then create the tangent linear model and adjoint of the code directly. This is because the adjoint of a discretized code is not necessarily identical to the discretization of the adjoint operator. We discretize (A.1.12) using finite differences and a scheme forward in time, centered in space, as discussed in Chapter 3:

$$u_i^{j+1} = u_i^j + \alpha(u_{i+1}^j - 2u_i^j + u_{i-1}^j) \tag{A.1.13}$$

where $\alpha = \sigma \Delta t / (\Delta x)^2$, $x_i = i\Delta x, i = 1, \ldots, I$, and $t_j = j\Delta t$. If we assume $u = u_b + \delta u$, where $u_b(t)$ is the basic solution, then the tangent linear model is

$$\begin{aligned} \delta u_i^{j+1} &= \delta u_i^j + \alpha(\delta u_{i+1}^j - 2\delta u_i^j + \delta u_{i-1}^j) \\ &= (1 - 2\alpha)\delta u_i^j + \alpha\delta u_{i+1}^j + \alpha\delta u_{i-1}^j \end{aligned} \tag{A.1.14}$$

or in matrix form

$$\delta u_i^{j+1} = \begin{pmatrix} 1 - 2\alpha & \alpha & \alpha \end{pmatrix} \begin{pmatrix} \delta u_i^j \\ \delta u_{i-1}^j \\ \delta u_{i+1}^j \end{pmatrix} \quad i = 1, \ldots, I \tag{A.1.15}$$

In this computation, there are four "active" variables δu_i^{j+1}, δu_i^j, δu_{i-1}^j, and δu_{i+1}^j but only one of them has been modified. In preparation for the computation of the adjoint, it is necessary to indicate explicitly that the other three active variables are *not* modified. The tangent linear model step is then

$$\begin{pmatrix} \delta u_{i-1}^j \\ \delta u_i^j \\ \delta u_{i+1}^j \\ \delta u_i^{j+1} \end{pmatrix} = \begin{pmatrix} 1 & 0 & 0 & 0 \\ 0 & 1 & 0 & 0 \\ 0 & 0 & 1 & 0 \\ \alpha & 1-2\alpha & \alpha & 0 \end{pmatrix} \begin{pmatrix} \delta u_{i-1}^j \\ \delta u_i^j \\ \delta u_{i+1}^j \\ \delta u_i^{j+1} \end{pmatrix} \quad i = 1, \ldots, I \tag{A.1.16}$$

Equation (A.1.16) is the same as (A.1.15) but includes the additional identities (no modification) needed for the adjoint model. The adjoint model is the transpose of the tangent linear model matrix acting on the adjoint variables, so that this step becomes

$$\begin{pmatrix} \delta^* u_{i-1}^j \\ \delta^* u_i^j \\ \delta^* u_{i+1}^j \\ \delta^* u_i^{j+1} \end{pmatrix} = \begin{pmatrix} 1 & 0 & 0 & \alpha \\ 0 & 1 & 0 & 1-2\alpha \\ 0 & 0 & 1 & \alpha \\ 0 & 0 & 0 & 0 \end{pmatrix} \begin{pmatrix} \delta^* u_{i-1}^j \\ \delta^* u_i^j \\ \delta^* u_{i+1}^j \\ \delta^* u_i^{j+1} \end{pmatrix} \quad i = 1, \ldots, I \tag{A.1.17}$$

where the stars represent adjoint variables.

Appendix A Coding and Checking the Tangent Linear and the Adjoint Models

This can be written line by line as

$$
\left.
\begin{array}{l}
\delta^* u^j_{i-1} = \delta^* u^j_{i-1} + \alpha \delta^* u^{j+1}_i \\[4pt]
\delta^* u^j_i = \delta^* u^j_i + (1 - 2\alpha)\delta^* u^{j+1}_i \\[4pt]
\delta^* u^j_{i+1} = \delta^* u^j_{i+1} + \alpha \delta^* u^{j+1}_i \\[4pt]
\delta^* u^{j+1}_i = 0
\end{array}
\quad i = 1,\ldots,1
\right\}
\tag{A.1.18}
$$

Note that the equation for the adjoint variable on the left-hand side of (A.1.15) is executed last.

A second example is adapted from Giering and Kaminski (1998). Consider the nonlinear nth step of the following algorithm:

$$
z^n = x^{n-1} \sin[(y^{n-1})^2]
\tag{A.1.19}
$$

where x, y, z are active variables. Using the chain rule, the tangent linear algorithm for this step is

$$
\delta z^n = \sin[(y^{n-1})^2]\delta x^{n-1} + x^{n-1} \cos[(y^{n-1})^2]2y^{n-1}\delta y^{n-1}
\tag{A.1.20}
$$

or in matrix form

$$
\begin{pmatrix} \delta x \\ \delta y \\ \delta z \end{pmatrix}^n =
\begin{pmatrix}
1 & 0 & 0 \\
0 & 1 & 0 \\
\sin[(y^{n-1})^2] & x^{n-1} \cos[(y^{n-1})^2]2y^{n-1} & 0
\end{pmatrix}
\begin{pmatrix} \delta x \\ \delta y \\ \delta z \end{pmatrix}^{n-1}
\tag{A.1.21}
$$

The adjoint operator is the transposed matrix of (A.1.21) acting on the adjoint variables:

$$
\begin{pmatrix} \delta x \\ \delta y \\ \delta z \end{pmatrix}^{n-1} =
\begin{pmatrix}
1 & 0 & \sin[(y^{n-1})^2] \\
0 & 1 & x^{n-1} \cos[(y^{n-1})^2]2y^{n-1} \\
0 & 0 & 0
\end{pmatrix}
\begin{pmatrix} \delta x \\ \delta y \\ \delta z \end{pmatrix}^{n}
\tag{A.1.22}
$$

The FORTRAN statement for the nonlinear forward step is:
```
Z=X*SIN(Y**2)
```
The TLM FORTRAN statement is
```
DZ=(SIN(Y**2))*DX+(X*COS(Y**2)*2*Y)*DY,
```
and the adjoint FORTRAN statements for this step are
```
ADX=ADX+ (SIN(Y**2))*ADZ
ADY=ADY+ (X*COS(Y**2)*2*Y)*ADZ
ADZ=0.0
```

A.1 Verification

Finally, we discuss how to verify the correctness of the tangent linear and adjoint codes (Navon et al., 1992). The verification of the tangent linear model is straight forward: For small increments in the initial conditions, the tangent linear model should reproduce the difference between two nonlinear integrations with quadratic errors:

$$
\delta x(t) = \mathbf{L}\delta x(0) = \mathcal{M}[x_0 + \delta x(0)] - \mathcal{M}(x_0) + O(||\delta x||^2)
\tag{A.1.23}
$$

A.2 Example of FORTRAN Code

Therefore, Navon et al. (1992) suggested computing the pattern correlation of the left- and right-hand sides of (A.1.23), as well as the relative error, using an appropriate norm. For sufficiently small perturbation amplitudes, the relative error should be proportional to the amplitude of the initial perturbation.

To verify the correctness of the adjoint code, Navon et al. (1992) used the identity

$$(\mathbf{L}\delta x_0)^T (\mathbf{L}\delta x_0) = (\delta x)^T \mathbf{L}^T (\mathbf{L}\delta x_0) \tag{A.1.24}$$

This check can be applied to every single subroutine or DO loop:

$$(AQ)^T (AQ) = Q^{*T} [A^T (AQ)] \tag{A.1.25}$$

Here Q represents the input of the original code, A represents either a single DO loop or a subroutine. The left-hand side involves only the TLM, whereas the right-hand side also involves the adjoint code. In practice, if the adjoint code is correct with respect to the tangent linear model, the identity (A.1.25) holds true up to machine accuracy.

A.2 Example of FORTRAN Code

The codes for a complete model for the nonlinear Burgers equation, its tangent linear model and adjoint model, kindly provided by Seon Ki Park, are presented at the end of this appendix. The forward model uses the leapfrog scheme for the advection term and the DuFort–Frankel scheme for the diffusion. The continuous Burgers equation is

$$\frac{\partial u}{\partial t} = -u \frac{\partial u}{\partial x} + \frac{1}{R} \frac{\partial^2 u}{\partial x^2} \tag{A.2.1}$$

and the finite differences used in the code start with a single forward step followed by the leapfrog/DuFort–Frankel scheme:

$$\begin{aligned}
u_i^{j+1} = u_i^{j-1} &- \frac{\Delta t}{\Delta x} [u_i^j (u_{i+1}^j - u_{i-1}^j)] \\
&+ \frac{1}{R} \frac{2\Delta t}{\Delta x^2} [u_{i+1}^j - (u_i^{j+1} - u_i^{j-1}) + u_i^j]
\end{aligned} \tag{A.2.2}$$

Here the index j represents the time step and i represents space (x). Note that in the adjoint code, the order of the substeps and all the DO loops is reversed.

```fortran
1   !>@file m_burger
2   !>@brief Module that contains FWD,TLM and ADJ for 1D burger equations.
3   !
4   !>@details Module that contains FWD,TLM and ADJ for 1D burger equations.
5   !         the orignal author is Seon Ki Park (03/24/98). Updated to the
6   !         Fortran-90 style.
7   module m_burger
8     implicit none
9
10    private
11    public :: burger, burger_tlm, burger_adj
12
```

Appendix A Coding and Checking the Tangent Linear and the Adjoint Models

```fortran
13    real,parameter :: R = 100.d0   !> Reynolds number (reciprocal of
          diffusion coefficient)
14    real,parameter :: dx = 1.0d0    !> space increment
15    real,parameter :: dt = 0.1d0    !> time increment
16    real,parameter :: dtdx = dt/dx
17    real,parameter :: dtdxsq = dt/(dx**2)
18    real,parameter :: c1 = (2.d0/R)*dtdxsq
19    real,parameter :: c0 = 1.d0/(1.d0+c1)
20
21  contains
22
23  subroutine burger(nx,n,ui,ub,uob,u,cost)
24    implicit none
25
26    integer,intent(in)  :: nx         !> # of grid pts
27    integer,intent(in)  :: n          !> # of time steps
28    real,   intent(out) :: cost       !> cost function
29    real,   intent(in)  :: uob(nx,n)  !> observations
30    real,   intent(out) :: ub(nx,n)   !> basic states (u)
31    real,   intent(out) :: u(nx,n)    !> model solutions (u)
32    real,   intent(in)  :: ui(nx)     !> initial conditions
33
34    integer :: i, j
35
36  ! initialize the cost function:
37    cost = 0.d0
38
39  ! set the initial conditions:
40    do i = 1, nx
41       u(i,1) = ui(i)
42    enddo
43
44  ! set the boundary conditions:
45    do j = 1, n
46       u(1, j) = 0.d0
47       u(nx,j) = 0.d0
48    enddo
49
50  ! integrate the model numerically:
51  ! FTCS for the 1st time step integration
52    do i = 2, nx-1
53       u(i,2) = u(i,1) - 0.5d0*dtdx*u(i,1)*(u(i+1,1)-u(i-1,1)) &
54                       + 0.5d0*c1*(u(i+1,1)-2.d0*u(i,1)+u(i-1,1))
55    enddo
56
57  ! Leap/frog/DuFort-Frankel afterwards
58    do j = 3, n
59       do i = 2, nx-1
60          u(i,j) = c0*(u(i,j-2) + c1*(u(i+1,j-1)-u(i,j-2)+u(i-1,j-1)) &
61                      - dtdx*u(i,j-1)*(u(i+1,j-1)-u(i-1,j-1)))
62       enddo
63    enddo
64
65  ! cost function:
66    do j = 1, n
67       do i = 1, nx
68          cost = cost + 0.5*(u(i,j)-uob(i,j))**2
```

A.2 Example of FORTRAN Code 289

```fortran
69          enddo
70       enddo
71
72    ! save nonlinear solutions to the basic fields:
73       do j = 1, n
74          do i = 1, nx
75             ub(i,j) = u(i,j)
76          enddo
77       enddo
78
79    endsubroutine burger
80
81
82    subroutine burger_tlm(nx,n,ui,ubasic,u)
83    !The tangent linear model of the burger's equations
84       implicit none
85
86       integer,intent(in) :: nx
87       integer,intent(in) :: n
88       real ui(nx)          ! initial conditions
89       real ubasic(nx,n)    ! basic states
90       real u(nx,n)         ! TLM solutions
91
92       integer :: i, j
93
94    ! set the initial conditions:
95       do i = 1, nx
96          u(i,1) = ui(i)
97       enddo
98
99    ! set the boundary conditions:
100      do j = 1, n
101         u(1 ,j) = 0.d0
102         u(nx,j) = 0.d0
103      enddo
104
105   ! FTCS for 1st time step integration
106      do i = 2, nx-1
107         !FWD: u(i,2) = u(i,1) - 0.5d0*dtdx*u(i,1)*(u(i+1 1)-u(i-1,1)) &
108         !                    + 0.5d0*c1*(u(i+1,1)-2.d0*u(i,1)+u(i-1,1))
109         u(i,2) = u(i,1) - 0.5d0*dtdx*( u(i,1)*(ubasic(i+1,1)-ubasic(i-1,1)) &
110                           +ubasic(i,1)*(u(i+1,1)-u(i-1,1)) ) &
111                   + 0.5d0*c1*(u(i+1,1)-2.d0*u(i,1) + u(i-1,1))
112      enddo
113
114   ! leap frog/DuFort-Frankel afterwards
115      do j = 3, n
116         do i = 2, nx-1
117            !FWD: u(i,j) = c0*(u(i,j-2) &
118            !                  + c1*(u(i+1,j-1)-u(i,j-2)+u(i-1,j-1)) &
119            !                  - dtdx*u(i,j-1)*(u(i+1,j-1)-u(i-1,j-1)))
120            u(i,j) = c0*( u(i,j-2) + c1*(u(i+1,j-1)-u(i,j-2)+u(i-1,j-1)) &
121                    - dtdx*(u(i,j-1)*(ubasic(i+1,j-1)-ubasic(i-1,j-1)) &
122                    + ubasic(i,j-1)*(u(i+1,j-1)-u(i-1,j-1))) )
123         enddo
124      enddo
125
```

Appendix A Coding and Checking the Tangent Linear and the Adjoint Models

```fortran
126    ! set the final value of u in ui:
127      do i = 1, nx
128        ui(i) = u(i,n)
129      enddo
130
131    endsubroutine burger_tlm
132
133
134    subroutine burger_adj(iforcing, nx, n, ui, ubasic, uob, u)
135    ! The adjoint model of the burger's equations.
136      implicit none
137
138      integer,intent(in)    :: iforcing ! index for forcing in the adjoint
139      integer,intent(in)    :: nx
140      integer,intent(in)    :: n
141      real,    intent(inout) :: ui(nx)         ! initial conditions
142      real,    intent(in)    :: ubasic(nx,n) ! basic states
143      real,    intent(in)    :: uob(nx,n)    ! observations
144      real,    intent(inout) :: u(nx,n)        ! model solutions
145
146      integer :: i, j
147
148    ! initialie adjoint variables:
149      do j = 1, n
150        do i = 1, nx
151          u(i,j) = 0.d0
152        enddo
153      enddo
154
155    ! set the final conditions:
156      if (iforcing .eq. 0) then
157        do i = 1, nx
158          u(i,n) = ui(i)
159          ui(i)  = 0.d0
160        enddo
161      else  ! add cost function part as a forcing
162        do i = 1, nx
163          u(i,n) = ubasic(i,n) - uob(i,n)
164          ui(i)  = 0.d0
165        enddo
166      endif
167
168    ! adjoint of Leap Frog/Dufort-Frankel
169      do j = n, 3, -1
170        do i = nx-1, 2, -1
171          !TL: u(i,j) = c0*(u(i,j-2) + c1*(u(i+1,j-1)-u(i,j-2)+u(i-1,j-1)) &
172          !              - dtdx*(u(i,j-1)*(ubasic(i+1,j-1)-ubasic(i-1,j-1)) &
173          !              + ubasic(i,j-1)*(u(i+1,j-1)-u(i-1,j-1))))
174          u(i-1,j-1) = u(i-1,j-1) + c0*(c1+dtdx*ubasic(i,j-1))*u(i,j)
175          u(i+1,j-1) = u(i+1,j-1) + c0*(c1-dtdx*ubasic(i,j-1))*u(i,j)
176          u(i,j-1) = u(i,j-1) - c0*dtdx*(ubasic(i+1,j-1)-ubasic(i-1,j-1))*u(
177                     i,j)
                       u(i,j-2) = u(i,j-2) + c0*(1.d0-c1)*u(i,j)
178          !u(i,j) = 0.d0
179        enddo
180        if (iforcing .eq. 1) then
181          do i = 1, nx
```

A.2 Example of FORTRAN Code

```fortran
182               u(i,j-1) = u(i,j-1) + ubasic(i,j-1) - uob(i,j-1)
183            enddo
184         endif
185      enddo
186
187 ! adjoint of FTCS
188      do i = nx-1, 2, -1
189         !TL: u(i,2) = u(i,1) &
190         !             - 0.5d0*dtdx*( u(i,1)*(ubasic(i+1,1)-ubasic(i-1,1))   &
191         !                          + ubasic(i,1)*(u(i+1,1)-u(i-1,1)) ) &
192         !             + 0.5d0*c1*(u(i+1,1)-2.d0*u(i,1) + u(i-1,1))
193         u(i-1,1) = u(i-1,1) + 0.5d0*(c1+dtdx*ubasic(i,1))*u(i,2)
194         u(i+1,1) = u(i+1,1) + 0.5d0*(c1-dtdx*ubasic(i,1))*u(i,2)
195         u(i,  1) = u(i,  1) + u(i,2)*(1.d0-c1-0.5d0*dtdx*(ubasic(i+1,1)-
             ubasic(i-1,1)))
196         !u(i,  2) = 0.d0
197      enddo
198
199      if (iforcing.eq.1) then
200         do i= 1, nx
201            u(i,1) = u(i,1) + ubasic(i,1) - uob(i,1)
202         enddo
203      endif
204
205 ! set the boundary conditions:
206      do i = 1, n
207         u(1, i) = 0.d0
208         u(nx,i) = 0.d0
209      enddo
210
211 ! set the final value of u in ui:
212      do i = 1, nx
213         ui(i) = ui(i) + u(i 1)
214      enddo
215
216 endsubroutine burger_adj
217
218 endmodule m_burger
```

Appendix B Postprocessing of Numerical Model Output to Obtain Station Weather Forecasts

If the numerical model forecasts are skillful, the forecast variables should be strongly related to the weather parameters of interest to the "person in the street" and for other important applications. These include precipitation (amount and type), surface wind, and surface temperature, visibility, cloud amount and type, etc. However, the model output variables are not optimal direct estimates of local weather forecasts. This is because models have biases, the bottom surface of the models is not a good representation of the actual orography, and models may not represent well the effect of local forcings important for local weather forecasts. In addition, models do not forecast some required parameters, such as visibility and probability of thunderstorms.

In order to optimize the use of numerical weather forecasts as guidance for human forecasters, it has been customary to use statistical methods to "post-process" the model forecasts and adapt them to produce local forecasts. In this appendix we discuss three of the methods that have been used for this purpose.

B.1   Model Output Statistics

This method, when applied under ideal circumstances, is the gold standard of NWP model output post processing (Glahn and Lowry, 1972; Carter et al., 1989). MOS[1] is essentially multiple linear regression, where the predictors h_{nj} are model forecast variables (e.g., temperature, humidity, or wind at any grid point, either near the surface or in the upper levels), and may also include other astronomical or geographical parameters (such as latitude, longitude and time of the year) valid at time t_n. The predictors could also include past observations. The predictand y_n is a station weather observation (e.g., maximum temperature or wind speed) valid at the same time as the forecast. Here, as in any statistical regression, the quality of the results improves with the quality and length of the training data set used to determine the regression coefficients b_j.

The *dependent* data set used for determining the regression coefficients is

$$\left. \begin{array}{ll} y_n = y(t_n) & n = 1, \ldots, N \\ h_{nj} = h_j(t_n) & n = 1, \ldots, N; j = 1, \ldots, J \end{array} \right\} \tag{B.1.1}$$

where we consider one predictand y_n as a function of time t_n and J predictors h_{nj}.

---

[1] I am grateful to J. Paul Dallavalle of the National Weather Service for information about MOS and Perfect Prog.

The linear regression (forecast) equation is

$$\hat{y}_n = b_0 + \sum_{j=1}^{J} b_j h_{nj} = \sum_{j=0}^{J} b_j h_{nj} \tag{B.1.2}$$

where for convenience the predictors associated with the constant term b_0 are defined as $h_{n0} \equiv 1$. In linear regression the coefficients b_j are determined by minimizing the sum of squares of the forecast errors over the training period (Wilks, 2011). The sum of squared errors is given by

$$SSE = \sum_{n=1}^{N} (y_n - \hat{y}_n)^2 = \sum_{n=1}^{N} e_n^2 \tag{B.1.3}$$

Taking the derivatives with respect to the coefficients b_j and setting them to zero we obtain:

$$\frac{\partial SSE}{\partial b_j} = 0 = \sum_{n=1}^{N} \left(y_n - \sum_{l=0}^{J} b_l h_{nl} \right) h_{nj} \quad j = 0, 1, \ldots, J \tag{B.1.4}$$

or

$$\sum_{n=1}^{N} \left[h_{jn}^T y_n - h_{jn}^T \sum_{l=0}^{J} h_{nl} b_l \right] h_{nj} = 0 \quad j = 0, \ldots, J \tag{B.1.5}$$

where $h_{jn}^T = h_{nj}$. Equations (B.1.5) are the "normal" equations for multiple linear regression that determine the linear regression coefficients b_j, $j = 0, \ldots, J$. In matrix form, they can be written as

$$\mathbf{H}^T \mathbf{H} \mathbf{b} = \mathbf{H}^T \mathbf{y} \quad or \quad \mathbf{b} = (\mathbf{H}^T \mathbf{H})^{-1} \mathbf{H}^T \mathbf{y} \tag{B.1.6}$$

where

$$\mathbf{H} = \begin{bmatrix} 1 & h_{11} & \ldots & h_{1J} \\ 1 & h_{21} & \ldots & h_{2J} \\ \vdots & \vdots & h_{nj} & \vdots \\ 1 & h_{N1} & \ldots & h_{NJ} \end{bmatrix} \quad \mathbf{b} = \begin{bmatrix} b_0 \\ b_1 \\ \vdots \\ b_J \end{bmatrix} \quad \mathbf{y} = \begin{bmatrix} y_1 \\ y_2 \\ \vdots \\ y_N \end{bmatrix} \tag{B.1.7}$$

are, respectively, the dependent sample predictor matrix (model output variables, geographical and astronomical parameters, etc.), the vector of regression coefficients, and the vector of predictands in the dependent sample. $\hat{\mathbf{y}} = \mathbf{Hb}, \mathbf{e} = \mathbf{y} - \mathbf{Hb}$ are the linear predictions and the prediction error, respectively, in the dependent sample. The *dependent* estimate of the error variance of the prediction is $s_e^2 = SSE/(N - J - 1)$ since the number of degrees of freedom is $N - J - 1$. This indicates that one should avoid overfitting the dependent sample by ensuring that $N \gg J$. For independent data, the expected error can be considerably larger than the dependent estimate s_e^2 because of the uncertainties in estimating the coefficients b_j. The best way to estimate the skill of MOS (or any statistical prediction) that can be expected when applied to *independent* data is to perform *cross-validation*. This can be done by reserving a portion (such as 10%) of the dependent data, deriving the regression coefficients from the other 90%, and then applying it to the unused 10%. The process can be repeated 10 times with different subsets of the dependent data to increase the confidence of the cross-validation, but this also increases the computational cost.

It is clear that for a MOS system to perform optimally, several conditions must be fulfilled:

(a) The training period should be as long as possible (at least several years).
(b) The model-based forecasting system should be kept unchanged to the maximum extent possible during the training period.
(c) After training, the MOS system should be applied to future model forecasts that also use the same unchanged model system.

These conditions, while favorable for the MOS performance, are not favorable for the continued improvement of the NWP model, since they require "frozen" models. The main advantage of MOS is that if the conditions stated above are satisfied, it achieves the best possible linear prediction. Another advantage is that it naturally takes into account the fact that forecast skill decreases with the forecast length, since the training sample will include, for instance, the information that a 1-day model prediction is on the average considerably more skillful than a 3-day prediction. The main disadvantage of MOS is that it is not easily adapted to an operational situation in which the model and data assimilation systems are frequently upgraded.

Typically, MOS equations have 10–20 predictors chosen by forward screening (Wilks, 2011). In the US NWS, the same MOS equations are computed for a few (4–10) relatively homogeneous regions in order to increase the size of the developmental database. In order to stratify the data into few but relatively homogeneous time periods, separate MOS equations are developed for the cool season (October–March) and the warm season (April–September). As shown in Table B.3.1, MOS can reduce very substantially the errors in the NWP model forecasts, especially at short lead times. At long lead times, the forecast skill is lost, so that the MOS forecast becomes a climatological forecast and the MOS forecast error variance asymptotes to the climatology error. The error variance of an individual NWP forecast, on the other hand, asymptotes to twice the climatological error variance, plus the square of the model bias (see Section 6.4.2).

Figure B.1.1 shows the evolution of the error in predicting the maximum temperature by the statistical guidance (MOS) and by the local human forecasters (LCL). The human forecasters skill in the 2-day forecast is now as good as the one-day forecast was in the 1970s. The human forecasters bring added value i.e., make better forecasts than the MOS statistical guidance, which in turn is considerably better than the direct NWP model output. Nevertheless, the long-term improvements are driven mostly by the improvements in the NWP model and data assimilation systems, as discussed in Chapter 1.

In summary, the forecast statistical guidance (and in particular MOS) adds value to the direct NWP model output by objectively interpreting model output to remove

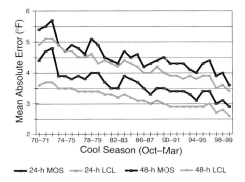

Figure B.1.1 Evolution of the mean absolute error of the MOS guidance and of the local official NWS forecasts (LCL) averaged over the USA. (Courtesy of J. Paul Dallavalle and Valery Dagostaro from the US NWS.)

systematic biases and quantifying uncertainty, predicting parameters that the model does not predict, and producing site-specific forecasts. It assists forecasters providing a first guess for the expected local conditions and allows convenient access to information on local model and climatology conditions.

B.2 Perfect Prog

Perfect Prog is an approach similar to MOS, except that the regression equations are derived using as predictors, observations or analyses (rather than forecasts) valid at the prediction time, as if the forecasts were perfect. If station observations are used as predictors in the dependent sample, it is not possible to use the same variable for a predictor as for the predictand (e.g., Boston's observed maximum surface temperature could not be used as predictor for the maximum temperature in Boston). However, if model analyses are used as "perfect" forecasts, one can use like variables as predictors. For obvious reasons, Perfect Prog has not been much used except for very short forecasts. Perhaps it would be possible now to use the long homogeneous reanalyses that have been completed (Kistler et al., 2001; Kalnay et al., 1996; Gibson et al., 1997) to derive very long and robust Perfect Prog statistics between model output and station data. After the regression between the reanalysis and station data is completed, the prediction of surface parameters could be done in two steps. In the first step, multiple regression would be used to predict the reanalysis field from model forecasts, which should be easier to achieve than predicting the station data directly, since a few parameters would be enough to represent the model bias and the decay of skill with time. In the second step, the Perfect Prog equations would be used to translate the predicted analysis into station weather parameters. In this approach, the disadvantage of Perfect Prog of not including the effective loss of skill associated with longer forecast lengths would be handled in the first step discussed earlier. This approach has yet to be thoroughly explored.

296   **Appendix B   Postprocessing of Numerical Model**

B.3   Adaptive Regression Based on a Simple Kalman Filter Approach

Adaptive regression based on Kalman filtering has also been widely used as a post-processor.[2] In MOS and in other statistical prediction methods such as nonlinear regression or neural networks, the regression coefficients are computed from the dependent sample and are not changed as new observations are collected until a new set of MOS equations are derived every 5 or 10 years. Because the regression coefficients are constant, the order of the observations is irrelevant in MOS, so that older data have as much influence as the newest observations used to derive the coefficients.

In adaptive regression, the Kalman filter equations (Section 5.6) are applied in a simple, sequential formulation to the multiple regression coefficients $\mathbf{b}_k = \mathbf{b}(t_k)$, whose values are *updated* every time step, rather than keeping them constant as in (B.1.2):

$$\hat{y}_k = \sum_{j=0}^{J} b_j(t_k) h_{kj} = \begin{bmatrix} 1 & h_{k1} & \dots & h_{kJ} \end{bmatrix} \begin{bmatrix} b_0 \\ b_1 \\ \vdots \\ b_J \end{bmatrix}_k = \mathbf{h}_k^T \mathbf{b}_k \qquad (B.3.1)$$

If we compare this equation with those in Chapter 5, we see that it has the form of a forecast of observations, $y_k^f = \mathbf{H}_k \mathbf{b}_k$, so that we can use the Kalman filter formulation with an "observation operator" $\mathbf{H} = \mathbf{h}_k^T$, a row vector in (B.1.7) corresponding to the time t_k. Recall that Kalman filtering consists of two steps (see Section 5.6). In the first step, starting from the analysis at time t_{k-1}, we forecast the values of the model variables (in this case the coefficients $\mathbf{b}_k$) and their error covariance at time t_k. In the second step, the Kalman weight matrix is derived, and, after obtaining the observations at time t_k, the model variables and error covariance are updated, giving the analysis at time t_k. In adaptive regression, the "forecast" or first guess of the regression coefficients at t_k is simply that they are the same as the (analysis) coefficients at t_{k-1}, and their error covariance is the same as that estimated in the previous time step, plus an additional error introduced by this "regression forecast model":

$$\left. \begin{array}{l} \mathbf{b}_k^f = \mathbf{b}_{k-1}^a \\ \mathbf{P}_k^f = \mathbf{P}_{k-1}^a + \mathbf{Q}_{k-1} \end{array} \right\} \qquad (B.3.2)$$

Here $\mathbf{Q}_{k-1} = \mathbf{q}_k \mathbf{q}_k^T$ is the "regression model" error covariance (a matrix of tunable coefficients that is diagonal if we assume that the errors of the different coefficients are not correlated).

The Kalman gain or weight vector for adaptive regression is given by

$$\mathbf{k}_k = \mathbf{P}_k^f \mathbf{h}_k (\mathbf{h}_k^T \mathbf{P}_k^f \mathbf{h}_k + r_k)^{-1} \qquad (B.3.3)$$

---

[2] I am very grateful to Joaquim Ballabrera for insightful comments and suggestions that improved this section.

B.3 Adaptive Regression Based on a Simple Kalman Filter Approach

Table B.3.1 RMS error in the forecast of the surface temperature at 00Z averaged for eight US stations. In dependent regression and Kalman filtering, the only predictor used was the direct model prediction of the temperature interpolated to the station. The MOS prediction has more than 10 predictors and several years of training

NWP (aviation model)	Dependent regression	Adaptive regression	MOS
5.36K	2.67K	3.07K	2.29K

Note that for a single predictand, the forecast error covariance $\mathbf{h}_k^T \mathbf{P}_k^f \mathbf{h}_k$ and the observational error covariance $\mathbf{R}_k = r_k$ are both scalars, and computing the Kalman gain matrix does not require a matrix inversion.

At time t_k the observed forecast error or innovation $e_k = y_k^o - \mathbf{h}_k^T \mathbf{b}_k^f$ is used to *update* the regression coefficients:

$$\left. \begin{aligned} \mathbf{b}_k^a &= \mathbf{b}_k^f + \mathbf{k}_k (y_k^o - \mathbf{h}_k^T \mathbf{b}_k^f) \\ \mathbf{P}_k^a &= (\mathbf{I} - \mathbf{k}_k \mathbf{h}_k^T) \mathbf{P}_k^f \end{aligned} \right\} \tag{B.3.4}$$

In summary, the adaptive regression algorithm based on Kalman filtering can be written as:

$$\left. \begin{aligned} y_k^f &= \mathbf{h}_k^T \mathbf{b}_{k-1}^a \\ \mathbf{P}_k^f &= \mathbf{P}_{k-1}^a + \mathbf{Q}_{k-1} \\ e_k &= y_k^o - y_k^f \\ w_k &= \mathbf{h}_k^T \mathbf{P}_k^f \mathbf{h}_k + r_k \\ \mathbf{k}_k &= \mathbf{P}_k^f \mathbf{h}_k w_k^{-1} \\ \mathbf{b}_k^a &= \mathbf{b}_{k-1}^a + \mathbf{k}_k e_k \\ \mathbf{P}_k^a &= \mathbf{P}_k^f - \mathbf{k}_k w_k \mathbf{k}_k^T \end{aligned} \right\} \tag{B.3.5}$$

where w_k is a temporary scalar defined for convenience. The two tuning parameters in the algorithm are r_k, the observational error covariance (a scalar), and $\mathbf{Q}_k$, the "regression model" error covariance (a diagonal matrix with one coefficient for the variance of each predictor if the errors are uncorrelated). Unlike regression, MOS, or neural networks, adaptive regression is *sequential*, and gives more weight to recent data than to older observations. The larger $\mathbf{Q}_k$, the faster older data will be forgotten. It also allows for observational errors r_k. This method can be generalized to several predictands, in which case the observation error covariance matrix may also include observational error correlations.

Table B.3.1 compares a simple Kalman filtering applied to the 24-hr surface temperature forecasts for July and August 1997 at 00Z, averaged for eight different US stations, using as a single predictor the global model output for surface temperature interpolated to each individual station. It was found that after only a few days of spin-up, starting with a climatological first guess, and with minimal tuning, the adaptive regression algorithm was able to reach a fairly steady error level substantially better than the numerical model error, and not much higher than regression on the dependent sample. Not surprisingly, MOS, using many more predictors and several years of training, provides an even better forecast than this simple AR.

In summary, Kalman filtering provides a simple algorithm for adaptive regression. It requires little training so that it is able to adapt rather quickly to changes in the model, and to long-lasting weather regimes. It is particularly good in correcting model biases. However, in general it is not as good as regression based on long dependent samples. A nice example about applying AR to the Lorenz-96 model is kindly provided by Zenginoğlu at https://github.com/anilzen/adaptive_regression/.

Bibliography

Kathleen T. Alligood, Tim D. Sauer, and James A. Yorke. *Chaos – An Introduction to Dynamical Systems*. Springer, 1st – corrected edition, November 1996. ISBN 0-387-94677-2.

Javier Amezcua, Eugenia Kalnay, and Paul D. Williams. The effects of the RAW filter on the climatology and forecast skill of the SPEEDY model. *Monthly Weather Review*, 139(2): 608–619, 2011. doi: 10.1175/2010MWR3530.1.

Brian Ancell and Gregory J. Hakim. Comparing adjoint- and ensemble-sensitivity analysis with applications to observation targeting. *Monthly Weather Review*, 135(12):4117–4134, December 2007. doi: 10.1175/2007MWR1904.1.

Jeffrey L. Anderson. An ensemble adjustment Kalman filter for data assimilation. *Monthly Weather Review*, 129(12):2884–2903, 2001. http://journals.ametsoc.org/doi/abs/10.1175/1520-0493(2001)129%3C2884:AEAKFF%3E2.0.CO%3B2.

Jeffrey L. Anderson. A local least squares framework for ensemble filtering. *Monthly Weather Review*, 131(4):634–642, April 2003. doi: 10.1175/1520-0493(2003)131<0634:ALLSFF>2.0.CO;2. https://journals.ametsoc.org/view/journals/mwre/131/4/1520-0493_2003_131_0634_allsff_2.0.co_2.xml. Place: Boston MA, USA Publisher: American Meteorological Society.

Erik Anderson and Heikki Järvinen. Variational quality control. *Quarterly Journal of the Royal Meteorological Society*, 125(554):697–722, 1999. http://onlinelibrary.wiley.com/doi/10.1002/qj.49712555416/abstract.

Erik Andersson, Jan Haseler, Per Undén, Philippe Courtier, Graeme Kelly, Drasko Vasiljevic, Cedo Brankovic, Catherine Gaffard, Anthony Hollingsworth, and Christian Jakob. The ECMWF implementation of three-dimensional variational assimilation (3D-Var). III: Experimental results. *Quarterly Journal of the Royal Meteorological Society*, 124(550): 1831–1860, 1998. http://onlinelibrary.wiley.com/doi/10.1002/qj.49712455004/abstract.

Akio Arakawa. Computational design for long-term numerical integration of the equations of fluid motion: Two-dimensional incompressible flow. Part I. *Journal of Computational Physics*, 1(1):119–143, 1966. www.sciencedirect.com/science/article/pii/0021999166900155.

Akio Arakawa. Adjustment mechanisms in atmospheric models. *Journal meteorological Society of Japan Series 2*, 75:45–69, 1997.

Akio Arakawa. The cumulus parameterization problem: Past, present, and future. *Journal of Climate*, 17(13):2493–2525, July 2004. doi: 10.1175/1520-0442(2004)017<2493:RATCPP>2.0.CO;2.

Akio Arakawa and Celal S. Konor. Vertical differencing of the primitive equations based on the Charney-Phillips grid in hybrid σ-p vertical coordinates. *Monthly Weather Review*, 124(3):511–528, 1996. http://journals.ametsoc.org/doi/abs/10.1175/1520-0493(1996)124%3C0511%3AVDOTPE%3E2.0.CO%3B2.

Bibliography

Akio Arakawa and Vivian R. Lamb. Computational design of the basic dynamical processes of the UCLA general circulation model. *Methods in Computational Physics*, 17:173–265, 1977. http://books.google.com/books?hl=en&lr=&id=nN_4561KTIIC&oi=fnd&pg=PA173&dq=arakawa+lamb+1977+ucla&ots=yJU84ij1es&sig=lYru9kLDsOA5q4COgzUJp30P6BE.

Akio Arakawa and Shrinivas Moorthi. Baroclinic instability in vertically discrete systems. *Journal of the Atmospheric Sciences*, 45(11):1688–1708, 1988. http://journals.ametsoc.org/doi/abs/10.1175/1520-0469(1988)045%3C1688%3ABIIVDS%3E2.0.CO%3B2.

Akio Arakawa and Wayne Howard Schubert. Interaction of a cumulus cloud ensemble with the large-scale environment, Part I. *Journal of the Atmospheric Sciences*, 31(3):674–701, 1974. http://journals.ametsoc.org/doi/abs/10.1175/1520-0469(1974)031%3C0674:IOACCE%3E2.0.CO;2.

Troy Arcomano, Istvan Szunyogh, Jaideep Pathak, Alexander Wikner, Brian R. Hunt, and Edward Ott. A machine learning-based global atmospheric forecast model. *Geophysical Research Letters*, 47(9):e2020GL087776, May 2020. doi: 10.1029/2020GL087776.

Richard Asselin. Frequency filter for time integrations. *Monthly Weather Review*, 100(6):487–490, 1972. http://journals.ametsoc.org/doi/abs/10.1175/1520-0493(1972)100%3C0487:FFFTI%3E2.3.CO;2.

R. Atlas, E. Kalnay, W. E. Baker, J. Susskind, D. Reuter, and M. Halem. Simulation studies of the impact of future observing systems on weather prediction. 1985.

R. Atlas, N. Wolfson, and J. Terry. The effect of SST and soil moisture anomalies on GLA model simulations of the 1988 U.S. summer drought. *Journal of Climate*, 6(11):2034–2048, November 1993. doi: 10.1175/1520-0442(1993)006<2034:TEOSAS>2.0.CO;2.

Eviatar Bach, Safa Motesharrei, Eugenia Kalnay, and Alfredo Ruiz-Barradas. Local atmosphere–ocean predictability: Dynamical origins, lead times, and seasonality. *Journal of Climate*, 32(21):7507–7519, November 2019. doi: 10.1175/JCLI-D-18-0817.1.

Wayman E. Baker, Stephen C. Bloom, John S. Woollen, Mark S. Nestler, Eugenia Brin, Thomas W. Schlatter, and Grant W. Branstator. Experiments with a three-dimensional statistical objective analysis scheme using FGGE data. *Monthly Weather Review*, 115(1):272–296, January 1987. doi: 10.1175/1520-0493(1987)115<0272:EWATDS>2.0.CO;2.

R. Balgovind, A. Dalcher, M. Ghil, and E. Kalnay. A stochastic-dynamic model for the spatial structure of forecast error statistics. *Monthly Weather Review*, 111(4):701–722, April 1983. doi: 10.1175/1520-0493(1983)111<0701:ASDMFT>2.0.CO;2.

D. M. Barker, W. Huang, Y-R. Guo, A. J. Bourgeois, and Q. N. Xiao. A three-dimensional variational data assimilation system for MM5: Implementation and initial results. *Monthly Weather Review*, 132(4):897–914, April 2004. doi: 10.1175/1520-0493(2004)132<0897:ATVDAS>2.0.CO;2.

Jan Barkmeijer, Franccois Bouttier, and Martin Van Gijzen. Singular vectors and estimates of the analysis-error covariance metric. *Quarterly Journal of the Royal Meteorological Society*, 124(549):1695–1713, July 1998. doi: 10.1002/qj.49712454916.

Stanley L. Barnes. A technique for maximizing details in numerical weather map analysis. *Journal of Applied Meteorology*, 3(4):396–409, August 1964. doi: 10.1175/1520-0450(1964)003<0396:ATFMDI>2.0.CO;2.

Stanley L. Barnes. Oklahoma thunderstorms on 29–30 April 1970. Part I: Morphology of a Tornadic storm. *Monthly Weather Review*, 106(5):673–684, May 1978. doi: 10.1175/1520-0493(1978)106<0673:OTOAPI>2.0.CO;2.

Anthony G. Barnston, Huug M. van den Dool, David R. Rodenhuis, Chester R. Ropelewski, Vernon E. Kousky, Edward A. O'Lenic, Robert E. Livezey, Stephen E. Zebiak, Mark A. Cane, Tim P. Barnett, Nicholas E. Graham, Ming Ji, and Ants Leetmaa. Long-lead seasonal forecasts – where do we stand? *Bulletin of the American Meteorological Society*, 75 (11):2097–2114, November 1994. doi: 10.1175/1520-0477(1994)C75<2097:LLSFDW>2.0. CO;2.

Richard Barrett, Michael Berry, Tony F. Chan, James Demmel, June Donato, Jack Dongarra, Victor Eijkhout, Roldan Pozo, Charles Romine, and Henk van der Vorst. *Templates for the Solution of Linear Systems: Building Blocks for Iterative Methods*. Other Titles in Applied Mathematics. Society for Industrial and Applied Mathematics, January 1994. doi: 10.1137/ 1.9781611971538.

J. R. Bates and A. McDonald. Multiply-Upstream, Semi-Lagrangian Advective Schemes: Analysis and application to a multi-level primitive equation model. *Monthly Weather Review*, 110 (12):1831–1842, December 1982. doi: 10.1175/1520-0493(1982)110<1831:MUSLAS>2.0. CO;2.

J. R. Bates, Y. Li, A. Brandt, S. F. McCormick, and J. Ruge. A global shallow-water numerical model based on the semi-lagrangian advection of potential vorticity. *Quarterly Journal of the Royal Meteorological Society*, 121(528):1981–2005, October 1995. doi: 10.1002/qj. 49712152810.

David S. Battisti and Anthony C. Hirst. Interannual variability in a tropical atmosphere–ocean model: Influence of the basic state, ocean geometry and nonlinearity. *Journal of the Atmospheric Sciences*, 46(12):1687–1712, June 1988. doi: 10.1175/1520-0469(1989)046<1687: IVIATA>2.0.CO;2.

Lennart Bengtsson, Michael Ghil, and Erland Källén, editors. *Dynamic Meteorology: Data Assimilation Methods*, volume 36 of *Applied Mathematical Sciences*. Springer New York, New York, 1981. doi: 10.1007/978-1-4612-5970-1.

R. Benoit, J. Côté, and J. Mailhot. Inclusion of a TKE boundary layer parameterization in the Canadian regional finite-element model. *Monthly Weather Review*, 117(8):1726–1750, August 1989. doi: 10.1175/1520-0493(1989)117<1726:IOATBL>2.0.CO;2.

Robert Benoit, Michel Desgagné, Pierre Pellerin, Simon Pellerin, Yves Chartier, and Serge Desjardins. The Canadian MC2: A semi-lagrangian, semi-implicit wideband atmospheric model suited for finescale process studies and simulation. *Monthly Weather Review*, 125(10): 2382–2415, October 1997. doi: 10.1175/1520-0493(1997)125<2382:TCMASL>2.0.CO;2.

Kenneth H. Bergman. Multivariate analysis of temperatures and winds using optimum interpolation. *Monthly Weather Review*, 107(11):1423–1444, November 1979. doi: 10.1175/ 1520-0493(1979)107<1423:MAOTAW>2.0.CO;2.

P. Bergthorsson, B. R. Döös, S. Fryklund, O. Haug, and R. Lindquist. Routine forecasting with the barotropic model. *Tellus*, 7(2):272–274, January 1955. doi: 10.3402/tellusa.v7i2.8775.

Páll Bergthórsson and Bo R. Döös. Numerical weather map analysis1. *Tellus*, 7(3):329–340, August 1955. doi: 10.1111/j.2153-3490.1955.tb01170.x.

A. K. Betts and M. J. Miller. A new convective adjustment scheme. Part II: Single column tests using GATE wave, BOMEX, ATEX and arctic air-mass data sets. *Quarterly Journal of the Royal Meteorological Society*, 112(473):693–709, July 1986. doi: 10.1002/qj.49711247308.

Craig H. Bishop and Zoltan Toth. Ensemble transformation and adaptive observations. *Journal of the Atmospheric Sciences*, 56(11):1748–1765, June 1999. doi: 10.1175/1520-0469(1999) 056<1748:ETAAO>2.0.CO;2.

Craig H. Bishop, Brian J. Etherton, and Sharanya J. Majumdar. Adaptive sampling with the ensemble transform Kalman filter. Part I: Theoretical aspects. *Monthly Weather Review*, 129 (3):420–436, March 2001. doi: 10.1175/1520-0493(2001)129<0420:ASWTET>2.0.CO;2.

J. Bjerknes. Atmospheric teleconnections from the equatorial pacific. *Monthly Weather Review*, 97(3):163–172, March 1969. doi: 10.1175/1520-0493(1969)097<0163:ATFTEP>2.3.CO;2.

V. Bjerknes. Das Problem der Wettervorhersage, betrachtet vom Standpunkte der Mechanik und der Physik. *Meteorologische Zeitschrift*, 21:1–7, 1904.

Thomas L. Black. The new NMC mesoscale eta model: Description and forecast examples. *Weather and Forecasting*, 9(2):265–278, June 1994. doi: 10.1175/1520-0434(1994) 009<0265:TNNMEM>2.0.CO;2.

Rainer Bleck and Stanley G. Benjamin. Regional weather prediction with a model combining terrain-following and isentropic coordinates. Part I: Model description. *Monthly Weather Review*, 121(6):1770–1785, June 1993. doi: 10.1175/1520-0493(1993)121<1770: RWPWAM>2.0.CO;2.

M. Bocquet and P. Sakov. Combining inflation-free and iterative ensemble Kalman filters for strongly nonlinear systems. *Nonlinear Processes in Geophysics*, 19(3):383–399, June 2012. doi: 10.5194/npg-19-383-2012.

Bert Bolin. Carl-Gustaf Rossby The stockholm period 1947–1957. *Tellus B: Chemical and Physical Meteorology*, 51(1):4–12, January 1999. doi: 10.3402/tellusb.v51i1.16255.

Massimo Bonavita, Lars Isaksen, and Elías Hólm. On the use of EDA background error variances in the ECMWF 4D-Var. *ECMWF Technical Memoranda*, 2012. www.ecmwf.int/node/8272. ECMWF.

Massimo Bonavita, Mats Hamrud, and Lars Isaksen. EnKF and hybrid gain ensemble data assimilation. Part II: EnKF and hybrid gain results. *Monthly Weather Review*, 143(12): 4865–4882, November 2015a. doi: 10.1175/MWR-D-15-0071.1.

Massimo Bonavita, Elias Hólm, Lars Isaksen, and Mike Fisher. The evolution of the ECMWF hybrid data assimilation system. *Quarterly Journal of the Royal Meteorological Society*, 142 (694):287–303, August 2015b. doi: 10.1002/qj.2652.

Massimo Bonavita, Rossella Arcucci, Alberto Carrassi, Peter Dueben, Alan J Geer, Bertrand Le Saux, Nicolas Longépé, Pierre-Philippe Mathieu, and Laure Raynaud. Machine learning for earth system observation and prediction. *Bulletin of the American Meteorological Society*, 102(4):E710–E716, 2021.

Jay P. Boris and David L. Book. Flux-corrected transport. I. SHASTA, a fluid transport algorithm that works. *Journal of Computational Physics*, 11(1):38–69, January 1973. doi: 10.1016/0021-9991(73)90147-2.

Philippe Bougeault. A non-reflective upper boundary condition for limited-height hydrostatic models. *Monthly Weather Review*, 111(3):420–429, March 1983. doi: 10.1175/1520-0493(1983)111<0420:ANRUBC>2.0.CO;2.

Sid-Ahmed Boukabara, Vladimir Krasnopolsky, Stephen G. Penny, Jebb Q. Stewart, Amy McGovern, David Hall, John E. Ten Hoeve, Jason Hickey, Hung-Lung Allen Huang, John K. Williams, and others. Outlook for exploiting artificial intelligence in the earth and environmental sciences. *Bulletin of the American Meteorological Society*, 102(5):E1016–E1032, 2021.

F. Bouttier. Arome: A new operational mesoscale NWP system for Meteo-France. *CAS/JSC WGNE Research Activities in Atmospheric and Oceanic Modelling*, 33:0503–0504, 2003.

F. Bouttier and F. Rabier. The operational implementation of 4D-Var. ECMWF Newsletter 78, ECMWF, 1997.

F. Bouttier, J. C. Derber, and M. Fisher. The 1997 revision of the Jb term in 3D/4D-Var. Technical Memoranda., 238, ECMWF, 1997.

Tim P. Boyer, Olga K. Baranova, Carla Coleman, Hernan E. Garcia, Alexandra Grodsky, Ricardo A. Locarnini, Alexey V. Mishonov, Christopher R. Paver, James R. Reagan, Dan Seidov, Igor V. Smolyar, Katharine W. Weathers, and Melissa M. Zweng. NOAA Atlas

NESDIS 87, World Ocean Database 2018 (Pre-release). *World Ocean Database*, 2018. www.ncei.noaa.gov/sites/default/files/2020-04/wod_intro.pdf.

Čedo Branković and T. N. Palmer. Seasonal skill and predictability of ECMWF PROVOST ensembles. *Quarterly Journal of the Royal Meteorological Society*, 126(567):2035–2067, July 2000. doi: 10.1002/qj.49712656704.

Arne M. Bratseth. Statistical interpolation by means of successive corrections. *Tellus A*, 38A (5):439–447, October 1986. doi: 10.1111/j.1600-0870.1986.tb00476.x.

William L. Briggs. *A Multigrid Tutorial*. SIAM, Philadelphia, PA, 1987.

Are Magnus Bruaset. *A Survey of Preconditioned Iterative Methods*. CRC Press, May 1995. ISBN 978-0-582-27654-3. Google-Books-ID: hubxyVnwW2kC.

Radmila Bubnová, Gwenaëlle Hello, Pierre Bénard, and Jean-François Geleyn. Integration of the fully elastic equations cast in the hydrostatic pressure terrain-following coordinate in the framework of the ARPEGE/Aladin NWP system. *Monthly Weather Review*, 123(2):515–535, February 1995. doi: 10.1175/1520-0493(1995)123<0515:IOTFEE>2.0.CO;2.

S. Buckeridge, M. J. P. Cullen, R. Scheichl, and M. Wlasak. A robust numerical method for the potential vorticity based control variable transform in variational data assimilation. *Quarterly Journal of the Royal Meteorological Society*, 137(657):1033–1094, April 2011. doi: 10.1002/qj.826.

M. Buehner, J. Morneau, and C. Charette. Four-dimensional ensemble-variational data assimilation for global deterministic weather prediction. *Nonlinear Processes in Geophysics*, 20 (5):669–682, September 2013. doi: 10.5194/npg-20-669-2013.

Mark Buehner, P. L. Houtekamer, Cecilien Charette, Herschel L. Mitchell, and Bin He. Intercomparison of variational data assimilation and the ensemble Kalman filter for global deterministic NWP. Part I: Description and single-observation experiments. *Monthly Weather Review*, 138(5):1550–1566, May 2010a. doi: 10.1175/2009MWR3157.1.

Mark Buehner, P. L. Houtekamer, Cecilien Charette, Herschel L. Mitchell, and Bin He. Intercomparison of variational data assimilation and the ensemble Kalman filter for global deterministic NWP. Part II: One-month experiments with real observations. *Monthly Weather Review*, 138(5):1567–1586, May 2010b. doi: 10.1175/2009MWR3158.1.

Mark Buehner, Ron McTaggart-Cowan, Alain Beaulne, Cécilien Charette, Louis Garand, Sylvain Heilliette, Ervig Lapalme, Stéphane Laroche, Stephen R. Macpherson, Josée Morneau, and Ayrton Zadra. Implementation of deterministic weather forecasting systems based on ensemble–variational data assimilation at environment Canada. Part I: The global system. *Monthly Weather Review*, 143(7):2532–2559, July 2015. doi: 10.1175/MWR-D-14-00354.1.

Mark Buehner, Ping Du, and Joël Bédard. A new approach for estimating the observation impact in ensemble–variational data assimilation. *Monthly Weather Review*, 146(2): 447–465, January 2018. doi: 10.1175/MWR-D-17-0252.1.

R. Buizza and T. N. Palmer. The singular-vector structure of the atmospheric global circulation. *Journal of the Atmospheric Sciences*, 52(9):1434–1456, May 1995. doi: 10.1175/ 1520-0469(1995)052<1434:TSVSOT>2.0.CO;2.

R. Buizza, M. Milleer, and T. N. Palmer. Stochastic representation of model uncertainties in the ECMWF ensemble prediction system. *Quarterly Journal of the Royal Meteorological Society*, 125(560):2887–2908, October 1999. doi: 10.1002/qj 49712556006.

Roberto Buizza. Sensitivity of optimal unstable structures. *Quarterly Journal of the Royal Meteorological Society*, 120(516):429–451, January 1994. doi: 10.1002/qj.49712051609.

Roberto Buizza. Potential forecast skill of ensemble prediction and spread and skill distributions of the ECMWF ensemble prediction system. *Monthly Weather Review*, 125(1):99–119, January 1997. doi: 10.1175/1520-0493(1997)125<0099:PFSOEP>2.0.CO;2.

Gerrit Burgers, Peter Jan van Leeuwen, and Geir Evensen. Analysis scheme in the ensemble Kalman filter. *Monthly Weather Review*, 126(6):1719–1724, June 1998. doi: 10.1175/1520-0493(1998)126<1719:ASITEK>2.0.CO;2.

D. M. Burridge. A split semi-implict reformulation of the Bushby-Timpson 10-level model. *Quarterly Journal of the Royal Meteorological Society*, 101(430):777–792, October 1975. doi: 10.1002/qj.49710143006.

P. Bénard, J. Mašek, and P. Smolíková. Stability of leapfrog constant-coefficients semi-implicit schemes for the fully elastic system of Euler equations: Case with orography. *Monthly Weather Review*, 133(5):1065–1075, May 2005. doi: 10.1175/MWR2907.1.

Dan G. Cacuci. Sensitivity theory for nonlinear systems. I. Nonlinear functional analysis approach. *Journal of Mathematical Physics*, 22(12):2794–2802, December 1981. doi: 10.1063/1.525186.

Robert F. Cahalan, William Ridgway, Warren J. Wiscombe, Thomas L. Bell, and Jack B. Snider. The albedo of fractal stratocumulus clouds. *Journal of the Atmospheric Sciences*, 51(16): 2434–2455, August 1994. doi: 10.1175/1520-0469(1994)051<2434:TAOFSC>2.0.CO;2.

Ming Cai, Eugenia Kalnay, and Zoltan Toth. Bred vectors of the Zebiak–Cane model and their potential application to ENSO predictions. *Journal of Climate*, 16(1):40–56, January 2003. doi: 10.1175/1520-0442(2003)016<0040:BVOTZC>2.0.CO;2.

Katherine Calvin and Ben Bond-Lamberty. Integrated human-earth system modeling – state of the science and future directions. *Environmental Research Letters*, 13(6):063006, June 2018. doi: 10.1088/1748-9326/aac642.

K. Campana. Use of cloud analyses to validate and improve model-diagnostic clouds at NMC. Shinfield Park, Reading, England, November 1994. ECMWF.

Mark A. Cane, Stephen E. Zebiak, and Sean C. Dolan. Experimental forecasts of El Niño. *Nature*, 321(6073):827–832, June 1986. doi: 10.1038/321827a0.

Peter M. Caplan and Glenn H. White. Performance of the national meteorological center's medium-range model. *Weather and Forecasting*, 4(3):391–400, September 1989. doi: 10.1175/1520-0434(1989)004<0391:POTNMC>2.0.CO;2.

R. E. Carbone, J. W. Wilson, T. D. Keenan, and J. M. Hacker. Tropical island convection in the absence of significant topography. Part I: Life cycle of diurnally forced convection. *Monthly Weather Review*, 128(10):3459–3480, October 2000. doi: 10.1175/1520-0493(2000)128<3459:TICITA>2.0.CO;2.

Carla Cardinali. Monitoring the observation impact on the short-range forecast. *Quarterly Journal of the Royal Meteorological Society*, 135(638):239–250, January 2009. doi: 10.1002/qj.366.

Carla Cardinali, Sergio Pezzulli, and Erik Andersson. Influence-matrix diagnostic of a data assimilation system. *Quarterly Journal of the Royal Meteorological Society*, 130(603):2767–2786, October 2004. doi: 10.1256/qj.03.205.

Alberto Carrassi, Marc Bocquet, Laurent Bertino, and Geir Evensen. Data assimilation in the geosciences: An overview of methods, issues, and perspectives. *Wiley Interdisciplinary Reviews: Climate Change*, 9(5):e535, September 2018. doi: 10.1002/wcc.535.

Gary M. Carter, J. Paul Dallavalle, and Harry R. Glahn. Statistical forecasts based on the national meteorological center's numerical weather prediction system. *Weather and Forecasting*, 4(3):401–412, September 1989. doi: 10.1175/1520-0434(1989)004<0401:SFBOTN>2.0.CO;2.

A. Caya, J. Sun, and C. Snyder. A Comparison between the 4DVAR and the ensemble Kalman filter techniques for radar data assimilation. *Monthly Weather Review*, 133(11):3081–3094, November 2005. doi: 10.1175/MWR3021.1.

C.-C. Chang and E. Kalnay. Applying prior correlations for ensemble-based spatial localization. *Nonlinear Processes in Geophysics*, 29(3):317–327, September 2022. doi: 10.5194/npg-29-317-2022.

J. G. Charney. On a physical basis for numerical prediction of large-scale motions in the atmosphere. *Journal of Meteorology*, 6(6):372–385, December 1949. doi: 10.1175/1520-0469(1949)006<0372:OAPBFN>2.0.CO;2.

J. G. Charney. Dynamic forecasting by numerical process. In *Compendium of Meteorology*, pages 470–482. American Meteorological Society, Boston, MA, 1951. doi: 10.1007/978-1-940033-70-9_40.

J. G. Charney. Numerical prediction of cyclogenesis. *Proceedings of the National Academy of Sciences of the United States of America*, 40(2):99–110, February 1954. ISSN 0027-8424.

J. G. Charney. Integration of the primitive and balance equations. In *Proceedings of the International Symposium on Numerical Weather Prediction*, Tokyo, Japan, 1962. Meteorological Society of Japan.

J. G. Charney and N. A. Phillips. Numerical integration of the quasi-geostrophic equations for barotropic and simple baroclinic flows. *Journal of Meteorology*, 10(2):71–99, April 1953. doi: 10.1175/1520-0469(1953)010<0071:NIOTQG>2.0.CO;2.

J. G. Charney and J. Shukla. Predictability of monsoons. In *Monsoon Dynamics*. Cambridge University Press, Cambridge England; New York, April 1981. ISBN 978-0-521-22497-0.

J. G. Charney, R. Fjörtoft, and J. von Neumann. Numerical integration of the barotropic vorticity equation. *Tellus*, 2(4):237–254, November 1950. doi: 10.1111/j.2153-3490.1950.tb00336.x.

J. G. Charney, R. G. Fleagle, H. Riehl, V. E. Lally, and D. Q. Wark. The feasibility of a global observation and analysis experiment. *Bulletin of the American Meteorological Society*, 47:200–220, 1966.

Jule G. Charney. On the scale of atmospheric motions. *Geofysiske Publikasjoner*, 17(2):3–17, 1948.

Jule G. Charney and Arnt Eliassen. On the growth of the hurricane depression. *Journal of the Atmospheric Sciences*, 21(1):68–75, January 1964. doi: 10.1175/1520-0469(1964)021<0068:OTGOTH>2.0.CO;2.

H. Charnock. Wind stress on a water surface. *Quarterly Journal of the Royal Meteorological Society*, 81(350):639–640, October 1955. doi: 10.1002/qj.49708135027.

Eric P. Chassignet and Zulema D. Garraffo. Viscosity Parameterization and the Gulf Stream Separation. Technical Report ADP013577, Miami University Florida Institute of Marine and Atmospheric Sciences, January 2001. www.dtic.mil/docs/citations/ADP013577.

Tse-Chun Chen. *Applications of ensemble forecast sensitivity to observations for improving numerical weather prediction*. Ph.D. Thesis, 2018.

Tse-Chun Chen and Eugenia Kalnay. Proactive quality control: Observing system simulation experiments with the Lorenz '96 model. *Monthly Weather Review*, 147(1):53–67, January 2019. doi: 10.1175/MWR-D-18-0138.1.

Tse-Chun Chen and Eugenia Kalnay. Proactive quality control: Observing system experiments using the NCEP global forecast system. *Monthly Weather Review*, June 2020. doi: 10.1175/MWR-D-20-0001.1.

Tse-Chun Chen, Eugenia Kalnay, and Daisuke Hotta. Use of EFSO for online data assimilation quality monitoring and proactive quality control. *CAS/JSC WGNE Research Activities in Atmospheric and Oceanic Modelling*, (47):9, 2017.

A. M. Clayton, A. C. Lorenc, and D. M. Barker. Operational implementation of a hybrid ensemble/4D-Var global data assimilation system at the Met Office. *Quarterly Journal of the Royal Meteorological Society*, 139(675):1445–1461, July 2013. doi: 10.1002/qj.2054.

Stephen E. Cohn and David F. Parrish. The behavior of forecast error covariances for a Kalman filter in two dimensions. *Monthly Weather Review*, 119(8):1757–1785, August 1990. doi: 10.1175/1520-0493(1991)119<1757:TBOFEC>2.0.CO;2.

Stephen E. Cohn and Ricardo Todling. Approximate data assimilation schemes for stable and unstable dynamics. *Journal of the Meteorological Society of Japan*, 74(1):63–75, 1996. doi: 10.2151/jmsj1965.74.1_63.

Stephen E. Cohn, Arlindo da Silva, Jing Guo, Meta Sienkiewicz, and David Lamich. Assessing the effects of data selection with the DAO physical-space statistical analysis system. *Monthly Weather Review*, 126(11):2913–2926, November 1998. doi: 10.1175/1520-0493(1998) 126<2913:ATEODS>2.0.CO;2.

G. P. Compo, J. S. Whitaker, P. D. Sardeshmukh, N. Matsui, R. J. Allan, X. Yin, B. E. Gleason, R. S. Vose, G. Rutledge, P. Bessemoulin, S. Brönnimann, M. Brunet, R. I. Crouthamel, A. N. Grant, P. Y. Groisman, P. D. Jones, M. C. Kruk, A. C. Kruger, G. J. Marshall, M. Maugeri, H. Y. Mok, Ø. Nordli, T. F. Ross, R. M. Trigo, X. L. Wang, S. D. Woodruff, and S. J. Worley. The twentieth century reanalysis project. *Quarterly Journal of the Royal Meteorological Society*, 137(654):1–28, January 2011. doi: 10.1002/qj.776.

M. Corazza, E. Kalnay, D. J. Patil, S.-C. Yang, R. Morss, M. Cai, I. Szunyogh, B. R. Hunt, and J. A. Yorke. Use of the breeding technique to estimate the structure of the analysis "errors of the day." *Nonlinear Processes in Geophysics*, 10(3):233–243, 2003. doi: 10.5194/npg-10-233-2003.

M. Corazza, E. Kalnay, and S. C. Yang. An implementation of the local ensemble Kalman filter in a quasi geostrophic model and comparison with 3D-Var. *Nonlinear Processes in Geophysics*, 14(1):89–101, 2007. doi: 10.5194/npg-14-89-2007.

Richard Courant and David Hilbert. *Methods of Mathematical Physics: Partial Differential Equations*, volume II. Interscience Publishers, Inc., New York, 1962.

P. Courtier, J.-N. Thépaut, and A. Hollingsworth. A strategy for operational implementation of 4D-Var, using an incremental approach. *Quarterly Journal of the Royal Meteorological Society*, 120(519):1367–1387, July 1994. doi: 10.1002/qj.49712051912.

P. Courtier, E. Andersson, W. Heckley, D. Vasiljevic, M. Hamrud, A. Hollingsworth, F. Rabier, M. Fisher, and J. Pailleux. The ECMWF implementation of three-dimensional variational assimilation (3D-Var). I: Formulation. *Quarterly Journal of the Royal Meteorological Society*, 124(550):1783–1807, July 1998. doi: 10.1002/qj.49712455002.

Philippe Courtier and Jean-Francois Geleyn. A global numerical weather prediction model with variable resolution: Application to the shallow-water equations. *Quarterly Journal of the Royal Meteorological Society*, 114(483):1321–1346, July 1988. doi: 10.1002/qj. 49711448309.

Philippe Courtier and Olivier Talagrand. Variational assimilation of meteorological observations with the direct and adjoint shallow-water equations. *Tellus A*, 42(5):531–549, October 1990. doi: 10.1034/j.1600-0870.1990.t01-4-00004.x.

George P. Cressman. An operational objective analysis system. *Monthly Weather Review*, 87 (10):367–374, October 1959. doi: 10.1175/1520-0493(1959)087<0367:AOOAS>2.0.CO;2.

M. J. P. Cullen. The unified forecast/climate model. *Meteorological Magazine*, 122(1449):81–94, 1993.

Jean Côté, Jean-Guy Desmarais, Sylvie Gravel, André Méthot, Alain Patoine, Michel Roch, and Andrew Staniforth. The operational CMC–MRB global environmental multiscale (GEM) model. Part II: Results. *Monthly Weather Review*, 126(6):1397–1418, June 1998a. doi: 10. 1175/1520-0493(1998)126<1397:TOCMGE>2.0.CO;2.

Jean Côté, Sylvie Gravel, André Méthot, Alain Patoine, Michel Roch, and Andrew Staniforth. The operational CMC–MRB global environmental multiscale (GEM) model. Part I: Design considerations and formulation. *Monthly Weather Review*, 126(6):1373–1395, June 1998b. doi: 10.1175/1520-0493(1998)126<1373:TOCMGE>2.0.CO;2.

Germund Dahlquist and Åke Björck. *Numerical Methods*. Prentice-Hall, 1974.

Amnon Dalcher and Eugenia Kalnay. Error growth and predictability in operational ECMWF forecasts. *Tellus A: Dynamic Meteorology and Oceanography*, 39(5):474–491, January 1987. doi: 10.3402/tellusa.v39i5.11774.

Amnon Dalcher, Eugenia Kalnay, and Ross N. Hoffman. Medium range lagged average forecasts. *Monthly Weather Review*, 116(2):402–416, February 1988. doi: 10.1175/1520-0493(1988)116<0402:MRLAF>2.0.CO;2.

Roger Daley. *Atmospheric Data Analysis*. Cambridge University Press, 1991. ISBN 978-0-521-45825-2. Google-Books-ID: RHM6pTMRTHwC.

Roger Daley and Edward Barker. NAVDAS: Formulation and diagnostics. *Monthly Weather Review*, 129(4):869–883, April 2001. doi: 10.1175/1520-0493(2001)129<0869:NFAD>2.0.CO;2.

H. C. Davies. A lateral boundary formulation for multi-level prediction models. *Quarterly Journal of the Royal Meteorological Society*, 102(432):405–418, April 1976. doi: 10.1002/qj.49710243210.

Huw C. Davies. Limitations of some common lateral boundary schemes used in regional NWP models. *Monthly Weather Review*, 111(5):1002–1012, May 1983. doi: 10.1175/1520-0493(1983)111<1002:LOSCLB>2.0.CO;2.

T. Davies, M. J. P. Cullen, A. J. Malcolm, M. H. Mawson, A. Staniforth, A. A. White, and N. Wood. A new dynamical core for the Met Office's global and regional modelling of the atmosphere. *Quarterly Journal of the Royal Meteorological Society*, 131(608):1759–1782, April 2005. doi: 10.1256/qj.04.101.

Helena Barbieri de Azevedo, Luis Goncalves, Eugenia Kalnay, and Matthew Wespetal. Dynamically weighted hybrid gain data assimilation: Perfect model experiments. *AGUFM*, 2018: IN43C–0913, 2018.

James W. Deardorff. Parameterization of the planetary boundary layer for use in general circulation models. *Monthly Weather Review*, 100(2):93–106, February 1972. doi: 10.1175/1520-0493(1972)100<0093:POTPBL>2.3.CO;2.

Dick P. Dee and Arlindo M. Da Silva. Data assimilation in the presence of forecast bias. *Quarterly Journal of the Royal Meteorological Society*, 124(545):269–295, January 1998. doi: 10.1002/qj.49712454512.

J. Derber and F. Bouttier. A reformulation of the background error covariance in the ECMWF global data assimilation system. *Tellus A: Dynamic Meteorology and Oceanography*, 51(2): 195–221, January 1999. doi: 10.3402/tellusa.v51i2.12316.

John C. Derber. A variational continuous assimilation technique. *Monthly Weather Review*, 117 (11):2437–2446, November 1989. doi: 10.1175/1520-0493(1989)117<2437:AVCAT>2.0.CO;2.

John C. Derber and Wan-Shu Wu. The use of TOVS cloud-cleared radiances in the NCEP SSI analysis system. *Monthly Weather Review*, 126(8):2287–2299, August 1998. doi: 10.1175/1520-0493(1998)126<2287:TUOTCC>2.0.CO;2.

John C. Derber, David F. Parrish, and Stephen J. Lord. The new global operational analysis system at the national meteorological center. *Weather and Forecasting*, 6(4):538–547, December 1991. doi: 10.1175/1520-0434(1991)006<0538:TNGOAS>2.0.CO;2.

308   **Bibliography**

R. E. Dickinson. Land surface. In *Climate System Modeling*, page 788. Cambridge University Press, 1992.

Geoffrey J. DiMego. The national meteorological center regional analysis system. *Monthly Weather Review*, 116(5):977–1000, May 1988. doi: 10.1175/1520-0493(1988)116<0977: TNMCRA>2.0.CO;2.

Geoffrey J. DiMego, Patricia A. Phoebus, and James E. McDonell. Data processing, and quality control for optimum interpolation analyses at the National Meteorological Center. Technical Report Office Note 306, U.S. Dept. of Commerce, National Oceanic and Atmospheric Administration, National Weather Service., Washington, D.C., 1985. https://repository.library.noaa.gov/view/noaa/11502.

K. K. Droegemeier. The numerical prediction of thunderstorms: Challenges, potential benefits and results from real-time operational tests. *Bulletin of the World Meteorological Organization*, 46(4):324–335, 1997.

Dale R. Durran. *Numerical Methods for Wave Equations in Geophysical Fluid Dynamics*. Springer Science & Business Media, 1999. ISBN 978-1-4757-3081-4. Google-Books-ID: JioBCAAAQBAJ.

Dale R. Durran. *Numerical Methods for Fluid Dynamics: With Applications to Geophysics*. Springer Science & Business Media, September 2010. ISBN 978-1-4419-6412-0. Google-Books-ID: ThMZrEOTuuUC.

Dale R. Durran, Ming-Jen Yang, Donald N. Slinn, and Randy G. Brown. Toward more accurate wave-permeable boundary conditions. *Monthly Weather Review*, 121(2):604–620, February 1993. doi: 10.1175/1520-0493(1993)121<0604:TMAWPB>2.0.CO;2.

Bo R. Döös and Max A. Eaton. Upper-air analysis over ocean areas1. *Tellus*, 9(2):184–194, 1957. doi: 10.1111/j.2153-3490.1957.tb01872.x.

W. Ebisuzaki and E. Kalnay. Ensemble experiments with a new lagged average forecasting scheme. Technical Report 15, World Meteorological Organization, Geneva, Switzerland, 1991.

Anthony William Fairbank Edwards. *Likelihood*. CUP Archive, 1984.

Martin Ehrendorfer and Joseph J. Tribbia. Optimal prediction of forecast error covariances through singular vectors. *Journal of the Atmospheric Sciences*, 54(2):286–313, January 1997. doi: 10.1175/1520-0469(1997)054<0286:OPOFEC>2.0.CO;2.

Amal El Akkraoui, Pierre Gauthier, Simon Pellerin, and Samuel Buis. Intercomparison of the primal and dual formulations of variational data assimilation. *Quarterly Journal of the Royal Meteorological Society*, 134(633):1015–1025, April 2008. doi: 10.1002/qj.257.

A. Eliassen, J. S. Sawyer, and Smagorinsky, J. Upper air network requirements for numerical weather prediction. Technical Report 29, World Meteorological Organization, 1954.

Arnt Eliassen. *The Quasi-Static Equations of Motion with Pressure as Independent Variable*, volume 17. Grøndahl & Sons boktr., I kommisjon hos Cammermeyers boghandel, 1949.

Arnt Eliassen and Elmer Raustein. A numerical integration experiment with a model atmosphere based on isentropic surfaces. *Meteorologiske Annaler*, 5:45–63, 1968.

Kerry Emanuel and D. J. Raymond. *The Representation of Cumulus Convection in Numerical Models*, volume 24 of *Meteorological Monographs*. American Meteorological Society, 1993. ISBN 978-1-935704-13-3. Google-Books-ID: utC9BwAAQBAJ.

Edward S. Epstein. Stochastic dynamic prediction. *Tellus*, 21(6):739–759, January 1969. doi: 10.3402/tellusa.v21i6.10143.

R. Errico and D. Baumhefner. Predictability experiments using a high-resolution limited-area model. *Monthly Weather Review*, 115(2):488–504, February 1987. doi: 10.1175/1520-0493(1987)115<0488:PEUAHR>2.0.CO;2.

Ronald M. Errico. What is an adjoint model? *Bulletin of the American Meteorological Society*, 78(11):2577–2591, November 1997. doi: 10.1175/1520-0477(1997)078<2577:WIAAM>2. 0.CO;2.

Ronald M. Errico and Tomislava Vukicevic. Sensitivity analysis using an adjoint of the PSU-NCAR mesoscale model. *Monthly Weather Review*, 120(8):1644–1660, August 1992. doi: 10.1175/1520-0493(1992)120<1644:SAUAAO>2.0.CO;2.

Erin Evans, Nadia Bhatti, Jacki Kinney, Lisa Pann, Malaquias Peña, Shu-Chih Yang, Eugenia Kalnay, and James Hansen. Rise undergraduates find that regime changes in Lorenz's model are predictable. *Bulletin of the American Meteorological Society*, 85(4):520–524, 2004.

Geir Evensen. Sequential data assimilation with a nonlinear quasi-geostrophic model using Monte Carlo methods to forecast error statistics. *Journal of Geophysical Research: Oceans*, 99(C5):10143–10162, May 1994. doi: 10.1029/94JC00572.

Geir Evensen. The ensemble Kalman Filter: Theoretical formulation and practical implementation. *Ocean Dynamics*, 53(4):343–367, November 2003. doi: 10.1007/s10236-003-0036-9.

Geir Evensen. *Data Assimilation: The Ensemble Kalman Filter*. Springer Science & Business Media, 2009.

Joel H. Ferziger and Milovan Perić. *Computational Methods for Fluid Dynamics*, volume 3. Springer, 2002.

Claude Fischer, Alain Joly, and François Lalaurette. Error growth and Kalman filtering within an idealized baroclinic flow. *Tellus A: Dynamic Meteorology and Oceanography*, 50(5):596–615, January 1998. doi: 10.3402/tellusa.v50i5.14561.

E. M. Fischer, S. Sippel, and R. Knutti. Increasing probability of record-shattering climate extremes. *Nature Climate Change*, 11(8):689–695, August 2021. doi: 10.1038/s41558-021-01092-9.

Martin Fischer and Antonio Navarra. GIOTTO: A coupled atmosphere-ocean general-circulation model: The tropical Pacific. *Quarterly Journal of the Royal Meteorological Society*, 126(567):1991–2012, July 2000. doi: 10.1002/qj.49712656702.

M. Fisher, M. Leutbecher, and G. A. Kelly. On the equivalence between Kalman smoothing and weak-constraint four-dimensional variational data assimilation. *Quarterly Journal of the Royal Meteorological Society*, 131(613):3235–3246, October 2005. doi: 10.1256/qj.04.142.

Michael Fisher and Philippe Courtier. Estimating the covariance matrices of analysis and forecast error in variational data assimilation. Technical report, European Centre for Medium-Range Weather Forecasts, 1995.

Thomas W. Flattery. Spectral models for global analysis and forecasting. Technical Report 243, Air Weather Service, US Air Force, 1971.

Clive A. J. Fletcher. Computational techniques for fluid dynamics. Volume 1-Fundamental and general techniques. *Springer-Verlag*, 1:409, 1988.

Michael S. Fox-Rabinovitz, Georgiy L. Stenchikov, Max J. Suarez, and Lawrence L. Takacs. A finite-difference GCM dynamical core with a variable-resolution stretched grid. *Monthly Weather Review*, 125(11):2943–2968, November 1997. doi: 10.1175/1520-0493(1997)125<2943:AFDGDC>2.0.CO;2.

J. M. Fritsch, J. Hilliker, J. Ross, and R. L. Vislocky. Model consensus. *Weather and Forecasting*, 15(5):571–582, October 2000. doi: 10.1175/1520-0434(2000)015<0571:MC>2.0. CO;2.

Bojie Fu and Yan Li. Bidirectional coupling between the Earth and human systems is essential for modeling sustainability. *National Science Review*, 3(4):397–398, December 2016. doi: 10.1093/nsr/nww094.

A. J. Gadd. A split explicit integration scheme for numerical weather prediction. *Quarterly Journal of the Royal Meteorological Society*, 104(441):569–582, 1978. doi: 10.1002/qj.49710444103.

L. Gandin. *Objective Analysis of Meteorological Fields*. Gidrometeorologicheskoe Izdatel'stvo (GIMIZ), Leningrad, 1963. Translated to English by the Israel Program for Scientific Translations.

Lev S. Gandin. Complex quality control of meteorological observations. *Monthly Weather Review*, 116(5):1137–1156, May 1988. doi: 10.1175/1520-0493(1988)116<1137:CQCOMO>2.0.CO;2.

J. R. Garratt. *The Atmospheric Boundary Layer*. Cambridge University Press, 1994.

Gregory Gaspari and Stephen E. Cohn. Construction of correlation functions in two and three dimensions. *Quarterly Journal of the Royal Meteorological Society*, 125(554):723–757, January 1999. doi: 10.1002/qj.49712555417.

W. Lawrence Gates, James S. Boyle, Curt Covey, Clyde G. Dease, Charles M. Doutriaux, Robert S. Drach, Michael Fiorino, Peter J. Gleckler, Justin J. Hnilo, Susan M. Marlais, Thomas J. Phillips, Gerald L. Potter, Benjamin D. Santer, Kenneth R. Sperber, Karl E. Taylor, and Dean N. Williams. An overview of the results of the atmospheric model intercomparison project (AMIP I). *Bulletin of the American Meteorological Society*, 80(1):29–55, January 1999. doi: 10.1175/1520-0477(1999)080<0029:AOOTRO>2.0.CO;2.

Ronald Gelaro and Yanqiu Zhu. Examination of observation impacts derived from observing system experiments (OSEs) and adjoint models. *Tellus A*, 61(2):179–193, March 2009. doi: 10.1111/j.1600-0870.2008.00388.x.

Ronald Gelaro, Rolf H. Langland, Simon Pellerin, and Ricardo Todling. The THORPEX observation impact intercomparison experiment. *Monthly Weather Review*, 138(11):4009–4025, November 2010. doi: 10.1175/2010MWR3393.1.

Joseph P. Gerrity, Ronald D. McPherson, and Paul D. Polger. On the efficient reduction, of truncation error in numerical weather prediction models. *Monthly Weather Review*, 100(8): 637–643, August 1972. doi: 10.1175/1520-0493(1972)100<0637:OTEROT>2.3.CO;2.

M. Ghil, S. Cohn, J. Tavantzis, K. Bube, and E. Isaacson. Applications of estimation theory to numerical weather prediction. In *Dynamic Meteorology: Data Assimilation Methods*, Applied Mathematical Sciences, pages 139–224. Springer, New York, 1981. doi: 10.1007/978-1-4612-5970-1_5.

Michael Ghil and Paola Malanotte-Rizzoli. Data assimilation in meteorology and oceanography. In Renata Dmowska and Barry Saltzman, editors, *Advances in Geophysics*, volume 33, pages 141–266. Elsevier, January 1991. doi: 10.1016/S0065-2687(08)60442-2.

J. K. Gibson, P. Kallberg, S. Uppala, A. Nomura, A. Hernandez, and E. Serrano. ECMWF reanalysis description. Technical Report 1, ECMWF, 1997.

Ralf Giering. Adjoint model compiler. *MPI report*, 1995.

Ralf Giering and Thomas Kaminski. Recipes for adjoint code construction. *ACM Transactions on Mathematical Software*, 24(4):437–474, December 1998. doi: 10.1145/293686.293695.

B. Gilchrist and G. P. Cressman. An experiment in objective analysis. *Tellus*, 6(4):309–318, January 1954. ISSN 0040-2826. doi: 10.3402/tellusa.v6i4.8762.

A. E. Gill. Some simple solutions for heat-induced tropical circulation. *Quarterly Journal of the Royal Meteorological Society*, 106(449):447–462, July 1980. doi: 10.1002/qj.49710644905.

Harry R. Glahn and Dale A. Lowry. The use of model output statistics (MOS) in objective weather forecasting. *Journal of Applied Meteorology (1962–1982)*, 11(8):1203–1211, 1972.

Gene H. Golub and Charles F. Van Loan. *Matrix Computations*. The Johns Hopkins University Press, Ltd., 3 edition, 1996.

Neil J. Gordon, David J. Salmond, and Adrian F. M. Smith. Novel approach to nonlinear/non-Gaussian Bayesian state estimation. In *IEE Proceedings F (Radar and Signal Processing)*, volume 140, pages 107–113. IET, 1993. Issue: 2.

Wojciech W. Grabowski. Coupling cloud processes with the large-scale dynamics using the cloud-resolving convection parameterization (CRCP). *Journal of the Atmospheric Sciences*, 58(9):978–997, May 2001. doi: 10.1175/1520-0469(2001)058<0978:CCPWTL>2.0.CO;2.

Wojciech W Grabowski and Piotr K Smolarkiewicz. CRCP: A cloud resolving convection parameterization for modeling the tropical convecting atmosphere. *Physica D: Nonlinear Phenomena*, 133(1):171–178, September 1999. doi: 10.1016/S0167-2789(99)00104-9.

Gabriele Gramelsberger. Conceiving meteorology as the exact science of the atmosphere: Vilhelm Bjerknes's paper of 1904 as a milestone, December 2009. www.ingentaconnect.com/content/schweiz/mz/2009/00000018/00000006/art00009.

Anne Greenbaum. *Iterative Methods for Solving Linear Systems*. SIAM, 1997.

Steven J. Greybush, Eugenia Kalnay, Takemasa Miyoshi, Kayo Ide, and Brian R. Hunt. Balance and ensemble Kalman filter localization techniques. *Monthly Weather Review*, 139(2):511–522, February 2011. doi: 10.1175/2010MWR3328.1.

Stephen M. Griffies and Alistair J. Adcroft. Formulating the equations of ocean models. In *Ocean Modeling in an Eddying Regime*, pages 281–317. American Geophysical Union (AGU), 2008. doi: 10.1029/177GM18.

Bertil Gustafsson, Heinz-Otto Kreiss, and Joseph Oliger. *Time Dependent Problems and Difference Methods*, volume 24. John Wiley & Sons, 1995. ISBN 0-471-50734-2.

N. Gustafsson and J. Bojarova. Four-dimensional ensemble variational (4D-En-Var) data assimilation for the high resolution limited area model (HIRLAM). *Nonlinear Processes in Geophysics*, 21(4):745–762, 2014. doi: 10.5194/npg-21-745-2014.

Nils Gustafsson. A review of methods for objective analysis. In Lennart Bengtsson, Michael Ghil, and Erland Källén, editors, *Dynamic Meteorology: Data Assimilation Methods*, pages 17–76. Springer New York, New York, 1981. doi: 10.1007/978-1-4612-5970-1_2.

Wolfgang Hackbusch. *Multi-grid Methods and Applications*. Springer-Verlag, Berlin, 1985.

L. A. Hageman and David M. Young. *Applied Iterative Methods*. Wiley, 1981.

Dale B. Haidvogel and Frank O. Bryan. Ocean general circulation modeling. In *Climate System Modeling*, page 788. Cambridge University Press, Cambridge, UK, 1992.

George J. Haltiner and Roger T. Williams. *Numerical Prediction and Dynamic Meteorology*. John Wiley and Sons, 1980.

Thomas M. Hamill and Chris Snyder. A hybrid ensemble Kalman filter–3D variational analysis scheme. *Monthly Weather Review*, 128(8):2905–2919, August 2000. doi: 10.1175/1520-0493(2000)128<2905:AHEKFV>2.0.CO;2.

Thomas M. Hamill, Chris Snyder, and Rebecca E. Morss. A comparison of probabilistic forecasts from bred, singular-vector, and perturbed observation ensembles. *Monthly Weather Review*, 128(6):1835–1851, June 2000. doi: 10.1175/1520-0493(2000)128<1835:ACOPFF>2.0.CO;2.

Thomas M. Hamill, Jeffrey S. Whitaker, and Chris Snyder. Distance-dependent filtering of background error covariance estimates in an ensemble Kalman filter. *Monthly Weather Review*, 129(11):2776–2790, November 2001. doi: 10.1175/1520-0493(2001)129<2776:DDFOBE>2.0.CO;2.

Mats Hamrud, Massimo Bonavita, and Lars Isaksen. EnKF and hybrid gain ensemble data assimilation. *ECMWF Technical Memoranda*, 2014. www.ecmwf.int/node/9766. Publisher: ECMWF.

Mats Hamrud, Massimo Bonavita, and Lars Isaksen. EnKF and hybrid gain ensemble data assimilation. Part I: EnKF implementation. *Monthly Weather Review*, 143(12):4847–4864, November 2015. doi: 10.1175/MWR-D-14-00333.1.

John Harlim. *Errors in the initial conditions for numerical weather prediction: A study of error growth patterns and error reduction with ensemble filtering*. PhD Thesis, 2006.

D. L. Hartmann, R. Buizza, and T. N. Palmer. Singular vectors: The effect of spatial scale on linear growth of disturbances. *Journal of the Atmospheric Sciences*, 52(22):3885–3894, November 1995. doi: 10.1175/1520-0469(1995)052<3885:SVTEOS>2.0.CO;2.

L. Hascoët and V. Pascual. The tapenade automatic differentiation tool: principles, model, and specification. *ACM Transactions On Mathematical Software*, 39(3), 2013.

Ross N. Hoffman. A four-dimensional analysis exactly satisfying equations of motion. *Monthly Weather Review*, 114(2):388–397, February 1986. doi: 10.1175/1520-0493(1986)114<0388: AFDAES>2.0.CO;2.

Ross N. Hoffman and Eugenia Kalnay. Lagged average forecasting, an alternative to Monte Carlo forecasting. *Tellus A*, 35A(2):100–118, March 1983. doi: 10.1111/j.1600-0870.1983. tb00189.x.

James E. Hoke and Richard A. Anthes. The initialization of numerical models by a dynamic-initialization technique. *Monthly Weather Review*, 104(12):1551–1556, December 1976. doi: 10.1175/1520-0493(1976)104<1551:TIONMB>2.0.CO;2.

A. Hollingsworth. An experiment in Monte Carlo forecasting. In *Proceedings from Workshop on Stochastic-Dynamic Forecasting*, pages 65–85, Shinfield Park, reading, UK, 1980.

A. Hollingsworth and P. Lönnberg. The statistical structure of short-range forecast errors as determined from radiosonde data. Part I: The wind field. *Tellus A*, 38A(2):111–136, March 1986. doi: 10.1111/j.1600-0870.1986.tb00460.x.

James Holton. *Introduction to Dynamic Meteorology*. International Geophysics (Book 48). Academic Press, Inc., 3rd edition, April 1992. www.textbooks.com/Introduction-to-Dynamic-Meteorology-3rd-Edition/9780123543554/James-R-Holton.php.

Song-You Hong and Eugenia Kalnay. Role of sea surface temperature and soil-moisture feedback in the 1998 Oklahoma–Texas drought. *Nature*, 408(6814):842–844, December 2000. doi: 10.1038/35048548.

Song-You Hong and Hua-Lu Pan. Nonlocal boundary layer vertical diffusion in a medium-range forecast model. *Monthly Weather Review*, 124(10):2322–2339, October 1996. doi: 10.1175/1520-0493(1996)124<2322:NBLVDI>2.0.CO;2.

John D. Horel and John M. Wallace. Planetary-scale atmospheric phenomena associated with the southern oscillation. *Monthly Weather Review*, 109(4):813–829, April 1981. doi: 10.1175/1520-0493(1981)109<0813:PSAPAW>2.0.CO;2.

Daisuke Hotta, Eugenia Kalnay, and Paul Ullrich. A semi-implicit modification to the Lorenz N-cycle scheme and its application for integration of meteorological equations. *Monthly Weather Review*, 144(6):2215–2233, January 2016. doi: 10.1175/MWR-D-15-0330.1.

Daisuke Hotta, Tse-Chun Chen, Eugenia Kalnay, Yoichiro Ota, and Takemasa Miyoshi. Proactive QC: A fully flow-dependent quality control scheme based on EFSO. *Monthly Weather Review*, 145(8):3331–3354, July 2017. doi: 10.1175/MWR-D-16-0290.1.

Dingchen Hou, Eugenia Kalnay, and Kelvin K. Droegemeier. Objective verification of the SAMEX '98 ensemble forecasts. *Monthly Weather Review*, 129(1):73–91, January 2001. doi: 10.1175/1520-0493(2001)129<0073:OVOTSE>2.0.CO;2.

P. L. Houtekamer and Herschel L. Mitchell. Data assimilation using an ensemble Kalman filter technique. *Monthly Weather Review*, 126(3):796–811, March 1998. doi: 10.1175/1520-0493(1998)126<0796:DAUAEK>2.0.CO;2.

P. L. Houtekamer and Herschel L. Mitchell. A sequential ensemble Kalman filter for atmospheric data assimilation. *Monthly Weather Review*, 129(1):123–137, January 2001. doi: 10.1175/1520-0493(2001)129<0123:ASEKFF>2.0.CO;2.

P. L. Houtekamer and Fuqing Zhang. Review of the ensemble Kalman filter for atmospheric data assimilation. *Monthly Weather Review*, 144(12):4489–4532, June 2016. doi: 10.1175/MWR-D-15-0440.1.

P. L. Houtekamer, Louis Lefaivre, Jacques Derome, Harold Ritchie, and Herschel L. Mitchell. A system simulation approach to ensemble prediction. *Monthly Weather Review*, 124(6): 1225–1242, June 1996. doi: 10.1175/1520-0493(1996)124<1225:ASSATE>2.0.CO;2.

James G. Howcroft. *Local Forecast Model, Present Status and Preliminary Verification*. US Department of Commerce, National Oceanic and Atmospheric Administration, National Weather Service, National Meteorological Center, Camp Spring, MD, 1971.

Francis D. Hughes. Skill of the medium range forecast group. 1987. https://repository.library .noaa.gov/view/noaa/11485.

B. R. Hunt, E. Kalnay, E. J. Kostelich, E. Ott, D. J. Patil, T. Sauer, I. Szunyogh, J. A. Yorke, and A. V. Zimin. Four-dimensional ensemble Kalman filtering. *Tellus A: Dynamic Meteorology and Oceanography*, 56(4):273–277, January 2004. doi: 10.3402/tellusa.v56i4.14424.

Brian R. Hunt. Efficient data assimilation for spatiotemporal chaos: A local ensemble transform Kalman filter. https://arxiv.org/abs/physics/0511236v1, 2005.

Brian R. Hunt, Eric J. Kostelich, and Istvan Szunyogh. Efficient data assimilation for spatiotemporal chaos: A local ensemble transform Kalman filter. *Physica D: Nonlinear Phenomena*, 230(1):112–126, June 2007. doi: 10.1016/j.physd.2006.11.008.

Kayo Ide, Philippe Courtier, Michael Ghil, and Andrew C. Lorenc. Unified notation for data assimilation: operational, sequential and variational (gtspecial issue it data assimilation in meteology and oceanography: theory and practice). *Journal of the Meteorological Society of Japan. Ser. II*, 75(1B):181–189, 1997.

Motohki Ikawa. *Description of a nonhydrostatic model developed at the Forecast Research Department of the MRI.*, volume 28. Technical Reports of the MRI, 1991.

IPCC. *Climate Change 2022: Impacts, Adaptation and Vulnerability*. Summary for Policymakers. Cambridge University Press, Cambridge, UK and New York, 2022.

Lars Isaksen, M. Bonavita, R Buizza, M. Fisher, J. Haseler, M. Leutbecher, and Laure Raynaud. Ensemble of data assimilations at ECMWF. Technical Memorandum 636, European Centre for Medium-Range Weather Forecasts, Shinfield Park, Reading, Berkshire RG2 9AX, England, December 2010.

David R. Jackett, Trevor J. McDougall, Rainer Feistel, Daniel G. Wright, and Stephen M. Griffies. Algorithms for density, potential temperature, conservative temperature, and the freezing temperature of seawater. *Journal of Atmospheric and Oceanic Technology*, 23(12): 1709–1728, December 2006. doi: 10.1175/JTECH1946.1.

Mark Z. Jacobson, Mark A. Delucchi, Mary A. Cameron, and Bethany A. Frew. Low-cost solution to the grid reliability problem with 100% penetration of intermittent wind, water, and solar for all purposes. *Proceedings of the National Academy of Sciences*, 112(49):15060, December 2015. doi: 10.1073/pnas.1510028112.

Mark Z. Jacobson, Mark A. Delucchi, Mary A. Cameron, and Brian V. Mathiesen. Matching demand with supply at low cost in 139 countries among 20 world regions with 100% intermittent wind, water, and sunlight (WWS) for all purposes. *Renewable Energy*, 123:236–248, August 2018. doi: 10.1016/j.renene.2018.02.009.

Ian N. James. *Introduction to Circulating Atmospheres*. Cambridge University Press, NY, 1994.

Bibliography

Z. I. Janjic. A nonhydrostatic model based on a new approach. *Meteorology and Atmospheric Physics*, 82(1–4):271–285, January 2003. doi: 10.1007/s00703-001-0587-6.

Zavisa I. Janjic, J. P. Gerrity Jr, and S. Nickovic. An alternative approach to nonhydrostatic modeling. *Monthly Weather Review*, 129(5):1164–1178, 2001.

Zaviša I. Janjić. A stable centered difference scheme free of two-grid-interval noise. *Monthly Weather Review*, 102(4):319–323, 1974.

Zaviša I. Janjić. The step-mountain coordinate: physical package. *Monthly Weather Review*, 118 (7):1429–1443, July 1990. doi: 10.1175/1520-0493(1990)118<1429:TSMCPP>2.0.CO;2.

Zaviša I. Janjić. The step-mountain eta coordinate model: further developments of the convection, viscous sublayer, and turbulence closure schemes. *Monthly Weather Review*, 122(5): 927–945, May 1994. doi: 10.1175/1520-0493(1994)122<0927:TSMECM>2.0.CO;2.

A. H. Jazwinski. *Stochastic Processes and Filtering Theory*. Academic Press, 1970.

Ming Ji, Arun Kumar, and Ants Leetmaa. An experimental coupled forecast system at the National Meteorological Center. *Tellus A*, 46(4):398–418, 1994.

Ming Ji, Ants Leetmaa, and Vernon E. Kousky. Coupled model predictions of ENSO during the 1980s and the 1990sat the National Centers for Environmental Prediction. *Journal of Climate*, 9(12):3105–3120, 1996.

Donald R. Johnson, Tom H. Zapotocny, Fred M. Reames, Bart J. Wolf, and R. Bradley Pierce. A comparison of simulated precipitation by hybrid isentropic-sigma and sigma models. *Monthly Weather Review*, 121(7):2088–2114, 1993.

J. Joiner and A. M. Da Silva. Efficient methods to assimilate remotely sensed data based on information content. *Quarterly Journal of the Royal Meteorological Society*, 124(549):1669–1694, July 1998. doi: 10.1002/qj.49712454915.

R. G. Jones, J. M. Murphy, and M. Noguer. Simulation of climate change over europe using a nested regional-climate model. I: Assessment of control climate, including sensitivity to location of lateral boundaries. *Quarterly Journal of the Royal Meteorological Society*, 121 (526):1413–1449, 1995.

Hann-Ming Henry Juang. The NCEP mesoscale spectral model: A revised version of the nonhydrostatic regional spectral model. *Monthly Weather Review*, 128(7):2329–2362, July 2000. doi: 10.1175/1520-0493(2000)128<2329:TNMSMA>2.0.CO;2.

Hann-Ming Henry Juang and Masao Kanamitsu. The NMC nested regional spectral model. *Monthly Weather Review*, 122(1):3–26, 1994.

Hann-Ming Henry Juang, Song-You Hong, and Masao Kanamitsu. The NCEP regional spectral model: An update. *Bulletin of the American Meteorological Society*, 78(10):2125–2143, 1997.

Heikki Järvinen, Erik Andersson, and François Bouttier. Variational assimilation of time sequences of surface observations with serially correlated errors. *Tellus A*, 51(4):469–488, August 1999. doi: 10.1034/j.1600-0870.1999.t01-4-00002.x.

Eigil Kaas, Annette Guldberg, Wilhelm May, and Michel Déqué. Using tendency errors to tune the parameterisation of unresolved dynamical scale interactions in atmospheric general circulation models. *Tellus A: Dynamic Meteorology and Oceanography*, 51(5):612–629, January 1999. doi: 10.3402/tellusa.v51i5.14481.

John S. Kain and J. Michael Fritsch. A one-dimensional entraining/detraining plume model and its application in convective parameterization. *Journal of the Atmospheric Sciences*, 47 (23):2784–2802, December 1990. doi: 10.1175/1520-0469(1990)047<2784:AODEPM>2. 0.CO;2.

Per Kallberg. Test of a boundary relaxation scheme in a barotropic model, ECMWF Res. Technical report, Dept Internal Report, 1977.

Rudolph E. Kalman and Richard S. Bucy. New results in linear filtering and prediction theory. *Journal of Basic Engineering*, 83(1):95–108, 1961.

Rudolph E. Kalman. A new approach to linear filtering and prediction problems. 1960.

E. Kalnay and M. Ham. Forecasting forecast skill in the Southern Hemisphere. In *Preprints of the 3rd International Conference on Southern Hemisphere Meteorology and Oceanography, Buenos Aires*, pages 13–17, 1989.

E. Kalnay and Z. Toth. The breeding method. In *Proceedings 1995 ECMWF Seminar on Predictability*, volume 1, pages 69–82, 1996.

E. Kalnay, M. Kanamitsu, R. Kistler, W. Collins, D. Deaven, L. Gandin, M. Iredell, S. Saha, G. White, J. Woollen, Y. Zhu, A. Leetmaa, R. Reynolds, M. Chelliah, W. Ebisuzaki, W. Higgins, J. Janowiak, K. C. Mo, C. Ropelewski, J. Wang, Roy Jenne, and Dennis Joseph. The NCEP/NCAR 40-year reanalysis project. *Bulletin of the American Meteorological Society*, 77(3):437–471, March 1996. doi: 10.1175/1520-0477(1996)077<0437:TNYRP>2.0.CO;2.

E. Balgovind Kalnay. Documentation of the GLAS fourth order general circulation model. Volume 2: Scalar code. Technical report, December 1983. https://ntrs.nasa.gov/search.jsp?R=19840015981.

Eugenia Kalnay and Zoltan Toth. Removing growing errors in the analysis cycle. In *Proceedings of the 10th Conference on Numerical Weather Prediction, 18–22 July, 1994, Portland, OR*, pages 212–215, Portland, OR, 1994. American Meteorological Society.

Eugenia Kalnay and Shu-Chih Yang. Accelerating the spin-up of ensemble Kalman filtering. *Quarterly Journal of the Royal Meteorological Society*, 136(651):1644–1651, July 2010. doi: 10.1002/qj.652.

Eugenia Kalnay, David L. T. Anderson, Andrew F. Bennett, Antonio J. Busalacchi, Stephen E. Cohn, Philippe Courtier, John Derber, Andrew C. Lorenc, David Parrish, and Thomas Schlatter. Data Assimilation in the Ocean and in the Atmosphere: What Should be Next? January 1997. http://ntrs.nasa.gov/search.jsp?R=20000037959.

Eugenia Kalnay, Stephen J. Lord, and Ronald D. McPherson. Maturity of operational numerical weather prediction: Medium range. *Bulletin of the American Meteorological Society*, 79 (12):2753–2769, December 1998. doi: 10.1175/1520-0477(1998)079<2753:MOONWP>2.0.CO;2.

Eugenia Kalnay, Matteo Corazza, and Ming Cai. Are bred vectors the same as Lyapunov vectors? page 6820, 2002.

Eugenia Kalnay, Hong Li, Takemasa Miyoshi, shu-chih Yang, and Joaquim Balabrera-Poy. 4-D-Var or ensemble Kalman filter? *Tellus A*, 59(5):758–773, October 2007a. doi: 10.1111/j.1600-0870.2007.00261.x.

Eugenia Kalnay, Hong Li, Takemasa Miyoshi, Shu-Chih Yang, and Joaquim Ballabrera-Poy. Response to the discussion on "4-D-Var or EnKF?" by Nils Gustafsson. *Tellus A: Dynamic Meteorology and Oceanography*, 59(5):778–780, January 2007b. doi: 10.1111/j.1600-0870.2007.00263.x.

Eugenia Kalnay, Yoichiro Ota, Takemasa Miyoshi, and Junjie Liu. A simpler formulation of forecast sensitivity to observations: Application to ensemble Kalman filters. *Tellus A: Dynamic Meteorology and Oceanography*, 64(1):18462, December 2012. doi: 10.3402/tellusa.v64i0.18462.

Eugenia Kalnay, Travis Sluka, Takuma Yoshida, Cheng Da, and Safa Mote. Towards strongly-coupled ensemble data assimilation with additional improvements from machine learning. *Nonlinear Processes in Geophysics Discussion*, 2023:1–31. January 2023. doi: 10.5194/npg-2023-1.

E. Bayliss Kalnay-Rivas. The 4th order GISS model of the global atmosphere. January 1977. https://ntrs.nasa.gov/search.jsp?R=19770066699.

Eugenia Kalnay-Rivas and Lee-Or Merkine. A simple mechanism for blocking. *Journal of the Atmospheric Sciences*, 38(10):2077–2091, 1981.

M. Kanamitsu, J.C. Alpert, K.A. Campana, P.M. Caplan, D.G. Deaven, M. Iredell, B. Katz, H.-L. Pan, J. Sela, and G.H. White. Recent changes implemented into the global forecast system at NMC. *Weather and Forecasting*, 6(3):425–435, September 1991. doi: 10.1175/1520-0434(1991)006<0425:RCIITG>2.0.CO;2.

James L. Kaplan and James A. Yorke. Chaotic behavior of multidimensional difference equations. In *Functional Differential Equations and Approximation of Fixed Points*, pages 204–227. Springer, 1979.

Akira Kasahara. Various vertical coordinate systems used for numerical weather prediction. *Monthly Weather Review*, 102(7):509–522, July 1974. doi: 10.1175/1520-0493(1974)102<0509:VVCSUF>2.0.CO;2.

Christian L. Keppenne. Data assimilation into a primitive-equation model with a parallel ensemble kalman filter. *Monthly Weather Review*, 128(6):1971–1981, June 2000. doi: 10.1175/1520-0493(2000)128<1971:DAIAPE>2.0.CO;2.

Marat Khairoutdinov, David Randall, and Charlotte DeMott. Simulations of the atmospheric general circulation using a cloud-resolving model as a superparameterization of physical processes. *Journal of the Atmospheric Sciences*, 62(7):2136–2154, July 2005. doi: 10.1175/JAS3453.1.

Marat F. Khairoutdinov and David A. Randall. A cloud resolving model as a cloud parameterization in the NCAR community climate system model: Preliminary results. *Geophysical Research Letters*, 28(18):3617–3620, 2001. doi: 10.1029/2001GL013552.

Jeffrey T. Kiehl. Atmospheric general circulation modeling. In *Climate System Modeling*, pages 319–370. Cambridge University Press, Cambridge, UK, 1992.

Young-Joon Kim and Akio Arakawa. Improvement of orographic gravity wave parameterization using a mesoscale gravity wave model. *Journal of the Atmospheric Sciences*, 52(11):1875–1902, June 1995. doi: 10.1175/1520-0469(1995)052<1875:IOOGWP>2.0.CO;2.

Robert Kistler, William Collins, Suranjana Saha, Glenn White, John Woollen, Eugenia Kalnay, Muthuvel Chelliah, Wesley Ebisuzaki, Masao Kanamitsu, and Vernon Kousky. The NCEP–NCAR 50–year reanalysis: Monthly means CD–ROM and documentation. *Bulletin of the American Meteorological Society*, 82(2):247–267, 2001.

Robert Edward Kistler. *A Study of Data Assimilation Techniques in an Autobarotropic, Primitive Equation, Channel Model*. PhD Thesis, Pennsylvania State University, 1974.

Daryl T. Kleist and Kayo Ide. An OSSE-based evaluation of hybrid variational– ensemble data assimilation for the NCEP GFS. Part I: System description and 3D-Hybrid results. *Monthly Weather Review*, 143(2):433–451, February 2015a. doi: 10.1175/MWR-D-13-00351.1.

Daryl T. Kleist and Kayo Ide. An OSSE-Based evaluation of hybrid variational– ensemble data assimilation for the NCEP GFS. Part II: 4DEnVar and hybrid variants. *Monthly Weather Review*, 143(2):452–470, February 2015b. doi: 10.1175/MWR-D-13-00350.1.

Daryl T. Kleist, David F. Parrish, John C. Derber, Russ Treadon, Wan-Shu Wu, and Stephen Lord. Introduction of the GSI into the NCEP global data assimilation system. *Weather and Forecasting*, 24(6):1691–1705, December 2009. doi: 10.1175/2009WAF2222201.1.

Daryl Timothy Kleist. *An evaluation of hybrid variational-ensemble data assimilation for the NCEP GFS*. PhD Thesis, 2012.

J. B. Klemp and D. K. Lilly. Numerical simulation of hydrostatic mountain waves. *Journal of Atmospheric Sciences*, 35(1):78–107, January 1978. doi: 10.1175/1520-0469(1978)035<0078:NSOHMW>2.0.CO;2.

J. B. Klemp, W. C. Skamarock and J. Dudhia. Conservative split-Explicit time integration methods for the compressible nonhydrostatic equations. *Monthly Weather Review*, 135(8): 2897–2913, August 2007. doi: 10.1175/MWR3440.1.

Joseph B. Klemp and Dale R. Durran. An upper boundary condition permitting internal gravity wave radiation in numerical mesoscale models *Monthly Weather Review*, 111(3):430–444, March 1983. doi: 10.1175/1520-0493(1983)111<0430:AUBCPI>2.0.CO;2.

Joseph B. Klemp and Robert B. Wilhelmson. The simulation of three-dimensional convective storm dynamics *Journal of the Atmospheric Sciences*, 35(6):1070–1096, June 1978. doi: 10.1175/1520-0469(1978)035<1070:TSOTDC>2.0.CO;2.

Thomas Knutson, Suzana J. Camargo, Johnny C. L. Chan, Kerry Emanuel, Chang-Hoi Ho, James Kossin, Mrutyunjay Mohapatra, Masaki Satoh, Masato Sugi, Kevin Walsh, and Liguang Wu. Tropical cyclones and climate change assessment: Part II: Projected response to anthropogenic warming. *Bulletin of the American Meteorological Society*, 101 (3):E303–E322, March 2020. doi: 10.1175/BAMS-D-18-0194.1.

Chiaki Kobayashi, Kiyoharu Takano, Shoji Kusunoki, Masato Sugi, and Akio Kitoh. Seasonal predictability in winter over eastern Asia using the JMA global model. *Quarterly Journal of the Royal Meteorological Society*, 126(567):2111–2123, 2000.

Dmitrii Kochkov, Jamie A. Smith, Ayya Alieva, Qing Wang, Michael P. Brenner, and Stephan Hoyer. Machine learning–accelerated computational fluid dynamics. *Proceedings of the National Academy of Sciences*, 118(21):e2101784118, May 2021. doi: 10.1073/pnas.2101784118.

Keiichi Kondo and Takemasa Miyoshi. Impact of removing covariance localization in an ensemble Kalman filter: Experiments with 10 240 members using an intermediate AGCM. *Monthly Weather Review*, 144(12):4849–4865, November 2016. doi: 10.1175/MWR-D-15-0388.1.

Randal D. Koster, Max J. Suarez, and Mark Heiser. Variance and predictability of precipitation at seasonal-to-interannual timescales. *Journal of Hydrometeorology*, 1(1):26–46, February 2000. doi: 10.1175/1525-7541(2000)001<0026:VAPOPA>2.0.CO;2.

Randal D. Koster, Zhichang Guo, Rongqian Yang, Paul A. Dirmeyer, Kenneth Mitchell, and Michael J. Puma. On the nature of soil moisture in land surface models. *Journal of Climate*, 22(16):4322–4335, August 2009. doi: 10.1175/2009JCLI2832.1.

Shunji Kotsuki, Steven J. Greybush, and Takemasa Miyoshi. Can we optimize the assimilation order in the serial ensemble Kalman filter? A study with the Lorenz- 96 model. *Monthly Weather Review*, 145(12):4977–4995, December 2017a. doi: 10.1175/MWR-D-17-0094.1.

Shunji Kotsuki, Takemasa Miyoshi, Koji Terasaki, Guo-Yuan Lien, and Eugenia Kalnay. Assimilating the global satellite mapping of precipitation data with the nonhydrostatic icosahedral atmospheric model (NICAM). *Journal of Geophysical Research: Atmospheres*, 122(2): 631–650, January 2017b. doi: 10.1002/2016JD025355.

Vladimir M. Krasnopolsky. *The Application of Neural Networks in the Earth System Sciences*, volume 46. Springer, 2013. ISBN 978-94-007-6073-8. www.springer.com/gp/book/9789400760721.

T. N. Krishnamurti, C. M. Kishtawal, Zhan Zhang, Timothy LaRow, David Bachiochi, Eric Williford, Sulochana Gadgil, and Sajani Surendran. Multimodel ensemble forecasts for weather and seasonal climate. *Journal of Climate*, 13(23):4196–4216, December 2000. doi: 10.1175/1520-0442(2000)013<4196:MEFFWA>2.0.CO;2.

Bibliography

Jürgen Kröger and Fred Kucharski. Sensitivity of ENSO characteristics to a new interactive flux correction scheme in a coupled GCM. *Climate Dynamics*, 36(1):119–137, January 2011. doi: 10.1007/s00382-010-0759-5.

Fred Kucharski, Franco Molteni, and Annalisa Bracco. Decadal interactions between the western tropical Pacific and the North Atlantic Oscillation. *Climate Dynamics*, 26(1):79–91, January 2006. doi: 10.1007/s00382-005-0085-5.

Fred Kucharski, Ning Zeng, and Eugenia Kalnay. A further assessment of vegetation feedback on decadal Sahel rainfall variability. *Climate Dynamics*, 40(5):1453–1466, 2012. doi: 10.1007/s00382-012-1397-x.

H. L. Kuo. Further studies of the parameterization of the Influence of cumulus convection on large-Scale flow. *Journal of the Atmospheric Sciences*, 31(5):1232–1240, July 1974. doi: 10.1175/1520-0469(1974)031<1232:FSOTPO>2.0.CO;2.

Yoshio Kurihara and Morris A. Bender. Use of a movable nested-mesh model for tracking a small vortex. *Monthly Weather Review*, 108(11):1792–1809, November 1980. doi: 10.1175/1520-0493(1980)108<1792:UOAMNM>2.0.CO;2.

Boris Khattatov, William Lahoz, and Richard Menard. *Data Assimilation*. Springer, 2010.

R. H. Langland, Z. Toth, R. Gelaro, I. Szunyogh, M. A. Shapiro, S. J. Majumdar, R. E. Morss, G. D. Rohaly, C. Velden, N. Bond, and C. H. Bishop. The North Pacific Experiment (NORPEX-98): Targeted observations for improved North American weather forecasts. *Bulletin of the American Meteorological Society*, 80(7):1363–1384, July 1999. doi: 10.1175/1520-0477(1999)080<1363:TNPENT>2.0.CO;2.

Rolf H. Langland and Nancy L. Baker. Estimation of observation impact using the NRL atmospheric variational data assimilation adjoint system. *Tellus A: Dynamic Meteorology and Oceanography*, 56(3):189–201, January 2004. doi: 10.3402/tellusa.v56i3.14413.

René Laprise. The Euler equations of motion with hydrostatic pressure as an independent variable. *Monthly Weather Review*, 120(1):197–207, January 1992. doi: 10.1175/1520-0493(1992)120<0197:TEEOMW>2.0.CO;2.

René Laprise, Daniel Caya, Guy Bergeron, and Michel Giguère. The formulation of the André Robert MC2 (mesoscale compressible community) model. *Atmosphere-Ocean*, 35(sup1):195–220, January 1997. doi: 10.1080/07055900.1997.9687348.

M. Latif, T. P. Barnett, M. A. Cane, M. Flügel, N. E. Graham, H. von Storch, J.-S. Xu, and S. E. Zebiak. A review of ENSO prediction studies. *Climate Dynamics*, 9(4–5):167–179, January 1994. doi: 10.1007/BF00208250.

Mojib Latif, D. Anderson, T. Barnett, M. Cane, R. Kleeman, A. Leetmaa, J. O'Brien, A. Rosati, and E. Schneider. A review of the predictability and prediction of ENSO. *Journal of Geophysical Research: Oceans*, 103(C7):14375–14393, 1998.

Kody Law, Andrew Stuart, and Kostas Zygalakis. *Data Assimilation*. Springer, Cham, 2015.

B. Legras and R. Vautard. A guide to Liapunov vectors. In *Proceedings 1995 ECMWF Seminar on Predictability*, volume 1, pages 143–156, 1996.

C. E. Leith. Spectral statistical-dynamical forecast experiments. In *The GARP Programme on Numerical Experimentation*, volume 7 of *GARP Working Group on Numerical Experimentation*, pages 445–467, Geneva, Switzerland, 1974a. World Meteorological Organization.

C. E. Leith. Theoretical skill of monte carlo forecasts. *Monthly Weather Review*, 102(6):409–418, June 1974b. doi: 10.1175/1520-0493(1974)102<0409:TSOMCF>2.0.CO;2.

Lance M. Leslie and R. James Purser. Three-dimensional mass-conserving semi- lagrangian scheme employing forward trajectories. *Monthly Weather Review*, 123(8):2551–2566, August 1995. doi: 10.1175/1520-0493(1995)123<2551:TDMCSL>2.0.CO;2.

John M. Lewis and John C. Derber. The use of adjoint equations to solve a variational adjustment problem with advective constraints. *Tellus A: Dynamic Meteorology and Oceanography*, 37 (4):309–322, January 1985. doi: 10.3402/tellusa.v37i4.11675.

John M. Lewis, Sivaramakrishnan Lakshmivarahan, and Sudarshan Dhall. *Dynamic Data Assimilation: A Least Square Approach*, volume 13. Cambridge University Press, 2006.

Yan Li, Eugenia Kalnay, Safa Motesharrei, Jorge Rivas, Fred Kucharski, Daniel Kirk-Davidoff, Eviatar Bach, and Ning Zeng. Climate model shows large-scale wind and solar farms in the Sahara increase rain and vegetation. *Science*, 361(6406):1019–1022, September 2018. doi: 10.1126/science.aar5629.

G.-Y. Lien, D. Hotta, E. Kalnay T. Miyoshi, and T.-C. Chen. Accelerating assimilation development for new observing systems using EFSO. *Nonlinear Processes in Geophysics*, 25(1): 129–143, 2018. doi: 10.5194/npg-25-129-2018.

Guo-Yuan Lien, Eugenia Kalnay, and Takemasa Miyoshi. Effective assimilation of global precipitation: Simulation experiments. *Tellus A: Dynamic Meteorology and Oceanography*, 65 (1):19915, December 2013. doi: 10.3402/tellusa.v65i0.19915.

Guo-Yuan Lien, Eugenia Kalnay, Takemasa Miyoshi, and George J. Huffman. Statistical properties of global precipitation in the NCEP GFS model and TMPA observations for data assimilation. *Monthly Weather Review*, 144(2):663–679, February 2016a. doi: 10.1175/ MWR-D-15-0150.1.

Guo-Yuan Lien, Takemasa Miyoshi, and Eugenia Kalnay. Assimilation of TRMM multisatellite precipitation analysis with a low-resolution NCEP global forecast system. *Monthly Weather Review*, 144(2):643–661, February 2016b. doi: 10.1175/MWR-D-15-0149.1.

D. K. Lilly and P. J. Kennedy Observations of a stationary mountain wave and its associated momentum flux and energy dissipation. *Journal of the Atmospheric Sciences*, 30(6):1135–1152, September 1973. doi: 10.1175/1520-0469(1973)030<1135:OOASMW>2.0.CO;2.

Douglas K. Lilly. On the computational stability of numerical solutions of time-dependent nonlinear geophysical fluid dynamics problems. *Monthly Weather Review*, 93(1):11–26, 1965.

Shian-Jiann Lin. A finite-volume integration method for computing pressure gradient force in general vertical coordinates. *Quarterly Journal of the Royal Meteorological Society*, 123 (542):1749–1762, July 1997. doi: 10.1002/qj.49712354214.

Shian-Jiann Lin. A "Vertically Lagrangian" finite-volume dynamical core for global models. *Monthly Weather Review*, 132(10):2293–2307, October 2004. doi: 10.1175/1520-0493 (2004)132<2293:AVLFDC>2.0.CO;2.

Shian-Jiann Lin and Richard B. Rood. Multidimensional flux-form semi-lagrangian transport schemes. *Monthly Weather Review*, 124(9):2046–2070, September 1996. doi: 10.1175/ 1520-0493(1996)124<2046:MFFSLT>2.0.CO;2.

Shian-Jiann Lin and Richard B. Rood. An explicit flux-form semi-lagrangian shallow-water model on the sphere. *Quarterly Journal of the Royal Meteorological Society*, 123(544): 2477–2498, October 1997. doi: 10.1002/qj.49712354416.

Richard S. Lindzen. Supersaturation of vertically propagating internal gravity waves. *Journal of the Atmospheric Sciences*, 45(4):705–711, February 1988. doi: 10.1175/1520-0469(1988) 045<0705:SOVPIG>2.0.CO;2.

Chengsi Liu, Qingnong Xiao, and Bin Wang. An ensemble-based four-dimensional variational data assimilation scheme. Part I: Technical formulation and preliminary test. *Monthly Weather Review*, 136(9) 3363–3373, September 2008. doi: 10.1175/2008MWR2312.1.

Chengsi Liu, Qingnong Xiao, and Bin Wang. An ensemble-based four-dimensional variational data assimilation scheme. Part II: Observing system simulation experiments with advanced research WRF (ARW). *Monthly Weather Review*, 137(5):1687–1704, May 2009. doi: 10. 1175/2008MWR2699.1.

Y. Liu, Z. Liu, S. Zhang, R. Jacob, F. Lu, X. Rong, and S. Wu. Ensemble-based parameter estimation in a coupled general circulation model. *Journal of Climate*, 27(18):7151–7162, September 2014a. doi: 10.1175/JCLI-D-13-00406.1.

Y. Liu, Z. Liu, S. Zhang, X. Rong, R. Jacob, S. Wu, and F. Lu. Ensemble-based parameter estimation in a coupled GCM using the adaptive spatial average method. *Journal of Climate*, 27(11):4002–4014, June 2014b. doi: 10.1175/JCLI-D-13-00091.1.

Stephen Lord, George Gayno, and Fanglin Yang. Analysis of an observing system experiment for the joint polar satellite system. *Bulletin of the American Meteorological Society*, 97(8): 1409–1425, August 2016. doi: 10.1175/BAMS-D-14-00207.1.

A. C. Lorenc. A global three-dimensional multivariate statistical interpolation scheme. *Monthly Weather Review*, 109(4):701–721, April 1981. doi: 10.1175/1520-0493(1981)109<0701: AGTDMS>2.0.CO;2.

A. C. Lorenc. Analysis methods for numerical weather prediction. *Quarterly Journal of the Royal Meteorological Society*, 112(474):1177–1194, October 1986. doi: 10.1002/qj. 49711247414.

Andrew C. Lorenc. Development of an operational variational assimilation scheme (gtSpecial issueltData assimilation in meteology and oceanography: Theory and practice). *Journal of the Meteorological Society of Japan. Ser. II*, 75(1B):339–346, 1997.

Andrew C. Lorenc. Modelling of error covariances by 4D-Var data assimilation. *Quarterly Journal of the Royal Meteorological Society*, 129(595):3167–3182, October 2003a. doi: 10.1256/qj.02.131.

Andrew C. Lorenc. The potential of the ensemble Kalman filter for NWP – a comparison with 4D-Var. *Quarterly Journal of the Royal Meteorological Society*, 129(595):3183–3203, October 2003b. doi: 10.1256/qj.02.132.

Andrew C. Lorenc and Richard T. Marriott. Forecast sensitivity to observations in the met office global numerical weather prediction system. *Quarterly Journal of the Royal Meteorological Society*, 140(678):209–224, January 2014. doi: 10.1002/qj.2122.

Andrew C. Lorenc, Neill E. Bowler, Adam M. Clayton, Stephen R. Pring, and David Fairbairn. Comparison of hybrid-4DEnVar and hybrid-4DVar data assimilation methods for global NWP. *Monthly Weather Review*, 143(1):212–229, January 2015. doi: 10.1175/ MWR-D-14-00195.1.

E. N. Lorenz. Atmospheric predictability experiments with a large numerical model. *Tellus*, 34 (6):505–513, 1982.

Edward N. Lorenz. Available potential energy and the maintenance of the general circulation. *Tellus*, 7(2):157–167, 1955.

Edward N. Lorenz. Energy and numerical weather prediction. *Tellus*, 12(4):364–373, 1960.

Edward N. Lorenz. Deterministic nonperiodic flow. *Journal of the Atmospheric Sciences*, 20(2): 130–141, March 1963a. doi: 10.1175/1520-0469(1963)020<0130:DNF>2.0.CO;2.

Edward N. Lorenz. The predictability of hydrodynamic flow. *Transactions of the New York Academy of Sciences*, 25(4 Series II):409–432, February 1963b. doi: 10.1111/j.2164-0947. 1963.tb01464.x.

Edward N. Lorenz. A study of the predictability of a 28-variable atmospheric model. *Tellus*, 17 (3):321–333, 1965.

Edward N. Lorenz. The predictability of a flow which possesses many scales of motion. *Tellus*, 21(3):289–307, 1969.

Edward N. Lorenz. Charney – a remarkable colleague. In Richard S. Lindzen, Edward N. Lorenz, and George W. Platzman, editors, *The Atmosphere – A Challenge: The Science of Jule Gregory Charney*, pages 89–91. American Meteorological Society, Boston, MA, 1990. doi: 10.1007/978-1-944970-35-2.

Edward N. Lorenz. *The Essence of Chaos*. The Jessie and John Danz lectures. University of Washington Press, Seattle, 1993. ISBN 978-0-295-97270-1 978-1-85728-454-6 978-1-85728-187-3. www.gbv.de/dms/bowker/toc/9781857284546.pdf.

Edward N. Lorenz. Predictability: A problem partly solved. In *Proc. Seminar on Predictability*, volume 1, 1996.

François Lott and Martin J. Miller. A new subgrid-scale orographic drag parametrization: Its formulation and testing. *Quarterly Journal of the Royal Meteorological Society*, 123(537): 101–127, 1997.

Jean-Francois Louis. A parametric model of vertical eddy fluxes in the atmosphere. *Boundary-Layer Meteorology*, 17(2):187–202, 1979.

Zhixin Lu, Brian R. Hunt, and Edward Ott. Attractor reconstruction by machine learning. *Chaos: An Interdisciplinary Journal of Nonlinear Science*, 28(6):061104, June 2018. doi: 10.1063/1.5039508.

Peter Lynch. *The Emergence of Numerical Weather Prediction: Richardson's Dream*. Cambridge University Press, November 2006. ISBN 978-0-521-85729-1. Google-Books-ID: EV5bZqOO7kkC.

Roland A. Madden. On predicting probability distributions of time-averaged meteorological data. *Journal of Climate*, 2(8):922–925, 1989.

Roland A. Madden and Paul R. Julian. Detection of a 40–50 day oscillation in the zonal wind in the tropical Pacific. *Journal of the Atmospheric Sciences*, 28(5):702–708, 1971.

Roland A. Madden and Paul R. Julian. Description of global-scale circulation cells in the tropics with a 40–50 day period. *Journal of the Atmospheric Sciences*, 29(6):1109–1123, 1972.

Linus Magnusson and Erland Källén. Factors influencing skill improvements in the ECMWF forecasting system. *Monthly Weather Review*, 141(9):3142–3153, September 2013. doi: 10.1175/MWR-D-12-00318.1.

Syukuro Manabe, Joseph Smagorinsky, and Robert F. Strickler. Simulated climatology of a general circulation model with a hydrologic cycle. *Monthly Weather Review*, 93(12):769–798, 1965.

Taroh Matsuno. Numerical integrations of the primitive equations by a simulated backward difference method. *Journal of the Meteorological Society of Japan. Ser. II*, 44(1):76–84, 1966a.

Taroh Matsuno. Quasi-geostrophic motions in the equatorial area. *Journal of the Meteorological Society of Japan. Ser. II*, 44(1):25–43, 1966b. doi: 10.2151/jmsj1965.44.1_25.

A. McDonald and Janerik Haugen. A two-time-level, three-dimensional semi-Lagrangian, semi-implicit, limited-area gridpoint model of the primitive equations. *Monthly Weather Review*, 120(11):2603–2621, 1992.

Aidan McDonald. *Lateral boundary conditions for operational regional forecast models; a review*. HIRLAM 4 Project, 1997.

N. A. McFarlane. The effect of orographically excited gravity wave drag on the general circulation of the lower stratosphere and troposphere. *Journal of the Atmospheric Sciences*, 44(14): 1775–1800, 1987.

Ronald D. McPherson, K. H. Bergman, R. E. Kistler, G. E. Rasch, and D. S. Gordon. The NMC operational global data assimilation system. *Monthly Weather Review*, 107(11):1445–1461, 1979.

J. C. McWilliams. Modeling the oceanic general circulation. *Annual Review of Fluid Mechanics*, 28(1):215–248, January 1996. doi: 10.1146/annurev.fl.28.010196.001243.

George L. Mellor and Tetsuji Yamada. A hierarchy of turbulence closure models for planetary boundary layers. *Journal of the Atmospheric Sciences*, 31(7):1791–1806, October 1974. doi: 10.1175/1520-0469(1974)031<1791:AHOTCM>2.0.CO;2.

George L. Mellor and Tetsuji Yamada. Development of a turbulence closure model for geophysical fluid problems. *Reviews of Geophysics*, 20(4):851–875, 1982.

F. Mesinger. A blocking technique for representation of mountains in atmospheric models. *Rivista di Meteorologia Aeronautica*, 44(1–4):195–202, 1984. http://cat.inist.fr/?aModele=afficheN&cpsidt=8419610.

Fedor Mesinger. Forward-backward scheme, and its use in a limited area model. *Contributions to Atmospheric Physics*, 50(1977):200–210, 1977.

Fedor Mesinger, Zaviša I. Janjić, Slobodan Ničković, Dušanka Gavrilov, and Dennis G. Deaven. The step-mountain coordinate: Model description and performance for cases of Alpine Lee Cyclogenesis and for a case of an Appalachian redevelopment. *Monthly Weather Review*, 116 (7):1493–1518, July 1988. doi: 10.1175/1520-0493(1988)116<1493:TSMCMD>2.0.CO;2.

Nicholas Metropolis and S. Ulam. The Monte Carlo method. *Journal of the American Statistical Association*, 44(247):335–341, September 1949. doi: 10.1080/01621459.1949.10483310.

Gerard Meurant. *Computer Solution of Large Linear Systems*. Elsevier, 1999.

M. J. Miller and A. J. Thorpe. Radiation conditions for the lateral boundaries of limited-area numerical models. *Quarterly Journal of the Royal Meteorological Society*, 107(453):615–628, July 1981. doi: 10.1002/qj.49710745310.

Robert N. Miller, Edward D. Zaron, and Andrew F. Bennett. Data assimilation in models with convective adjustment. *Monthly Weather Review*, 122(11):2607–2613, November 1994. doi: 10.1175/1520-0493(1994)122<2607:DAIMWC>2.0.CO;2.

Frank J. Millero and Alain Poisson. International one-atmosphere equation of state of seawater. *Deep Sea Research Part A. Oceanographic Research Papers*, 28(6):625–629, June 1981. doi: 10.1016/0198-0149(81)90122-9.

Herschel L. Mitchell, P. L. Houtekamer, and Gérard Pellerin. Ensemble size, balance, and model-error representation in an ensemble Kalman filter*. *Monthly Weather Review*, 130 (11):2791–2808, November 2002. doi: 10.1175/1520-0493(2002)130<2791:ESBAME>2.0.CO;2.

K. Miyakoda and J. Sirutis. Comparative global prediction experiments on parameterized subgrid-scale vertical eddy transports. *Control of Atmospheric Physics*, 50:445–487, 1977.

K. Miyakoda, L. Umscheid, D. H. Lee, J. Sirutis, R. Lusen, and F. Pratte. The near-real-time, global, four-dimensional analysis experiment during the GATE period, Part I. *Journal of the Atmospheric Sciences*, 33(4):561–591, April 1976. doi: 10.1175/1520-0469(1976) 033<0561:TNRTGF>2.0.CO;2.

T. Miyoshi and Q. Sun. Control simulation experiment with Lorenz's butterfly attractor. *Nonlinear Processes in Geophysics*, 29(1):133–139, March 2022. doi: 10.5194/npg-29-133-2022.

Takemasa Miyoshi. *Ensemble Kalman filter experiments with a primitive-equation global model*. PhD Thesis, University of Maryland, College Park, MD, 2005.

Takemasa Miyoshi, Eugenia Kalnay, and Hong Li. Estimating and including observation-error correlations in data assimilation. *Inverse Problems in Science and Engineering*, 21(3):387–398, April 2013. doi: 10.1080/17415977.2012.712527.

Takemasa Miyoshi, Keiichi Kondo, and Toshiyuki Imamura. The 10,240-member ensemble Kalman filtering with an intermediate AGCM. *Geophysical Research Letters*, 41(14): 5264–5271, July 2014. doi: 10.1002/2014GL060863.

Takemasa Miyoshi, Shunji Kotsuki, Keiichi Kondo, and Roland Potthast. Local particle filter implemented with minor modifications to the LETKF code. In *NG13A – Advances in Data Assimilation, Predictability, and Uncertainty Quantification I*, San Franciso, CA, 2019.

Chin-Hoh Moeng and John C. Wyngaard. Evaluation of turbulent transport and dissipation closures in second-order modeling. *Journal of the Atmospheric Sciences*, 46(14):2311–2330, July 1988. doi: 10.1175/1520-0469(1989)046<2311:EOTTAD>2.0.CO;2.

F. Molteni. Atmospheric simulations using a GCM with simplified physical parametrizations. I: Model climatology and variability in multi-decadal experiments. *Climate Dynamics*, 20 (2):175–191, January 2003. doi: 10.1007/s00382-002-0268-2.

F. Molteni, R. Buizza, T. N. Palmer, and T. Petroliagis. The ECMWF ensemble prediction system: Methodology and validation. *Quarterly Journal of the Royal Meteorological Society*, 122(529):73–119, January 1996. doi: 10.1002/qj.49712252905.

Franco Molteni and T. N. Palmer. Predictability and finite-time instability of the northern winter circulation. *Quarterly Journal of the Royal Meteorological Society*, 119(510):269–298, January 1993. doi: 10.1002/qj.49711951004.

A. S. Monin and A. M. Obukhov. Basic laws of turbulent mixing in the atmosphere near the ground. *Trudy Geofizicheskogo Instituta, Akademiya Nauk SSSR*, 24(151):163–187, 1954.

Rebecca E. Morss, Kerry A. Emanuel, and Chris Snyder. Idealized adaptive observation strategies for improving numerical weather prediction. *Journal of the Atmospheric Sciences*, 58 (2):210–232, January 2001. doi: 10.1175/1520-0469(2001)058<0210:IAOSFI>2.0.CO;2.

Safa Mote, Jorge Rivas, and Eugenia Kalnay. A novel approach to carrying capacity: From a priori prescription to a posteriori derivation based on underlying mechanisms and dynamics. *Annual Review of Earth and Planetary Sciences*, 48(1):657–683, May 2020. doi: 10.1146/annurev-earth-053018-060428.

Safa Motesharrei, Jorge Rivas, and Eugenia Kalnay. Human and nature dynamics (HANDY): Modeling inequality and use of resources in the collapse or sustainability of societies. *Ecological Economics*, 101:90–102, May 2014. doi: 10.1016/j.ecolecon.2014.02.014.

Safa Motesharrei, Jorge Rivas, Eugenia Kalnay, Ghassem R. Asrar, Antonio J. Busalacchi, Robert F. Cahalan, Mark A. Cane, Rita R. Colwell, Kuishuang Feng, Rachel S. Franklin, Klaus Hubacek, Fernando Miralles-Wilhelm, Takemasa Miyoshi, Matthias Ruth, Roald Sagdeev, Adel Shirmohammadi, Jagadish Shukla, Jelena Srebric, Victor M. Yakovenko, and Ning Zeng. Modeling sustainability: Population, inequality, consumption, and bidirectional coupling of the Earth and Human Systems. *National Science Review*, 3(4):470–494, December 2016. doi: 10.1093/nsr/nww081.

Steven L. Mullen and David P. Baumhefner. Monte carlo simulations of explosive cyclogenesis. *Monthly Weather Review*, 122(7):1548–1567, July 1994. doi: 10.1175/1520-0493(1994) 122<1548:MCSOEC>2.0.CO;2.

Aarne Männik and Rein Room. *Nonhydrostatic Adiabatic Kernel for HIRLAM: Part II: Anelastic, Hybrid-coordinate, Explicit-Eulerian Model*, volume 49. HIRLAM Technical Reports, 2001.

Richard Ménard and Roger Daley. The application of Kalman smoother theory to the estimation of 4DVAR error statistics. *Tellus A*, 48(2):221–237, 1996.

I. M. Navon and David M Legler. Conjugate-gradient methods for large-scale minimization in meteorology. *Monthly Weather Review*, 115(8):1479–1502, August 1987. doi: 10.1175/1520-0493(1987)115<1479:CGMFLS>2.0.CO;2.

I. M. Navon, X. Zou, J. Derber, and J. Sela. Variational data assimilation with an adiabatic version of the NMC spectral model. *Monthly Weather Review*, 120(7):1433–1446, July 1992. doi: 10.1175/1520-0493(1992)120<1433:VDAWAA>2.0.CO;2.

Bibliography

J. D. Neelin, M. Latif, and F. Jin. Dynamics of coupled ocean-atmosphere models: The tropical problem. *Annual Review of Fluid Mechanics*, 26(1):617–659, January 1994. doi: 10.1146/annurev.fl.26.010194.003153.

Lars Nerger, Tijana Janjić, Jens Schröter, and Wolfgang Hiller. A unification of ensemble square root Kalman filters. *Monthly Weather Review*, 140(7):2335–2345, July 2012. doi: 10.1175/MWR-D-11-00102.1.

N. K. Nichols. Mathematical Concepts of Data Assimilation. In William Lahoz, Boris Khattatov, and Richard Menard, editors, *Data Assimilation: Making Sense of Observations*, pages 13–39. Springer Berlin Heidelberg, Berlin, Heidelberg, 2010. doi: 10.1007/978-3-540-74703-1_2.

Adrienne Norwood, Eugenia Kalnay, Kayo Ide, Shu-Chih Yang, and Christopher Wolfe. Lyapunov, singular and bred vectors in a multi-scale system: An empirical exploration of vectors related to instabilities. *Journal of Physics A: Mathematical and Theoretical*, 46(25):254021, June 2013. doi: 10.1088/1751-8113/46/25/254021.

Yoshimitsu Ogura and Norman A. Phillips. Scale analysis of deep and shallow convection in the atmosphere. *Journal of the Atmospheric Sciences*, 19(2):173–179, March 1962. doi: 10.1175/1520-0469(1962)019<0173:SAODAS>2.0.CO;2.

J. Oliger and A. Sundström. Theoretical and practical aspects of some initial boundary value problems in fluid dynamics. *SIAM Journal on Applied Mathematics*, 35(3):419–446, November 1978. doi: 10.1137/0135035.

I. Orlanski. A simple boundary condition for unbounded hyperbolic flows. *Journal of Computational Physics*, 21(3):251–269, July 1976. doi: 10.1016/0021-9991(76)90023-1.

Steven A. Orszag. On the elimination of aliasing in finite-difference schemes by filtering high-wavenumber components. *Journal of the Atmospheric Sciences*, 28(6):1074–1074, September 1971. doi: 10.1175/1520-0469(1971)028<1074:OTEOAI>2.0.CO;2.

Yoichiro Ota, John C. Derber, Eugenia Kalnay, and Takemasa Miyoshi. Ensemble-based observation impact estimates using the NCEP GFS. *Tellus A: Dynamic Meteorology and Oceanography*, 65(1):20038, December 2013. doi: 10.3402/tellusa.v65i0.20038.

Edward Ott, Brian R. Hunt, Istvan Szunyogh, Matteo Corazza, Eugenia Kalnay, D. J. Patil, and James A. Yorke. Exploiting Local Low Dimensionality of the Atmospheric Dynamics for Efficient Ensemble Kalman Filtering. https://arxiv.org/abs/physics/0203058v1, 2002.

Edward Ott, Brian R. Hunt, Istvan Szunyogh, Aleksey V. Zimin, Eric J. Kostelich, Matteo Corazza, Eugenia Kalnay, D. J. Patil, and James A. Yorke. A local ensemble Kalman filter for atmospheric data assimilation. *Tellus A: Dynamic Meteorology and Oceanography*, 56(5):415–428, January 2004. doi: 10.3402/tellusa.v56i5.14462.

T. N. Palmer, G. J. Shutts, and R. Swinbank. Alleviation of a systematic westerly bias in general circulation and numerical weather prediction models through an orographic gravity wave drag parametrization. *Quarterly Journal of the Royal Meteorological Society*, 112(474):1001–1039, October 1986. doi: 10.1002/qj.49711247406.

T. N. Palmer, R. Gelaro, J. Barkmeijer, and R. Buizza. Singular vectors, metrics, and adaptive observations. *Journal of the Atmospheric Sciences*, 55(4):633–653, February 1998. doi: 10.1175/1520-0469(1998)055<0633:SVMAAO>2.0.CO;2.

T. N. Palmer, F. Molteni, R. Mureau, R. Buizza, P. Chapelet, and J. Tribbia. Ensemble prediction. In *Proceedings of the ECMWF Seminar on Validation of Models Over Europe*, volume 1, pages 21–66, 1993.

H.-L. Pan and L. Mahrt. Interaction between soil hydrology and boundary-layer development. *Boundary-Layer Meteorology*, 38(1–2):185–202, January 1987. doi: 10.1007/BF00121563.

Hua-Lu Pan. A simple parameterization scheme of evapotranspiration over land for the NMC medium-range forecast model. *Monthly Weather Review*, 118(12):2500–2512, December 1990. doi: 10.1175/1520-0493(1990)118<2500:ASPSOE>2.0.CO,2.

Hua-Lu Pan and Wan-Shu Wu. Implementing a mass flux convection parameterization package for the NMC medium-range forecast model. 1995. https://repository.library.noaa.gov/view/noaa/11429.

R. A. Panofsky. Objective weather-map analysis. *Journal of Meteorology*, 6(6):386–392, December 1949. doi: 10.1175/1520-0469(1949)006<0386:OWMA>2.0.CO;2.

Seon Ki Park and Liang Xu. *Data Assimilation for Atmospheric, Oceanic and Hydrologic Applications*, volume 2. Springer Science & Business Media, 2013.

David F. Parrish and John C. Derber. The national meteorological center's spectral statistical-interpolation analysis system. *Monthly Weather Review*, 120(8):1747–1763, August 1992. doi: 10.1175/1520-0493(1992)120<1747:TNMCSS>2.0.CO;2.

D. J. Patil, B. R. Hunt, E. Kalnay, J. A. Yorke, and E. Ott. Local low dimensionality of atmospheric dynamics. *Physical Review Letters*, 86(26 Pt 1):5878–5881, June 2001. doi: 10.1103/PhysRevLett.86.5878.

S. G. Penny, E. Kalnay, J. A. Carton, B. R. Hunt, K. Ide, T. Miyoshi, and G. A. Chepurin. The local ensemble transform Kalman filter and the running-in-place algorithm applied to a global ocean general circulation model. *Nonlinear Processes in Geophysics*, 20(6):1031–1046, November 2013. doi: 10.5194/npg-20-1031-2013.

Stephen G. Penny. The hybrid local ensemble transform Kalman filter. *Monthly Weather Review*, 142(6):2139–2149, May 2014. doi: 10.1175/MWR-D-13-00131.1.

Stephen G. Penny. Mathematical foundations of hybrid data assimilation from a synchronization perspective. *Chaos: An Interdisciplinary Journal of Nonlinear Science*, 27(12):126801, December 2017. doi: 10.1063/1.5001819.

Stephen G. Penny and Takemasa Miyoshi. A local particle filter for high-dimensional geophysical systems. *Nonlinear Processes in Geophysics*, 23(6):391–405, November 2016. doi: https://doi.org/10.5194/npg-23-391-2016.

Stephen G. Penny, David W. Behringer, James A. Carton, and Eugenia Kalnay. A hybrid global ocean data assimilation system at NCEP. *Monthly Weather Review*, 143(11):4660–4677, November 2015. doi: 10.1175/MWR-D-14-00376.1.

Donald J. Perkey and Carl W. Kreitzberg. A time-dependent lateral boundary scheme for limited-area primitive equation models. *Monthly Weather Review*, 104(6):744–755, June 1976. doi: 10.1175/1520-0493(1976)104<0744:ATDLBS>2.0.CO;2.

Anders Persson. Early operational Numerical Weather Prediction outside the USA: an historical introduction Part III: Endurance and mathematics – British NWP, 1948–1965. *Meteorological Applications*, 12(4):381–413, 2005a. doi: 10.1017/S1350482705001933.

Anders Persson. Early operational Numerical Weather Prediction outside the USA: An historical Introduction. Part 1: Internationalism and engineering NWP in Sweden, 1952–69. *Meteorological Applications*, 12(2):135–159, 2005b. doi: 10.1017/S1350482705001593.

Anders Persson. Early operational Numerical Weather Prediction outside the USA: An historical introduction: Part II: Twenty countries around the world. *Meteorological Applications*, 12(3): 269–289, 2005c. doi: 10.1017/S1350482705001751.

M. Peña, E. Kalnay, and M. Cai. Statistics of locally coupled ocean and atmosphere intraseasonal anomalies in Reanalysis and AMIP data. *Nonlinear Processes in Geophysics*, 10(3): 245–251, June 2003. doi: 10.5194/npg-10-245-2003.

Dinh Tuan Pham. Stochastic methods for sequential data assimilation in strongly nonlinear systems. *Monthly Weather Review*, 129(5):1194–1207, May 2001. doi: 10.1175/1520-0493(2001)129<1194:SMFSDA>2.0.CO;2.

S. G. Philander. *El Nino, La Nina, and the Southern Oscillation*. Academic Press, New York, 1990. https://books.google.com/books?id=9fwrkW_B1YYC.

N. A. Phillips. A coordinate system having some special advantages for numerical forecasting. *Journal of Meteorology*, 14(2):184–185, April 1957. doi: 10.1175/1520-0469(1957) 014<0184:ACSHSS>2.0.CO;2.

N. A. Phillips. Principles of large scale numerical weather prediction. In *Dynamic Meteorology*, pages 1–96. Springer, Dordrecht, 1973. doi: 10.1007/978-94-010-2599-7_1.

N. A. Phillips. The emergence of quasi-geostrophic theory. In Richard S. Lindzen, Edward N. Lorenz, and George W. Platzman, editors, *The Atmosphere – A Challenge: The Science of Jule Gregory Charney*, pages 177–206. American Meteorological Society, Boston, MA, 1990. ISBN 1-878220-03-9.

Norman A. Phillips. An example of non-linear computational instability. In *The Atmosphere and the Sea in Motion*, page 516. The Rockefeller Institute Press, New York, 1959.

Norman A. Phillips. The equations of motion for a shallow rotating atmosphere and the "Traditional Approximation." *Journal of the Atmospheric Sciences*, 23(5):626–628, September 1966. doi: 10.1175/1520-0469(1966)023<0626:TEOMFA>2.0.CO;2.

Norman A. Phillips. *The nested grid model*. NOAA. Tech. Report NWS 22, 1979. www.weather .gov/media/owp/oh/hdsc/docs/TR22.pdf.

Norman A. Phillips. The spatial statistics of random geostrophic modes and first-guess errors. *Tellus A*, 38A(4):314–332, August 1986. doi: 10.1111/j.1600-0870.1986.tb00418.x.

Norman A. Phillips. Carl-gustaf rossby: His times, personality, and actions. *Bulletin of the American Meteorological Society*, 79(6):1097–1112, June 1998. doi: 10.1175/ 1520-0477(1998)079<1097:CGRHTP>2.0.CO;2.

R. A. Pielke, W. R. Cotton, R. L. Walko, C. J. Tremback, W. A. Lyons, L. D. Grasso, M. E. Nicholls, M. D. Moran, D. A. Wesley, T. J. Lee, and J. H. Copeland. A comprehensive meteorological modeling system – RAMS. *Meteorology and Atmospheric Physics*, 49(1–4): 69–91, March 1992. doi: 10.1007/BF01025401.

Carlos Pires, Robert Vautard, and Oliver Talagrand. On extending the limits of variational assimilation in nonlinear chaotic systems. *Tellus A*, 48(1):96–121, January 1996. doi: 10.1034/j.1600-0870.1996.00006.x.

Jonathan Poterjoy. A localized particle filter for high-dimensional nonlinear systems. *Monthly Weather Review*, 144(1):59–76, January 2016. doi: 10.1175/MWR-D-15-0163.1.

Jonathan Poterjoy and Jeffrey L. Anderson. Efficient assimilation of simulated observations in a high-dimensional geophysical system using a localized particle filter. *Monthly Weather Review*, 144(5):2007–2020, April 2016. doi: 10.1175/MWR-D-15-0322.1.

Roland Potthast, Anne Walter, and Andreas Rhodin. A localized adaptive particle filter within an operational NWP framework. *Monthly Weather Review*, 147(1):345–362, January 2019. doi: 10.1175/MWR-D-18-0028.1.

Zhao-Xia Pu and Eugenia Kalnay. Targeting observations with the quasi-inverse linear and adjoint NCEP global models: Performance during FASTEX. *Quarterly Journal of the Royal Meteorological Society*, 125(561):3329–3337, October 1999. doi: 10.1002/qj.49712556110.

Zhao-Xia Pu, Eugenia Kalnay, David Parrish, Wanshu Wu, and Zoltan Toth. The use of bred vectors in the NCEP global 3D variational analysis system. *Weather and Forecasting*, 12(3): 689–695, September 1997a. doi: 10.1175/1520-0434(1997)012<0689:TUOBVI>2.0.CO;2.

Zhao-Xia Pu, Eugenia Kalnay, Joseph Sela, and Istvan Szunyogh. Sensitivity of forecast errors to initial conditions with a quasi-inverse linear method. *Monthly Weather Review*, 125(10): 2479–2503, October 1997b. doi: 10.1175/1520-0493(1997)125<2479:SOFETI>2.0.CO;2.

Zhao-Xia Pu, Stephen J. Lord, and Eugenia Kalnay. Forecast sensitivity with dropwindsonde data and targeted observations. *Tellus A: Dynamic Meteorology and Oceanography*, 50(4): 391–410, January 1998. doi: 10.3402/tellusa.v50i4.14536.

R. Purser. A new approach to optimal assimilation of meteorological data by iterative Bayesian analysis. Clearwater, FL, June 1984.

R. J. Purser and L. M. Leslie. An efficient interpolation procedure for high-order three-dimensional semi-lagrangian models. *Monthly Weather Review*, 119(10):2492–2498, October 1991. doi: 10.1175/1520-0493(1991)119<2492:AEIPFH>2.0.CO;2.

R. James Purser and Lance M. Leslie. Generalized Adams–Bashforth time integration schemes for a semi-Lagrangian model employing the second-derivative form of the horizontal momentum equations. *Quarterly Journal of the Royal Meteorological Society*, 122(531): 737–763, April 1996. doi: 10.1002/qj.49712253109.

R. James Purser, Wan-Shu Wu David F. Parrish, and Nigel M. Roberts. Numerical aspects of the application of recursive filters to variational statistical analysis. Part II: Spatially inhomogeneous and anisotropic general covariances. *Monthly Weather Review*, 131(8):1536–1548, 2003a.

R. James Purser, Wan-Shu Wu, David F. Parrish, and Nigel M. Roberts. Numerical aspects of the application of recursive filters to variational statistical analysis. Part I: Spatially homogeneous and isotropic Gaussian covariances. *Monthly Weather Review*, 131(8):1524–1535, 2003b.

William M. Putman and Shian-Jiann Lin. Finite-volume transport on various cubed-sphere grids. *Journal of Computational Physics*, 227(1):55–78, November 2007. doi: 10.1016/j.jcp.2007.07.022.

F. Rabier, E. Klinker, P. Courtier, and A. Hollingsworth. Sensitivity of forecast errors to initial conditions. *Quarterly Journal of the Royal Meteorological Society*, 122(529):121–150, January 1996. doi: 10.1002/qj.49712252906.

F. Rabier, A. McNally, E. Andersson, P. Courtier, P. Undén, J. Eyre, A. Hollingsworth, and F. Bouttier. The ECMWF implementation of three-dimensional variational assimilation (3D-Var). II: Structure functions. *Quarterly Journal of the Royal Meteorological Society*, 124 (550):1809–1829, July 1998. doi: 10.1002/qj.49712455003.

F. Rabier, H. Järvinen, E. Klinker, J.-F. Mahfouf, and A. Simmons. The ECMWF operational implementation of four-dimensional variational assimilation. I: Experimental results with simplified physics. *Quarterly Journal of the Royal Meteorological Society*, 126:1143–1170, 2000. doi: 10.1002/qj.49712656415.

Florence Rabier and Philippe Courtier. Four-dimensional assimilation in the presence of baroclinic instability. *Quarterly Journal of the Royal Meteorological Society*, 118(506):649–672, July 1992. doi: 10.1002/qj.49711850604.

David Randall, Marat Kharoutdinov, Akio Arakawa, and Wojciech Grabowski. Breaking the cloud parameterization deadlock. *Bulletin of the American Meteorological Society*, 84(11): 1547–1564, November 2003. doi: 10.1175/BAMS-84-11-1547.

David A. Randall. *General circulation model development: Past, present, and future*, volume 70. Academic Press, July 2000. ISBN 978-0-08-050723-1. Google-Books-ID: vnYeHl6AvgkC.

M. Rančić, R. J. Purser, and F. Mesinger. A global shallow-water model using an expanded spherical cube: Gnomonic versus conformal coordinates. *Quarterly Journal of the Royal Meteorological Society*, 122(532):959–982, April 1996. doi: 10.1002/qj.49712253209.

Carolyn A. Reynolds, Peter J. Webster, and Eugenia Kalnay. Random error growth in NMC's global forecasts. *Monthly Weather Review*, 122(6):1281–1305, June 1994. doi: 10.1175/1520-0493(1994)122<1281:REGING>2.0.CO;2.

Lewis Fry Richardson. *Prediction by Numerical Process*. Cambridge, The University Press, 1922. http://archive.org/details/weatherpredictio00richrich.

A. Robert. The integration of a spectral model of the atmosphere by the implicit method. In *Proc. WMO/IUGG Symposium on NWP, Tokyo, Japan Meteorological Agency*, volume 7, pages 19–24, 1969.

Andre J. Robert. The integration of a low order spectral form of the primitive meteorological equations. *Journal of the Meteorological Society of Japan. Ser. II*, 44(5):237–245, 1966. doi: 10.2151/jmsj1965.44.5_237.

André Robert. A semi-lagrangian and semi-implicit numerical integration scheme for the primitive meteorological equations. *Journal of the Meteorological Society of Japan. Ser. II*, 60 (1):319–325, 1982. doi: 10.2151/jmsj1965.60.1_319.

André Robert, Tai Loy Yee, and Harold Ritchie. A semi-lagrangian and semi-implicit numerical integration scheme for multilevel atmospheric models. *Monthly Weather Review*, 113(3): 388–394, March 1985. doi: 10.1175/1520-0493(1985)113<0388:ASLASI>2.0.CO;2.

Clive D. Rodgers. *Inverse Methods for Atmospheric Sounding*, Volume 2 of *Series on Atmospheric, Oceanic and Planetary Physics*. World Scientific, July 2000. doi: 10.1142/3171.

R. Room, A. Mannik, and A. Luhamaa. *Nonhydrostatic adiabatic kernel for HIRLAM Part IV Semi-implicit semi-Lagrangian scheme*, volume 65. HIRLAM Technical Reports, 2006.

Hoffman N. Ross and Robert Atlas. Future observing system simulation experiments. *Bulletin of the American Meteorological Society*, 97(9):1601–1616, 2016. ISSN 0003-0007.

C. G. Rossby. Planetary flow patterns in the atmosphere. *Quarterly Journal of the Royal Meteorological Society*, 66:68–87, 1939a.

Carl G. Rossby. Relation between variations in the intensity of the zonal circulation of the atmosphere and the displacements of the semi-permanent centers of action. *Journal of Marine Research*, 2:38–55, 1939b.

Nicole Rostaing, StéPhane Dalmas, and André Galligo. Automatic differentiation in Odyssée. *Tellus A: Dynamic Meteorology and Oceanography*, 45(5):558–568, January 1993. doi: 10. 3402/tellusa.v45i5.15060.

William F. Ruddiman. The anthropogenic greenhouse era began thousands of years ago. *Climatic Change*, 61(3):261–293, December 2003. doi: 10.1023/B:CLIM.0000004577. 17928.fa.

William F. Ruddiman. *Plows, Plagues, and Petroleum: How Humans Took Control of Climate*. Princeton University Press, 2005. ISBN 0-691-12164-8. Google-Books-ID: kTSDb2efRmYC.

D. Ruelle. *Chaotic Evolution and Strange Attractors*. Cambridge University Press, September 1989. ISBN 978-0-521-36830-8. Google-Books-ID: PXm43Y5NaJEC.

Alfredo Ruiz-Barradas, Eugenia Kalnay, Malaquías Peña, Amir E. BozorgMagham, and Safa Motesharrei. Finding the driver of local ocean–atmosphere coupling in reanalyses and CMIP5 climate models. *Climate Dynamics*, 48(7):2153–2172, April 2017. doi: 10.1007/ s00382-016-3197-1.

Robert Sadourny, Akio Arakawa, and Yale Mintz. Integration of the nondivergent barotropic vorticity equation with an icosahedral-hexagonal grid for the sphere. *Monthly Weather Review*, 96(6):351–356, June 1968. doi: 10.1175/1520-0493(1968)096<0351:IOTNBV>2. 0.CO;2.

Kazuo Saito, Teruyuki Kato, Hisaki Eito, and Chiashi Muroi. *Documentation of the meteorological research institute/numerical prediction division unified nonhydrostatic model*, volume 42. Technical Report. MRI, 2001.

Kazuo Saito, Tsukasa Fujita, Yoshinori Yamada, Jun-ichi Ishida, Yukihiro Kumagai, Kohei Aranami, Shiro Ohmori, Ryoji Nagasawa, Saori Kumagai, Chiashi Muroi, Teruyuki Kato, Hisaki Eito, and Yosuke Yamazaki. The operational JMA nonhydrostatic mesoscale model. *Monthly Weather Review*, 134(4):1266–1298, April 2006. doi: 10.1175/MWR3120.1.

Kazuo Saito, Jun-ichi Ishida, Kohei Aranami, Tabito Hara, Tomonori Segawa, Masami Narita, and Yuuki Honda. Nonhydrostatic atmospheric models and operational development at JMA. *Journal of the Meteorological Society of Japan. Ser. II*, 85B:271–304, 2007. doi: 10.2151/jmsj.85B.271.

Pavel Sakov, Dean S. Oliver, and Laurent Bertino. An iterative EnKF for strongly nonlinear systems. *Monthly Weather Review*, 140(6):1988–2004, February 2012. doi: 10.1175/MWR-D-11-00176.1.

Barry Saltzman. Finite amplitude free convection as an initial value problem – I. *Journal of the Atmospheric Sciences*, 19(4):329–341, July 1962. doi: 10.1175/1520-0469(1962)019<0329:FAFCAA>2.0.CO;2.

Ahmed H. Sameh and Vivek Sarin. Hybrid parallel linear system solvers. *International Journal of Computational Fluid Dynamics*, 12(3–4):213–223, January 1999. doi: 10.1080/10618569908940826.

Y. Sasaki. An objective analysis based on the variational method. *Journal of the Meteorological Society of Japan. Ser. II*, 36(3):77–88, 1958. doi: 10.2151/jmsj1923.36.3_77. www.jstage.jst.go.jp/article/jmsj1923/36/3/36_3_77/_article.

Yoshikazu Sasaki. Some basic formalisms in numerical variational analysis. *Monthly Weather Review*, 98(12):875–883, December 1970. doi: 10.1175/1520-0493(1970)098<0875:SBFINV>2.3.CO;2.

Masaki Satoh, Hirofumi Tomita, Hisashi Yashiro, Hiroaki Miura, Chihiro Kodama, Tatsuya Seiki, Akira T. Noda, Yohei Yamada, Daisuke Goto, Masahiro Sawada, Takemasa Miyoshi, Yosuke Niwa, Masayuki Hara, Tomoki Ohno, Shin-ichi Iga, Takashi Arakawa, Takahiro Inoue, and Hiroyasu Kubokawa. The non-hydrostatic icosahedral atmospheric model: Description and development. *Progress in Earth and Planetary Science*, 1:18, October 2014. doi: 10.1186/s40645-014-0018-1.

Tim Sauer. How long do numerical chaotic solutions remain valid? *Physical Review Letters*, July 1997.

Thomas W. Schlatter. Some experiments with a multivariate statistical objective analysis scheme. *Monthly Weather Review*, 103(3):246–257, March 1975. doi: 10.1175/1520-0493(1975)103<0246:SEWAMS>2.0.CO;2.

Paul S. Schopf and Max J. Suarez. Vacillations in a coupled ocean–atmosphere model. *Journal of the Atmospheric Sciences*, 45(3):549–566, February 1988. doi: 10.1175/1520-0469(1988)045<0549:VIACOM>2.0.CO;2.

C. Schraff, H. Reich, A. Rhodin, A. Schomburg, K. Stephan, A. Periáñez, and R. Potthast. Kilometre-scale ensemble data assimilation for the COSMO model (KENDA). *Quarterly Journal of the Royal Meteorological Society*, 142(696):1453–1472, April 2016. doi: 10.1002/qj.2748.

Joseph G. Sela. Spectral modeling at the national meteorological center. *Monthly Weather Review*, 108(9):1279–1292, September 1980. doi: 10.1175/1520-0493(1980)108<1279:SMATNM>2.0.CO;2.

P. J. Sellers. Biophysical models of land surface processes. *Climate System Modeling*, pages 451–490, 1992.

Ralph Shapiro. Smoothing, filtering, and boundary effects. *Reviews of Geophysics*, 8(2):359–387, May 1970. doi: 10.1029/RG008i002p00359.

J. Shukla, L. Marx, D. Paolino, D. Straus, J. Anderson, J. Ploshay, D. Baumhefner, J. Tribbia, C. Brankovic, T. Palmer, Y. Chang, S. Schubert, M. Suarez, and E. Kalnay. Dynamical seasonal prediction. *Bulletin of the American Meteorological Society*, 81(11):2593–2606, November 2000. doi: 10.1175/1520-0477(2000)081<2593:DSP>2.3.CO;2.

Frederick G. Shuman. History of numerical weather prediction at the national meteorological center. *Weather and Forecasting*, 4(3):286–296, September 1989. doi: 10.1175/1520-0434(1989)004<0286:HONWPA>2.0.CO;2.

Frederick G. Shuman and John B. Hovermale. An operational six-layer primitive equation model. *Journal of Applied Meteorology*, 7(4):525–547, August 1968. doi: 10.1175/1520-0450(1968)007<0525:AOSLPE>2.0.CO;2.

A. J. Simmons and D. M. Burridge. An energy and angular-momentum conserving vertical finite-difference scheme and hybrid vertical coordinates. *Monthly Weather Review*, 109(4):758–766, April 1981. doi: 10.1175/1520-0493(1981)109<0758:AEAAMC>2.0.CO;2.

A. J. Simmons, R. Mureau, and T. Petroliagis. Error growth and estimates of predictability from the ECMWF forecasting system. *Quarterly Journal of the Royal Meteorological Society*, 121(527):1739–1771, October 1995. doi: 10.1002/qj.49712152711.

William C. Skamarock and Joseph B. Klemp. A time-split nonhydrostatic atmospheric model for weather research and forecasting applications. *Predicting Weather, Climate and Extreme Events*, 227(7):3465–3485, March 2008. doi: 10.1016/j.jcp.2007.01.037.

William C. Skamarock, Joseph B. Klemp, Jimy Dudhia, David O. Gill, Dale M. Barker, Wei Wang, and Jordan G. Powers. *A description of the advanced research WRF version 2*, volume 468 of *NCAR Tech. Note*. National Center For Atmospheric Research Boulder Co Mesoscale and Microscale . . . , 2005.

William C. Skamrock. Truncation error estimates for refinement criteria in nested and adaptive models. *Monthly Weather Review*, 117(4):872–886, April 1989. doi: 10.1175/1520-0493(1989)117<0872:TEEFRC>2.0.CO;2.

J. M. Slingo. The development and verification of a cloud prediction scheme for the Ecmwf model. *Quarterly Journal of the Royal Meteorological Society*, 113(477):899–927, July 1987. doi: 10.1002/qj.49711347710.

Piotr K. Smolarkiewicz. A simple positive definite advection scheme with small implicit diffusion. *Monthly Weather Review*, 111(3):479–486, March 1983. doi: 10.1175/1520-0493(1983)111<0479:ASPDAS>2.0.CO;2.

Piotr K. Smolarkiewicz. A fully multidimensional positive definite advection transport algorithm with small implicit diffusion. *Journal of Computational Physics*, 54(2):325–362, May 1984. doi: 10.1016/0021-9991(84)90121-9.

Piotr K. Smolarkiewicz and Wojciech W. Grabowski. The multidimensional positive definite advection transport algorithm: Nonoscillatory option. *Journal of Computational Physics*, 86(2):355–375, February 1990. doi: 10.1016/0021-9991(90)90105-A.

Chris Snyder, Thomas Bengtsson, Peter Bickel, and Jeff Anderson. Obstacles to high-dimensional particle filtering. *Monthly Weather Review*, 136(12):4629–4640, December 2008. doi: 10.1175/2008MWR2529.1.

Matthias Sommer and Martin Weissmann. Ensemble-based approximation of observation impact using an observation-based verification metric. *Tellus A: Dynamic Meteorology and Oceanography*, 68(1):27885, December 2016. doi: 10.3402/tellusa.v68.27885.

Gilles Sommeria. Three-dimensional simulation of turbulent processes in an undisturbed trade wind boundary layer. *Journal of the Atmospheric Sciences*, 33(2):216–241, February 1976. doi: 10.1175/1520-0469(1976)033<0216:TDSOTP>2.0.CO;2.

S-T. Soong and Y. Ogura. Response of tradewind cumuli to large-scale processes. *Journal of Atmospheric Sciences*, 37(9):2035–2050, September 1980. doi: 10.1175/1520-0469(1980) 037<2035:ROTCTL>2.0.CO;2.

Colin Sparrow. *The Lorenz Equations: Bifurcations, Chaos, and Strange Attractors*. Springer Science & Business Media, December 2012. ISBN 978-1-4612-5767-7. Google-Books-ID: ttviBwAAQBAJ.

John D. Stackpole. The NMC 9-Layer Global Primitive Equation Model on a latitude-longitude grid. 1978. https://repository.library.noaa.gov/view/noaa/12745.

A. Staniforth, A. White, N. Wood, J. Thuburn, M. Zerroukat, E. Cordero, and et al. *Unified Model Documentation Paper, 15, Joy of UM 5.3 – Model formulation*. UKMO, 2002.

Andrew Staniforth and Jean Côté. Semi-lagrangian integration schemes for atmospheric models – a review. *Monthly Weather Review*, 119(9):2206–2223, September 1990. doi: 10.1175/ 1520-0493(1991)119<2206:SLISFA>2.0.CO;2.

Andrew N. Staniforth and Roger W. Daley. A finite-element formulation for the vertical discretization of sigma-coordinate primitive equation models. *Monthly Weather Review*, 105 (9):1108–1118, September 1977. doi: 10.1175/1520-0493(1977)105<1108:AFEFFT>2.0. CO;2.

David R. Stauffer and Nelson L. Seaman. Use of four-dimensional data assimilation in a limited-area mesoscale model. Part I: Experiments with synoptic-scale data. *Monthly Weather Review*, 118(6):1250–1277, June 1990. doi: 10.1175/1520-0493(1990)118<1250: UOFDDA>2.0.CO;2.

Will Steffen, Regina Angelina Sanderson, Peter D. Tyson, Jill Jäger, Pamela A. Matson, Berrien Moore III, Frank Oldfield, Katherine Richardson, Hans Joachim Schellnhuber, Billie L. Turner, and Robert J. Wasson. *Global Change and the Earth System: A Planet Under Pressure*. Springer-Verlag, Berlin; Heidelberg; New York, January 2006. ISBN 978-3-540-26607-5.

Will Steffen, Wendy Broadgate, Lisa Deutsch, Owen Gaffney, and Cornelia Ludwig. The trajectory of the Anthropocene: The great acceleration. *The Anthropocene Review*, January 2015. doi: 10.1177/2053019614564785.

David J. Stensrud. *Parameterization Schemes: Keys to Understanding Numerical Weather Prediction Models*. Cambridge University Press, May 2007. ISBN 978-0-521-86540-1. Google-Books-ID: lMXSpRwKNO8C.

Roland B. Stull. *An Introduction to Boundary Layer Meteorology | Roland B. Stull | Springer*. 1988. ISBN 978-94-009-3027-8. www.springer.com/us/book/9789027727688.

Max J. Suarez and Paul S. Schopf. A delayed action oscillator for ENSO. *Journal of the Atmospheric Sciences*, 45(21):3283–3287, November 1988. doi: 10.1175/1520-0469(1988) 045<3283:ADAOFE>2.0.CO;2.

I. Szunyogh, Z. Toth, R. E. Morss, S. J. Majumdar, B. J. Etherton, and C. H. Bishop. The effect of targeted dropsonde observations during the 1999 winter storm reconnaissance program. *Monthly Weather Review*, 128(10):3520–3537, October 2000. doi: 10.1175/ 1520-0493(2000)128<3520:TEOTDO>2.0.CO;2.

Istvan Szunyogh, Eugenia Kalnay, and Zoltan Toth. A comparison of Lyapunov and optimal vectors in a low-resolution GCM. *Tellus A: Dynamic Meteorology and Oceanography*, 49 (2):200–227, January 1997. doi: 10.3402/tellusa.v49i2.14467.

Lawrence L. Takacs. A two-step scheme for the advection equation with minimized dissipation and dispersion errors. *Monthly Weather Review*, 113(6):1050–1065, June 1985. doi: 10.1175/ 1520-0493(1985)113<1050:ATSSFT>2.0.CO;2.

Eugene S. Takle, William J. Gutowski, Raymond W. Arritt, Zaitao Pan, Christopher J. Anderson, Renato Ramos da Silva, Daniel Caya, Shyh-Chin Chen, F. Giorgi, Jens Hesselbjerg Christensen, Song-You Hong, Hann-Ming Henry Juang, Jack Katzfey, William M. Lapenta, Rene Laprise, Glen E. Liston, Philippe Lopez, John McGregor, Roger A. Pielke, and John O. Roads. Project to intercompare regional climate simulations (PIRCS): Description and initial results. *Journal of Geophysical Research: Atmospheres*, 104(D16):19443–19461, August 1999. doi: 10.1029/1999JD900352.

Olivier Talagrand. The use of adjoint equations in numerical modelling of the atmospheric circulation. *Automatic Differentiation of Algorithms: Theory, Implementation, and Application*, 169:180, 1991. Publisher: SIAM Philadelphia, PA.

Olivier Talagrand. Assimilation of observations, an introduction (special issue data assimilation in meteorology and oceanography: Theory and practice). *Journal of the Meteorological Society of Japan. Ser. II*, 75(1B):191–209, 1997. doi: 10.2151/jmsj1965.75.1B_191.

Olivier Talagrand and Philippe Courtier. Variational assimilation of meteorological observations with the adjoint vorticity equation. I: Theory. *Quarterly Journal of the Royal Meteorological Society*, 113(478):1311–1328, October 1987. doi: 10.1002/qj.49711347812.

M. C. Tapp and P. W. White. A non-hydrostatic mesoscale model. *Quarterly Journal of the Royal Meteorological Society*, 102(432):277–296, April 1976. doi: 10.1002/qj.49710243202.

Yasuo Tatsumi. An economical explicit time integration scheme for a primitive model. *Journal of the Meteorological Society of Japan. Ser. II*, 61(2):269–288, 1983. doi: 10.2151/jmsj1965.61.2_269.

Yasuo Tatsumi. A spectral limited-area model with time-dependent lateral boundary conditions and its application to a multi-level primitive equation model. *Journal of the Meteorological Society of Japan. Ser. II*, 64(5):637–664, 1986. doi: 10.2151/jmsj1965.64.5_637.

Mark Taylor, Joseph Tribbia, and Mohamed Iskandarani. The spectral element method for the shallow water equations on the sphere. *Journal of Computational Physics*, 130(1):92–108, January 1997. doi: 10.1006/jcph.1996.5554.

Clive Temperton and David L. Williamson. Normal mode initialization for a multilevel gridpoint model. Part I: Linear aspects. *Monthly Weather Review*, 109(4):729–743, April 1981. doi: 10.1175/1520-0493(1981)109<0729:NMIFAM>2.0.CO;2.

Sidney Teweles and Hermann B. Wobus. Verification of prognostic charts. *Bulletin of the American Meteorological Society*, 35(10):455–463, December 1954. doi: 10.1175/1520-0477-35.10.455.

H. J. Thiébaux and M. A. Pedder. *Spatial Objective Analysis: With Applications in Atmospheric Science*. London; Orlando: Academic Press, 1987. ISBN 978-0-12-686930-9. http://trove.nla.gov.au/version/21852782.

Philip D. Thompson. *Numerical Weather Analysis and Prediction*. New York: Macmillan, 1961. http://archive.org/details/numericalweather00thom.

Philip Duncan Thompson. Charney and the revival of numerical weather prediction. In Richard S. Lindzen, Edward N. Lorenz, and George W. Platzman, editors, *The Atmosphere – A Challenge: The Science of Jule Gregory Charney*, pages 93–199. American Meteorological Society, Boston, MA, 1990. doi: 10.1007/978-1-944970-35-2.

Jean-Noéul Thépaut and Philippe Courtier. Four-dimensional variational data assimilation using the adjoint of a multilevel primitive-equation model. *Quarterly Journal of the Royal Meteorological Society*, 117(502):1225–1254, October 1991. doi: 10.1002/qj.49711750206.

Jean-Noël Thépaut, Ross N. Hoffman, and Philippe Courtier. Interactions of dynamics and observations in a four-dimensional variational assimilation. *Monthly Weather Review*, 121

(12):3393–3414, December 1993. doi: 10.1175/1520-0493(1993)_21<3393:IODAOI>2.0. CO;2.

Michael K. Tippett, Jeffrey L. Anderson, Craig H. Bishop, Thomas M. Hamill, and Jeffrey S. Whitaker. Ensemble square root filters. *Monthly Weather Review*, 131(7):1485–1490, July 2003. doi: 10.1175/1520-0493(2003)131<1485:ESRF>2.0.CO;2.

Hendrik L Tolman. User manual and system documentation of WAVEWATCH III TM version 3.14. *Technical Note, MMAB Contribution*, 276:220, 2009.

Ryan D. Torn and Gregory J. Hakim. Ensemble-based sensitivity analysis. *Monthly Weather Review*, 136(2):663–677, February 2008. doi: 10.1175/2007MWR2132.1.

Ryan D. Torn, Gregory J. Hakim, and Chris Snyder. Boundary conditions for limited-area ensemble Kalman filters. *Monthly Weather Review*, 134(9):2490–2502, September 2006. doi: 10.1175/MWR3187.1.

Zoltan Toth and Eugenia Kalnay. Ensemble forecasting at NMC: The generation of perturbations. *Bulletin of the American Meteorological Society*, 74(12):2317–2330, December 1993. doi: 10.1175/1520-0477(1993)074<2317:EFANTG>2.0.CO;2.

Zoltan Toth and Eugenia Kalnay. Ensemble forecasting at NCEP. In *Proc. Seminar on Predictability*, volume 2, pages 39–61, 1996.

Zoltan Toth and Eugenia Kalnay. Ensemble forecasting at NCEP and the breeding method. *Monthly Weather Review*, 125(12):3297–3319, December 1997. doi: 10.1175/1520-0493(1997)125<3297:EFANAT>2.0.CO;2.

M. Steven Tracton and Eugen a Kalnay. Operational ensemble prediction at the National Meteorological Center: Practical aspects. *Weather and Forecasting*, 8(3):379–398, September 1993. doi: 10.1175/1520-0434(1993)008<0379:OEPATN>2.0.CO;2.

Kevin E. Trenberth. *Climate System Modeling*. Cambridge University Press, 1992. ISBN 978-0-521-43231-3. Google-Books-ID: EDClFW7JWrQC.

Anna Trevisan and Roberto Legnani. Transient error growth and local predictability: a study in the Lorenz system. *Tellus A*, 47(1):103–117, January 1995. doi: 10.1034/j.1600-0870.1995. 00006.x.

Anna Trevisan and Francesco Pancotti. Periodic orbits, Lyapunov vectors, and singular vectors in the Lorenz system. *Journal of the Atmospheric Sciences*, 55(3):390–398, February 1998. doi: 10.1175/1520-0469(1998)055<0390:POLVAS>2.0.CO;2.

Yannick Trémolet. Accounting for an imperfect model in 4D-Var. *Quarterly Journal of the Royal Meteorological Society: A Journal of the Atmospheric Sciences, Applied Meteorology and Physical Oceanography*, 132(621):2483–2504, 2006.

S. Valcke, V. Balaji, A. Craig, C. DeLuca, R. Dunlap, R. W. Ford, R. Jacob, J. Larson, R. O'Kuinghttons, G. D. Riley, and M. Vertenstein. Coupling technologies for Earth System Modelling. *Geoscientific Model Development*, 5(6):1589–1596, December 2012. doi: 10.5194/gmd-5-1589-2012.

H. M. Van den Dool. Searching for analogues, how long must we wait? *Tellus A: Dynamic Meteorology and Oceanography*, 46(3):314–324, January 1994. doi: 10.3402/tellusa.v46i3. 15481.

P. J. van Leeuwen. Nonlinear data assimilation in geosciences an extremely efficient particle filter. *Quarterly Journal of the Royal Meteorological Society*, 136(653):1991–1999, October 2010. doi: 10.1002/qj.699.

Peter Jan van Leeuwen. Particle filtering in geophysical systems. *Monthly Weather Review*, 137 (12):4089–4114, December 2009. doi: 10.1175/2009MWR2835.1.

Peter Jan van Leeuwen. A consistent interpretation of the stochastic version of the ensemble Kalman filter. *Quarterly Journal of the Royal Meteorological Society*, 146(731):2815–2825, July 2020. doi: 10.1002/qj.3819.

Peter Jan van Leeuwen, Hans R. Künsch, Lars Nerger, Roland Potthast, and Sebastian Reich. Particle filters for high-dimensional geoscience applications: A review. *Quarterly Journal of the Royal Meteorological Society*, 145(723):2335–2365, 2019. doi: 10.1002/qj.3551.

Sanita Vetra-Carvalho, Peter Jan van Leeuwen, Lars Nerger, Alexander Barth, M. Umer Altaf, Pierre Brasseur, Paul Kirchgessner, and Jean-Marie Beckers. State-of-the-art stochastic data assimilation methods for high-dimensional non-Gaussian problems. *Tellus A: Dynamic Meteorology and Oceanography*, 70(1):1–43, January 2018. doi: 10.1080/16000870.2018. 1445364.

Sir Gilbert Thomas Walker. *World Weather III*. Edward Stanford, 1928.

J. M. Wallace, E. M. Rasmusson, T. P. Mitchell, V. E. Kousky, E. S. Sarachik, and H. Storch. On the structure and evolution of ENSO-related climate variability in the tropical Pacific: Lessons from TOGA. *Journal of Geophysical Research: Oceans*, 103(C7):14241–14259, June 1998. doi: 10.1029/97JC02905.

John M. Wallace, Stefano Tibaldi, and Adrian J. Simmons. Reduction of systematic forecast errors in the ECMWF model through the introduction of an envelope orography. *Quarterly Journal of the Royal Meteorological Society*, 109(462):683–717, October 1983. doi: 10.1002/qj.49710946202.

Shizhang Wang, Ming Xue, Alexander D. Schenkman, and Jinzhong Min. An iterative ensemble square root filter and tests with simulated radar data for storm-scale data assimilation. *Quarterly Journal of the Royal Meteorological Society*, 139(676):1888–1903, October 2013a. doi: 10.1002/qj.2077.

Xuguang Wang, Chris Snyder, and Thomas M. Hamill. On the theoretical equivalence of differently proposed ensemble–3DVAR hybrid analysis schemes. *Monthly Weather Review*, 135 (1):222–227, January 2007. doi: 10.1175/MWR3282.1.

Xuguang Wang, David Parrish, Daryl Kleist, and Jeffrey Whitaker. GSI 3DVar-based ensemble–variational hybrid data assimilation for NCEP global forecast system: Single–resolution experiments. *Monthly Weather Review*, 141(11):4098–4117, June 2013b. doi: 10.1175/MWR-D-12-00141.1.

Klaus M. Weickmann, Glenn R. Lussky, and John E. Kutzbach. Intraseasonal (30–60 Day) fluctuations of outgoing longwave radiation and 250 mb streamfunction during northern winter. *Monthly Weather Review*, 113(6):941–961, June 1985. doi: 10.1175/1520-0493(1985) 113<0941:IDFOOL>2.0.CO;2.

M. C. Wheeler. Tropical Meteorology | Equatorial Waves. In James R. Holton, editor, *Encyclopedia of Atmospheric Sciences*, pages 2313–2325. Academic Press, Oxford, January 2003. doi: 10.1016/B0-12-227090-8/00414-0.

Jeffrey S. Whitaker and Thomas M. Hamill. Ensemble data assimilation without perturbed observations. *Monthly Weather Review*, 130(7):1913–1924, July 2002. doi: 10.1175/1520-0493(2002)130<1913:EDAWPO>2.0.CO;2.

Jeffrey S. Whitaker, Gilbert P. Compo, Xue Wei, and Thomas M. Hamill. Reanalysis without radiosondes using ensemble data assimilation. *Monthly Weather Review*, 132(5):1190–1200, May 2004. doi: 10.1175/1520-0493(2004)132<1190:RWRUED>2.0.CO;2.

A. Wiin-Nielsen. The birth of numerical weather prediction. *Tellus A*, 43(4):36–52, August 1991. doi: 10.1034/j.1600-0870.1991.t01-2-00006.x.

Daniel S. Wilks. *Statistical Methods in the Atmospheric Sciences*, volume 100. Academic press, 2011.

Paul D. Williams. A proposed modification to the Robert–Asselin time filter. *Monthly Weather Review*, 137(8):2538–2546, August 2009. doi: 10.1175/2009MWR2724.1.

David L. Williamson. Integration of the barotropic vorticity equation on a spherical geodesic grid. *Tellus*, 20(4):642–653, November 1968. doi: 10.1111/j.2153-3490.1968.tb00406.x.

David L. Williamson and Rene Laprise. Numerical approximations for global atmospheric general circulation models. *Numerical Modeling of Global Atmosphere in the Climate System*, 127:219, 2000. Publisher: Kluwer Academic Publishers.

Richard L. Wobus and Eugenia Kalnay. Three years of operational prediction of forecast skill at NMC. *Monthly Weather Review*, 123(7):2132–2148, July 1995. doi: 10.1175/1520-0493(1995)123<2132:TYOOPO>2.0.CO;2.

Wan-Shu Wu, R. James Purser and David F. Parrish. Three-dimensional variational analysis with spatially inhomogeneous covariances. *Monthly Weather Review*, 130(12):2905–2916, 2002.

M. Xue, K. K. Droegemeier, V. Wong, A. Shapiro, and K. Brewster. ARPS 4.0 user's guide. *Center for Advanced Prediction Systems, The University of Oklahoma*, 380, 1995.

Ming Xue, Donghai Wang, Jidong Gao, Keith Brewster, and Kelvin K. Droegemeier. The advanced regional prediction system (ARPS), storm-scale numerical weather prediction and data assimilation. *Meteorology & Atmospheric Physics*, 82, 2003.

Shu-Chih Yang, Debra Baker, Hong Li, Katy Cordes, Morgan Huff, Geetika Nagpal, Ena Okereke, Josue Villafañe, Eugenia Kalnay, and Gregory S. Duane. Data assimilation as synchronization of truth and model: Experiments with the three-variable Lorenz system*. *Journal of the Atmospheric Sciences*, 63(9):2340–2354, September 2006. doi: 10.1175/JAS3739.1.

Shu-Chih Yang, Eugenia Kalnay, Brian Hunt, and Neill E. Bowler. Weight interpolation for efficient data assimilation with the local ensemble transform Kalman filter. *Quarterly Journal of the Royal Meteorological Society*, 135(638):251–262, January 2009. doi: 10.1002/qj.353.

Shu-Chih Yang, Eugenia Kalnay, and Brian Hunt. Handling nonlinearity in an ensemble Kalman filter: Experiments with the three-variable Lorenz model. *Monthly Weather Review*, 140(8): 2628–2646, March 2012. doi: 10.1175/MWR-D-11-00313.1.

Shu-Chih Yang, Kuan-Jen Lin, Takemasa Miyoshi, and Eugenia Kalnay. Improving the spin-up of regional EnKF for typhoon assimilation and forecasting with Typhoon Sinlaku (2008). *Tellus A: Dynamic Meteorology and Oceanography*, 65(1):20804, December 2013. doi: 10.3402/tellusa.v65i0.20804.

Weiyu Yang and Ionel Michael Navon. Documentation of the Tangent Linear Model and Its Adjoint of the Adiabatic Version of the NASA GEOS-1 C-Grid GCM, Version 5.2. Technical report, National Aeronautics and Space Administration, Goddard Space Flight Center, 1996.

Kao-San Yeh, Jean Côté, Sylvie Gravel, André Méthot, Alain Patoine, Michel Roch, and Andrew Staniforth. The CMC–MRB global environmental multiscale (GEM) model. Part III: Nonhydrostatic formulation. *Monthly Weather Review*, 130(2):339–356, February 2002. doi: 10.1175/1520-0493(2002)130<0339:TCMGEM>2.0.CO;2.

Takuma Yoshida. *Covariance localization in strongly coupled data assimilation*. PhD Thesis, 2019.

Takuma Yoshida and Eugenia Kalnay. Correlation-cutoff method for covariance localization in strongly coupled data assimilation. *Monthly Weather Review*, 146(9):2881–2889, August 2018. doi: 10.1175/MWR-D-17-0365.1.

Steven T. Zalesak. Fully multidimensional flux-corrected transport algorithms for fluids. *Journal of Computational Physics*, 31(3):335–362, June 1979. doi: 10.1016/0021-9991(79)90051-2.

Bibliography

Stephen E. Zebiak and Mark A. Cane. A model El Ni&ntilde–Southern oscillation. *Monthly Weather Review*, 115(10):2262–2278, October 1987. doi: 10.1175/1520-0493(1987) 115<2262:AMENO>2.0.CO;2.

N. Zeng, A. Mariotti, and P. Wetzel. Terrestrial mechanisms of interannual CO2 variability. *Global Biogeochemical Cycles*, 19(1), March 2005. doi: 10.1029/2004GB002273.

Ning Zeng. Glacial-interglacial atmospheric CO_2 change – The glacial burial hypothesis. *Advances in Atmospheric Sciences*, 20(5):677–693, September 2003. doi: 10.1007/ BF02915395.

Da-Lin Zhang, Hai-Ru Chang, Nelson L. Seaman, Thomas T. Warner, and J. Michael Fritsch. A two-way interactive nesting procedure with variable terrain resolution. *Monthly Weather Review*, 114(7):1330–1339, July 1986. doi: 10.1175/1520-0493(1986)114<1330: ATWINP>2.0.CO;2.

Fuqing Zhang, Zhiyong Meng, and Altug Aksoy. Tests of an ensemble Kalman filter for mesoscale and regional-scale data assimilation. Part I: Perfect model experiments. *Monthly Weather Review*, 134(2):722–736, February 2006. doi: 10.1175/MWR3101.1.

Qingyun Zhao and Frederick H. Carr. A prognostic cloud scheme for operational NWP models. *Monthly Weather Review*, 125(8):1931–1953, August 1997. doi: 10.1175/1520-0493(1997) 125<1931:APCSFO>2.0.CO;2.

Xiaqiong Zhou, Yuejian Zhu, Dingchen Hou, and Daryl Kleist. A comparison of perturbations from an ensemble transform and an ensemble Kalman filter for the NCEP global ensemble forecast system. *Weather and Forecasting*, 31(6):2057–2074, December 2016. doi: 10.1175/ WAF-D-16-0109.1.

X. Zou, I. M. Navon, and F. X. Ledimet. An optimal nudging data assimilation scheme using parameter estimation. *Quarterly Journal of the Royal Meteorological Society*, 118(508): 1163–1186, October 1992. doi: 10.1002/qj.49711850808.

Milija Zupanski. Regional four-dimensional variational data assimilation in a Quasi-operational forecasting environment. *Monthly Weather Review*, 121(8):2396–2408, August 1993. doi: 10.1175/1520-0493(1993)121<2396:RFDVDA>2.0.CO;2.

Milija Zupanski. Maximum likelihood ensemble filter: Theoretical aspects. *Monthly Weather Review*, 133(6):1710–1726, June 2005. doi: 10.1175/MWR2946.1.

Index

3D-Var, 156, 165, 173
4D-LETKF, 214
4D-Var, 156, 187
4DEnVar, 211

absolute acceleration, 35
absolute velocity, 36
absolute vorticity, 56
absolutely stable, 95
absolutely unstable, 96
Adams-Bashford, 93
adjoint, 231
adjoint model, 165, 197
adjustment processes, 140
advection equation, 76
alternating direction implicit, ADI, 125
amplification factor, 87, 88, 97
analysis cycle, 153
analysis error covariance, 170
analysis increment, 179
analysis weights interpolation, 207
anomaly correlation, 265
apparent force, 36
Arakawa Jacobian, 111, 115
Arctic Oscillation, AO, 273
assimilation window, 189
Atmospheric Models Intercomparison Project, AMIP, 271
attractor, 227
autonomous, 226

background error covariance, 164, 165
background field, 153
backward Lyapunov vectors, 239
baroclinic instabilities, 257, 266
basic state, 41
Bayes theorem, 162
beneficial observations, 218
best linear unbiased estimation, BLUE, 167
β-plane, 51, 59
bifurcation point, 227
boundary value problem, 114
Boussinesq approximation, 68
bred vector dimension, 255

bred vectors, 182, 198, 253
breeding, 253
Brunt–Vaïsälä frequency, 44
bucket model, 147
bulk parameterization, 145

centered differences, 83
chaotic bounded flow, 229
chaotic solutions, 226
chaotic system, 229
characteristic Lyapunov vectors, 239
Charney–Phillips grid, 118
climate change, 273
climate models, 139
Cloud Resolving Convective Parameterization, CRCP, 148
Cloud System Resolving Model, CSRM, 148
computational group velocity, 101
computational mode, 90
computational stability, 83
conditionally stable, 96
conjugate gradient, 174, 192
conservation of energy, 38
conservation of mass, 37
conservation of momentum, 35
conservation of potential vorticity, 56
continuity equation, 37
Control Simulation Experiment, CSE, 281
control variable, 174, 190
Controlling Chaos in Control Simulation Experiments, 280
convective instabilities, 257
Coriolis force, 36
cost function, 160, 173, 175, 189, 191, 208, 212
Courant number, 83
Courant-Friedrichs-Lewy (CFL) condition, 84
covariance hybrid, 208
covariance localization, 208
Crank–Nicholson, 96
cumulus parameterization, 148
curse of dimensionality, 222

data assimilation, 150
decadal variability, 272

Index

delayed oscillator, 270
deterministic EnKF, 203
detrimental observations, 218
dimension of the phase space, 226
double sweep method, 122
dry convective adjustment, 143
dynamic system, 226
dynamical processes, 142

elliptic equations, 77
Ensemble Adjustment Kalman Filter, EAKF, 203
Ensemble Forecast Sensitivity to Observations,
    EFSO, 217
Ensemble Kalman Filter, EnKF, 156, 199
Ensemble of Data Assimilations, EDA, 194
ensemble space, 205
Ensemble Square Root Filter, EnSRF, 204
Ensemble Transform Kalman Filter, ETKF, 204
enstrophy-conserving, 120
envelope orography, 146
equatorial Rossby waves, 74
equatorially trapped waves, 72
equivalent depth, 55, 57
error covariance matrices, 169
error growth in a perfect model , 264
error growth in an imperfect model, 267
error of representativeness, 171
errors of the day, 187, 188, 208, 249
eta coordinates, 65
Euclidean norm, 231
Euler-backward (Matsuno) scheme, 87
Eulerian time derivative, 37
Extended Kalman Filter, EKF, 165, 199, 202
external waves, 48

filter degeneracy, 222
filtering approximations, 50, 54
final singular vectors, 233, 241
finite difference equation, FDE, 80
finite differences, 80
finite element method, 103
finite volume methods, 118
first guess, 153
first LLV, 238
first order closure, 142
flow, 226
flow relaxation scheme, 129
flux corrected transport, FCT, 102
forecast errors of the day, 164
Forecast Sensitivity to Observations, FSO, 216
forward Lyapunov vectors, 239
forward model, 158
four-dimensional data assimilation, 4DDA, 153
f-plane, 41, 46
fractal, 229
frequency dispersion relationship, FDR, 42, 43, 45,
    48, 51, 52, 57, 59, 74

gain hybrid, 209
Galerkin approach, 103
Gauss–Seidel (successive relaxation) method, 124
Gaussian elimination, 122
general vertical coordinates, 61
generalized inverse, 196
geopotential, 37
geostrophic mode, 45
(global) Lyapunov exponents, 237
global weather forecast models, 139
gradient of the cost function, 174, 175, 192, 212
gravity waves, 185
gravity-wave parameterization, 146
Green's theorem, 119
grey zone, 135, 144
group velocity, 42
growth rate of errors, 263

Hamiltonian systems, 226
Helmholtz linear equation, 121
Hessian, 174, 185
Hessian of the PSAS, 185
hindcast, 271
Horizontally Explicit-Vertically Implicit, HE-VI,
    132
Horizontally Implicit-Vertically Implicit, HI-VI,
    133
Human System, 277
Hybrid Forecast Sensitivity to Observations, HFSO,
    217
hydrostatic approximation, 46, 52
hydrostatic pressure, 134
hyperbolic equations, 77

identical twin experiment, 224
ill-posed, 77
implicit scheme, 96
implicit time schemes, 94
importance of the particles, 221
incremental form, 174, 196
inertia Lamb waves, 45
inertia oscillations, 44
inertia-gravity waves, 45, 53, 60, 74
initial conditions, 150
initial singular vectors, 232, 241
initialization, 185
inner loop, 194
innovation, 158
inter-variable correlation, 177
internal waves, 49
isentropic coordinates, 67
isentropic potential vorticity, 56
isothermal atmosphere, 46

Jacobi simultaneous relaxation method,
    124
Jacobian of the observation operator,
    169

Kalman filter, 156
Kalman gain matrix, 165, 170
Kaplan-Yorke dimension, 229
Kelvin waves, 71, 75
Krylov subspace methods, 125
K-theory, 70, 142

Lagged Average Forecasting, LAF, 247
Lagrangian scheme, 106
Lagrangian time derivative, 37
Lamb waves, 43, 44, 60
Lanczos algorithm, 235
Lanczos method, 126
Laplace equation, 105
Laplace tidal equation, 47
lateral boundary conditions, 126
leading Lyapunov Vectors, 182
leapfrog scheme, 81, 90
least square, 156
Legendre polynomials, 105
likelihood, 160, 162
linear observation operator, 169
local (finite time) Lyapunov exponent, 238
Local Ensemble Transform Kalman Filter, LETKF, 199, 204
local Lyapunov vector, 238
Local Particle Filter, LPF, 222
local polynomial interpolation, 152
local stability, 227
Lorenz grid, 118
Lorenz N-cycle, 92
LU decomposition, 122
Lyapunov exponents, 227

Madden Julian Oscillation, MJO, 266
maximum likelihood approach, 160
Maximum Likelihood Ensemble Filter, 204
mesoscale models, 139
minimization algorithms, 174
model coupler, 70
model error, 163, 195
model space, 158
moist convective adjustment, 143
Monte Carlo Forecasting, 245
Montgomery potential, 67
multigrid methods, 125
multiscale modeling framework, MMF, 149
multisystem ensemble, 263

nature run, NR, 216, 281
Newtonian relaxation, 155
NMC method, 165, 174, 182
no-cost smoother, 214
non-hydrostatic models, 132
nonlinear computational instability, NCI, 104, 108
nonlinear model, 165
nonlinear normal mode initialization, 186

nonlinear observation operator, 169
normal mode, 54
North Atlantic Oscillation, NAO, 273
nudging, 155, 222

objective analysis, 151, 156
observation error, 158
observation operator, 158
observation space, 158
observational increment, 159
Observing Systems Simulation Experiment, OSSE, 211, 279, 281
optimal interpolation, OI, 150, 162, 171, 184
orbit, 226
outer loop, 186, 194

Pacific Decadal Oscillation, PDO, 273
parabolic equations, 77
partial differential equations, PDE, 76
Particle Filter, 219
perturbation equation, 41, 58
perturbed observations EnKF, 203
phase space, 226
phase speed (velocity), 42, 43, 60
Physical-space Statistical Analysis System, PSAS, 184
plane wave solutions, 42, 46
planetary boundary layer, PBL, 148
pne-way nested, 127
Poisson equation, 126
pole problem, 105
posterior probability, 162
potential predictability, 268
potential temperature, 38
potential vorticity equation, 59
practical salinity units, psu, 69
precision, 157
precondition, 174, 175
predictability of the first kind', 268
predictability of the second kind, 268
predictor-corrector scheme, 93
pressure coordinates, 64
pressure gradient force, 36
primitive equations, 45, 61, 64
prior information, 153
prior probability, 162
Proactive quality control, PQC, 218

quadratic function, 167
quadratically conserving schemes, 110
quasi-Boussinesq (anelastic) approximation, 46, 51
quasi-compressible approximation, 133
quasi-geostrophic Approximation, 50
quasi-geostrophic balance, 57
quasi-Newton, 174, 192
Quasi-Outer Loop, QOL, 215

Index

recursive filter, 175, 176
relaxation time scale, 155
reliability of the forecast, 250
resampling, 220
Reynolds averaging, 141
rigid top, 64
Robert-Asselin-Williams (RAW) filter, 91
Rossby number, 60
Rossby waves, 51
rotating frame of reference, 36
Runge-Kutta, 93
Running in Place, RIP, 215

scaled lagged average forecasting, SLAF, 249
second-order closure, 143
Semi-Implicit, Semi-Lagrangian, SI-SL, 133
semi-implicit scheme, 97
semi-Lagrangian scheme, 106
sequential method, 156
shallow water equations, SWE, 45, 55, 58, 71
Shapiro filter, 110
sigma coordinates, 65
simplification operator, 196
single observation test, 178
singular vectors, 259
sound waves, 42
Southern Oscillation, 270
spatial error correlation, 176
SPEEDY, 147
spin-up, 202
split explicit, 132
sponge layer, 128
square-root EnKF, 203
stable system, 229
staggered grids, 115
stationary points, 227
statistical interpolation, 156
stochastic EnKF, 203
Stochastic-dynamic Forecasting, 244
Storm-scale models, 139
strange attractors, 227, 229
strong constraint, 190

strongly coupled data assimilation, 203
subgrid-scale mountains, 146
subgrid-scale processes, 139, 142
subspace spanned by B, 178
successive corrections method, SCM, 150, 153
successive overrelaxation method, SOR, 124
superensemble, 263
superobservation, superob, 171
superparameterization, 148

tangent linear model, 165, 197, 231
targeted observations, 258
tempering, 246
tendency modification scheme, 129
thermal wind, 65
thinning, 171
TOGA, 270–272
total derivative, 37
total potential energy, 55
trajectory, 226
transfer coefficients, 145
transform matrix, 205, 206
transform method, 104
transformation of the control variable, 175
triangular truncation, 105
truncation error, 82
turbulent diffusion process, 142
two-way interactive, 131

upstream scheme, 81, 86

variational approach, 160
variational quality control, 186
virtual temperature, 38
von Neumann stability criterion, 85
vorticity equation, 56

weak constraint, 195
well-posed, 77
wind/height correlation, 180

zeroth order closure, 142